NANOPLASMONICS - FUNDAMENTALS AND APPLICATIONS

Edited by **Grégory Barbillon**

Nanoplasmonics - Fundamentals and Applications
http://dx.doi.org/10.5772/65150
Edited by Grégory Barbillon

Contributors

Yasushi Oshikane, Karlo Queiroz Costa, Janilson Souza, Victor Dmitriev, Mohamed Hameed, Salah Obayya, Mohamed Hussein, Youssef El-Toukhy, Nasrin Hooshmand, Montasir Qasymeh, Jingzhi Wu, Yanhong Wang, Chengpu Liu, Qing Zhang, Jun You, Dmitry Kuzmin, Igor Bychkov, Vladimir Shavrov, Vasily Temnov, Hiroaki Matsui, Nikolaos Kalfagiannis, Demosthenes Koutsogeorgis, Elefterios Lidorikis, Panos Patsalas, Guiru Gu, Xuejun Lu, Thitikorn Kemsri, Yingjie Zhang, Joachim Herrmann, Anton Husakou, Marios Mattheakis, George Tsironis, Efthimios Kaxiras, Jose L Hueso, Carlos J. Bueno-Alejo, Adriana Arca-Ramos, Yuling Wang, Dan Li, Xuezhi Zheng, Guy A. E. Vandenbosch, Victor V. Moshchalkov, Jeong-Woo Choi, Waleed El-Said, Hyeon-Yeol Cho, Pedro J. Rivero, Javier Goicoechea, Francisco Arregui, Michael Leuenberger, Hari Paudel, Muhammad Waqas Shabbir

Published by InTech

Janeza Trdine 9, 51000 Rijeka, Croatia

Publishing Process Manager Edi Lipovic
Technical Editor SPi Global
Cover InTech Design team

First published June, 2017
Printed in Croatia
Legal deposit, Croatia: National and University Library in Zagreb

Additional hard copies can be obtained from orders@intechopen.com

Nanoplasmonics - Fundamentals and Applications, Edited by Grégory Barbillon
p. cm.
Print ISBN 978-953-51-3277-6
Online ISBN 978-953-51-3278-3

For EU product safety concerns:
IN TECH d.o.o., Prolaz Marije Krucifikse Kozulić 3, 51000 Rijeka, Croatia,
info@intechopen.com or visit our website at intechopen.com.

Meet the editor

Grégory Barbillon completed his PhD degree in Physics (2007) with distinction at the University of Technology of Troyes (France). Then, he obtained his Habilitation (HDR) in Physics (2013) at the University of Paris-Sud (Orsay, France). Actually, he is a contractual researcher at the Centre for Nanoscience and Nanotechnology (C2N) located at Paris-Saclay University (Orsay, France). His research interests were focused on plasmonics, nanophotonics, surface-enhanced Raman scattering, biosensors, near-field optics, biophotonics, nanotechnology, fluorescence, and nanofabrication.

Contents

Preface

Nanoplasmonics is a young topic of research, which is part of nanophotonics and nano-optics. Nanoplasmonics concerns to the investigation of electron oscillations in metallic nanostructures and nanoparticles (NPs). Surface plasmons have optical properties, which are very interesting. For instance, surface plasmons have the unique capacity to confine light at the nanoscale. Moreover, surface plasmons are very sensitive to the surrounding medium and the properties of the materials on which they propagate. In addition to the above, the surface plasmon resonances can be controlled by adjusting the size, shape, periodicity, and materials' nature. All these optical properties can enable a great number of applications, such as biosensors, optical modulators, photodetectors, and photovoltaic devices. This book is intended for a broad audience and provides an overview of some of the fundamental knowledges and applications of nanoplasmonics.

The book is divided into 19 chapters with the following scope:

Section 1 of this book entitled "Fundamentals" presents six chapters. The first two chapters investigate graphene-based plasmonics. Indeed, Mattheakis et al. investigate plasmonic metamaterial based on graphene in order to obtain tunable wave propagation properties such as epsilon-near-zero behavior. Then, Kuzmin et al. study in detail two ways of breaking the degeneracy of the plasmonic modes with the opposite azimuthal rotations for graphene-coated nanowires. In addition, Paudel et al. present interesting descriptions of the plasmonic properties of metal nanostructures of different geometries and their applications in modern nanotechnology (Chapter 3). In Chapter 4, Kim et al. present a theoretical method for the calculation of the transient nonlinearity in dielectric composites doped with metal nanoparticles and some applications using this approach. Then, Zheng et al. are focused on understanding the electromagnetic response of plasmonic antennas through a classical computational electromagnetic algorithm (Chapter 5). To finish this section "Fundamentals," Hooshmand investigates the interdimer separation influence on the localized surface plasmon resonance (LSPR) of homo- and heterodimers composed of silver and gold nanocubes by using the method of discrete-dipole approximation (DDA).

Section 2 describes some applications of nanoplasmonics to thermal and infrared (IR) domains. Firstly, the work by Wu et al. investigates the thermal effects of plasmonic nanostructures, which can be employed for energy conversion, thermal management, and optical trapping. In addition, Matsui et al. explore plasmonic responses of ITO nanoparticles (NPs) in the infrared (IR) domain and describe IR plasmonic applications of ITO NP sheets for solar-thermal shielding technology. To conclude this section, Gu et al. present a work on the description of surface plasmon resonance excitation for a plasmonic 2D hole array structure and its application in IR detector enhancement.

In Section 3, some applications of nanoplasmonics to optoelectronics are presented. Firstly, the work by Zhang et al. introduces a new methodology for designing plasmonic metamaterial devices such as Butterworth filter and metal-insulator-metal absorber. In the next chapter, Qasymesh highlights the potential of functioning electro-optical materials in nanoplasmonic waveguides to achieve novel ultracompact and efficient devices. Then, Costa et al. present an impedance-matching analysis of two plasmonic nanocircuits connected to cylindrical nanoantennas. Finally, El-Toukhy et al. investigate novel designs of tapered plasmonic nanoantennas for energy-harvesting applications.

Section 4 displays some applications of nanoplasmonics to biology and chemistry. In the first chapter of this section, Bueno-Alejo et al. introduce plasmonics devoted to photocatalytic applications in different environments. In the second chapter of Section 4, Wang et al. interest to the investigation of plasmonic biosensors based on surface-enhanced Raman scattering (SERS) for protein biomarker detection. Then, El-Said et al. present analyses of living cells by SERS. To finish this section, Rivero et al. investigate LSPR optical fibers for biosensing.

Section 5 concludes this book by describing some of the fabrication techniques dedicated to plasmonics. Firstly, Kalfagiannis et al. present the technique of laser annealing as an interesting platform for plasmonic nanostructuring. Lastly, the work by Oshikane investigates the fabrication and properties of plasmonic thin film of titanium nitride by pulsed laser deposition.

For making this book on nanoplasmonics, I had the pleasure to collaborate with authors of high quality coming from worldwide, and I thank them for that. This book dedicated to nanoplasmonics could serve as support for the students or researchers who are interested in this field.

Dr. HDR. Grégory Barbillon
Paris-Saclay University,
France

Section 1

Fundamentals

Chapter 1

Graphene and Active Metamaterials: Theoretical Methods and Physical Properties

Marios Mattheakis, Giorgos P. Tsironis and
Efthimios Kaxiras

Additional information is available at the end of the chapter

http://dx.doi.org/10.5772/67900

Abstract

The interaction of light with matter has triggered the interest of scientists for a long time. The area of plasmonics emerges in this context through the interaction of light with valence electrons in metals. The random phase approximation in the long wavelength limit is used for analytical investigation of plasmons in three-dimensional metals, in a two-dimensional electron gas, and finally in the most famous two-dimensional semi-metal, namely graphene. We show that plasmons in bulk metals as well as in a two-dimensional electron gas originate from classical laws, whereas quantum effects appear as non-local corrections. On the other hand, graphene plasmons are purely quantum modes, and thus, they would not exist in a "classical world." Furthermore, under certain circumstances, light is able to couple with plasmons on metallic surfaces, forming a surface plasmon polariton, which is very important in nanoplasmonics due to its subwavelength nature. In addition, we outline two applications that complete our theoretical investigation. First, we examine how the presence of gain (active) dielectrics affects surface plasmon polariton properties and we find that there is a gain value for which the metallic losses are completely eliminated resulting in lossless plasmon propagation. Second, we combine monolayers of graphene in a periodic order and construct a plasmonic metamaterial that provides tunable wave propagation properties, such as epsilon-near-zero behavior, normal, and negative refraction.

Keywords: random phase approximation, graphene, gain dielectrics, plasmonic metamaterial

1. Introduction

The interaction of light with matter has triggered the interest of scientists for a long time. The area of plasmonics emerges in this context through the interaction of light with electrons in metals, while a plasmon is the quantum of the induced electronic collective oscillation. In three-dimensional (3D) metals as well as in a two-dimensional electron gas (2DEG), the plasmon arises classically through a depolarized electromagnetic field generated through Coulomb long-range interaction of valence electrons and crystal ions [1]. Under certain circumstances, light is able to couple with plasmons on metallic surfaces, forming a surface plasmon polariton (SPP) [2–4]. The SPPs are very important in nanoplasmonics and nanodevices, due to their subwavelength nature, that is, because their spatial scale is smaller than that of corresponding free electromagnetic modes. In addition to classical plasmons, purely quantum plasmon modes exist in graphene, the famous two-dimensional (2D) semimetal. Since we need the Dirac equation to describe the electronic structure of graphene, the resulting plasmons are purely quantum objects [5–8]. As a consequence, graphene is quite special from this point of view, possessing exceptional optical properties, such as ultra-subwavelength plasmons stemming from the specifics of the light-matter interaction [7–10].

In this chapter, we present basic properties of plasmons, both from a classical standpoint but also quantum mechanically using the random phase approximation approach. Plasmons in 3D metals as well as in 2DEG originate from classical laws, whereas quantum effects appear as non-local corrections [11–13]. In addition, we point out the fundamental differences between volume (bulk), surface, and two-dimensional plasmons. We show that graphene plasmons are a purely quantum phenomenon and that they would not exist in a "classical world." We then outline two applications that complete our theoretical investigation. First, we examine how the presence of gain (active) dielectrics affects SPP properties and we find that there is a gain value for which the metallic losses are completely eliminated resulting in lossless SPP propagation [3]. Second, we combine monolayers of graphene in a periodic order and construct a plasmonic metamaterial that provides tunable wave propagation properties, such as epsilon-near-zero behavior, normal, and negative refraction [9].

2. Volume and surface plasmons in three-dimensional metals

2.1. Free collective oscillations: plasmons

Plasma is a medium with equal concentration of positive and negative charges, of which at least one charge type is mobile [1]. In a classical approach, metals are considered to form plasma made of ions and electrons. The latter are only the valence electrons that do not interact with each other forming an ideal negatively charged free electron gas [1, 14]. The positive ions, that is, atomic nuclei, are uniformly distributed forming a constant background of positive charge. The background positive charge is considered to be fixed in space, and as a result, it does not respond to any electronic fluctuation or any external field while the electron gas is free to move. In equilibrium, the electron density (plasma sea) is also distributed uniformly at any point preserving the overall electrical neutrality of the system. Metals support free and collective longitudinal charge oscillation with well-defined natural frequency, called the plasma frequency ω_p. The quanta of these charge oscillations are *plasmons*, that is, quasi-particles with energy $E_p = \hbar\omega_p$, where $\hbar$ is the reduced Plank constant.

http://dx.doi.org/10.5772/67900

We assume a plasma model with electron (and ion) density n. A uniform charge imbalance δn is established in the plasma by displacing uniformly a zone of electrons (e.g., a small slab in Cartesian coordinates) by a small distance $\mathbf{x}$ (**Figure 1**). The uniform displacement implies that all electrons oscillate in phase [2]; this is compatible with a long wavelength approximation ($\lambda_p/\alpha \to \infty$, where λ_p is the plasmon wavelength and α is the crystal lattice constant); in this case, the associated wavenumber $|\mathbf{q}|$ (**Figure 1(b)**) is very small compared with Fermi wavenumber k_F, viz. $q/k_F \to 0$ [7]. Longitudinal oscillations including finite wave vector $\mathbf{q}$ will be taken into account later in the context of quantum mechanics. The immobilized ions form a constant charge density indicated by en, where e is the elementary charge. Let $\mathbf{x}(t)$ denote the position of the displaced electronic slab at time t with charge density given by $-e\delta n(t)$. Due to the electron displacement, an excess positive charge density is created that is equal to $e\delta n(t)$, which in equilibrium, $\delta n = 0$, reduces to zero. Accordingly, an electric field is generated and interacts with the positive background via Coulomb interaction, forcing the electron cloud to move as a whole with respect to the immobilized ions, forming an electron density oscillation, that is, the plasma oscillation. The polarized electric field is determined by the first Maxwell equation as

$$\nabla \cdot \mathbf{E} = 4\pi e\delta n, \tag{1}$$

in CGS units.[1] The displacement $\mathbf{x}(t)$ in the electronic gas produces an electric current density $\mathbf{J} = -e(n+\delta n)\dot{\mathbf{x}} \approx -en\dot{\mathbf{x}}$ (since $\delta n/n \to 0$), related to the electron charge density via the continuity equation $\nabla \cdot \mathbf{J} = -e\partial_t \delta n$. After integration in time, we obtain

$$\delta n = n\nabla \cdot \mathbf{x} \tag{2}$$

Combining Eqs. (1) and (2), we find the electric field that is induced by the electron charge displacement, that is,

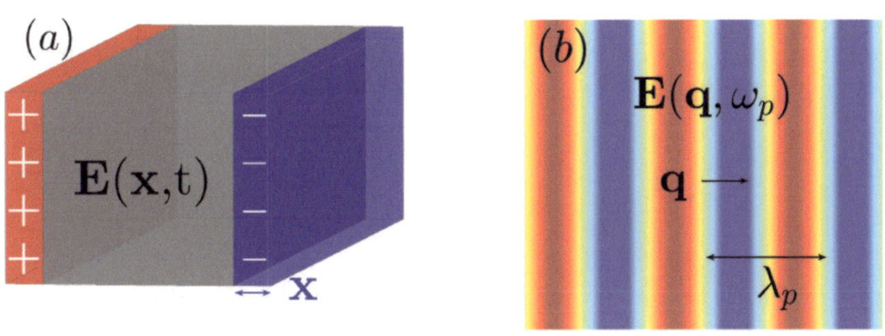

Figure 1. (a) A charge displacement is established by displacing uniformly a slab of electrons at a small distance $\mathbf{x}$, creating a polarized electric field in the solid. (b) A plasma longitudinal oscillation electric field in the bulk of a solid. The arrow indicates the direction of displacement of electrons and of the wavevector $\mathbf{q}$, while the double-faced arrow shows the plasmon wavelength λ_p.

[1]For SI units, we make the substitution $1/\varepsilon_0 = 4\pi$.

$$\mathbf{E} = 4\pi e n \mathbf{x}. \tag{3}$$

Newtonian mechanics states that an electron with mass m in an electric field $\mathbf{E}$ obeys the equation $m\ddot{\mathbf{x}} = -e\mathbf{E}$, yielding finally the equation of motion

$$m\ddot{\mathbf{x}} + 4\pi e^2 n \mathbf{x} = 0, \tag{4}$$

indicating that electrons form a collective oscillation with plasma frequency

$$\omega_p(0) = \sqrt{\frac{4\pi e^2 n}{m}}. \tag{5}$$

where $\omega_p(0) \equiv \omega_p(\mathbf{q} = 0)$. The energy $E_p = \hbar\omega_p$ is the minimum energy necessary for exciting a plasmon. Typical values of plasmon energy E_p at metallic densities are in the range of $2 - 20$ eV.

Having shown that an electron gas supports free and collective oscillation modes, we proceed to investigate the dynamical dielectric function $\varepsilon(\mathbf{q}, \omega)$ of the free electron gas. The dielectric function is the response of the electronic gas to an external electric field and determines the electronic properties of the solid [1, 11, 15]. We consider an electrically neutral homogeneous electronic gas and introduce a weak space-time-varying external charge density $\rho_{\text{ext}}(\mathbf{x},t)$ [14]. Our goal is to investigate the longitudinal response of the system as a result of the external perturbation. In free space, the external charge density produces an electric displacement field $\mathbf{D}(\mathbf{x},t)$ determined by the divergence relation $\nabla \cdot \mathbf{D} = 4\pi\rho_{\text{ext}}$. Moreover, the system responds and generates additional charges (induced charges) with density $\rho_{\text{ind}}(\mathbf{x},t)$ creating a polarization field $\mathbf{P}(\mathbf{x},t)$ defined by the expression $\nabla \cdot \mathbf{P} = -\rho_{\text{ind}}$ [1]. Because of the polarization, the total charge density inside the electron gas will be $\rho_{\text{tot}} = \rho_{\text{ext}} + \rho_{\text{ind}}$, leading to the screened electric field $\mathbf{E}$, determined by $\nabla \cdot \mathbf{E} = 4\pi\rho_{\text{tot}}$. The fundamental relation $\mathbf{D} = \mathbf{E} + 4\pi\mathbf{P}$ is derived after combining the aforementioned field equations.

The dielectric function is introduced as the linear optical response of the system. According to the linear response theory and taking into account the non-locality in time and space [2, 14], the total field depends linearly on the external field, if the latter is weak. In the most general case, we have

$$\mathbf{D}(\mathbf{x}, t) = \int d\mathbf{x}' \int_{-\infty}^{\infty} dt' \varepsilon(\mathbf{x} - \mathbf{x}', t - t')\mathbf{E}(\mathbf{x}', t'), \tag{6}$$

where we have implicitly assumed that all length scales are significantly larger than the crystal lattice, ensuring homogeneity. Thence, the response function depends only on the differences between spatial and temporal coordinates [2, 8]. In Fourier space, the convolutions turn into multiplications and the fields are decomposed into individual plane-wave components of the wavevector $\mathbf{q}$ and angular frequency ω. Thus, in the Fourier domain, Eq. (6) reads

$$\mathbf{D}(\mathbf{q}, \omega) = \varepsilon(\mathbf{q}, \omega)\mathbf{E}(\mathbf{q}, \omega). \tag{7}$$

For notational convenience, we designate the Fourier-transformed quantities with the same symbol as the original while they differ in the dependent variables. The Fourier transform of an arbitrary field $\mathbf{F}(\mathbf{r},t)$ is given by $\mathbf{F}(\mathbf{r},t) = \int \mathbf{F}(\mathbf{q},\omega)e^{i(\mathbf{q}\cdot\mathbf{r}-\omega t)}d\mathbf{q}dt$ where ω, $\mathbf{q}$ represent the Fourier transform quantities. Hence, the Fourier transform of the divergence equations of **D** and **E** yields

$$-i\mathbf{q}\cdot\mathbf{D}(\mathbf{q},\omega) = 4\pi\rho_{\text{ext}}(\mathbf{q},\omega) \tag{8}$$

$$-i\mathbf{q}\cdot\mathbf{E}(\mathbf{q},\omega) = 4\pi\rho_{\text{tot}}(\mathbf{q},\omega). \tag{9}$$

In longitudinal oscillations, the electron displacement field is in the direction of **q** (**Figure 1(b)**), thus, $\mathbf{q}\cdot\mathbf{D} = qD$ and $\mathbf{q}\cdot\mathbf{E} = qE$, where $D(\mathbf{q},\omega)$ and $E(\mathbf{q},\omega)$ refer to longitudinal fields. Combining Eqs. (7)–(9) yields

$$\rho_{\text{tot}}(\mathbf{q},\omega) = \frac{\rho_{\text{ext}}(\mathbf{q},\omega)}{\varepsilon(\mathbf{q},\omega)}. \tag{10}$$

Interestingly enough, in the absence of external charges, $\rho_{\text{ext}}(\mathbf{q},\omega) = 0$, Eq. (10) states that non-zero amplitudes of charge oscillation exist, that is, $\rho_{\text{tot}}(\mathbf{q},\omega) \neq 0$, under the condition

$$\varepsilon(\mathbf{q},\omega) = 0. \tag{11}$$

In other words, in the absence of any external perturbation, free collective charge oscillations exist with dispersion relation $\omega(\mathbf{q})$ that satisfies condition (11). These are plasmon modes, and consequently, Eq. (11) is referred as *plasmon condition*. Furthermore, condition (11) leads to $\mathbf{E} = -4\pi\mathbf{P}$, revealing that at plasmon frequencies the electric field is a pure depolarization field [1, 2].

We note that due to their longitudinal nature, plasmon waves cannot couple to any transverse wave such as electromagnetic waves; as a result, volume plasmons cannot be excited by light. On the other hand, moving charged particles can be used for exciting plasmons. For instance, an electron beam passing through a thin metal excites plasmons by transferring part of its energy to the plasmon excitation. As a result, plasmons do not decay directly via electromagnetic radiation but only through energy transfer to electron-hole excitation (Landau damping) [2, 8, 14].

2.2. Dynamical dielectric function

Based on the plasmon condition (11), the problem has been reduced in the calculation of the dynamical dielectric function $\varepsilon(\mathbf{q}, \omega)$. Further investigation of $\varepsilon(\mathbf{q}, \omega)$ reveals the plasmon dispersion relation as well as the Landau-damping regime, that is, where plasmons decay very fast exciting electron-hole pairs [8]. Classically, in the long wavelength limit, the dielectric response $\varepsilon(0, \omega)$ can be calculated in the context of the plasma model [1, 11]. Let us consider the plasma model of Eq. (4) subjected to a weak and harmonic time-varying external field $\mathbf{D}(t) = \mathbf{D}(\omega)e^{-i\omega t}$; Eq. (4) is modified to read

$$m\ddot{\mathbf{x}}(t) + 4\pi e^2 n\mathbf{x}(t) = -e\mathbf{D}(t). \tag{12}$$

Assuming also a harmonic in time electron displacement, that is, $\mathbf{x}(t) = \mathbf{x}(\omega)e^{-i\omega t}$, the Fourier transform of Eq. (12) yields

$$(-m\omega^2 + 4\pi e^2 n)\mathbf{x}(\mathbf{q},\omega) = -e\mathbf{D}(\mathbf{q},\omega). \tag{13}$$

Introducing Eq. (3) in Eq. (13) and using the relation (7), we derive the spatially local dielectric response

$$\varepsilon(0,\omega) = 1 - \frac{\omega_p(0)^2}{\omega^2}, \tag{14}$$

where the plasma frequency $\omega_p(0)$ is defined in Eq. (5). Eq. (14) verifies that the plasmon condition (11) is satisfied at the plasma frequency. The dielectric function (14) coincides with the Drude model permittivity.

Further investigation of the dynamical dielectric function can be performed using quantum mechanics. An explicit form of $\varepsilon(\mathbf{q},\omega)$ including screening effect has been evaluated in the context of the *random phase approximation* (RPA) [8, 12–14] and is given by

$$\varepsilon(\mathbf{q},\omega) = 1 - v_c(\mathbf{q})\chi_0(\mathbf{q},\omega) \tag{15}$$

where $v_c(\mathbf{q})$ is the Fourier transform of the Coulomb potential and $\chi_0(\mathbf{q},\omega)$ is the polarizability function, known as Lindhard formula [8, 12–14]. The Coulomb potential in two and three dimensions, respectively, reads

$$v_c(\mathbf{q}) = \begin{cases} \dfrac{2\pi e^2}{|\mathbf{q}|\varepsilon_b} & \text{(2D)} \\ \dfrac{4\pi e^2}{|\mathbf{q}|^2\varepsilon_b} & \text{(3D)} \end{cases} \tag{16}$$

where ε_b represents the background lattice dielectric constant of the system.

In RPA approach, the dynamical conductivity $\sigma(\mathbf{q},\omega)$ reads [8]

$$\sigma = \frac{i\omega e^2}{q^2}\chi_0(\mathbf{q},\omega), \tag{17}$$

revealing the fundamental relation between $\varepsilon(\mathbf{q},\omega)$ and $\sigma(\mathbf{q},\omega)$ that also depends on system dimensions; we have finally

$$\varepsilon(\mathbf{q},\omega) = 1 + i\frac{q^2 v_c}{\omega e^2}\sigma(\mathbf{q},\omega). \tag{18}$$

In the random phase approximation, the most important effect of interactions is that they produce electronic screening, while the electron-electron interaction is neglected. The polarizability of a non-interacting electron gas is represented by Lindhard formula as follows:

$$\chi_0(\mathbf{q},\omega) = -\frac{2}{V}\sum_{\mathbf{k}} \frac{f(\epsilon_{\mathbf{k}+\mathbf{q}}) - f(\epsilon_{\mathbf{k}})}{\hbar\omega - (\epsilon_{\mathbf{k}+\mathbf{q}} - \epsilon_{\mathbf{k}}) + i\hbar\eta} \tag{19}$$

where factor 2 is derived by spin degeneracy (summation over the two possible values of spin $s = \uparrow,\downarrow$) [8, 13, 14]. The summation is over all the wavevectors $\mathbf{k}$, V is the volume, $i\hbar\eta$ represents a small imaginary number to be brought to zero after the summation, and $\epsilon_{\mathbf{k}}$ is the kinetic energy for the wave vector $\mathbf{k}$. The carrier distribution f is given by Fermi-Dirac distribution $f(\epsilon_{\mathbf{k}}) = \left(\exp\left[\beta(\epsilon_{\mathbf{k}} - \mu)\right] + 1\right)^{-1}$, where μ is the chemical potential and $\beta = 1/k_B T$ with Boltzmann's constant denoted by k_B and T is the absolute temperature. Equation (19) describes processes in which a particle in state $\mathbf{k}$, which is occupied with probability $f(\epsilon_{\mathbf{k}})$, is scattered into state $\mathbf{k} + \mathbf{q}$, which is empty with probability $1 - f(\epsilon_{\mathbf{k+q}})$. Eqs. (15)–(19) consist of the basic equations for a detailed investigation of charge density fluctuations and the screening effect, electron-hole pair excitation, and plasmons. With respect to condition (11), the roots of Eq. (15) determine the plasmon modes. Moreover, the poles of χ_0 account for electron-hole pair excitation defining the Plasmon-damping regime [12–14].

For an analytical investigation, we split the summation of Eq. (19) in two parts. We make an elementary change of variables $\mathbf{k} + \mathbf{q} \to -\mathbf{k}$, in the term that includes $f(\epsilon_{\mathbf{k+q}})$, and assume that the kinetic energy is symmetric with respect to the wavevector, that is, $\epsilon_{\mathbf{k}} = \epsilon_{-\mathbf{k}}$. Therefore, formula (19) yields

$$\chi_0(\mathbf{q},\omega) = \frac{2}{V}\left(\sum_{\mathbf{k}} \frac{f(\epsilon_{\mathbf{k}})}{\hbar z - (\epsilon_{\mathbf{k+q}} - \epsilon_{\mathbf{k}})} - \sum_{\mathbf{k}} \frac{f(\epsilon_{\mathbf{k}})}{\hbar z + (\epsilon_{\mathbf{k+q}} - \epsilon_{\mathbf{k}})}\right) \tag{20}$$

where $z = \omega + i\eta$. At zero temperature, the chemical potential is equal to Fermi energy, that is, $\mu = E_F$ [8, 11, 14], and the Fermi-Dirac distribution is reduced to Heaviside step function, thus, $f(\epsilon_{\mathbf{k}})|_{T=0} = \Theta(E_F - \epsilon_{\mathbf{k}})$. The kinetic energy of each electron of mass m in state $\mathbf{k}$ is given by

$$\epsilon_{\mathbf{k}} = \frac{\hbar^2 |\mathbf{k}|^2}{2m}, \tag{21}$$

hence

$$\epsilon_{\mathbf{k+q}} - \epsilon_{\mathbf{k}} = \frac{\hbar^2}{2m}\left(|\mathbf{q}|^2 + 2\mathbf{k}\cdot\mathbf{q}\right). \tag{22}$$

At zero temperature, because of the Heaviside step function, the only terms that survive in summation (20) are those with $|\mathbf{k}| < k_F$, where k_F is the Fermi wavenumber and related to Fermi energy by equation (21) as $k_F = (2mE_F/\hbar^2)^{1/2}$. Subsequently, we obtain for the Lindhard formula

$$\chi_0(\mathbf{q},\omega) = \frac{4}{V}\sum_{|\mathbf{k}|<k_F} \frac{\epsilon_{\mathbf{k+q}} - \epsilon_{\mathbf{k}}}{(\hbar z)^2 - (\epsilon_{\mathbf{k+q}} - \epsilon_{\mathbf{k}})^2} \tag{23}$$

Summation turns into integration by using $V^{-1}\sum_{|\mathbf{k}|}(\ldots) \to (2\pi)^{-3}\int d^3\mathbf{k}(\ldots)$, hence

$$\chi_0(\mathbf{q},\omega) = \frac{4}{(2\pi)^3}\int d^3\mathbf{k}\frac{\epsilon_{\mathbf{k+q}} - \epsilon_{\mathbf{k}}}{(\hbar z)^2 - (\epsilon_{\mathbf{k+q}} - \epsilon_{\mathbf{k}})^2} \tag{24}$$

where the imaginary part in z guarantees the convergence of the integrals around the poles $\hbar\omega = \pm(\epsilon_{\mathbf{k+q}} - \epsilon_{\mathbf{k}})$. The poles of χ_0 determine the Landau-damping regime where plasmons decay into electron-hole pairs excitation. In particular, the damping regime is a continuum bounded by the limit values of $(\epsilon_{\mathbf{k+q}} - \epsilon_{\mathbf{k}})$; $\mathbf{k}$ takes its maximum absolute value $|\mathbf{k}| = k_F$ and the inner product takes the extreme values $k_F\,\hat{\mathbf{k}}\cdot\mathbf{q} = \pm k_F|\mathbf{q}|$.

$$\frac{\hbar q}{2m}(q - 2k_F) < \omega < \frac{\hbar q}{2m}(q + 2k_F), \tag{25}$$

where $q = |\mathbf{q}|$. The Landau-damping continuum (electron-hole excitation regime) is demonstrated in **Figure 2** by the shaded area.

Introducing relation (22) into Eq. (24) and changing to spherical coordinates (r,θ,ϕ), where $r = |\mathbf{k}|$ and θ are the angle between $\mathbf{k}$ and $\mathbf{q}$, we obtain

$$\chi_0(q,\omega) = \frac{2k_F^4 q}{(2\pi)^3 m z^2}\int_0^{2\pi} d\phi \int_0^1 dx\, x^2 \int_0^{\pi} d\theta \frac{\left(\frac{q}{k_F} + 2x\cos\theta\right)\sin\theta}{1 - \left(\frac{v_F q}{z}\right)^2\left(\frac{q}{2k_F} + x\cos\theta\right)^2}. \tag{26}$$

where $x = r/k_F$ is a dimensionless variable and $v_F = \hbar k_F/m$ is the Fermi velocity. In the non-static ($\omega \gg v_F q$) and long wavelength ($q \ll k_F$) limits, we can expand the integral in a power series of q. Keeping up to q^3 orders, we evaluate integral (26) and set the imaginary part of z zero, that is, $z = \omega$. That leads to a third-order approximation polarizability function

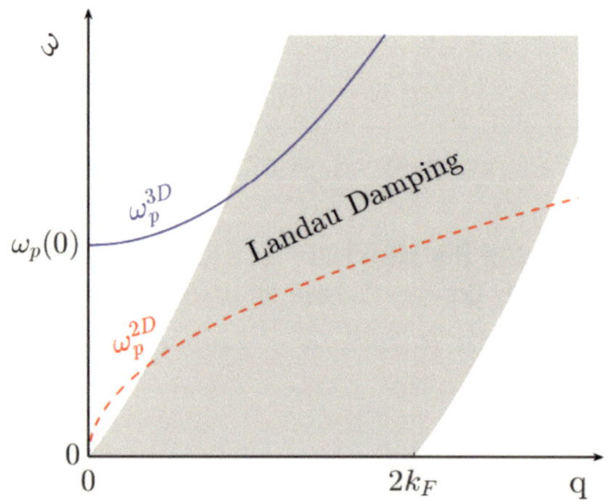

Figure 2. Dispersion relation of plasmons in the bulk of three-dimensional solid (blue solid line) and in two-dimensional electron gas (dashed red curve) plasmons. The shaded region demonstrates the Landau-damping regime where plasmons decay to electron-hole pairs excitation.

http://dx.doi.org/10.5772/67900

$$\chi_0(q,\omega) = \frac{k_F^3 q^2}{3\pi^2 m\omega^2}\left(1 + \frac{3v_F^2 q^2}{5\omega^2}\right), \tag{27}$$

which, in turn, yields the dielectric function by using formula (14) and the three-dimensional Coulomb interaction (16), hence

$$\varepsilon(q \to 0,\omega) = 1 - \frac{\omega_p(0)^2}{\omega^2}\left(1 + \frac{3}{5}\left(\frac{v_F q}{\omega}\right)^2\right), \tag{28}$$

where vacuum is assumed as background ($\varepsilon_d = 1$) and we use the relation 0 [1, 15] where n is the electron density. The result (28) is reduced to simple Drude dielectric function (14) for $q = 0$.

The plasmon condition (11) determines the q-dependent plasmon dispersion relation $\omega_p(q)$. Demanding $\varepsilon(q,\omega) = 0$, Eq. (28) yields approximately

$$\omega_p(q) \approx \omega_p(0)\left(1 + \frac{3}{10}\left(\frac{v_F q}{\omega_p(0)}\right)^2\right). \tag{29}$$

Interestingly enough, the leading term of plasma frequency (29) does not include any quantum quantity, such as v_F, which appears as non-local correction in sub-leading terms. That reveals that plasmons in 3D metals are purely classical modes. Moreover, a gap, that is, $\omega_p(0)$, appears in the plasmon spectrum of three-dimensional metals. The plasmon dispersion relation (29) is shown in **Figure 2**.

In the random phase approximation, the electrons do not scatter, that is, collision between electrons and crystal impurities is not taken into account. As a consequence, the dielectric function is calculated to be purely real; this is nevertheless an unphysical result as can be seen clearly at zero frequency where the dielectric function is not well defined, that is, $\varepsilon(q, 0) = \infty$. The problem is cured by introducing a relaxation time τ in the denominator of the dielectric function as follows:

$$\varepsilon(q \to 0,\omega) = 1 - \frac{\omega_p^2(q)}{\omega(\omega + \mathrm{i}/\tau)} \tag{30}$$

We can phenomenologically prove expression (30) by using the simple plasma model. In particular, we modify the equation of motion (12) to a damped-driven harmonic oscillator by assuming that the motion of electron is damped via collisions occurring with a characteristic frequency $\gamma = 1/\tau$ [2]; this approach immediately leads to the dielectric response (30). Typically values of relaxation time τ are of the order 10^{-14} s, at room temperature. The relaxation time is determined experimentally. In the presence of τ, the dielectric function (15) is well defined at $\omega = 0$, where the real part of permittivity has a peak with width τ^{-1} known as *Drude peak*. Furthermore, it can be shown that equation (30) satisfies the Kramers-Kronig relations (sum rules) [1, 14, 15].

2.3. Surface plasmon polariton

A new guided collective oscillation mode called *surface plasmon* arises in the presence of a boundary. Surface plasmon is a surface electromagnetic wave that propagates along an interface between a conductor (metal) and an insulator (dielectric). This guided mode couples to electromagnetic waves resulting in a polariton. Surface plasmon polaritons (SPPs) occur at frequencies close to but smaller than plasma frequency. These surface modes show exceptional properties for applications of nanophotonics, specifically they constitute a class of nanophotonics themselves, namely nanoplasmonics. The basic property is the subwavelength nature, that is, the wavelength of SPPs is smaller than electromagnetic radiation at the same frequency and in the same medium [2, 3, 9].

Let us consider a waveguide formed by a planar interface at $z = 0$ consisting of two semi-infinite nonmagnetic media (permeability $\mu = 1$) with dielectric functions ε_1 and ε_2 as **Figure 3a** denotes. The dielectric functions are assumed to be local in space (non-q−dependent) and nonlocal in time (ω dependence), hence $\varepsilon_{1,2} = \varepsilon_{1,2}(\omega)$. Assuming harmonic in time dependence in the form $\mathbf{u}(\mathbf{r},t) = \mathbf{u}(\mathbf{r})e^{-\mathrm{i}\omega t}$, the Maxwell equations (in CGS units) in the absence of external charges and currents read

$$\nabla \cdot (\varepsilon_j \mathbf{E}_j) = 0 \qquad \nabla \times \mathbf{E}_j = \mathrm{i}k_0 \mathbf{H}_j \tag{31}$$

$$\nabla \cdot (\mathbf{H}_j) = 0 \qquad \nabla \times \mathbf{H}_j = -\mathrm{i}\varepsilon_j k_0 \mathbf{E}_j \tag{32}$$

where $k_0 = \omega/c$ is the free space wavenumber and the index j denotes the media as $j = 1$ for $z < 0$ and $j = 2$ for $z > 0$. Combining Eqs. (31) and (32), the fields are decoupled into two separate Helmholtz equations [2, 4] as

$$\left[\nabla^2 + k_0^2 \varepsilon_j\right] \begin{pmatrix} \mathbf{E}_j(\mathbf{r}) \\ \mathbf{H}_j(\mathbf{r}) \end{pmatrix} = 0 \tag{33}$$

where $\mathbf{r} = (x,y,z)$. For simplicity, let us assume surface electromagnetic waves propagating along one direction, chosen to be the x direction (**Figure 3b**), and show no spatial variations in the perpendicular in-plane direction, hence $\partial_y \mathbf{u} = 0$. Under this assumption, we are seeking

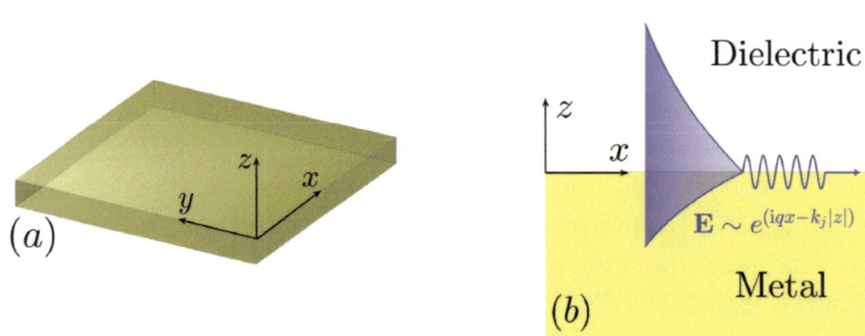

Figure 3. A planar interface is formed between a metal and a dielectric where surface plasmon polaritons (SPPs) propagate in (a) three- and (b) two-dimensional representation. (b) A schematic illustration of the SPP field.

http://dx.doi.org/10.5772/67900

electromagnetic waves of the form $\psi_j(\mathbf{r}) = \psi_j(z)e^{iq_jx}$, where $\psi_j = (\mathbf{E}_j,\mathbf{H}_j)^T$ and q will be the plasmon propagation constant. Substituting the aforementioned ansatz into Helmholtz equation (33), we obtain the guided electromagnetic modes equation [2]

$$\left[\frac{\partial^2}{\partial z^2} + (k_0^2\varepsilon_j - q_j^2)\right]\begin{pmatrix}\mathbf{E}_j(z)\\ \mathbf{H}_j(z)\end{pmatrix} = 0. \tag{34}$$

Surface waves are waves that have been trapped at the interface ($z = 0$) and decay exponentially away from it $\left(\psi_j(z) \sim e^{-\kappa_j|z|}\text{for } k_j > 0\right)$. Consequently, propagating wave solutions along z is not desired. In turn, we derive the *surface wave condition*

$$\kappa_j = \sqrt{q_j^2 - k_0^2\varepsilon_j} \in \mathbb{R}. \tag{35}$$

In order to determine the spatial field profiles and the SPP dispersion relation, we need to find explicit expressions for each field component of **E** and **H**. This can be achieved by solving the curl equations (31) and (32), which naturally lead to two self-consistent set of coupled governing equations. Each set corresponds to one of the fundamental polarizations, namely transverse magnetic (TM) (*p*-polarized waves) and transverse electric (TE) (*s*-polarized waves), hence

Transverse magnetic (TM)		**Transverse electric (TE)**	
E_{jz}	$= -\frac{q_j}{k_0\varepsilon_j}H_{jy}$	H_{jz}	$= \frac{q}{k_0}E_{jy}$
E_{jx}	$= -\frac{i}{k_0\varepsilon_j}\frac{\partial}{\partial z}H_{jy}$	H_{jx}	$= \frac{i}{k_0}\frac{\partial}{\partial z}E_{jy}$
$\frac{\partial^2}{\partial z^2}H_{jy}$	$-(q_j^2 - k_0^2\varepsilon_j)H_{jy} = 0$	$\frac{\partial^2}{\partial z^2}E_{jy}$	$-(q_j^2 - k_0^2\varepsilon_j)E_{jy} = 0$

We focus on transverse magnetic (TM) polarization, in which the magnetic field **H** is parallel to the interface. Since the planar interface extends along (x,y) plane, the TM fields read $\mathbf{E}_j = (E_{jx}, 0, E_{jz})$ and $\mathbf{H}_j = (0, H_{jy}, 0)$. Solving the TM equations for surface waves, we obtain for each half plane

$$\begin{gathered} z < 0 \quad (j = 1) \\ H_y = A_1 e^{iq_1x}e^{k_1z} \end{gathered} \tag{36}$$

$$E_x = -\frac{ik_1A_1}{k_0\varepsilon_1}e^{iq_1x}e^{k_1z} \tag{37}$$

$$E_z = -\frac{q_1A_1}{k_0\varepsilon_1}e^{iq_1x}e^{k_1z} \tag{38}$$

$$\begin{gathered} z > 0 \quad (j = 2) \\ H_y = A_2 e^{iq_2x}e^{-k_2z} \end{gathered} \tag{39}$$

$$E_x = \frac{ik_2 A_2}{k_0 \varepsilon_2} e^{iq_2 x} e^{-k_2 z} \tag{40}$$

$$E_z = -\frac{q_2 A_2}{k_0 \varepsilon_2} e^{iq_2 x} e^{-k_2 z} \tag{41}$$

where k_j is related to q_j by Eq. (35). The boundary conditions imply that the parallel to interface components of electric (E_x) and magnetic (H_y) fields must be continuous. Accordingly, we demand Eqs. (36) = (39) and Eqs. (37) = (40) at $z = 0$, hence we find the system of equations

$$\begin{pmatrix} e^{iq_1 x} & -e^{iq_2 x} \\ \frac{k_1}{\varepsilon_1} e^{iq_1 x} & \frac{k_2}{\varepsilon_2} e^{iq_2 x} \end{pmatrix} \begin{pmatrix} A_1 \\ A_2 \end{pmatrix} = 0, \tag{42}$$

which has a solution only if the determinant is zero. As an outcome, we obtain the so-called *surface plasmon polariton condition*

$$\frac{k_1}{\varepsilon_1} + \frac{k_2}{\varepsilon_2} = 0. \tag{43}$$

Condition (43) states that the interface must consist of materials with opposite signed permittivities, since surface wave condition requires the real part of both k_1 and k_2 to be non-negative numbers. For that reason, interface between metals and dielectrics may support surface plasmons, since metals show negative permittivity at frequencies smaller than plasma frequency [2]. Furthermore, boundary conditions demand the continuity of the normal to the interface electric displacement ($D_{jz} = \varepsilon_j E_{jz}$) yielding the continuity of the plasmon propagation constant $q_1 = q_2 = q$ [4]. In turn, by combining Eq. (35) with Eq. (43) we obtain the dispersion relation for the surface plasmon polariton

$$q(\omega) = \frac{\omega}{c} \sqrt{\frac{\varepsilon_1 \varepsilon_2}{\varepsilon_1 + \varepsilon_2}} \tag{44}$$

where $\varepsilon_{1,2}$ are, in general, complex functions of ω. For a metal-dielectric interface, it is more convenient to use the notation $\varepsilon_1 = \varepsilon_d$ and $\varepsilon_2 = \varepsilon_m$ for dielectric and metal permittivity, respectively. In long wavelengths, the SPP wavenumber is close to the light line in dielectric, viz. $q \simeq k_0 \sqrt{\varepsilon_d}$, and the waves are extended over many wavelengths into the dielectrics [2, 4]; these waves are known as Sommerfeld-Zenneck waves and share similarities with free surface electromagnetic modes [2]. On the other hand, at the limit $q \to \infty$, Eq. (44) asymptotically leads to the condition

$$\varepsilon_d + \varepsilon_m = 0 \tag{45}$$

indicating the *nonretarded* surface plasmon limit [4]. In the vicinity of the nonretarded limit, Eq. (35) yields $k_j \simeq q \gg k_0$. Furthermore, in the nonretarded limit the phase velocity $v_{\text{ph}} = \omega / q$ is tending to zero unveiling the electrostatic nature characterized by the surface plasmon [2, 3]. As a result, at the same frequency v_{ph} is much smaller than the speed of light and, thus, the SPP wavelength (λ_{sp}) is always smaller than the light wavelength (λ_{ph}), that is, $\lambda_{sp} < \lambda_{ph}$, revealing the subwavelength nature of surface plasmon polaritons [2, 4]. In addition, due to the fact that SPP phase velocity is always smaller than the phase velocity of propagating electromagnetic waves, SPPs cannot radiate and, hence, they are well-defined surface propagating electromagnetic waves. Demanding $q \to \infty$ in the dielectric function (30), we find the so-called *surface plasmon frequency* ω_{sp}, which is the upper frequency limit that SPPs occur

$$\omega_{sp} = \sqrt{\frac{\omega_p^2}{1+\varepsilon_d} - \gamma^2} \simeq \frac{\omega_p}{\sqrt{1+\varepsilon_d}}, \tag{46}$$

indicating that SPPs always occur at frequencies smaller than bulk plasmons.

If we follow the same procedure for transverse electric polarized fields, in which the electric field is parallel to interface and the only non-zero electromagnetic field components are E_y,H_x, and H_z, we will find the condition $k_1 + k_2 = 0$ [2]. This condition is satisfied only for $k_1 = k_2 = 0$ unveiling that s-polarized surface modes do not exist. Consequently, surface plasmon polaritons are always TM electromagnetic waves.

Due to metallic losses, SPPs decay exponentially along the interface restricting the propagation length. Mathematically speaking, losses are described by the small imaginary part in the complex dielectric function of metal $\varepsilon_m = -\varepsilon_{m'} - \mathrm{i}\varepsilon''_{m}$, where ε'_m,$\varepsilon''_m > 0$. Consequently, the SPPs propagation constant (44) becomes complex, that is, $q = q' + \mathrm{i}q''$, where the imaginary part accounts for losses of SPPs energy. In turn, the effective propagation length L, which shows the rate of change of the energy attenuation of SPPs [2, 3], is determined by the imaginary part $\mathrm{Im}[q]$ as $L^{-1} = 2\mathrm{Im}[q]$.

Gain materials rather than passive regular dielectrics have been used to reduce the losses in SPP propagation. Gain materials are characterized by a complex permittivity function, that is, $\varepsilon_d = \varepsilon'_d = +\mathrm{i}\varepsilon''_d$, with ε'_d,$\varepsilon''_d > 0$, where ε''_d is a small number compared to ε'_d and accounts for gain. As a result, gain dielectric gives energy to the system counterbalancing the metal losses. We investigate the SPP dispersion relation (44) in the presence of gain and loss materials, and find an explicit formula for gain ε''_d where the SPP wavenumber is reduced to real function, resulting in lossless SPPs propagation. In addition, we find an upper limit that values of gain are allowed. In this critical gain, the purely real SPP propagation constant becomes purely imaginary, destroying the SPP modes.

The dispersion relation (44) can also be written as $q = k_0 n_{sp}$ [3], where n_{sp} is the plasmon effective refractive index given by

$$n_{sp} = \sqrt{\frac{\varepsilon_d \varepsilon_m}{\varepsilon_d + \varepsilon_m}}. \tag{47}$$

We are seeking for a gain ε''_d such that the effective index n_{sp} becomes real. Substituting the complex function describing the dielectric and metal into Eq. (47), the function n_{sp} is written in the ordinary complex form as [3]

$$n_{sp} = \sqrt{\frac{\sqrt{x^2+y^2}+x}{2}} + \mathrm{i}\,\mathrm{sgn}(y)\sqrt{\frac{\sqrt{x^2+y^2}-x}{2}}, \tag{48}$$

where sgn(y) is the discontinuous signum function [3] and

$$x = \frac{\varepsilon_d'|\varepsilon_m|^2 - \varepsilon_m'|\varepsilon_d|^2}{|\varepsilon_d+\varepsilon_m|^2} \tag{49}$$

$$y = \frac{\varepsilon_d''|\varepsilon_m|^2 - \varepsilon_m''|\varepsilon_d|^2}{|\varepsilon_d+\varepsilon_m|^2} \tag{50}$$

with $|z_*|$ denoting the norm of the complex number z_*. The poles in Eqs. (49) and (50) correspond to the nonretarded surface plasmon limit (45).

Considering the plasmon effective index n_{sp} in Eq. (48) in the (x, y) plane, we observe that lossless SPP propagation ($\mathrm{Im}[n_{sp}] = \mathrm{Im}[q] = 0$) is warranted when the conditions $y = 0$ and $x > 0$ are simultaneously satisfied. Let us point out that for $y = 0$ and $x < 0$, although the imaginary part in Eq. (48) vanishes due to the signum function, its real part becomes imaginary, that is, $n_{\mathrm{sp}} = \mathrm{i}\sqrt{|x|}$, which does not correspond to propagation waves. Solving Eq. (50) for $y = 0$ with respect to gain ε_d'' and avoiding the nonretarded limit (45), that is, $\varepsilon_d \neq -\varepsilon_m$, we obtain two exact solutions [3] as follows:

$$\varepsilon_d''\pm = \frac{|\varepsilon_m|^2}{2\varepsilon_m''}\left(1 \pm \sqrt{1 - \left(\frac{2\varepsilon_d'\varepsilon_m''}{|\varepsilon_m|^2}\right)^2}\right). \tag{51}$$

Due to the fact that ε_d is real, we read from Eq. (51) that [3].

$$|\varepsilon_m|^2 \geqslant 2\varepsilon_d'\varepsilon_m''. \tag{52}$$

Using inequality (52), we read for the solution ε_{d+} of Eq. (51) that $\varepsilon''_{d+} \geqslant \varepsilon_d'$. This is a contradiction since the ε_d'' is defined to be smaller than ε_d'. Thus, ε_{d+} does not correspond to a physically relevant gain.

Solving, on the other hand, Eq. (49) for $x > 0$, with respect to the dielectric gain ε_d'', we determine a critical value ε_c distinguishing the regimes of lossless and prohibited SPP propagation [3], namely

$$\varepsilon_c = \varepsilon_d'\sqrt{\frac{|\varepsilon_m|^2}{\varepsilon_m'\varepsilon_d'} - 1}, \tag{53}$$

hence, Eq. (53) sets an upper limit in values of gain. The appearance of critical gain can be understood as follows: In Eq. (51) the gain ε_{d-} becomes equal to critical gain ε_c when $\varepsilon_d + \varepsilon_m = 0$ [3], where the last item is the nonretarded limit where $q \to \infty$. Specifically, the surface plasmon exists when the metal is characterized by the Drude dielectric function of Eq. (30), $\varepsilon_d''- = \varepsilon_c$ at $\omega = \omega_{sp}$, corresponding to a maximum frequency [3].

In order to represent the above theoretical findings, we use the dielectric function of Eq. (30) to calculate the SPP dispersion relation for an interface consisting of silver with $\omega_p(0) = 13.67$ PHz and $\gamma = 0.1018$ PHz, and silica glass with $\varepsilon_d' = 1.69$ and for gain $\varepsilon_d'' = \varepsilon_{d-}$ determined by Eq. (51). We represent in **Figure 3a** the SPP dispersion relation of Eq. (44) for lossless case ($\varepsilon_d'' = \varepsilon_d''-$), where the lossless gain is denoted by the inset image in **Figure 3a**. We indicate the real and imaginary of normalized SSP dispersion q/k_p ($k_p \equiv \omega_p/c$), with respect to the normalized frequency ω/ω_p. We observe, indeed, that for $\omega < \omega_{\rm sp}$ the imaginary part of q vanishes, whereas for $\omega > \omega_{\rm sp}$ the SPPs wavenumber is purely imaginary. Subsequently, in the vicinity of $\omega = \omega_{\rm sp}$ a phase transition from lossless to prohibited SPPs propagation is expected [3].

We also solve numerically the full system of Maxwell equations (31) and (32) in a two-dimensional space for transverse magnetic polarization. The numerical experiments have been performed by virtue of the multi-physics commercial software COMSOL and the frequency ω is confined in the range $[0.3\omega_p, 0.75\omega_p]$ with the integration step $\Delta\omega = 0.01\omega_p$. In the same range, the lossless gain is calculated by Eq. (51), to be $[8 \cdot 10^{-3}, 8 \cdot 10^{-2}]$. For the excitation of SPPs on the metallic surface, we use the near-field technique [2, 3, 9, 10]. For this purpose, a circular electromagnetic source of radius $R = 20$ nm has been located 100 nm above the metallic surface acting as a point source, since the wavelength λ of EM waves is much larger, that is, $\lambda >> R$ [2, 3]. In **Figure 4b**, we demonstrate, in a log-linear scale, the propagation length L, with respect to ω, subject in lossless gain ε_{d-} (blue line and open circles). For the sake of comparison, we plot $L(\omega)$ in the absence of gain (green line and filled circles). The solid lines represent the theoretical predictions obtained by the definition of L, whereas the circles

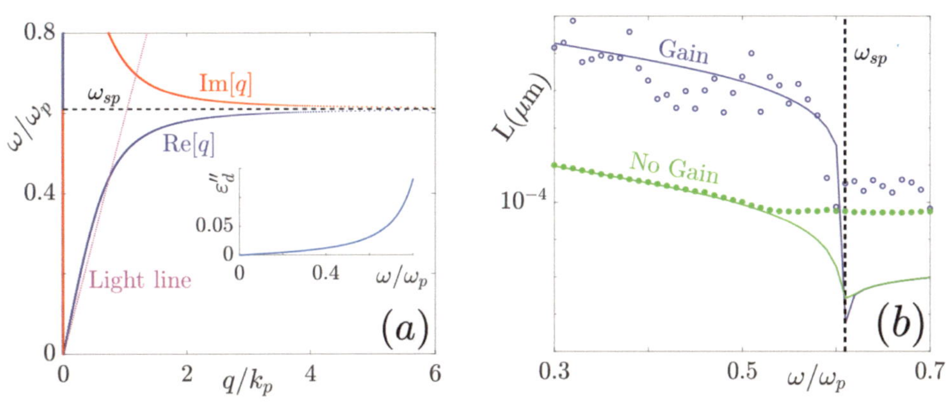

Figure 4. (a) The surface plasmon polariton (SPP) dispersion relation $q(\omega)$ in the presence of a gain material with gain corresponds to lossless SPP propagation. Re[q] and Im[q] are indicated by blue and red lines, respectively. The horizontal dashed black line denotes the SPP frequency ($\omega_{\rm sp} = 0.61\omega_p$) where an interchanging between Re[q] and Im[q] appears. The dotted magenta line indicates the light line in the dielectric. (Inset) Demonstration of the gain leads to lossless SPP propagation. (b) Theoretical (solid lines) and numerical (circles) prediction of SPP propagation length L in the presence (blue) and in the absence (green) of gain dielectric showing a phase transition that happens at $\omega_{\rm sp}$ (vertical dashed black line). Deviations between theoretical and numerical predications for $\omega > \omega_{\rm sp}$ correspond to quasi-bound EM modes. The $k_p = \omega_p/c$ is used as normalized unit of wavenumbers and ω_p as normalized unit for frequencies.

indicate numerical results. For the numerical calculations, the characteristic propagation length has been estimated by the inverse of the slope of the $\mathrm{Log}(I)$, where I is the magnetic intensity along the interface [2–4]. The black vertical dashed line denotes the SPP resonance frequency ω_{sp}, in which the phase transition appears. The graphs in **Figure 4b** indicate that in the presence of the lossless gain, SPPs may travel for very long, practically infinite, distances. Approaching the resonance frequency ω_{sp}, L decreases rapidly leading to a steep phase transition on the SPPs propagation. The deviations between theoretical and numerical results in **Figure 4** for frequencies near and greater than ω_{sp} are attributed to the fact that in the regime $\omega_{sp} < \omega < \omega_p$, there are quasi-bound EM modes [2, 3], where EM waves are evanescent along the metal-dielectric interface and radiate perpendicular to it. Consequently, the observed EM field for $\omega > \omega_{sp}$ corresponds to radiating modes [3].

3. Two-dimensional plasmons

In this section, we investigate plasmons in a two-dimensional electron gas (2DEG), where the electron sea is free to move only in two dimensions, tightly confined in the third. The reduced dimensions of electron confinement and Coulomb interaction cause crucial differences in plasmons excitation spectrum. For instance, plasmon spectrum in a 2DEG is gapless in contrast with three-dimensional case [13]. For the sake of completeness, we first discuss briefly plasmons in a regular 2DEG characterized by the usual parabolic dispersion relation (21) for a two-dimensional wavevector **k** lies in the plane of 2DEG. Thence, we focus on plasmons in a quite special two-dimensional material, viz. graphene. Graphene is a gapless two-dimensional semi-metal with linear dispersion relation. The linear energy spectrum offers great opportunity to describe graphene with chiral Dirac Hamiltonian for massless spin-1/2 fermions [7, 8, 10]. Furthermore, graphene can be doped with several methods, such as chemical doping [7], by applying an external voltage [10], or with lithium intercalation [16]. The doping shifts the Fermi level toward the conduction bands making graphene a great metal. The advantage to describe graphene electronic properties with massless carriers Dirac equation leads to exceptional optical and electronic properties, like very high electric conductivity and ultra-subwavelength plasmons [6–8, 10].

3.1. Dynamical dielectric function of 2D metals

In order to determine the plasmon spectrum of a two-dimensional electron gas, first of all we calculate the dielectric function in the context of random phase approximation (15) with v_q being the two-dimensional Coulomb interaction of Eq. (16). In the Lindhard formula (23), V and **k** denote a two-dimensional volume and wave-vector, respectively. First, we investigate a 2DEG described by the parabolic dispersion relation (21). The electrons are assumed to occupy a single band ignoring interband transitions, that is, transitions to higher bands. Thus, there is no orbital degeneracy ($g_v = 1$) resulting in the two-dimensional Fermi wavenumber $k_F = \sqrt{2\pi n}$, where n is the carrier (electrons) density [13, 17]. Turning summation (23) into integral by the substitution $V^{-1}\sum_{|\mathbf{k}|}(\ldots) = (2\pi)^{-2}\int d^2\mathbf{k}(\ldots)$, we obtain the Lindhard formula in integral form

$$\chi_0(\mathbf{q},\omega) = \frac{4}{(2\pi)^2}\int d^2|\mathbf{k}| \frac{\varepsilon_{\mathbf{k}+\mathbf{q}} - \varepsilon_{\mathbf{k}}}{(\hbar z)^2 - (\varepsilon_{\mathbf{k}+\mathbf{q}} - \varepsilon_{\mathbf{k}})^2} \tag{54}$$

The singe particle excitation continuum is still defined by expression (25), since the kinetic energy is considered to have the same form as in 3D case, even though the 2D Fermi wave-number has been modified. Transforming to polar coordinate system (r,θ) and using relation (22), integral (54) reads

$$\chi_0(q,\omega) = \frac{2k_F^3 q}{(2\pi)^2 m z^2}\int_0^1 dx\, x \int_0^{2\pi} d\theta \frac{\frac{q}{k_F} + 2x\cos\theta}{1 - \left(\frac{v_F q}{z}\right)^2\left(\frac{q}{2k_F} + x\cos\theta\right)^2} \tag{55}$$

where x is a dimensionless variable defined as $x = r/k_F$. Previously, since we are interested in long wavelength limit ($q \ll k_F$), we expand the integrand of Eq. (55) around $q = 0$. Keeping up to first orders of q, integral (55) yields

$$\chi_0(q,\omega) = \frac{k_F^2 q^2}{2\pi m\omega^2} \tag{56}$$

where $z \to \omega$ by sending the imaginary part of z to zero. The dielectric function is determined by the formula of Eq. (15) for 2D Coulomb interaction of Eq. (16), hence

$$\varepsilon(q,\omega) = 1 - \frac{2\pi n e^2 q}{m\omega^2} \tag{57}$$

The 2DEG plasmon dispersion relation is determined by Eq. (11) to be

$$\omega_p^{2D}(q) = \sqrt{\frac{2\pi n e^2 q}{m}} \tag{58}$$

related with volume plasmons dispersion relation by $\omega_p^{2D}(q) = \omega_p\sqrt{q/2}$. In contrast to three-dimensional electron gas where plasmon spectrum is gapped, in two-dimensional case the plasmon frequency depends on $\sqrt{q}$ making the plasmon spectrum gapless. In **Figure 2**, the 2D plasmon dispersion relation (58) is demonstrated together with three-dimensional case. Furthermore, it is worth pointing out the similarity between the plasmon dispersion relation of 2DEG of Eq. (58) and SPP of Eq. (44), that is, both show $\sqrt{q}$ dependence.

Let us now investigate the most special two-dimensional electron gas, namely graphene. At the limit where the excitation energy is small compared to E_F, the dispersion relation of graphene, viz. the relation between kinetic energy $\epsilon_{\mathbf{k}}^s$ and momentum $\mathbf{p} = \hbar\mathbf{k}$, is described by two linear bands as

$$\epsilon_{\mathbf{k}}^s = s\hbar v_F|\mathbf{k}| \tag{59}$$

where $s = \pm 1$ indicates the conduction (+1) and valence (-1) band, respectively, v_F is the two-dimensional Fermi velocity which is constant for graphene and equal to $v_F = 10^6$ m/s [7, 8, 10, 16, 18]. Because of valley degeneracy $g_v = 2$, the Fermi momentum is modified to read

$k_F = \sqrt{2\pi n/g_v} = \sqrt{\pi n}$ [8, 18]. The Fermi energy, given by $E_F = \hbar v_F k_F$, becomes zero in the absence of doping ($n = 0$). As a consequence, the E_F crosses the point where the linear valence and conduction bands touch each other, namely at the Dirac point, giving rise to the semi-metal character of the undoped graphene [7, 15, 16, 18]. The Lindhard formula of Eq. (19) needs to be generalized to include both intra- and interband transitions (valley degeneracy) as well as the overlap of states, hence

$$\chi_0(\mathbf{q},\omega) = -\frac{g_s g_v}{V} \sum_{s,s'} \sum_{\mathbf{k}} \frac{f(\epsilon^{s'}_{\mathbf{k}+\mathbf{q}}) - f(\epsilon^{s}_{\mathbf{k}})}{\hbar\omega - (\epsilon^{s'}_{\mathbf{k}+\mathbf{q}} - \epsilon^{s}_{\mathbf{k}}) + i\hbar\eta} F_{ss'}(\mathbf{k},\mathbf{k}+\mathbf{q}) \tag{60}$$

where the factors $g_s = g_v = 2$ account to spin and valley degeneracy, respectively. The Lindhard formula has been modified to contain two extra summations $\left(\sum_{s=-1}^{1} \sum_{s'=-1}^{1}\right)$ corresponding to valley degeneracy for the two bands of Eq. (59). In addition, the overlap of states function $F_{ss'}(\mathbf{k},\mathbf{k}+\mathbf{q})$ has been introduced and defined by $F_{ss'}(\mathbf{k},\mathbf{k}+\mathbf{q}) = (1 + ss' \cos\psi)/2$, where ψ is the angle between $\mathbf{k}$ and $\mathbf{k}+\mathbf{q}$ vectors [5, 18]. The term $\cos\psi$ can be expressed in $|\mathbf{k}|$, $|\mathbf{k}+\mathbf{q}|$ and θ terms, and subsequently the overlap function is written as [8]

$$F_{ss'}(\mathbf{k},\mathbf{k}+\mathbf{q}) = \frac{1}{2}\left(1 + ss' \frac{|\mathbf{k}| + |\mathbf{q}|\cos\theta}{|\mathbf{k}+\mathbf{q}|}\right). \tag{61}$$

In long wavelength limit, we approximately obtain

$$|\mathbf{k}+\mathbf{q}| = |\mathbf{k}|\left(1 + \frac{|\mathbf{q}|\cos\theta}{|\mathbf{k}|} + \frac{|\mathbf{q}|^2 \sin^2\theta}{2|\mathbf{k}|^2}\right). \tag{62}$$

In this limit, we obtain for the graphene dispersion relation (59) the general form

$$\epsilon^{s}_{\mathbf{k}+\mathbf{q}} - \epsilon^{s'}_{\mathbf{k}} = s\hbar v_F \left(\frac{s-s'}{s}|\mathbf{k}| + |\mathbf{q}|\left(\cos\theta + \frac{|\mathbf{q}|}{2|\mathbf{k}|}\sin^2\theta\right)\right). \tag{63}$$

In turn, the plasmon-damping regimes are determined by the poles of polarizability (60) by substituting expression (63). Due to the valley degeneracy, there are two damping regimes corresponding, respectively, to intraband ($s = s'$)

$$\omega < v_F q \tag{64}$$

and interband ($s = -s'$)

$$v_F(2k_F - q) < \omega < v_F(2k_F + q). \tag{65}$$

electron-hole pair excitations [8] demonstrated in **Figure 5** by shaded areas.

Substituting the long wavelength limit expression (62) in the overlap function (61), the latter reads

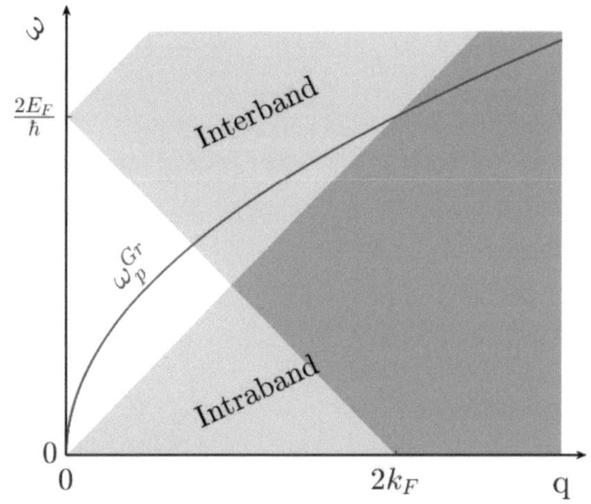

Figure 5. Blue solid line indicates the dispersion relation of graphene plasmons (ω_p^{Gr}). The shaded regimes represent the intra- and interband Landau damping where plasmon decays to electron-hole pairs excitation.

$$F_{ss'}(\mathbf{k},\mathbf{k}+\mathbf{q}) = \begin{cases} 1 - \dfrac{q^2}{4k^2}\sin^2\theta & \simeq 1 \quad s = s' \text{ (intraband)} \\ \dfrac{q^2}{4k^2}\sin^2\theta & \simeq 0 \quad s \neq s' \text{ (interband)} \end{cases} \tag{66}$$

Equation (66) states that in long wavelength limit, the interband contribution can be neglected [5], hence, the Lindhard formula (60) is simplified to

$$\chi_0(\mathbf{q}\to 0,\omega) = -\frac{4}{V}\sum_{\mathbf{k}}\left\{\frac{f(\epsilon^+_{\mathbf{k}+\mathbf{q}}) - f(\epsilon^+_{\mathbf{k}})}{\hbar z - (\epsilon^+_{\mathbf{k}+\mathbf{q}} - \epsilon^+_{\mathbf{k}})} + \frac{f(\epsilon^-_{\mathbf{k}+\mathbf{q}}) - f(\epsilon^-_{\mathbf{k}})}{\hbar z - (\epsilon^-_{\mathbf{k}+\mathbf{q}} - \epsilon^-_{\mathbf{k}})}\right\}. \tag{67}$$

As it has already been mentioned, in zero temperature limit, the Fermi-Dirac distribution $f(\epsilon^\pm_{\mathbf{k}})$ is simplified to Heaviside step function $\Theta(k_F \mp |\mathbf{k}|)$. In this limit, the second term in the right hand of Eq. (67) is always zero, since $\Theta(k_F + |\mathbf{k}|) = \Theta(k_F + |\mathbf{k}+\mathbf{q}|) = 1$, which reflects that all states in the valence band are occupied. Making again the elementary transformation $\mathbf{k}+\mathbf{q} \to -\mathbf{k}$ in the term of Eq. (67) that includes $f(\epsilon^+_{\mathbf{k}+\mathbf{q}})$, we obtain

$$\chi_0(\mathbf{q}\to 0,\omega) = \frac{8}{V}\sum_{|\mathbf{k}|<k_F}\frac{\epsilon^+_{\mathbf{k}+\mathbf{q}} - \epsilon^+_{\mathbf{k}}}{(\hbar z)^2 - (\epsilon^+_{\mathbf{k}+\mathbf{q}} - \epsilon^+_{\mathbf{k}})^2}. \tag{68}$$

Turning summation (68) into integral, we read

$$\chi_0(\mathbf{q}\to 0,\omega) = \frac{8}{(2\pi)^2}\int d^2|\mathbf{k}|\frac{\epsilon^+_{\mathbf{k}+\mathbf{q}} - \epsilon^+_{\mathbf{k}}}{(\hbar z)^2 - (\epsilon^+_{\mathbf{k}+\mathbf{q}} - \epsilon^+_{\mathbf{k}})^2}. \tag{69}$$

Transforming to polar coordinates for $r = |\mathbf{k}|$ and using relation (63), we obtain the integral

$$\chi_0(\mathbf{q},\omega) = \frac{2E_F k_F q}{\pi^2\hbar^2\omega^2}\int_0^1 dx \int_0^{2\pi} \frac{x\cos\theta + \frac{q}{2k_F}\sin^2\theta}{1-\left(\frac{v_F q}{\omega}\right)^2\left(\cos\theta + \frac{q}{2k_F x}\sin^2\theta\right)^2} d\theta, \tag{70}$$

where $x = r/k_F$, $q = |\mathbf{q}|$ and $\eta = 0 \Rightarrow z = \omega$. In non-static $(\omega \gg v_F q)$ and long wavelength $(q \ll k_F)$ limits, we expand the integrator of Eq. (69) in series of q. Keeping up to first power of q/k_F, we obtain

$$\chi_0(\mathbf{q}\to 0,\omega) = \frac{2E_F k_F q}{\pi^2\hbar^2\omega^2}\int_0^1 dx \int_0^{2\pi} \left(x\cos\theta + \frac{q}{2k_F}\sin^2\theta\right) d\theta. \tag{71}$$

The evaluation of integral (71) is trivial and leads to the polarizability function of graphene

$$\chi_0(\mathbf{q}\to 0,\omega) = \frac{E_F}{\pi\hbar^2}\frac{q^2}{\omega^2}. \tag{72}$$

Using the RPA formula (15), we obtain the long wavelength dielectric function of graphene

$$\varepsilon(q,\omega) = 1 - \frac{2e^2 E_F}{\hbar^2\omega^2} q \tag{73}$$

indicating that at low energies doped graphene is described by a Drude-type dielectric function with plasma frequency depending straightforward on the doping amount, namely the Fermi energy level E_F. The plasma frequency of graphene monolayer is determined by condition (11) and reads

$$\omega_p^{Gr}(q) = \sqrt{\frac{2e^2 E_F}{\hbar^2} q} \tag{74}$$

indicating the $q^{1/2}$ dependence likewise plasmons at a regular 2DEG. The most important result is the presence of $\hbar$ in the denominator of Eq. (74), which reveals that plasmons in graphene are purely quantum modes, that is, there are no classical plasmons in doped graphene. In addition, graphene plasmon frequency is proportional to $n^{1/4}$, which is different from classical 2D plasmon behavior where $\omega_p^{2D} \sim n^{1/2}$ [7, 18]. This is a direct consequence of the quantum relativistic nature of graphene, since Fermi energy is defined differently in any case, namely $E_F \sim k_F \sim n^{1/2}$ in graphene, whereas, $E_F \sim k_F^2 \sim n$ in 2DEG case. In **Figure 3(a)**, we represent the plasmon dispersion relation in doped graphene.

3.2. Graphene plasmonic metamaterial

Multilayers of plasmonic materials have been used for designing metamaterials providing electromagnetic propagation behavior not found under normal circumstances like negative refraction and epsilon-near-zero (ENZ) [9, 19, 20]. The bottleneck in creating plasmonic devices with

http://dx.doi.org/10.5772/67900

any desirable characteristic has been the limitations of typical 3D solids in producing perfect interfaces for the confinement of electrons and the features of dielectric host. This may no longer be a critical issue. The advent of truly two-dimensional materials like graphene (a metal), transition-metal dichalcogenides (TMDC's, semiconductors), and hexagonal boron nitride (hBN, an insulator) makes it possible to produce structures with atomic-level control of features in the direction perpendicular to the stacked layers [9, 21]. This is ushering a new era in manipulating the properties of plasmons and designing devices with extraordinary behavior.

Here, we propose a systematic method for constructing epsilon-near-zero (ENZ) metamaterials by appropriate combination on 2D materials. The aforementioned metamaterials exhibit interesting properties like diffractionless EM wave propagation with no phase delay [9]. We show analytically that EM wave propagation through layered heterostructures can be tuned dynamically by controlling the operating frequency and the doping level of the 2D metallic layers. Specifically, we find that multilayers of a plasmonic 2D material embedded in a dielectric host exhibit a plasmonic Dirac point (PDP), namely a point in wavenumber space where two linear coexisting dispersion curves cross each other, which, in turn, leads to an effective ENZ behavior [9]. To prove the feasibility of this design, we investigate numerically EM wave propagation in periodic plasmonic structures consisting of 2D metallic layers lying on yz plane in the form of graphene, arranged periodically along the x axis and possessing surface conductivity σ_s. The layers are embedded in a uniaxial dielectric host in the form of TMDC or hBN multilayers of thickness d and with uniaxial relative permittivity tensor $\overline{\overline{\varepsilon}}_d$ with diagonal components $\varepsilon_x \neq \varepsilon_y = \varepsilon_z$. We explore the resulting linear, elliptical, and hyperbolic EM dispersion relations which produce ENZ effect, ordinary and negative diffraction, respectively.

We solve the analytical problem under TM polarization, with the magnetic field parallel to the y direction which implies that there is no interaction of the electric field with ε_y. We consider a magnetically inert (relative permeability $\mu = 1$) lossless host ($\varepsilon_x, \varepsilon_z \in \mathbb{R}$). For monochromatic harmonic waves in time, the Maxwell equations lead to three equations connecting the components of the $\mathbf{E}$ and $\mathbf{H}$ fields. For the longitudinal component [9, 19], $E_z = (i\eta_0/k_0\varepsilon_z)(\partial H_y/\partial x)$ where $\eta_0 = \sqrt{\mu_0/\varepsilon_0}$ is the free space impedance. Defining the vector of the transversal field components as $\psi = (E_x, H_y)^T$ gives [9]

$$i\frac{\partial}{\partial z}\psi = k_0\eta_0 \begin{pmatrix} 0 & 1 + \frac{1}{k_0^2}\frac{\partial}{\partial_x}\frac{1}{\varepsilon_z}\frac{\partial}{\partial_x} \\ \frac{\varepsilon_x}{\eta_0^2} & 0 \end{pmatrix} \psi \tag{75}$$

Assuming EM waves propagating along the z axis, viz. $\psi(x,z) = \psi(x)e^{ik_z z}$, Eq. (75) leads to an eigenvalue problem for the wavenumber k_z of the plasmons along z [9, 19]. The metallic 2D planes are assumed to carry a surface current $J_s = \sigma_s E_z$, which acts as a boundary condition in the eigenvalue problem. Furthermore, infinite number of 2D metals are considered to be arranged periodically, along x axis, with structural period d. The magnetic field reads $H_y^-(x)e^{ik_z z}$ for $-d < x < 0$ and $H_y^+(x)e^{ik_z z}$ for $0 < x < d$ on either side of the metallic plane at $x = 0$, with boundary conditions $H_y^+(0) - H_y^-(0) = \sigma_s E_z(0)$ and $\partial_x H_y^+(0) = \partial_x H_y^-(0)$. Due to the periodicity, we use Bloch theorem along x as $H_y^+(x) = H_y^-(x-d)e^{ik_x d}$, with Bloch wavenumber k_x. As a result, we arrive at the dispersion relation [9, 19, 20]:

$$F(k_x,k_z) = \cos(k_x d) - \cosh(\kappa d) + \frac{\xi\kappa}{2}\sinh(\kappa d) = 0 \tag{76}$$

where $\kappa^2 = (\varepsilon_z/\varepsilon_x)(k_z^2 - k_0^2\varepsilon_x)$ expresses the anisotropy of the host medium and $\xi = -(\mathrm{i}\sigma_s\eta_0/k_0\varepsilon_z)$ coincides with the so-called "plasmonic thickness" which determines the SPP decay length [9, 19, 20]. In particular, ξ is twice the SPP penetration length and defines the maximum distance between two metallic layers where the plasmons are strongly interacting [9, 19, 20]. We point out that for lossless 2D metallic planes σ_s is purely imaginary and ξ is purely real (for $\varepsilon_z \in \mathbb{R}$). At the center of the first Brillouin zone $k_x = 0$, the equation has a trivial solution [19] for $\kappa = 0 \Rightarrow k_z = k_0\sqrt{\varepsilon_x}$ which corresponds to the propagation of x-polarized fields travelling into the host medium with refractive index $\sqrt{\varepsilon_x}$ without interacting with the 2D planes which are positioned along z axis [22]. Near the Brillouin zone center ($k_x/k_0 \ll 1$ and $\kappa \simeq 0$) and under the assumption of a very dense grid ($d \to 0$), we obtain $k_x d \ll 1$ and $\kappa d \ll 1$, we Taylor expand the dispersion equation (76) to second order in d, hence

$$\frac{k_z^2}{\varepsilon_x} + \frac{d}{(d-\xi)\varepsilon_z}k_x^2 = k_0^2. \tag{77}$$

The approximate relation (77) is identical to that of an equivalent homogenized medium described by dispersion: $k_z^2/\varepsilon_x^{\mathrm{eff}} + k_x^2/\varepsilon_z^{\mathrm{eff}} = k_0^2$ [9, 21]. Subsequently, from a metamaterial point of view, the entire system is treated as a homogeneous anisotropic medium with effective relative permittivities given by

$$\varepsilon_x^{\mathrm{eff}} = \varepsilon_x\,,\;\; \varepsilon_z^{\mathrm{eff}} = \varepsilon_z + \mathrm{i}\frac{\eta_0\sigma_s}{k_0 d} = \varepsilon_z\frac{d-\xi}{d}. \tag{78}$$

We read from Eq. (78) the capability to control the behavior of the overall structure along the z direction. For instance, the choice $d = \varepsilon_z/(\varepsilon_z - \varepsilon_x)\xi$ leads to an isotropic effective medium with $\varepsilon_z^{\mathrm{eff}} = \varepsilon_x^{\mathrm{eff}}$ [9].

For the lossless case ($\mathrm{Im}[\xi] = 0$), we identify two interesting regimes, viz. the strong plasmon coupling for $d < \xi$ and the weak plasmon coupling for $d > \xi$. In both cases, plasmons develop along z direction at the interfaces between the conducting planes and the dielectric host. In the strong coupling case ($d < \xi$), plasmons of adjacent interfaces interact strongly with each other. As a consequence, the shape of the supported band of Eq. (77), in the (k_x,k_z) plane, is hyperbolic (dashed red line in **Figure 6(a)**) and the system behaves as a hyperbolic metamaterial [9, 19, 22] with $\varepsilon_x^{\mathrm{eff}} > 0$, $\varepsilon_z^{\mathrm{eff}} < 0$. On the other hand, in the weak plasmon coupling ($d > \xi$), the interaction between plasmons of adjacent planes is very weak. In this case, the shape of the dispersion relation (77) on the (k_x,k_z) plane is an ellipse (dotted black line in **Figure 6(a)**) and the systems act as an ordinary anisotropic media with $\varepsilon_z^{\mathrm{eff}},\varepsilon_x^{\mathrm{eff}} > 0$ [9]. We note that in the case $\xi < 0$ the system does not support plasmons and the supported bands are always ellipses [9]. When either the 2D medium ($\mathrm{Re}[\sigma_s] \neq 0$) or the host material is lossy, a similar separation holds by replacing ξ by $\mathrm{Re}[\xi]$.

http://dx.doi.org/10.5772/67900

The most interesting case is the linear dispersion, where k_z is linearly dependent on k_x and dk_x/dk_z is constant for a wide range of k_z [9, 19]. When this condition holds, the spatial harmonics travel with the same group velocity into the effective medium [9, 19]. To engineer our structure to exhibit a close-to-linear dispersion relation, we inspect the approximate version of Eq. (77): a huge coefficient for k_x will make k_0^2 on the right-hand-side insignificant; if $\xi = d$, the term proportional to k_x^2 increases without bound yielding a linear relation between k_z and k_x. With this choice, $\sigma_s = -i(k_0 d\varepsilon_z/\eta_0)$, and substituting in the exact dispersion relation Eq. (76), we find that $(k_x,k_z) = (0,k_0\sqrt{\varepsilon_x})$ becomes a saddle point for the transcendental function $F(k_x,k_z)$ giving rise to the conditions for the appearance of two permitted bands, namely two lines on the (k_x,k_z) plane across which $F(k_x,k_z) = 0$. This argument connects a mathematical feature, the saddle point of the dispersion relation, with a physical feature, the crossing point of the two coexisting linear dispersion curves, the plasmonic Dirac point [9] (solid blue line in **Figure 6(a)**). From a macroscopic point of view, the choice $\xi = d$ makes the effective

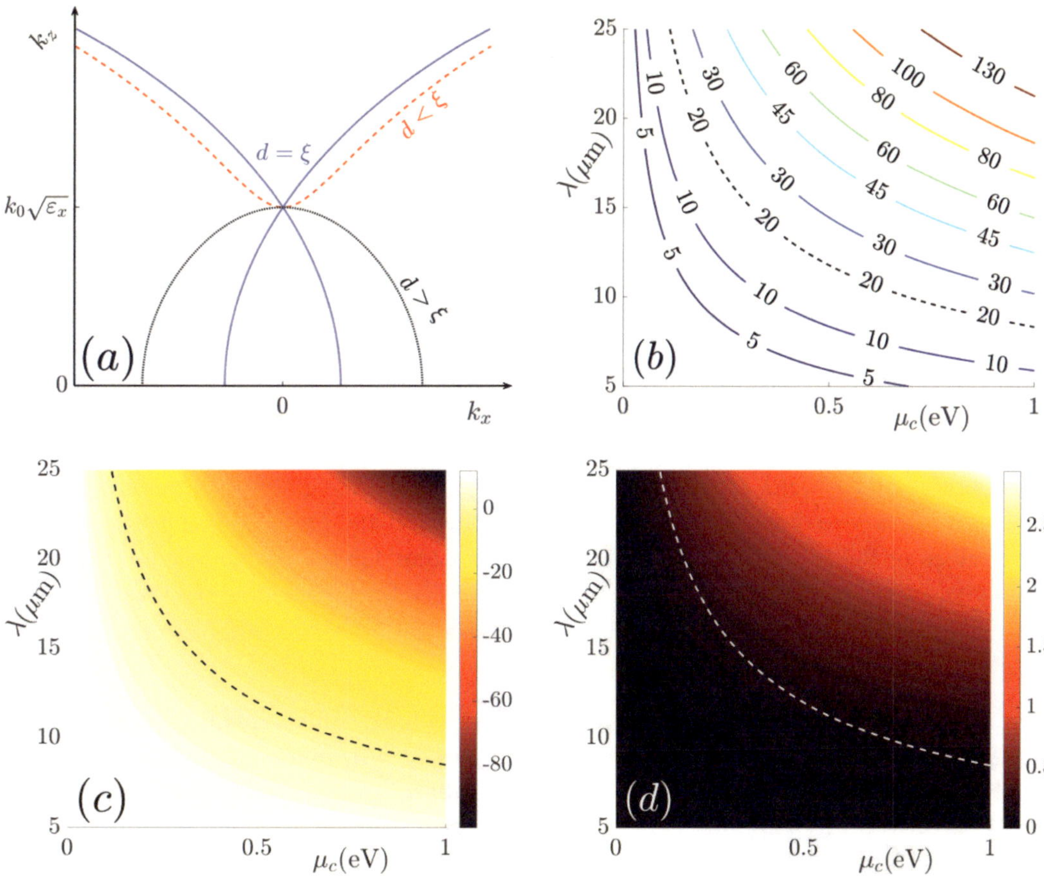

Figure 6. (a) The three supported dispersion plasmonic bands in (k_x,k_z) plane: hyperbolic (dashed red), elliptical (dotted black), and linear (solid blue) where plasmonic Dirac point (PDP) appears. (b) Combinations of graphene doping μ_c and free-space operational wavelengths λ leading to epsilon-near-zero (ENZ) behavior (PDP in dispersion relation) for several lattice periods d (in nm). (c) Real and (d) imaginary parts of the effective permittivity $\varepsilon_z^{\text{eff}}$ for the choice $d = 20\,\text{nm}$ (dashed line in (b)); dashed curves indicate the ENZ regime.

permittivity along the z direction vanish, as is evident from Eq. (78). As a result, the existence of a PDP makes the effective medium behave like an ENZ material in one direction ($\varepsilon_z^{\text{eff}} = 0$).

The plasmonic length ξ is, typically, restricted in few nanometers ($\xi < 100$ nm). Regular dielectrics always present imperfections in nanoscales, hence, the use of regular materials as dielectric hosts is impractical. Furthermore, graphene usually exfoliates or grows up on other 2D materials. Because of the aforementioned reasons, it is strongly recommended that the dielectric host to be also a 2D material with atomic scale control of the thickness d and no roughness. For instance, one could build a dielectric host by stacking 2D layers of materials molybdenum disulfide (MoS_2) [23] with essentially perfect planarity, complementing the planarity of graphene.

Substituting the graphene dielectric function (73) into formula (18), we calculate the two-dimensional Drude-type conductivity of graphene [6, 19, 21]

$$\sigma_s(\omega) = \frac{\mathrm{i}e^2 \mu_c}{\pi \hbar^2 (\omega + \mathrm{i}/\tau)}, \tag{79}$$

where μ_c is the tunable chemical potential equal to Fermi energy E_F and τ is the transport-scattering time of the electrons [6, 19] introduced in the same manner as in Eq. (30). In what follows, we use bulk MoS_2, which at THz frequencies is assumed lossless with a diagonal permittivity tensor of elements, $\varepsilon_x \cong 3.5$ (out of plane) and $\varepsilon_y = \varepsilon_z \cong 13$ (in plane) [23].

The optical losses of graphene are taken into account using $\tau = 0.5$ ps [19]. Since the optical properties of the under-investigated system can be controlled by tuning the doping amount, the operating frequency or the structural period, in **Figure 6(b)**, we show proper combinations of μ_c and operational wavelength in free space λ which lead to a PDP for several values of lattice density distances ($d = \text{Re}[\xi]$ in nm) [9]. To illustrate, for a reasonable distance between successive graphene planes of $d = 20$ nm, the real (**Figure 6(c)**) and imaginary (**Figure 6(d)**) effective permittivity values that can be emulated by this specific graphene-MoS_2 architecture determine the device characteristics at different frequencies and graphene-doping levels. Positive values of $\text{Re}[\varepsilon_z^{\text{eff}}]$ are relatively moderate and occur for larger frequencies and lower doping levels of graphene; on the other hand, $\text{Im}[\varepsilon_z^{\text{eff}}]$ is relatively small in the ENZ region as indicated by a dashed line in both graphs [9]. On the other hand, losses become larger as $\text{Re}[\varepsilon_z^{\text{eff}}]$ gets more negative.

To examine the actual electromagnetic field distribution in our graphene-MoS_2 configuration, we simulate the EM wave propagation through two finite structures consisting of 40 and 100 graphene planes with $\text{Re}[\xi] = 20.8\,\text{nm}$ and for operational wavelength in vacuum $\lambda = 12\,\text{m}$ ($f = 25$ THz $= 0.1$ eV). In order to have a complete picture of the propagation properties, we excite the under-investigating structures with a 2D dipole magnetic source as well as with a TM plane wave source. In particular, the 40-layered structure is excited by a 2D magnetic dipole source, which is positioned close to one of its two interfaces and oriented parallel to them, denoted by a white dot in **Figure 7(a)–(c)**. On the other hand, the 100-layered configuration is excited by a plane source, which is located below the multilayer and is rotated by 20°

http://dx.doi.org/10.5772/67900

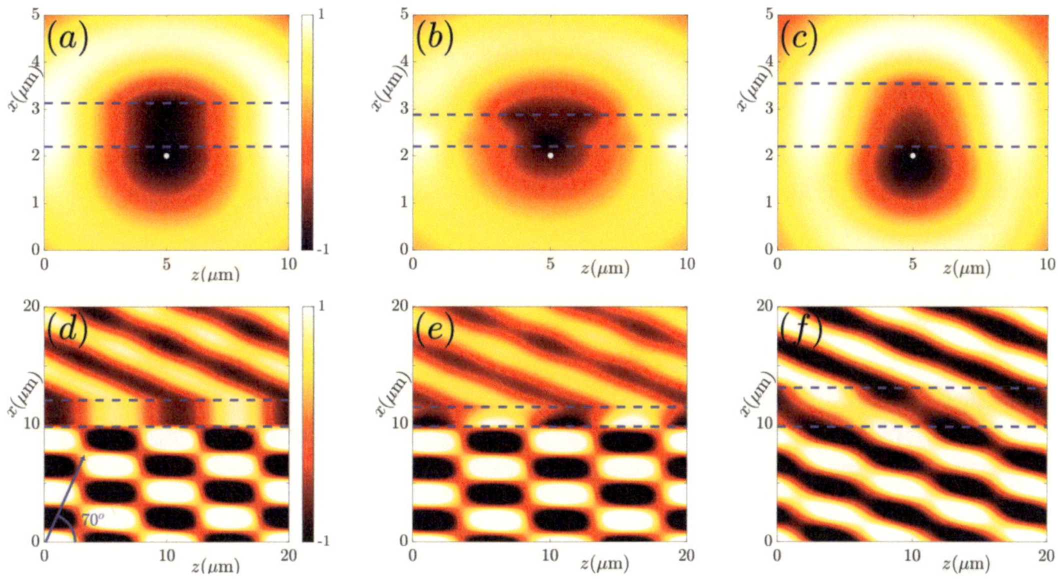

Figure 7. Spatial distribution of the magnetic field (color map) of graphene-MoS_2 multilayer structure located between the blue dashed lines and embedded in MoS_2 background. In (a)–(c), the metamaterial consists of 40 graphene sheets and excited by a magnetic dipole (white dot). In (d)–(f), the structure is composed by 100 graphene layers and excited by a TM plane wave source located at $y = 0$ and rotated 20° with respect to the interface. (a), (d) $d = \mathrm{Re}[\xi]$ (ENZ behavior). (b), (e) $d = 0.7\mathrm{Re}[\xi]$ hyperbolic metamaterial. (c), (f) $d = 1.5\mathrm{Re}[\xi]$ elliptical medium, where $\mathrm{Re}[\xi] = 20.8\,\mathrm{nm}$. Due to high reflections in (d), (e), we observe pattern formation of stationary waves below the metamaterial.

with respect to the interface; the blue arrow in **Figure 7(d)** indicates the direction of the incident wave. The normalized to one spatial distribution of the magnetic field value is shown in **Figure 7** in color representation, where the volume containing the graphene multilayers is between the dashed blue lines. To minimize the reflections, the background region is filled with a medium of the same dielectric properties as MoS_2. In **Figure 7(a** and **d)**, the system is in the critical case ($d = \mathrm{Re}[\xi]$), where the waves propagate through the graphene sheets without dispersion as in an ENZ medium. In **Figure 7(b** and **e)**, the interlayer distance is $d = 0.7\mathrm{Re}[\xi]$ (strong plasmon-coupling regime) and the system shows negative (anomalous) diffraction. In **Figure 7(c** and **f)** $d = 1.5\mathrm{Re}[\xi]$ (weak plasmon-coupling regime) and the EM wave show ordinary diffraction through the graphene planes [9].

4. Conclusion

In summary, we have studied volume and surface plasmons beyond the classical plasma model. In particular, we have described electronic excitations in solids, such as plasmons and their damping mechanism, viz. electron-hole pairs excitation, in the context of the quantum approach random phase approximation (RPA), a powerful self-consistent theory for determining the dielectric function of solids including screening non-local effect. The dielectric function and, in turn, the plasmon dispersion relation have been calculated for a bulk metal, a two-dimensional electron gas (2DEG) and for graphene, the famous two-dimensional semi-metal.

The completely different dispersion relation between plasmon in three- and two-dimensional metals has been pointed out. Furthermore, we have highlighted the fundamental difference between plasmons in a regular 2DEG and in doped graphene, indicating that plasmons in graphene are purely quantum modes, in contrast to plasmons in 2DEG, which originate from classical laws. Moreover, the propagation properties of surface plasmon polariton (SPP), a guided collective oscillation mode, have been also investigated. For the completeness of our theoretical investigation, we have outlined two applications. First, we have examined SPPs properties along an interface consisting of a bulk metal and an active (gain) dielectric. We have found that there is a gain value for which the metallic losses have been completely eliminated resulting in lossless SPP propagation. Second, we have investigated a plasmonic metamaterial composed of doped graphene monolayers. We have shown that depending on operating frequency, doping amount, and interlayer distance between adjacent graphene layers, the wave propagation properties present epsilon-near-zero behavior, normal, and negative refraction, providing a metamaterial with tunable optical properties.

Acknowledgements

We acknowledge discussions with D. Massatt and E. Manousakis and partial support by the European Union under programs H2020-MSCA-RISE-2015-691209-NHQWAVE and by the Seventh Framework Programme (FP7-REGPOT-2012-2013-1) under grant agreement no. 316165. We also acknowledge support by EFRI 2-DARE NSF Grant No. 1542807 (M.M); ARO MURI Award No. W911NF14-0247 (E.K.). We used computational resources on the Odyssey cluster of the FAS Research Computing Group at Harvard University.

Author details

Marios Mattheakis[1,2]*, Giorgos P. Tsironis[2] and Efthimios Kaxiras[1,3]

*Address all correspondence to: mariosmat@g.harvard.edu

1 School of Engineering and Applied Sciences, Harvard University, Cambridge, MA, USA

2 Department of Physics, University of Crete, Heraklion, Greece

3 Department of Physics, Harvard University, Cambridge, MA, USA

References

[1] Kittel C. Introduction to Solid State Physics. 8th ed. USA: John Wiley & Sons, Inc; 2005. 680 p.

[2] Maier SA. Plasmonics: Fundamentals and Applications. New York: Springer; 2007. 223 p.

[3] Mattheakis M, Oikonomou T, Molina MI, Tsironis GP. Phase transition in PT symmetric active plasmonic systems. IEEE J. Sel. Top. Quantum Electron. 2016;**22**:5000206. DOI: 10.1109/JSTQE.2015.2490018

[4] Pitarke JM, Silkin VM, Chulkov EV, Echenique PM. Theory of surface plasmons and surface-plasmon-polaritons. Rep. Pro. Phys. 2007;**70**:1–87. DOI: 10.1088/0034-4885/70/1/R01

[5] Kenneth W, Shung K. Dielectric function and plasmon structure of stage-1 intercalated graphite. Phys. Rev. B. 1986;**34**:979–993. DOI: 10.1103/PhysRevB.34.979

[6] Jablan M, Buljan H, Soljačić M. Plasmonics in graphene at infrared frequencies. Phys. Rev. B. 2009;**80**:245435. DOI: 10.1103/PhysRevB.80.245435

[7] Grigorenko AN, Polini M, Novoselov KS. Graphene plasmonics. Nat. Photon. 2012;**6**:749–758. DOI: 10.1038/NPHOTON.2012.262

[8] Gonçalves PAD, Peres NMR. An Introduction to Graphene Plasmonics. World Scientific Publishing, Singapore; 2016. 431 p.

[9] Mattheakis M, Valagiannopoulos CA, Kaxiras E. Epsilon-near-zero behavior from plasmonic Dirac point: Theory and realization using two-dimensional materials. Phys. Rev. B. 2016;**94**:201404(R). DOI: 10.1103/PhysRevB.94.201404

[10] Fei Z, Rodin AS, Andreev GO, Bao W, McLeod AS, Wagner M, Zhang LM, Zhao Z, Thiemens M, Dominguez G, Fogler MM, Castro Neto AH, Lau CN, Keilmann F, and Basov DN. Gate-tuning of graphene plasmons revealed by infrared nano-imaging. Nature. 2012;**487**:82–85. DOI: 10.1038/nature11253

[11] March NH, Parrinello M. Collective Effects in Solids and Liquids. Bristol: Adam Hilger LTD; 1982. pp. 4–45.

[12] Isihara A. Electron Liquids. 2nd ed. Berlin, Germany: Springer; 1998. pp. 21–36.

[13] Monarkha Y, Kono K. Two-Dimensional Coulomb Liquids and Solids. Berlin, Germany: Springer; 2004. pp. 65–103.

[14] Psaltakis GC. Quantum Many-Particle Systems. Heraklion: Crete University Press; 2012. 707 p. (In Greek)

[15] Kaxiras E. Atomic and Electronic Structure of Solids. New York, USA: Cambridge University Press; 2003. 676 p.

[16] Shirodkar SN, Mattheakis M, Cazeaux P, Narang P, Soljačic′ M, Kaxiras E. Visible quantum plasmons in lithium-intercalcated multilayer graphene (to be published in 2017). arXiv: 1703.01558

[17] Stern F. Polarizability of a two-dimensional electron gas. Phys. Rev. Let. 1967;**18**:546–548. DOI: 10.1103/PhysRevLett.18.546

[18] Hwang EH, Das Sarma S. Dielectric function, screening, and plasmons in two-dimensional graphene. Phys. Rev. B. 2007;**75**:205418. DOI: 10.1103/PhysRevB.75.205418

[19] Wang B, Zhang X, García Vidal FJ, Yuan X, Teng J. Strong coupling of surface plasmon polaritons in monolayer graphene sheet arrays. Phys. Rev. Lett. 2012;**109**:073901. DOI: 10.1103/PhysRevLett.109.073901

[20] Wang B, Zhang X, Loh KP, Teng J. Tunable broadband transmission and phase modulation of light through graphene multilayers. J. Appl. Phys. 2014;**115**:213102. DOI: 10.1063/1.4880336

[21] Nefedov IS, Valagiannopoulos CA, Melnikov LA. Perfect absorption in graphene multilayers. J. Optics. 2013;**15**:114003. DOI: 10.1088/2040-8978/15/11/114003

[22] Valagiannopoulos CA, Mirmoosa MS, Nefedov IS, Tretyakov SA, Simovski CR. Hyperbolic-metamaterial antennas for broadband enhancement of dipole emission to free space. J. Appl. Phys. 2014;**116**:163106. DOI: 10.1063/1.4900528

[23] Defo RK, Fang S, Shirodkar SN, Tritsaris GA, Dimoulas A, Kaxiras E. Strain dependence of band gaps and exciton energies in pure and mixed transition-metal dichalcogenides. Phys. Rev. B. 2016;**94**:155310. DOI: 10.1103/PhysRevB.94.155310

Chapter 2

Magneto-Plasmonics and Optical Activity in Graphene-Based Nanowires

Dmitry A. Kuzmin, Igor V. Bychkov,
Vladimir G. Shavrov and Vasily V. Temnov

Additional information is available at the end of the chapter

http://dx.doi.org/10.5772/67417

Abstract

Nowadays, graphene plasmonics shows a great number of features unusual for traditional (metal-based) plasmonics from high localization and large propagation distance of surface plasmon-polaritons (SPPs) through the existence of both TE- and TM-polarized SPPs to the possibility of controlled SPPs by graphene chemical potential (or, equivalently, by gate voltage or chemical doping). Cylindrical graphene-based plasmonic structures have some advantages in contrast to planar geometry: absence of edge losses, existence of high-order azimuthal modes, etc. In this work, we discuss some ways to obtain an optical activity in cylindrical graphene-based plasmonic structures and its possible applications to SPPs manipulation.

Keywords: surface plasmon-polaritons, graphene, optical activity, magneto-plasmonics, metasurface

1. Introduction

Nowadays, it is evident that graphene is a very promising material for many optics, photonics and plasmonics applications [1–3]. Graphene layers (single layer as well as two- and multi-layer waveguides) may support highly localized electromagnetic waves, i.e. surface plasmon-polaritons (SPPs), both TE and TM polarized [4–9]. Tight confinement and large propagation length of plasmons make it possible to observe strong light-matter interactions in graphene-based structures [10]. Practically, only graphene ribbons of finite weight may be used. Unfortunately, the edges of such ribbons lead to undesirable increase in losses [11]. The possible way to solve this problem is the use of cylindrical 2D surfaces [12]. Graphene-based cylindrical waveguides may operate in single- and multi-mode regimes in the frequency range from THz to mid-IR [13–15]. They may support TE-polarized plasmons [16], similar to the single graphene layer [4].

For realizing any plasmonic devices, one should have the instrument for manipulating by plasmon-polaritons. This goal may be achieved, for example, by the combination of plasmonic and optically active materials [17–22]. Among other optically active materials, the use of magnetic ones leads to cross-coupling between magnetic properties of materials and optical fields: different mechanisms may lead to optically induced magnetic fields [23–26] and excitation of localized plasmons may lead to a major increase in magneto-optical effects [27–31].

It is well known that magnetic field (or magnetization) in cylindrical optical fibres may lead to the rotation of the energy distribution (i.e. speckle-pattern) into cross-section of the fibre [32–35]. The nature of this effect is magnetic field-induced breaking of degeneracy of the modes with opposite signs of azimuthal mode index (i.e. rotating in opposite azimuthal directions). Recently, we have shown that in graphene-coated optical fibre one may control such rotation by both magnetic field and chemical potential of the graphene [36], but for observable rotation it is necessary that the fibre length should be of a few centimetres. Recently, we have shown that in case of magneto-active nanowire covered by graphene layer one may achieve the rotation of some plasmonic modes by up to ~100° on the scale of about 500 nm at mid-infrared frequencies [37]. Tuning carrier concentration in graphene by chemical doping or gate voltage allows controlling of SPP properties and notably the rotation angle of high-order azimuthal modes.

In this chapter, we summarize our previous results and discuss some magnetic-free ways to obtain similar effects. Our results may open the door for the application of straintronic control in plasmonics and the design of one-way propagation plasmonic devices.

The chapter consists of introduction, three sections and conclusions. In Section 2, we review basic properties of SPPs propagating in cylindrical graphene-based plasmonic waveguides. We discuss conditions of propagation of both TM and TE fundamental modes and TM-like high-order SPPs. This section also covers some features of effective magnetic field induced via inverse Faraday effect. In Section 3, we show the possibility of rotation of SPPs, which are supported by graphene-coated gyrotropic nanowire. Section 4 discusses the similar effects of the spiral graphene-based plasmonic waveguide.

2. Surface plasmon-polaritons in graphene-covered nanowires

Cylindrical graphene-based plasmonic waveguides of different configurations have been well investigated in the literature [13–16]. It has been shown that such a waveguide may support high-order azimuthal plasmonic modes and may work in the single-mode regime.

Let us consider a nanowire with the dielectric permittivity $\varepsilon_{\text{wire}}$ covered by a graphene layer. This structure is embedded in the medium with permittivity ε_{out}. The radius of nanowire is R. We will use cylindrical coordinates (r, ϕ, z). The nanowire axis is supposed to be the z-axis. We describe graphene by a 2D conductivity σ_g [38], which depends on the temperature T, the angular frequency ω, the scattering rate Γ and the chemical potential (or the Fermi energy) $\mu_{\text{ch}} \approx \hbar v_F(\pi n)^{1/2}$, where $v_F \approx 10^6$ m/s is the Fermi velocity. For example, $n \approx 8 \times 10^{13}$ cm^{-2} corresponds $\mu_{\text{ch}} \approx 1$ eV. We use a standard model of graphene surface conductivity calculated within the local random phase approximation with the dominant Drude term at SPP energies below the Fermi level [39, 40].

We will consider monochromatic plasmons propagating along the nanowire axis (*z*-axis) by putting electric and magnetic fields **E, H** $\propto \exp[i(\beta z - \omega t)]$ into Maxwell's equations ($\beta = \beta' + i\beta''$ is a complex propagation constant, ω is a circular frequency). Components of electric and magnetic fields in the cylindrical coordinates (r, ϕ, z) inside the nanowire (i.e. at $0 < r < R$) are the following:

$$\begin{aligned}
E^{\text{in}}_{r,m} &= -ig^{-2}\{\beta g A_m I'_m(gr) + i\omega m\mu_0 r^{-1} B_m I_m(gr)\}, \\
E^{\text{in}}_{\phi,m} &= -ig^{-2}\{i\beta m r^{-1} A_m I_m(gr) - \omega\mu_0 g B_m I'_m(gr)\}, \\
E^{\text{in}}_{z,m} &= A_m I_m(gr), \\
H^{\text{in}}_{r,m} &= -ig^{-2}\{\beta g B_m I'_m(gr) - i\omega m\varepsilon_{in} r^{-1} A_m I_m(gr)\}, \\
H^{\text{in}}_{\phi,m} &= -ig^{-2}\{i\beta m r^{-1} B_m I_m(gr) - \omega\varepsilon_{in} g A_m I'_m(gr)\}, \\
H^{\text{in}}_{z,m} &= B_m I_m(gr).
\end{aligned} \tag{1}$$

In Eq. (1), a multiplier $\exp[i(m\phi + \beta z - \omega t)]$ has been omitted. Equations describing fields outside the nanowire (at $r > R$) $\mathbf{E^{out}}$, $\mathbf{H^{out}}$ may be derived from Eq. (1) by substitutions:

$$g \to p, I_m(gr) \to K_m(pr), A_m \to C_m, B_m \to D_m, \ \varepsilon_{\text{in}} \to \ \varepsilon_{\text{out}}. \tag{2}$$

In all formulae, $I_m(x)$ and $K_m(x)$ are modified Bessel's functions of the first and second types, respectively; operation 'means differentiation with respect to the argument; $g^2 = \beta^2 - \omega^2\varepsilon_{\text{in}}\mu_0$, $p^2 = \beta^2 - \omega^2\varepsilon_{\text{out}}\mu_0$, where μ_0 is the magnetic constant. Parameters p and g are related to the field confinement of mode. Constants A_m, B_m, C_m and D_m are determined by boundary conditions and mode normalization procedure. The boundary conditions at $r = R$ are the following: $E^{\text{in}}_{z,m} = E^{\text{out}}_{z,m}$, $E^{\text{in}}_{\varphi,m} = E^{\text{out}}_{\varphi,m}$, $H^{\text{out}}_{z,m} - H^{\text{in}}_{z,m} = -\sigma_g E^{\text{in}}_{\varphi,m}$ and $H^{\text{out}}_{\varphi,m} - H^{\text{in}}_{\varphi,m} = \sigma_g E^{\text{in}}_{z,m}$.

The characteristic (or dispersion) equation for the mth plasmonic mode may be obtained from boundary conditions with fields expressions (1) and (2).

For $m = 0$, electromagnetic waves may be classified into TE and TM modes. High-order modes have all non-zero components of electric and magnetic fields.

Dispersion relation for TE-polarized fundamental mode ($m = 0$) reads [16]

$$i\omega\mu_0\sigma_g = \frac{gI_0(gR)}{I_1(gR)} + \frac{pK_0(pR)}{K_1(pR)}. \tag{3}$$

Both terms on the right-hand side are positive; so, we may conclude that condition $\text{Im}[\sigma_g] < 0$ is necessary. But this condition is not sufficient. Let us suppose that $|\text{Re}[\sigma_g]| \ll |\text{Im}[\sigma_g]|$. This condition is satisfied near the inter-band transition (i.e. when $1.667 < \hbar\omega/\mu_{\text{ch}} < 2$). Cut-off limit (i.e. $p \to 0$) leads to critical coupling among the core radius, frequency and dielectric permittivity. The limit $g_{cr}R \ll 1$, which corresponds to small core radius, small difference of permittivities of the inner and the outer mediums, or low frequency, leads to the condition of $\omega\mu_0|\text{Im}[\sigma_g]|R = 2$. Estimations of critical radius for the frequency range 100–600 THz (near-infrared to visible light) give $R \sim 1$ μm.

High-order TE-like SPPs modes may propagate in the structures with much larger radius, and, thus, are of interest for practical plasmonic applications.

TM-SPPs have been investigated in details in [13, 14]. Dispersion relation of TM fundamental plasmonic modes is

$$\frac{\varepsilon_{\text{out}} K_0(pR)}{pK_1(pR)} + \frac{\varepsilon_{in} I_0(gR)}{gI_1(gR)} + i\frac{\sigma_g}{\omega} = 0. \tag{4}$$

This mode exists for any radius values and frequencies of electromagnetic wave, when $\text{Im}[\sigma_g] > 0$.

The modes with index $|m| > 0$ exist above the cut-off frequency. The number of supported modes at the fixed vacuum wavelength λ_0 may be estimated as $\text{Re}[i2\pi R(\varepsilon_{\text{wire}}+\varepsilon_{\text{out}})c/(\sigma_g\lambda_0)]$. An increase of the core permittivity leads to an increase in the number of supported modes.

Comparison of the SPPs characteristics of graphene-covered nanowire and gold-coated nanowire in the frequency range of 30–50 THz shows [13] that the effective mode index of SPPs mode in graphene-covered nanowire is much larger than that of gold-coated nanowire, indicating that plasmon mode in graphene-covered nanowire has a much shorter SPP wavelength and better mode confinement. SPPs mode in graphene-covered nanowire has a much smaller mode area: the mode energy of graphene-covered nanowire is mainly localized inside the nanowire, while the mode energy of gold-coated nanowire resides outside the Au coating.

Field expressions (1) with (2) allow one to calculate the inverse magneto-plasmonic effect [41]: i.e. the effective magnetic field induced by propagating SPPs due to the inverse Faraday effect $\mathbf{H}_{\text{eff}} = \alpha \text{Im}[\mathbf{E}\times\mathbf{E}^*]$. For mode with $m = 0$ SPPs may be classified into TE and TM modes. It is easy to show that TE mode cannot produce any magnetic field, while TM mode can induce azimuthal magnetic field. Magnetic field components induced by any single mode do not depend on the azimuthal angle ϕ. For modes with $m \neq 0$, all components of magnetic field are non-zero. Change in the propagation direction leads to a change in the $\mathbf{H}_{\text{eff}}$ rotation direction. Modes with greater $|m|$ induces longitudinal component of magnetic field as well. The value of this component increases with an increase in mode number. We have found also that in two-mode regime it is possible to induce azimuthally periodic magnetic field distribution. This distribution may be rotated by these two mode phase shift controls.

3. Gyrotropic graphene-coated nanowires

Let us consider a gyrotropic nanowire covered by a graphene layer (see **Figure 1**). We will use cylindrical coordinates (r, ϕ, z). The nanowire axis is supposed to be the z-axis, which coincides with the gyration axis. Such situation may be realized, for example, in magnetic nanowires magnetized along the wire axis. The electrodynamic properties of the nanowire may be described by the following dielectric permittivity tensor:

$$\hat{\varepsilon}_{\text{wire}} = \varepsilon_0 \begin{pmatrix} \varepsilon_\perp & -i\varepsilon_a & 0 \\ i\varepsilon_a & \varepsilon_\perp & 0 \\ 0 & 0 & \varepsilon_\parallel \end{pmatrix} \tag{5}$$

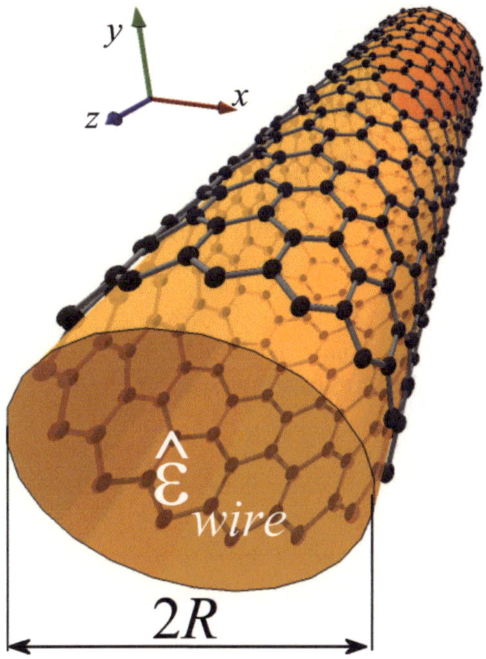

Figure 1. Geometry of the problem.

Here, ε_0 is the electric constant (we will use SI units throughout the chapter). Graphene layer may be described by 2D conductivity σ_g, which depends on the temperature T, the angular frequency ω, the scattering rate Γ and the chemical potential μ_{ch}. It may be calculated in local random phase approximation [39, 40], for example. We will take into account the presence of graphene only as a specific boundary condition [38]. We will suppose that the outer medium is the air, i.e. it has dielectric permittivity $\varepsilon_{out} = \varepsilon_0$.

Characteristics of SPP modes propagating in graphene-covered non-gyrotropic nanowire have been investigated in details [13–16]. In Section 2, it has been discussed that plasmonic modes in such structure may induce a complex distribution of magnetic field via inverse Faraday effect. Here, we suppose that intensity of plasmonic modes under consideration is small enough, and one may neglect the inverse Faraday effect inside magnetic nanowire.

Now, one has to solve Maxwell's equations inside each medium. We suppose that electromagnetic wave has harmonical time dependence and propagates along the z-axis, i.e. $\mathbf{E}$, $\mathbf{H} \sim \exp[-i\omega t + i\beta z]$, where $\beta = \beta' + i\beta''$ is a complex propagation constant. Electromagnetic field distribution inside magnetic nanowire with permittivity tensor (Eq. (5)) may be expressed similarly to that of circular microwave waveguides and optical fibres filled by gyrotropic medium [42–44]. Field outside the nanowire has a usual form (see Eqs. (1) and (2)). These fields should satisfy the boundary conditions at $r = R$: $E^{in}_{z,\,m} = E^{out}_{z,\,m}$, $E^{in}_{\varphi,m} = E^{out}_{\varphi,m}$, $H^{out}_{z,\,m} - H^{in}_{z,\,m}$

$= -\sigma_g E^{\mathrm{in}}_{\varphi,m}$ and $H^{\mathrm{out}}_{\varphi,m} - H^{\mathrm{in}}_{\varphi,m} = \sigma_g E^{\mathrm{in}}_{z,m}$. So, we will have the dispersion equation. Solving this equation, one will obtain β for each azimuthal mode index m.

Standard characteristics of SPP are the SPP wavelength $\lambda_{SPP} = 2\pi/\beta'$, and propagation length $L_{SPP} = (\beta'')^{-1}$. When L_{SPP} becomes less than λ_{SPP} for chosen m, the corresponding SPP mode becomes overdamped and cannot propagate in the structure.

Analytical analysis shows that dispersion equation has terms with the first and third powers of the mode index m. This leads to non-reciprocity for modes with the opposite azimuthal propagation direction, i.e. modes with different signs of m will propagate with slightly different velocities.

Let us suppose that at $z = 0$ one has a field distribution with azimuthal dependence $\sim\cos(m\phi)$. Such distribution may be described by superposition of two modes with $m = \pm|m|$, which are excited without phase shift:

$$E_i = \tilde{E}_{i,+m}(r) \exp\left[im\phi\right] exp[i\beta_{+m}z] + \tilde{E}_{i,-m}(r) \exp\left[-im\phi\right] exp[i\beta_{-m}z] \tag{6}$$

where $\tilde{E}_{i,\pm m}(r)$ are the radial distributions of the field, $i = r, \varphi, z$. Indeed, due to the difference between propagation constants $\beta_{\pm m}$, this distribution will differ for opposite signs of m, but in the first approach we will suppose that $\tilde{E}_{i,+m}(r) \approx \tilde{E}_{i,-m}(r)$. Correctness of such assumption will further be approved by numerical calculations of field distributions. Different values of propagation speed will lead to phase shift at chosen $z = z_0$ and, thus, to the rotation of field distribution on the angle $(\beta'_{-m} - \beta'_{+m})z_0/2m$. This formula is similar to the one for microwave waveguide filled by gyrotropic medium [43]. For the characterization of rotation angle, we will use the specific rotation angle for each mode defined as following:

$$\theta = \frac{\beta'_{-m} - \beta'_{+m}}{2m}. \tag{7}$$

Propagation length will also differ for modes with opposite signs of m. This may lead to the fact that at certain value of z, one of the modes becomes negligibly small. At such length, defined by the condition

$$|\beta''_{-m} - \beta''_{+m}|z_0 >> 1, \tag{8}$$

the initial azimuthal intensity distribution becomes spatially homogeneous.

For numerical solution of dispersion equation and investigation of field distributions, we will use the following parameters: linear frequency of electromagnetic wave $f = \omega/2\pi = 100$ THz (wavelength in vacuum $\lambda_0 = 3$ μm), nanowire radius $R = 50$ nm (quantum effects in graphene structures should be taken into account at the size of the structure less than ≈20 nm [45]). For simplicity, we will assume that $\varepsilon_\perp = \varepsilon_{||} = \varepsilon$. The value of ε will be set at 2. We will consider the room temperatures ($T = 300$ K) and graphene scattering rate $\Gamma = 0.1$ meV. Due to the fact that rotation may be observed only for modes that depend on azimuthal angle, we will consider the modes with $|m| \neq 0$.

Figure 2 shows the electric field distribution of some high-order modes at z = 100 nm for different values of ε_a. Graphene chemical potential is μ_{ch} = 1 eV. Dash-dot lines show the calculated position of maximum. One can see that the calculated rotation angles are in good agreement with numerical modelling. Change in the sign of gyrotropy ε_a (i.e. change in magnetization or magnetic field direction) leads to opposite rotation of field distribution. The difference between radial distributions of the fields with opposite signs of m has no evident effect. For high-index modes in giant gyrotropy case, condition (8) is not satisfied, and one can see the blurring of distribution we have mentioned above.

A change in graphene conductivity (or its chemical potential) may lead to greater difference in propagation constants of the modes with opposite signs of m. This may be used for adjusting the rotation angle, similarly to graphene-covered optical fibre [36]. Dependences of the specific rotation angle, calculated by Eq. (7) for some lower modes, are shown in **Figure 3**. This figure also contains SPP wavelength and propagation length. One can see that specific rotation angle reaches a maximum at certain chemical potential, for which values are different for each mode. For lower modes, the maximum corresponds to higher chemical potential values. These maximal values are indicated by dashed lines. Maximal rotation angle decreases when mode number increases. An increase in gyrotropy $|\varepsilon_a|$ leads to some shift of the maximum to lower chemical potentials. For positive values of gyrotropy, the specific rotation angles are negative but equal to the absolute value in the case of negative gyrotropy. For the graphs of SPP

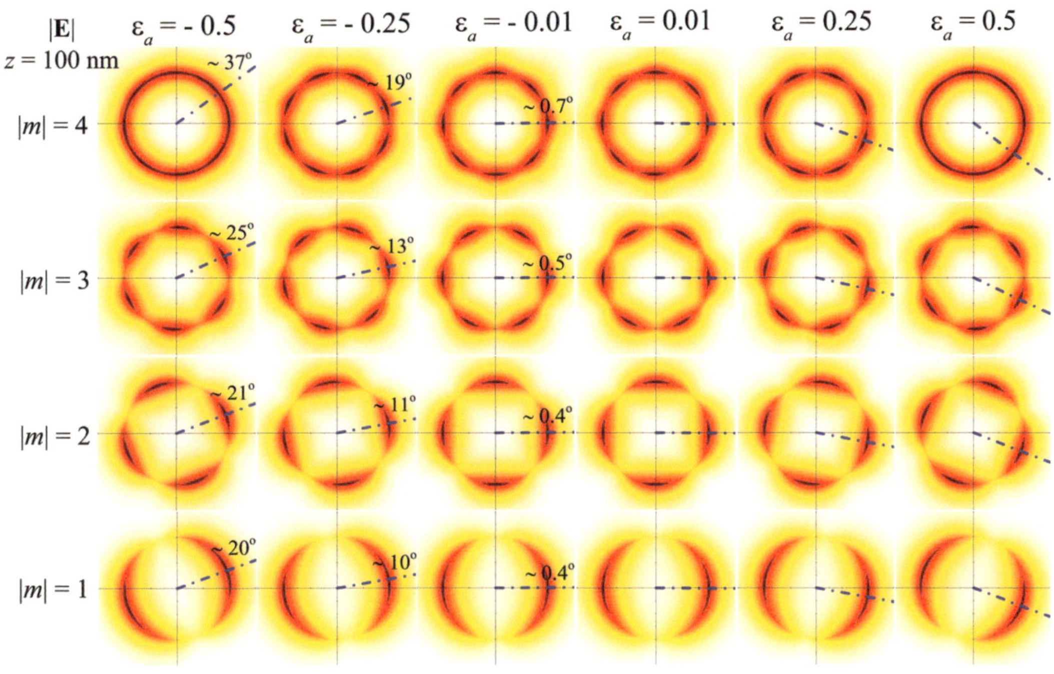

Figure 2. Electric field distribution of some lower modes at z = 100 nm for different values of ε_a. Dash-dot lines show the calculated position of the first maximum. Calculated rotation angles are shown as well. For $\varepsilon_a > 0$ rotation angles are the same as for corresponding case of $\varepsilon_a < 0$, but distribution rotates in opposite direction. Graphene chemical potential is μ_{ch} = 1 eV.

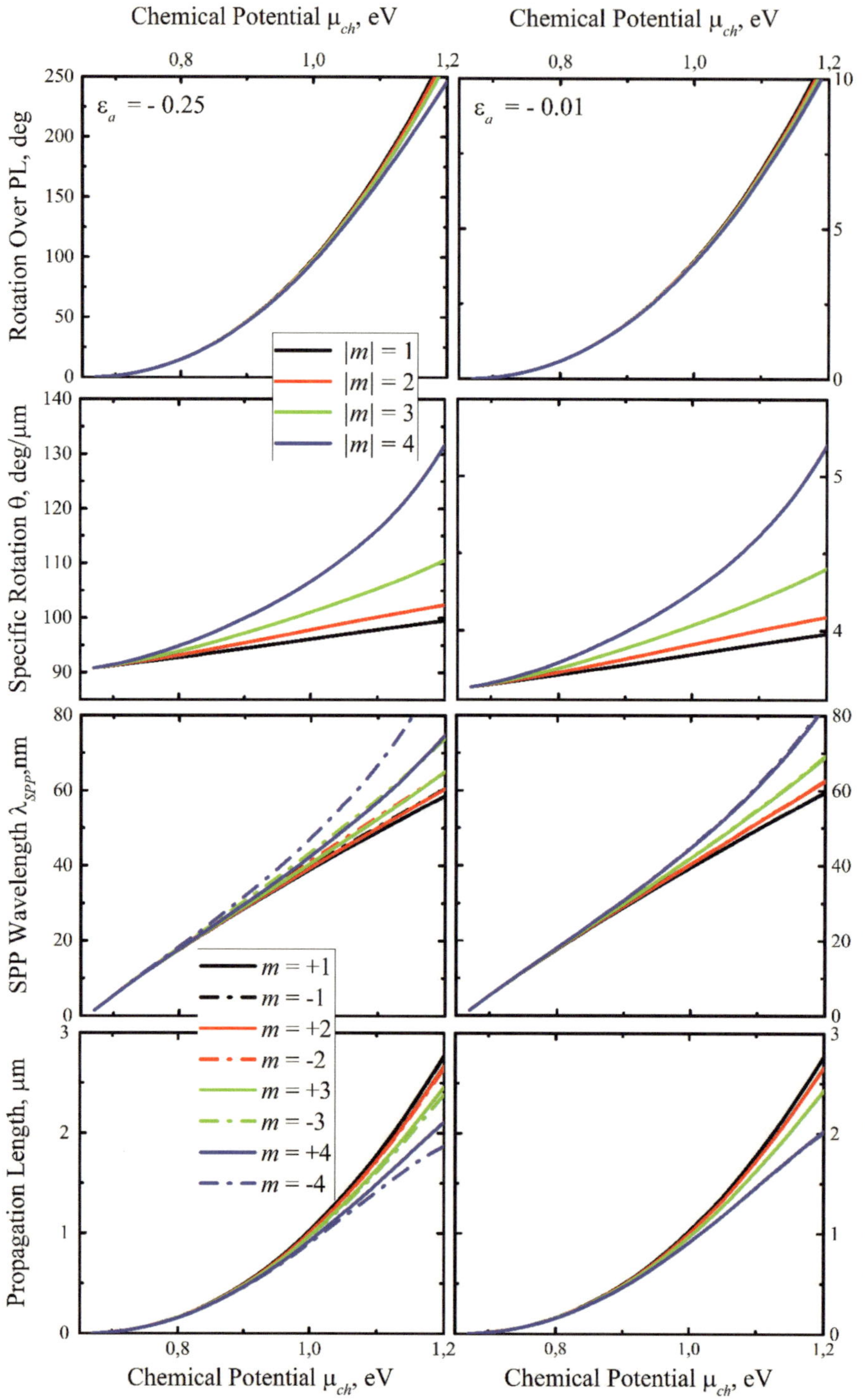

Figure 3. Dependences of the specific rotation angle, SPP wavelength and propagation length versus chemical potential of graphene. The vertical dashed lines show the position of maximum of specific rotation angle for corresponding mode.

wavelength and propagation length change, the sign of ε_a leads to the exchange of the lines for $m > 0$ and $m < 0$.

In general, the specific rotation angle may be adjusted approximately twice by changing the chemical potential of graphene.

It should be noted that the maximal specific rotation is observed near the inflection point of dependence of SPP wavelength versus chemical potential for the mode with higher wavelength. At such chemical potential values, corresponding modes have propagation length less than SPP wavelength, i.e. when the modes become evanescent. One can also see that for maximal rotation angles propagation lengths of the modes with opposite sign of m differ significantly. Thus, condition (8) plays a crucial role.

Propagation characteristics of the modes depend on the permittivity of nanowire, its radius and frequency of electromagnetic wave. All these values may be used for achieving the maximal rotation of desirable mode, but this question needs to be investigated separately.

Rotation angle linearly depends on the length of nanowire. So, the maximal rotation may be reached at the propagation length of SPPs. But one has to keep in mind condition (8) to avoid a non-desirable blurring of distribution.

It should be noted that for practical application of the effect under investigation, high values of ε_a are needed. It takes values $\varepsilon_a \sim 0.001$–0.01 at wavelengths approximately equal to be considered here to be used in magneto-optics materials frequently [29]. Faraday rotation angle and Verdet constant are used frequently used for characterizing gyrotropic materials. Faraday rotation angle may be defined as follows [43]: $\theta_F = z\omega[(\varepsilon + \varepsilon_a)^{1/2} - (\varepsilon - \varepsilon_a)^{1/2}]/(2c) = BVz$, where B is the external magnetic induction and V is the Verdet constant. For $\varepsilon_a \ll \varepsilon$, gyrotopy is proportional to BV. The high values of Verdet constant in the THz frequency range (0.1–10 THz) have some semiconductors: (Cd,Mn)Te (10^3 rad T^{-1} m^{-1}) [46], InSb (10^4 rad T^{-1} m^{-1}) [47], HgTe (10^6 rad T^{-1} m^{-1}) [48]. In such materials, the values of ε_a may be reached, which are necessary for practical application, but proposed structure should be redesigned for THz frequencies (radius of the core should be increased). We should note that at THz frequencies propagation length of SPPs is much greater than in infrared domain as considered here. Thus, it is possible to achieve a greater rotation angles but at greater scales.

4. Spiral graphene-based waveguides

Let us consider a dielectric cylinder (core of the waveguide) with dielectric permittivity $\varepsilon_{in} = \varepsilon^r{}_{in}\varepsilon_0$ (ε_0 is the electric constant) and radius R, which is coiled by graphene strip (see **Figure 4a** and **b**). Such cylinder is embedded in the dielectric medium with dielectric permittivity $\varepsilon_{out} = \varepsilon^r{}_{out}\varepsilon_0$. Both mediums will be considered as non-magnetic ($\mu_{in} = \mu_{out} = \mu_0$). We will use cylindrical coordinates (r, ϕ, z). Let us suppose that the z-axis coincides with the cylinder axis.

The net of cylindrical surface may be represented as meta-surface formed by graphene strips with width W and spacer width G (see **Figure 4c**). Such meta-surface may be described by 2D conductivity tensor $\hat{\sigma}$ of which components depend on the tilt angle θ [49, 50]:

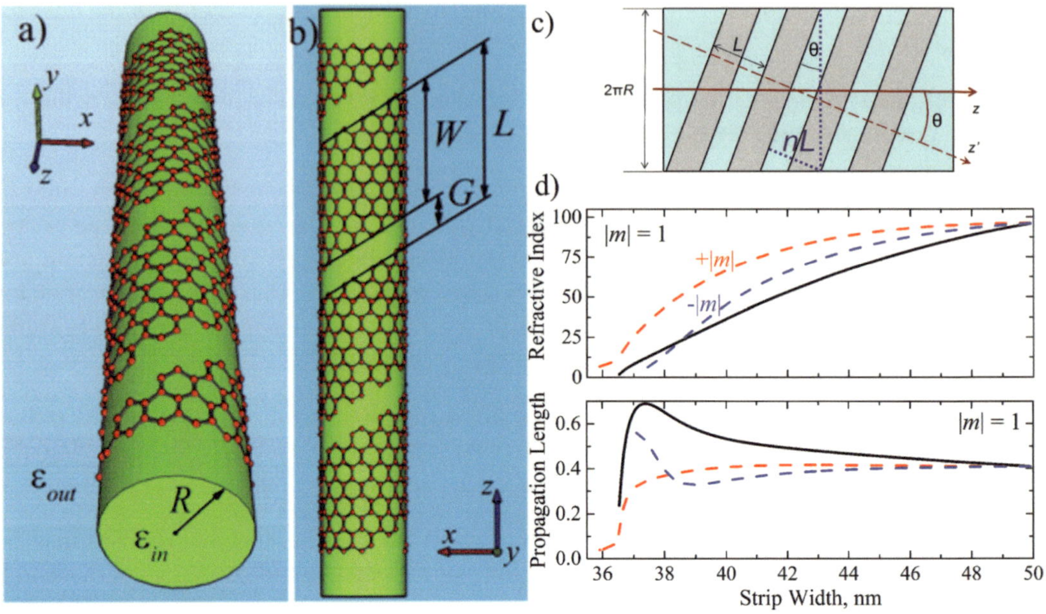

Figure 4. Geometry of the problem. Perspective (a) and top projection (b) views of the part of the meta-tube. Schemes illustrating the coupling between the tilt angle θ and tilt number n (c) and influence of the graphene strip width W on dispersion characteristic of the plasmonic modes with $m = \pm 1$ (d), $n = 0$ for the solid line corresponds to the non-chiral structure formed by graphene rings, $n = 9$ for dashed lines approximately corresponds to the tilt angle 45°.

$$\begin{aligned}
\hat{\sigma}_g &= \begin{pmatrix} \sigma_{\phi\phi} & \sigma_{\phi z} \\ \sigma_{z\phi} & \sigma_{zz} \end{pmatrix} \\
\sigma_{\phi\phi} &= \frac{L\sigma_g\sigma_C \sin^2\theta}{W\sigma_C + G\sigma_g} + \frac{W}{L}\sigma_g \cos^2\theta, \\
\sigma_{zz} &= \frac{L\sigma_g\sigma_C \cos^2\theta}{W\sigma_C + G\sigma_g} + \frac{W}{L}\sigma_g \sin^2\theta, \\
\sigma_{\phi z} &= \sigma_{z\phi} = \left(\frac{W}{L}\sigma_g - \frac{L\sigma_g\sigma_C}{W\sigma_C + G\sigma_g}\right)\sin\theta\cos\theta \\
\sigma_C &= -i\omega\varepsilon_0\varepsilon\left(\frac{L}{\pi}\right)\ln\left[\csc\left(\frac{\pi G}{2L}\right)\right]
\end{aligned} \tag{9}$$

In Eq. (9), σ_g is the graphene conductivity, which may be divided into intra- and inter-band contributions $\sigma_g = \sigma_{\text{intra}} + \sigma_{\text{inter}}$, while σ_C is the effective meta-surface conductivity caused by capacitive coupling between graphene strips. Such representation is valid when the wavelength of electromagnetic wave is much longer than the periodicity of the structure.

For fixed periodicity of the surface $L = W + G$, the tilt angle may get discrete values: $\theta_n = \arcsin[nL/2\pi R]$, where n is an integer number, which may be interpreted as the count of graphene spirals (or the count of spacers in the meta-surface between two enclosed strip edges). We will call this number the 'tilt number'.

For investigation of electrodynamics of such structure, one should solve Maxwell's equations inside each medium taking into account the boundary conditions: $E_{z,}^{\text{in}} = E_{z}^{\text{out}}$, $E_{\varphi}^{\text{in}} = E_{\varphi}^{\text{out}}$, $H_{z}^{\text{out}} - H_{z}^{\text{in}} = -\sigma_{\phi\phi}E_{\phi}^{\text{in}} - \sigma_{\phi z}E_{z}^{\text{in}}$ and $H_{\phi}^{\text{out}} - H_{\phi}^{\text{in}} = \sigma_{zz}E_{z}^{\text{in}} + \sigma_{z\phi}E_{\phi}^{\text{in}}$. Considering the waves propagating along the cylinder axis, one may put electric and magnetic fields **E**, **H** ~ $\exp[-i\omega t + ihz + im\phi]$, where ω is the circular frequency, h is the propagation constant and m is the azimuthal mode index.

The calculations show that propagation constants for the modes propagating along the z-axis with opposite azimuthal rotation direction (we will denote these propagation constants as $h_{+|m|}$ and $h_{-|m|}$) are different, similarly to the plasmonic modes in gyrotropic graphene-covered nanowires (see Section 3). Thus, if one has azimuthal field distribution at an input of the waveguide $z = 0$ proportional to $\cos[m\phi]$, at the output $z = z_0$ one will have rotated field distribution. Rotation angle may be calculated as $\psi = z_0(\text{Re}[h_{-|m|}] - \text{Re}[h_{+|m|}])/2|m|$, while $z_0|\text{Im}[h_{-|m|}] - \text{Im}[h_{+|m|}]| << 1$. To characterize the rotation effect, we will also use the specific rotation angle $\psi_0 = \psi/z_0$.

Figure 4(d) shows the dispersion characteristics (mode refractive index $\text{Re}[h_m]/k_0$ and the propagation length $1/2\text{Im}[h_m]$) for the structure formed by graphene strip with the periodicity $L = 50$ nm, core radius $R = 100$ nm for the frequency of electromagnetic wave $f = 50$ THz via graphene strip width for the modes with $m = \pm 1$ and two values of the tilt angles ($n = 0$ corresponds to the non-chiral structure formed by graphene rings, $n = 9$ approximately corresponds to the tilt angle 45°). We assume that $\varepsilon^r_{\text{in}} = 3$, and $\varepsilon^r_{\text{out}} = 1$. For the fixed periodicity L, a decrease in strip width leads to an increase in the spacer width. One can see that there is a critical strip width when the structure cannot support plasmonic mode. This critical width is different for a different mode index m. For non-zero tilt number modes with the opposite azimuthal rotation directions (i.e. $\pm|m|$) have different critical width. The mode rotating in the same direction as chirality of the structure exists at lower graphene width but has low propagation length, while the opposite mode has a bulk behaviour when graphene width reaches the critical value.

Figure 5 shows the specific rotation angle for the spiral waveguide with the periodicity $L = 50$ nm, graphene strip width $W = 45$ nm and the spacer width $G = 5$ nm via the tilt number of the structure. Other parameters are the same as for **Figure 2**. One can see that at $n < n_{cr} \approx 10$, specific rotation angle of lower modes is greater than that of higher modes. At $n > n_{cr}$, the opposite behaviour take a place. Maximal specific rotation reaches up to few hundreds of degrees, and it is observed near the tilt angle of about $\pi/4$, when non-diagonal components of conductivity are maximal, similarly to the TE-TM coupling for the fundamental plasmonic mode. The maximums of higher modes are shifted to greater tilt angles. We should note that the rotation angle strictly depends on the graphene chemical potential, which may be controlled by gate voltage or chemical doping. The maximal values of the specific rotation angle are much higher than that of gyrotropic graphene-covered nanowires for the similar set of parameters (see Section 3). Negative tilt numbers (and tilt angles) correspond to the opposite chirality of the structure. Such situation will be observed for the backward waves propagating in the structure, and the specific rotation angle will have negative values. Thus, the structure is strictly non-reciprocal. This property may be used for one-way wave propagation if we have graphene strip width lower than the critical one for the negative azimuthal mode index: mode

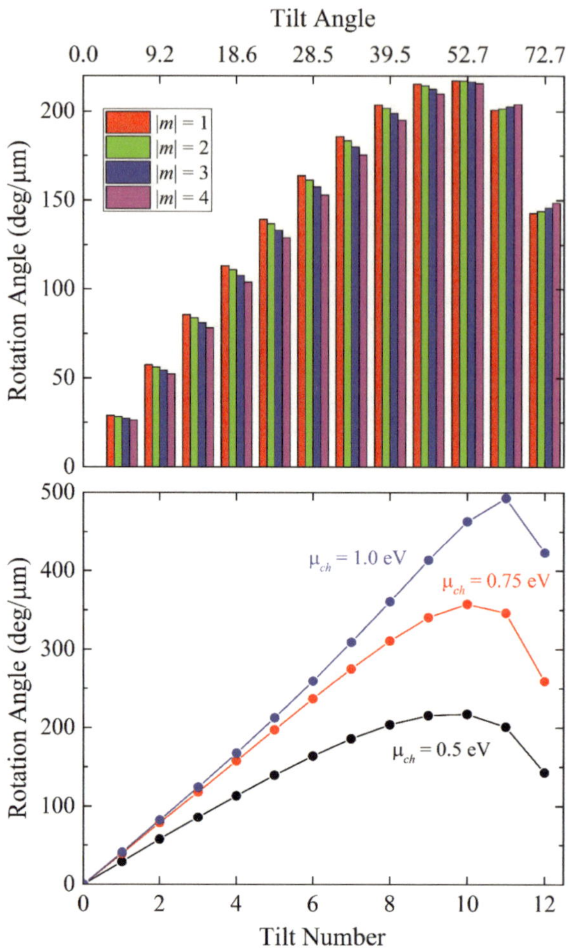

Figure 5. Specific rotation angle for the spiral waveguide with the periodicity $L = 50$ nm, graphene strip width $W = 45$ nm and the spacer width $G = 5$ nm via the tilt number (or the tilt angle) of the structure. The upper panel shows the specific rotation angle for first four modes for graphene chemical potential μ_{ch} = 0.5 eV. The lower panel shows the specific rotation angle for the mode with $|m| = 1$ for different values of graphene chemical potential.

with the positive azimuthal index will still propagate along the structure, while in the opposite direction only the mode with the negative azimuthal index will propagate.

One may see that even at maximal possible tilt angle the specific rotation angle is still high enough. This fact makes the concept of surface plasmon-polaritons control by the shear strains very promising for practical applications. Let us imagine that we have a non-chiral structure formed by nanowire longitudinally covered by graphene strips (see **Figure 6**). This situation corresponds to the tilt angle $\theta = 90°$. Non-diagonal components of the surface conductivity tensor will be equal to zero, and no rotation of high-order plasmonic modes will be observed. If shear strains are applied to such a structure, the spiral waveguide will be formed with the tilt angle defined by the strain value and field distribution at the output of the waveguide will be rotated.

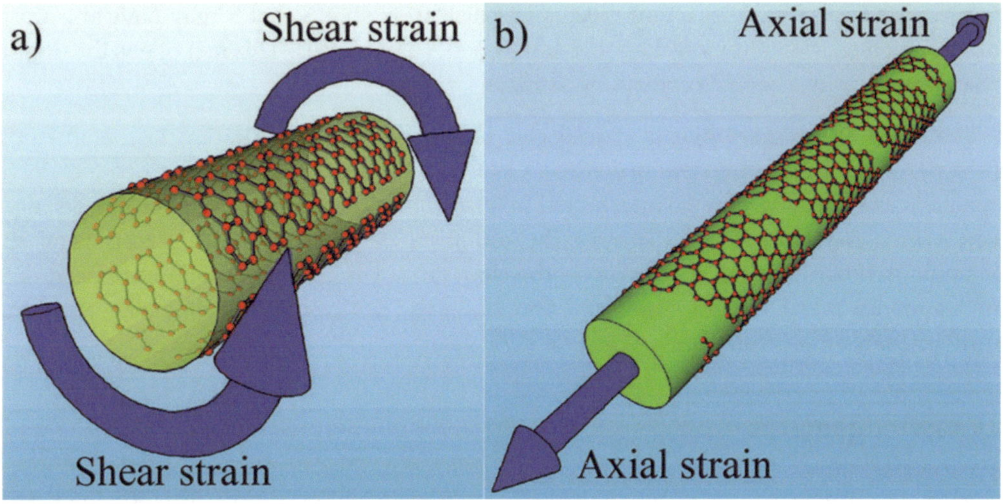

Figure 6. The schemes illustrating formation of the spiral waveguide by shear strains (a) and control of spiral waveguide parameters by axial strains (b).

Another way to control plasmons in the structure under investigation by external strains is to apply an axial strain to the structure. An axial strain will lead to the change in the spacer width G and the periodicity of the structure L. From Eq. (9), we may see that the relation W/L affects significantly the meta-surface conductivity. In the situation under discussion W/L may be controlled by the axial strain. It is seen from **Figure 4(d)** that even a change in the strip width on some nanometres leads to a significant difference in the propagation constants of counterrotating azimuthal modes. Thus, proposed structure should be very sensitive to an axial stresses.

5. Concluding remarks

In this chapter, we have investigated in detail two ways of breaking of the degeneracy of the plasmonic modes with the opposite azimuthal rotations for graphene-coated nanowires: by external magnetic field and by surface spiral structure. This breaking of the degeneracy may lead to a giant spatial rotation of high-order plasmonic modes, and to the redistribution of the intensity of electromagnetic wave.

The open problem is the self-consistent problem of SPPs propagation in plasmonic magnetic nanowires. Magnetization of the nanowire, in general, will lead to the change in SPPs properties, while SPPs themselves will induce an effective magnetic field, which will change the magnetization of the nanowire. This effect should be taken into account especially for nonlinear SPPs. To the best of our knowledge, this problem is still unsolved.

The effects discussed in Section 4 are caused by the off-diagonal components of the surface conductivity tensor. In addition to the considered structure, some similar effect may be observed for the nanowires covered by strained graphene layer. Recent investigation of graphene conductivity under a non-mechanical distortion shows that it may have anisotropic conductivity tensor with the off-diagonal components as well [51]. This fact opens the door for further investigations of SPPs control by artificial strains.

The predicted effects may play a crucial role in the polarization rotation in metamaterials consisting of multiple proposed structures. On the other hand, redistribution of the electromagnetic wave intensity may be interpreted in terms of local change in the photonic density of states, which may be used to control the radiation of quantum dots placed near such plasmonic nanowire. These results open the door to novel plasmonic applications ranging from nanowire-based Faraday isolators and one-way devices to the magnetic and strain control in quantum-optical applications.

Acknowledgements

The work was supported in part by Stratégie internationale NNN-Telecom de la Région Pays de La Loire, Alexander von Humboldt Stiftung, President of Russian Federation (project # MK-1653.2017.2), Russian Foundation for Basic Research (grants ## 16-37-00023, 16-07-00751 and 16-29-14045), and Act 211 Government of the Russian Federation (contract No 02.A03.21.0011).

Author details

Dmitry A. Kuzmin[1]*, Igor V. Bychkov[1], Vladimir G. Shavrov[2] and Vasily V. Temnov[3]

*Address all correspondence to: kuzminda@csu.ru

1 Chelyabinsk State University, Chelyabinsk, Russian Federation

2 Kotel'nikov Institute of Radio-Engineering and Electronics of RAS, Moscow, Russian Federation

3 Institut des Molécules et Matériaux du Mans, Université du Maine, France

References

[1] Bao Q., Loh K.P. Graphene photonics, plasmonics, and broadband optoelectronic devices. ACS Nano. 2012;**6**(5):3677–3694. DOI: 10.1021/nn300989g

[2] Grigorenko A.N., Polini M., Novoselov K.S. Graphene plasmonics. Nature Photonics. 2012;**6**:749–758. DOI: 10.1038/nphoton.2012.262

[3] Garcia de Abajo F.J. Graphene plasmonics: challenges and opportunities. ACS Photonics. 2014;**1**(3):135–152. DOI: 10.1021/ph400147y

[4] Mikhailov S.A., Ziegler K. New electromagnetic mode in graphene. Physical Review Letters. 2007;**99**:016803. DOI: 10.1103/PhysRevLett.99.016803

[5] Hanson G.W. Quasi-transverse electromagnetic modes supported by a graphene parallel-plate waveguide. Journal of Applied Physics. 2008;**104**(8):084314. DOI: 10.1063/1.3005881

[6] Bludov Y.V., Ferreira A., Peres N.M.R., Vasilevskiy M. A primer on surface plasmon-polaritons in graphene. International Journal of Modern Physics B. 2013;**27**(10):1341001. DOI: 10.1142/S0217979213410014

[7] Kotov O.V., Kol'chenko M.A., Lozovik Y.E. Ultrahigh refractive index sensitivity of TE-polarized electromagnetic waves in graphene at the interface between two dielectric media. Optics Express. 2013;**21**(11):13533–13546. DOI: 10.1364/OE.21.013533

[8] Buslaev P.I., Iorsh I.V., Shadrivov I.V., Belov P.A., Kivshar Y.S. Plasmons in waveguide structures formed by two graphene layers. JETP Letters. 2013;**97**(9):535–539. DOI: 10.1134/S0021364013090063

[9] Smirnova D.A., Iorsh I.V., Shadrivov I.V., Kivshar Y.S. Multilayer graphene waveguides. JETP Letters. 2014;**99**(8):456–460. DOI: 10.1134/S002136401408013X

[10] Koppens F.H.L., Chang D.E., García de Abajo F.J. Graphene plasmonics: a platform for strong light–matter interactions. Nano Letters. 2011;**11**(8):3370–3377. DOI: 10.1021/nl201771h

[11] Yan H., Low T., Zhu W., Wu Y., Freitag M., Li X., Guinea F., Avouris P., Xia F. Damping pathways of mid-infrared plasmons in graphene nanostructures. Nature Photonics. 2013;**7**:394–399. DOI: 10.1038/nphoton.2013.57

[12] Soto Lamata I., Alonso-Gonzalez P., Hillenbrand R., Nikitin A.Y. Plasmons in cylindrical 2D materials as a platform for nanophotonic circuits. ACS Photonics. 2015;**2**(2):280–286. DOI: 10.1021/ph500377u

[13] Gao Y., Ren G., Zhu B., Liu H., Lian Y., Jian Sh. Analytical model for plasmon modes in graphene-coated nanowire. Optics Express. 2014;**22**(20): 24322–24331. DOI: 10.1364/OE.22.024322

[14] Gao Y., Ren G., Zhu B., Wang J., Jian Sh. Single-mode graphene-coated nanowire plasmonic waveguide. Optics Letters. 2014;**39**(20):5909–5912. DOI: 10.1364/OL.39.005909

[15] Correas-Serrano D., Gomez-Diaz J.S., Alu A., Alvarez-Melcon A. Electrically and magnetically biased graphene-based cylindrical waveguides: analysis and applications as reconfigurable antennas. IEEE Transactions on Terahertz Science and Technology. 2015;**5**(6): 951–960. DOI: 10.1109/TTHZ.2015.2472985

[16] Kuzmin D.A., Bychkov I.V., Shavrov V.G., Kotov L.N. Transverse-electric plasmonic modes of cylindrical graphene-based waveguide at near-infrared and visible frequencies. Scientific Reports. 2016;**6**: 26915. DOI: 10.1038/srep26915

[17] Krasavin A.V., Zheludev N.I. Active plasmonics: controlling signals in Au/Ga waveguide using nanoscale structural transformations. Applied Physics Letters. 2014;**84**(8):1416. DOI: 10.1063/1.1650904

[18] Fedutik Y., Temnov V.V., Schöps O., Woggon U., Artemyev M.V. Exciton-plasmon-photon conversion in plasmonic nanostructures. Physical Review Letters. 2007;**99**:136802. DOI: 10.1103/PhysRevLett.99.136802

[19] Temnov V.V., Armelles G., Woggon U., Guzatov D., Cebollada A., Garcia-Martin A., et al. Active magneto-plasmonics in hybrid metal–ferromagnet structures. Nature Photonics. 2010;**4**:107–111. DOI: 10.1038/nphoton.2009.265

[20] LeBlanc S.J., McClanahan M.R., Jones M., Moyer P.J. Enhancement of multiphoton emission from single CdSe quantum dots coupled to gold films. Nano Letters. 2013;**13** (4):1662–1669. DOI: 10.1021/nl400117h

[21] Abbasi F., Davoyan A.R., Engheta N. One-way surface states due to nonreciprocal light-line crossing. New Journal of Physics. 2015;**17**:063014. DOI: 10.1088/1367-2630/17/6/063014

[22] Kuzmin D.A., Bychkov I.V., Shavrov V.G. Magnetic field control of plasmon polaritons in graphene-covered gyrotropic planar waveguide. Optics Letters. 2015;**40**(11):2557–2560. DOI: 10.1364/OL.40.002557

[23] Kurkin M.I., Bakulina N.B., Pisarev R.V. Transient inverse Faraday effect and ultrafast optical switching of magnetization. Physical Review B. 2008;**78**:134430. DOI: 10.1103/PhysRevB.78.134430

[24] Mentink J.H., Hellsvik J., Afanasiev D.V., Ivanov B.A., Kirilyuk A., Kimel A.V., et al. Ultrafast spin dynamics in multisublattice magnets. Physical Review Letters. 2012;**108**: 057202. DOI: 10.1103/PhysRevLett.108.057202

[25] Kirilyuk A., Kimel A.V., Rasing T. Laser-induced magnetization dynamics and reversal in ferrimagnetic alloys. Reports on Progress in Physics. 2013;**76**(2):026501. DOI: 10.1088/0034-4885/76/2/026501

[26] Kurkin M.I., Orlova N.B. Femtosecond magnetooptics and ultrafast magnetization reversal of ferromagnetic. Journal of Magnetism and Magnetic Materials. 2014;**361**:224–231. DOI: 10.1016/j.jmmm.2014.02.079

[27] Belotelov V.I., Doskolovich L.L., Zvezdin A.K. Extraordinary magneto-optical effects and transmission through metal-dielectric plasmonic systems. Physical Review Letters. 2007;**98**:077401. DOI: 10.1103/PhysRevLett.98.077401

[28] Belotelov V.I., Akimov I.A., Pohl M., Kotov V.A., Kasture S., Vengurlekar A.S., et al. Enhanced magneto-optical effects in magnetoplasmonic crystals. Nature Nanotechnology. 2011;**6**:370–376. DOI: 10.1038/nnano.2011.54

[29] Kreilkamp L.E., Belotelov V.I., Chin J.Y., Neutzner S., Dregely D., Wehlus T., et al. Waveguide-plasmon polaritons enhance transverse magneto-optical Kerr effect. Physical Review X. 2013;**3**:041019. DOI: 10.1103/PhysRevX.3.041019

[30] Khokhlov N.E., Prokopov A.R., Shaposhnikov A.N., Berzhansky V.N., Kozhaev M.A., Andreev S.N., et al. Photonic crystals with plasmonic patterns: novel type of the heterostructures for enhanced magneto-optical activity. Journal of Physics D: Applied Physics. 2015;**48**(9):095001. DOI: 10.1088/0022-3727/48/9/095001

[31] Razdolski I., Parchenko S., Stupakiewicz A., Semin S., Stognij A., Maziewski A., et al. Second-harmonic generation from a magnetic buried interface enhanced by an interplay of surface plasma resonances. ACS Photonics. 2015;**2**(1):20–26. DOI: 10.1021/ph500382u

[32] Baranova N.B., Zel'dovich B.Y. Rotation of a ray by a magnetic field. JETP Letters. 1994;**59**(10):681–684.

[33] Darsht M.Y., Zhirgalova I.V., Zel'dovich B.Y., Kundikova N.D. Observation of a "magnetic" rotation of the speckle of light passed through an optical fiber. JETP Letters. 1994;**59**(11):763–765.

[34] Ardasheva L.I., Sadykova M.O., Sadykov N.R., Chernyakov V.E. Rotation of the speckle pattern in a low-mode optical fiber in a longitudinal magnetic field. Journal of Optical Technology. 2002;**69**(7):451. DOI: 10.1364/JOT.69.000451

[35] Ardasheva L.I., Kundikova N.D., Sadykova M.O., Sadykov N.R., Chernyakov V.E. Speckle-pattern rotation in a few-mode optical fiber in a longitudinal magnetic field. Optics and Spectroscopy. 2003;**95**(4):645–651. DOI: 10.1134/1.1621451

[36] Kuzmin D.A., Bychkov I.V., Shavrov V.G. Influence of graphene coating on speckle-pattern rotation of light in gyrotropic optical fiber. Optics Letters. 2015;**40**(6):890–893. DOI: 10.1364/OL.40.000890

[37] Kuzmin A., Bychkov I.V., Shavrov V.G., Temnov V.V. Giant Faraday rotation of high-order plasmonic modes in graphene-covered nanowires. Nano Letters. 2016;**16**(7):4391–4395. DOI: 10.1021/acs.nanolett.6b01517

[38] Hanson G.W. Dyadic Green's functions and guided surface waves for a surface conductivity model of graphene. Journal of Applied Physics. 2008;**103**(6):064302. DOI: 10.1063/1.2891452

[39] Falkovsky L.A., Varlamov A.A. Space-time dispersion of graphene conductivity. The European Physical Journal B. 2007;**56**(4):281–284. DOI: 10.1140/epjb/e2007-00142-3

[40] Falkovsky L.A. Optical properties of graphene and IV–VI semiconductors. Physics-Uspekhi. 2008;**51**(9):887–897. DOI: 10.1070/PU2008v051n09ABEH006625

[41] Kuzmin D.A., Bychkov I.V., Shavrov V.G., Temnov V.V., Lee H.I., Mok J. Plasmonically induced magnetic field in graphene-coated nanowires. Optics Letters. 2016;**41**(2):396–399. DOI: 10.1364/OL.41.000396

[42] Suhl H., Walker L.R. Topics in guided-wave propagation through gyromagnetic media: Part I—The completely filled cylindrical guide. Bell System Technical Journal. 1954;**33**(5):579–659. DOI: 10.1002/j.1538-7305.1954.tb02358.x

[43] Gurevich A.G., Melkov G.A. Magnetization oscillations and waves. Boca Raton: CRC Press; 1996. 445 p.

[44] Cojocaru E. Modes in dielectric or ferrite gyrotropic slab and circular waveguides, longitudinally magnetized, with open and completely or partially filled wall. Journal of the Optical Society of America B. 2010;**27**(10):1965–1977. DOI: 10.1364/JOSAB.27.001965

[45] Thongrattanasiri S., Manjavacas A., García de Abajo F.J. Quantum Finite-Size Effects in Graphene Plasmons. ACS Nano. 2012;**6**(2):1766–1775. DOI: 10.1021/nn204780e

[46] Gaj J.A., Galazka R.R., Nawrocki M. Giant exciton Faraday rotation in Cd1-xMnxTe mixed crystals. Solid State Communications. 1978;**25**(3):193–195. DOI: 10.1016/0038-1098 (78)91477-1

[47] Palik E.D., Furdyna J.K. Infrared and microwave magnetoplasma effects in semiconductors. Reports on Progress in Physics. 1970;**33**(3):1193. DOI: 10.1088/0034-4885/33/3/307

[48] Shuvaev A.M., Astakhov G.V., Pimenov A., Brüne C., Buhmann H., Molenkamp L.W. Giant magneto-optical Faraday effect in HgTe thin films in the terahertz spectral range. Physical Review Letters. 2011;**106**:107404. DOI: 10.1103/PhysRevLett.106.107404

[49] Gomez-Diaz J.S., Tymchenko M., Alù A. Hyperbolic plasmons and topological transitions over uniaxial metasurfaces. Physical Review Letters. 2015;**114**:233901. DOI: 10.1103/PhysRevLett.114.233901

[50] Gomez-Diaz J.S., Tymchenko M., Alù A. Hyperbolic metasurfaces: surface plasmons, light-matter interactions, and physical implementation using graphene strips. Optical Materials Express. 2015;**5**(10):2313–2329. DOI: 10.1364/OME.5.002313

[51] Oliva-Leyva M., Naumis G.G. Effective Dirac Hamiltonian for anisotropic honeycomb lattices: optical properties. Physical Review B. 2016;**93**:035439. DOI: 10.1103/PhysRevB.93.035439

Chapter 3

Nanoplasmonics in Metallic Nanostructures and Dirac Systems

Hari P. Paudel, Alireza Safaei and
Michael N. Leuenberger

Additional information is available at the end of the chapter

http://dx.doi.org/10.5772/67689

Abstract

In this book chapter, we review some of the progress made in nanoplasmonics and related optoelectronics phenomena in the field of two-dimensional (2D) materials and the recent 3D Weyl semimetals. We give a brief overview of plasmonics for three-dimensional (3DEG) and two-dimensional electron gases and draw comparisons with graphene, 3D topological insulators, 3D Weyl semimetals, and nanoplasmonics in nanogeometries. We discuss the decay of plasmons into electron-hole pairs and the subsequent thermalization and cooling of the hot carriers. We present our recent results in the fields of plasmonics in different nanostructures made of noble metals, such as Silver, and plasmonics in Dirac systems such as graphene and 3D topological insulators. We show a possibility of dynamically shifting the plasmon resonances in hybrid metal-semiconductor nanostructures. Plasmonics in 3D topological insulator and 3D Weyl semimetals have been least explored in nanoplasmonics although it can provide a variety of interesting physical phenomena involving spin plasmonics and chirality. Due to the inherent large spin-orbit coupling, locked spin-momentum oscillations can exist under special conditions and in the presence of an external laser field. We explore symmetric and antisymmetric modes in a slab of 3D TIs and present their dependences on the thickness of the slab.

Keywords: surface plasmon polaritons, metal, Dirac fermions

1. Introduction

Electromagnetic properties of metal-dielectric interfaces have attracted a vast amount of research efforts. Ever since, the work of Mie [1] for small particles and Ritchie [2] for flat interfaces, a wide variety of scientists ranging from physicist, chemists, material scientists to biologists have explored plasmonics-based phenomena and their potential applications in practical life. In nanostructures, under the right circumstances, light waves propagating at metal-dielectric interface excite collective modes of electrons at the metal surface, resulting in the generation of charge density waves called surface plasmons (SPs), which can be divided into localized modes called localized surface plasmons (LSPs) and propagating modes called surface plasmon polaritons (SPPs), propagating along the interface like ripples across the surface of water with an effective wavelength much less than that of incident electromagnetic wave. Free electrons respond collectively by oscillating in resonance with the light waves. In optics, scientists have investigated methods to use plasmonics for concentrating, channeling, and changing the phase of light using subwavelength metallic structures. This would lead to miniaturized plasmonic circuits with length scales much smaller than those in current use [3–13]. A creatively engineered metal-dielectric interface can generate surface plasmons with the same frequency as the outside electromagnetic waves but with several times shorter wavelength. This interesting phenomenon can be utilized in a way in which surface plasmons carry information in microprocessors faster than current electronic transistors [14]. Plasmonics holds promise for a higher information density than conventional electronics [15]. While this proposed application needs still to be proven feasible, metallic nanostructures much smaller than the wavelength of light have already been successfully built for amplifying signals in surface enhanced Raman spectroscopy (SERS), providing a powerful method to detect a single molecule [16–27]. Plasmonically enhanced electric fields are already being used for sensing biomolecules [28–36]. The efficient heating property of plasmonic fields can be used for photothermal cancer treatment [37, 38] and also for thermally assisted magnetic recording [39–41]. Plasmonic lasers are able to achieve ultrafast dynamics with sub-wavelength mode confinement [42–50].

Metallic nanoparticles can be made in different sizes and shapes, and the distance among them can be controlled as well. These parameters can be used to tune the plasmon resonance frequency [51–62]. The use of optically excited plasmons as a tunable frequency source that can be mixed with a laser through Raman scattering enables dynamical shifting of the wavelength of light in a controlled manner [63]. Fluegel et al. [64] used a continuous laser beam of a few microwatts power to excite carriers in a carefully engineered narrow GaAs quantum well. These photogenerated carriers are selectively passed through a thick barrier of AlAs into a wide GaAs quantum well in which the two-dimensional electron gas (2DEG) supports high charge density waves with collective electron motion normal to the layer. A signal laser beam operating at a different wavelength undergoes inelastic Raman scattering from the plasmon-phonon (longitudinal optical phonon) modes in the two-dimensional (2D) quantum well. The result is that a signal with 13–15 nm redshifted frequency is generated. Plasmonics can also be exploited in optical tweezers to confine nanoparticles to small dimensions [63]. Grigorenko et al. [65, 66] have made electromagnetically coupled gold pillars. A trapping beam simultaneously excites gap plasmon (GP) modes in the gap between the pillars when they are used as a substrate, resulting in an enhanced field at the trapping site.

In a metallic nanoshell containing a core of dielectric material, due to different dielectric environments in the core and outside of the nanoshell, plasmons are excited in the inner and outer surface of the metallic nanoshell, which can be tuned by varying the ratio of inner to the outer radii of the nanoshell and by altering the dielectric materials [67, 68]. **Figure 1** shows that the plasmon resonance energy can be tuned over a wide range by varying the material and/or shaping parameters. **Figure 1** compares the plasmon resonance wavelengths of gold, silver, graphene, and topological insulator nanostructures. For a particle diameter much smaller than the wavelength of light, the light-matter interaction leads to an oscillating homogeneous polarization of the particle volume, resulting in an oscillating dipole field. For spherical Au and Ag nanoparticles, the dipole plasmon resonance occurs in the visible part of the spectrum. In particle ensembles, additional shifts are noticed due to electromagnetic coupling between LSP modes. For example, in case of a dimer, plasmons can be viewed as bonding and anti-bonding combinations, in analogy to molecular orbitals, i.e., hybridization of the individual nanoparticle LSPs occurs, giving rise to resonance shifts following the $1/d^3$ interaction between two classical dipoles [69]. The observed shifts in LSP resonance and the linewidth narrowing can be understood by means of the Mie scattering theory [1]. Due to the coupling of LSPs, the arrays support one longitudinal and two transverse modes of propagating polarization waves. The transport of energy along such a chain is the main idea for transmitting information using plasmonics [70, 71]. Large heat dissipation has so far slowed down the progress in this field.

Nevertheless, we proposed a method of guiding electromagnetic waves along a chain below the diffraction limit in a controlled manner using a chain of nanoshells [72, 73]. The dynamic control over the plasmon resonances and their coupling gave rise to the idea of an optically controlled plasmonic switch. In this chapter, we present a concept of dynamical control over the plasmon resonances that can be obtained by controlling the dielectric environment of the LSPs using a pump probe technique. In Ref. [74], we show that a shift up to 125 nm can be achieved in an Ag core-TiO_2 coated nanostructure.

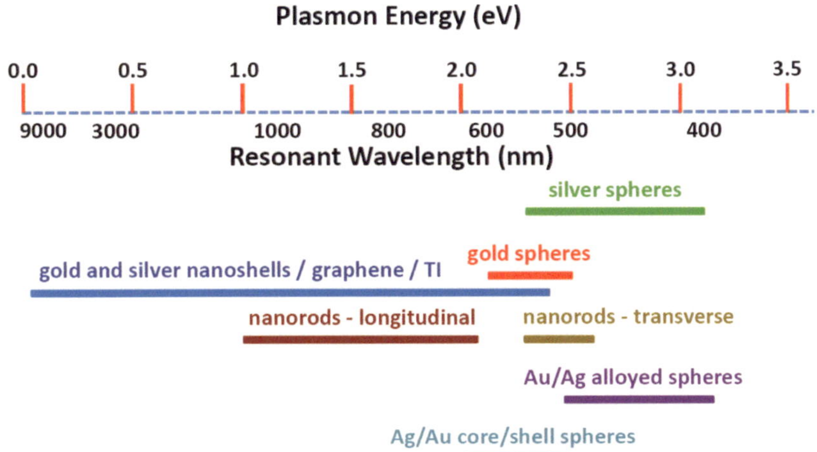

Figure 1. Comparison of the plasmon resonance wavelength spectra of gold, silver, graphene, and topological insulator (TI) nanostructures. The lifetime of plasmon for gold and silver ($\tau = 1 - 100\,fs$) is shorter than that of graphene and TI ($\tau = 10 - 100\,fs$).

In addition to that, we also review plasmonics in Dirac systems. Like in the case of a two-dimensional electron gas (2DEG), the surface plasmon mode in 2D materials such as graphene is tunable by changing the gate voltage through shifting the Fermi energy. Exciting surface plasmons give rise to light absorption enhancement in graphene, which can be utilized for photodetectors based on surface plasmon polaritons [75], optical switching of infrared plasmon polariton [76], and THz plasmonic lasing [77]. Plasmonics in Dirac systems show interesting features due to massless electrons around the Dirac nodes. In particular, we focus on surface plasmons in graphene, in Bismuth-based 3D topological insulators (3DTIs), and in 3D Weyl semimetals. Graphene, 3DTIs, and 3D Weyl semimetals are interesting due to their special electronic and optical properties arising from the linear dispersion relation around the Dirac cones in the Brillouin zone [78]. Around these points, energy dispersion of electrons can be described by a low energy Dirac Hamiltonian: $H_G(\mathbf{k}) = h(\mathbf{k}) \cdot \boldsymbol{\sigma}$, where $h(\mathbf{k}) = [h_x(\mathbf{k}), h_y(\mathbf{k}), 0]$ and $\boldsymbol{\sigma} = [\sigma_x, \sigma_y, \sigma_z]$. For a small wave vector, $\mathbf{q} = \mathbf{k} - \mathbf{K}$, we have $h(\mathbf{k}) = \hbar\upsilon_{Fq}$, showing that the electronic states are helical in nature, therefore allowing for "one-way traffic." This helical nature of electronic states significantly enhances the charge mobility, resulting in reduced plasmonic loss, and therefore high-quality factor in doped graphene.

Due to the excitation of surface plasmons, it is now possible to engineer the behavior of light on nanometer length scales and to increase the light-matter interaction [79–82]. This interaction is an outcome of the near-field enhancement close to the metal surface, which also leads to plasmon damping through radiative decay and through nonradiative decay inside the material, due to Landau damping, i.e., creation of electron-hole pairs via interband or intraband transitions, electron-phonon interaction, and boundary effects. The intraband transition happens in the conduction band and the interband transition occurs between other bands (such as the d-band) and the conduction band, as shown schematically in **Figure 2a** [83–85].

The radiative decay part of plasmon damping is due to the direct photon emission by coherent electron oscillation. As the size of the nanoparticle increases, the radiative decay of the plasmon is more significant. For larger nanoparticle elements, the radiative decay component is the main reason of plasmon resonance broadening and weakening of the dipole strength. In contrast, decreasing the size of the nanoparticle lets the nonradiative component dominate the plasmon decay. For applications in information technology, a slow dephasing of optical polarization by electron oscillation is essential, which is characterized by the dephasing time (T_2). The dephasing time can be found experimentally by measuring the plasmon linewidth $\Gamma = 2\hbar/T_2$, where T_2 is related to the time constant of the inelastic decay of the plasmons T_1 and elastic scattering time constant T^* by the authors of Refs. [82–84].

$$\frac{1}{T_2} = \frac{1}{2T_1} + \frac{1}{T^*}. \tag{1}$$

T_1 is a combination of time constants related to the radiative and nonradiative decay processes $T_1 = T_{1,r} + T_{1,nr}$. Nonradiative decay of surface plasmon through Landau damping gives rise to electron-hole pair generation. When the energy of these excited carriers is much larger than the thermally excited electron-hole pairs at ambient temperature, they are called hot carriers. This phenomenon of hot-carrier generation finds lots of applications in energy harvesting, photo

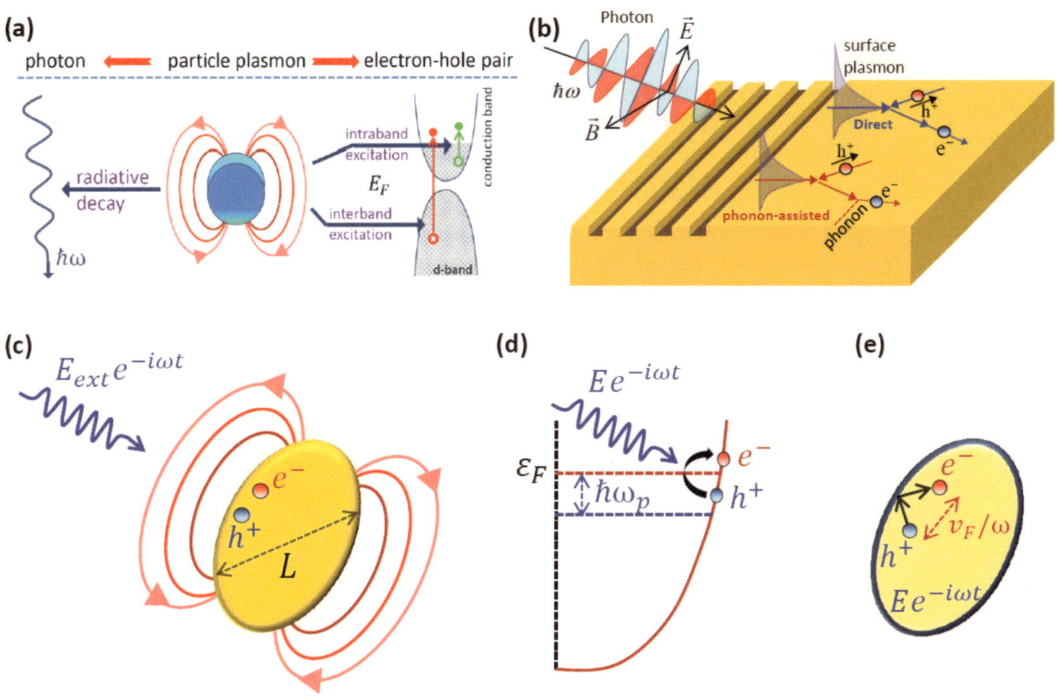

Figure 2. (a) Different types for decay of plasmon. The plasmon is damped by radiative (left) or nonradiative decay through Landau damping (right). (b) Schematic representation of the surface plasmon excitation followed by damping through hot-carrier generation. Schematics for the surface-assisted excitation of hot electron-hole (e-h) pair with energy $\hbar\omega_p$. (c) An external electric field incident on a metal nanoparticle with characteristic size of L excites surface plasmon. (d) It leads to geometry-assisted hot e-h excitation (e) followed by momentum relaxation through surface scattering in a region with size of the order of v_F/ω.

detection, and photocatalysis. The hot carriers can be used directly to drive chemical reactions [86] in metals and to drive photocurrents in a semiconductor for the photovoltaic effect or for photoelectrochemical processes [87, 88]. The energy distribution of the generated hot carriers is sensitive to the detailed band structure of the metal. For example, the energy of the excited holes in copper and gold is higher than for electrons; however, in aluminum and silver, the electron and hole energies are almost equal. The advantage of the asymmetry between the produced hot electron and holes by the surface plasmon helps to collect more hot carriers before inelastic scattering occurs. The energy distribution of the carriers generated by decay of plasmons, the subsequent elastic and inelastic scatterings and their transport are important for the design of the hot-carrier-based devices. The inelastic scatterings due to the Coulomb interaction lead to the thermalization of the hot carriers among each other, thereby reaching their carrier equilibrium temperature. The inelastic scatterings due to the electron-phonon interaction result in the cooling of the carriers, thereby reaching the equilibrium temperature of the lattice of the material. For example, in graphene, thermalization takes place on a time scale of 100 fs, while cooling is much slower and happens on a time scale of 10 ps.

This book chapter is organized as follows: in Section 2, we discuss the SP resonances in hybrid metal-semiconductor nanostructures. By altering the dielectric environment of nanostructures dynamically using pump-probe techniques, we show that it is possible to shift the SP resonance wavelength. In Section 3, we discuss in detail the plasmon excitations and their damping pathways in a three-dimensional electron gas (3DEG). We discuss both the radiative and nonradiative damping mechanisms of SPs in 3DEGs. In Section 4, we give a brief overview of plasmons in a two-dimensional electron gas (2DEG). In Section 5, we present the size dependent properties of the SPs in nanostructures. Graphene plasmonics and losses are discussed in Section 6. Section 7 is dedicated to the description of the SPs in Dirac systems. We focus on the SPs in 3DTI materials and 3D Weyl semimetals, and we discuss graphene plasmons as a limiting case of the 3DTI plasmons in the limit where the thickness of the 3DTI slab $d \rightarrow 0$. Finally, we wrap up the book chapter with the conclusion in Section 8.

2. Surface plasmon resonances in metal nanostructures

A nanoparticle shows tunable optical properties under controlled variation of its geometry. In a pure Ag spherical nanoparticle in vacuum, for example, the plasmon resonance occurs at 320 nm. These plasmon modes are shifted if the nanoparticle is coated with dielectric materials. It has been shown that with increasing shell thickness, the local electric field enhancement factor peak increases and redshifts for $\varepsilon_2 > \varepsilon_3$, whereas the local field enhancement factor peak decreases and blueshifts nonlinearly for $\varepsilon_2 < \varepsilon_3$, where ε_2 and ε_3 are dielectric functions of the shell and the surrounding materials, respectively [89]. The electric field enhancement factor is defined by $E_F = |E/E_0|^2$. It can be easily calculated by means of the finite-difference time domain (FDTD) technique. **Figure 3** shows the example of an Au nanoparticle with a radius of 50 nm in air with index of refraction $n = 1$.

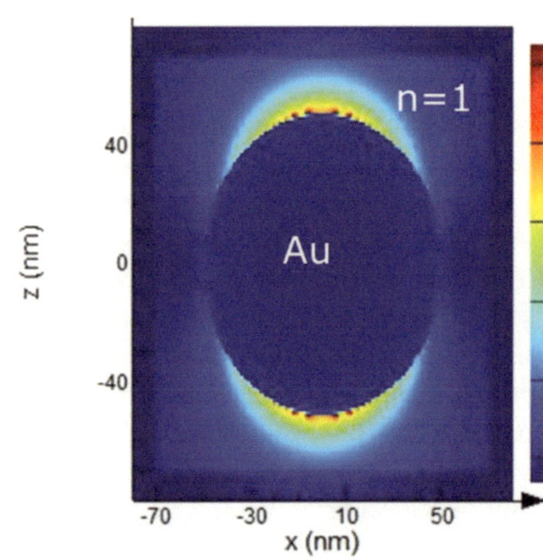

Figure 3. The resonant excitation of the Au nanoparticle causes large local electric fields close to the particle surface, which is obtained by means of finite-difference time domain (FDTD) calculations. Here, $E_F(x,y)$ is shown in the xz plane at 494 nm.

http://dx.doi.org/10.5772/67689

The quasi-static approximation provides a good estimate for a nanoparticle size of around 1/10 or smaller of the incident light wavelength. For larger nanoparticles, due to the finite speed of light, retardation effects lead to a redshift of the plasmon resonance [90]. In Ref. [91], authors have found an analytical expression for a spheroid that takes into account the depolarization factors and that gives a good approximation for nanogeometries of size up to 150 nm. **Figure 4** shows our results for the local field enhancement in the presence of an Ag nanocube. As expected from electrostatics, the largest enhancement occurs at the vertices of the Ag nanocube.

The optical resonances of a nanoshell exhibit enhanced sensitivity to its local dielectric environment relative to the solid nanoparticle, as shown in **Figure 5**. For a particle diameter less than the wavelength of light, the light-matter interaction leads to an oscillating homogeneous

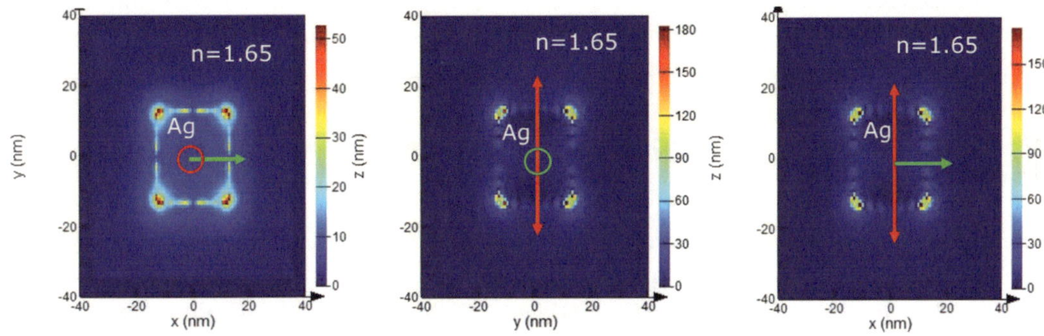

Figure 4. Local field enhancement E_F for an Ag nanocube of 20 nm size. The surrounding medium has a refractive index of $n = 1.65$, which is the case for boron nitride (BN). The green arrow points in the propagation direction of the electromagnetic wave, and the red double arrow shows the polarization of the electromagnetic wave.

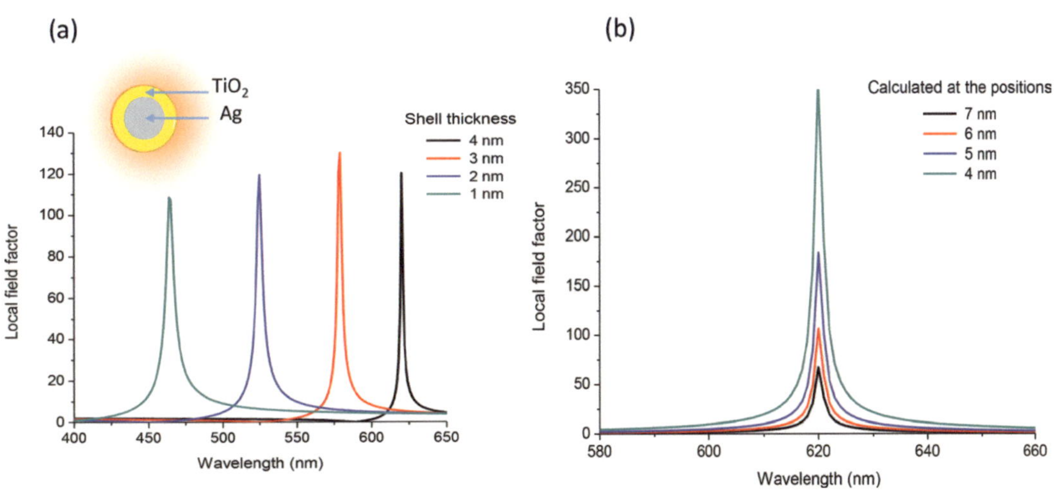

Figure 5. Surface plasmons (SPs) field enhancement in a hybrid nanoshell with a Ag core and TiO_2 shell of size of 15 nm at different shell thickness using the quasi-static field approximation (a), and different positions inside the shell (b) [see Ref. [74]. The inset in **Figure 5a** shows a scheme of a core-shell nanoparticle.

polarization of the particle volume, resulting in a dipole field. **Figure 5a** and **b** shows the dependence of plasmon resonances on the shell thickness and the size of a hybrid metal-semiconductor nanostructure of 15 nm diameter with an Ag core coated by TiO_2 shell obtained in the quasi-static approximation. The expressions are given in Ref. [67].

Plasmon resonances in a nanoshell can be tuned dynamically by letting a pump laser pulse of energy equal to the band gap or above generate electron-hole pairs in a semiconducting material surrounding the nanoshell. A probe laser pulse at a plasmon resonance frequency is used to excite plasmons on the metal surface. The generation of free electron-hole pairs alters the dielectric function of the surrounding semiconducting material. Due to the reduced dielectric function caused by the excitation of the electron-hole pairs, the excitation of surface plasmons by a probe pulse requires a higher energy. The frequency of the probe pulse is smaller than that of the pump pulse ensuring that no excitons are excited in the semiconductor during the probing. The change in the dielectric function of the surrounding medium due to the pump pulse can be calculated using Fermi's golden rule:

$$\mathrm{Im}(\varepsilon(\omega)) = \frac{4\pi^2 e^2}{m^2\omega^2} \sum_{k,c,v} |\mathbf{e}\cdot\mathbf{P}_{c,v}|^2 f(E_v(\mathbf{k}))[1-f(E_c(\mathbf{k}))]\delta(E_c(\mathbf{k})-E_v(\mathbf{k})-\hbar\omega), \tag{2}$$

where $\mathbf{P}_{c,v}$ is a transition dipole matrix element between conduction and the valence band, $\mathbf{e}$ is the direction of the polarization and $f(E)$ is the Fermi-Dirac distribution function. At zero temperature, the factor $f(E_v(\mathbf{k}))[1-f(E_c(\mathbf{k}))] = 1$ before the optical excitation and becomes less than 1 after optical excitation.

Figure 6 shows shifts in the resonance peak of the surface plasmons occurring at around 620 nm before the generation of excitons in a nanoshell structure (as shown in inset of **Figure 5a**) with diameter of 15 nm. After the pump pulse, depending on the density of the excitations, the

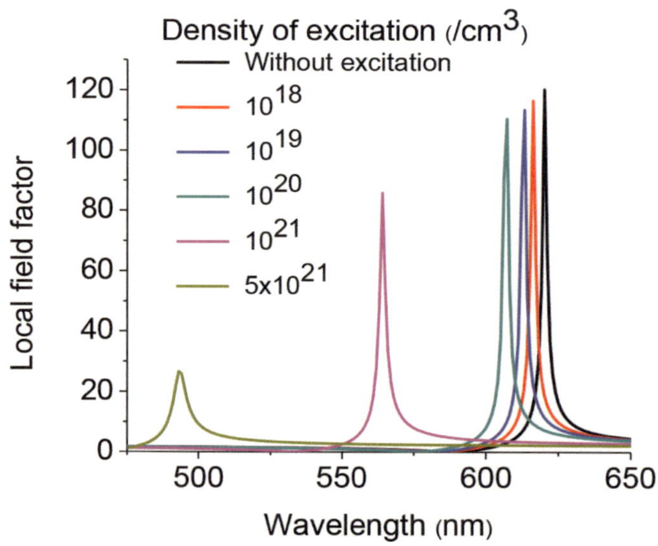

Figure 6. Local field enhancement factor E_F on the surface of a nanoshell of 15 nm diameter. The core is made of Ag with a diameter of 7 nm. At higher density of excitation, the resonance peak is gradually blueshifted.

plasmon resonance peaks are excited by the probe pulse shift. The larger the density of excited free electron-hole pairs in the semiconductors, the larger is the blueshift of the plasmon resonance peak. For a density of excitation of 5×10^{21} cm^{-3}, a resonance shift of up to 125 nm can be achieved.

3. Plasmon excitation and damping for a three-dimensional electron gas (3DEG)

In this section, we discuss the plasmon excitation and their damping pathways in 3D materials made of metal. The plasmonic damping pathways in 3D materials include radiative decay, Landau damping, and resistive loss, as depicted in **Figure 7**. During Landau damping, plasmon quasi-particles lose their energy by exciting hot electron-hole pairs via direct interband or phonon/geometry-assisted intraband transitions. In the case of geometry-assisted intraband transition, the translational symmetry is broken due to electric field confinement or boundaries of the material [92]. In the case of resistive loss, single carriers, electrons or holes that are the building blocks of the plasmon quasi-particle, are kicked out of the phase-coherent collective plasma oscillation through electron-electron or electron-phonon scattering, giving rise to plasmon damping. In 2D materials, the plasmons follow similar damping pathways. **Figure 7** shows the stages of the plasmon decay, the initial nonequilibrium configuration after the excitation, the thermalization, and the cooling of the hot carriers. Some of these damping

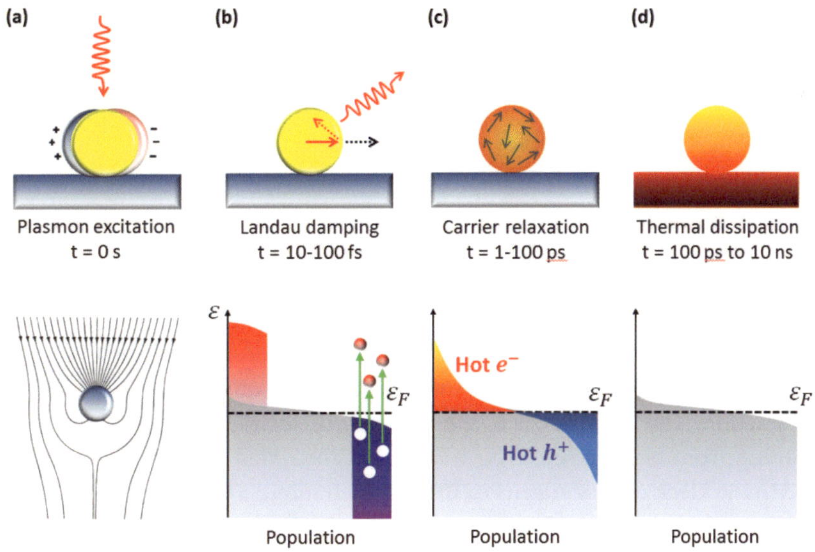

Figure 7. The stages of energy relaxation of excited plasmon. (a) The incident electric field on a graphene nanostructure excites the surface plasmon. (b) The plasmon quasi-particles lose their energies by different scattering processes to excite hot carriers, which have nonthermal distribution on the order of tens to hundreds of femtoseconds. (c) Electron-electron and other scattering processes redistribute the hot carriers and thermalized them in the order of one to hundreds of picoseconds to generate ling-lived hot carrier distribution. (d) Cooling stages and energy transferring to the substrate happen by phonon and geometry-assisted scattering processes over a longer time scale (hundreds of picoseconds to tens of nanoseconds) for relaxation of the hot carrier distribution to the equilibrium.

pathways can be used to inject hot carriers into other materials. For example, at a metal-semiconductor interface, hot electron-hole pairs can be separated by means of the Schottky barrier for the purpose of energy harvesting. In metal-graphene or metal-MoS_2 junctions, the surface plasmons can generate hot electron-hole pairs, thereby injecting electrons/holes into n/p-doped 2D materials, giving rise to hot carrier-induced doping [93] or even insulator-to-metal phase transitions [94].

The initial distribution of the hot carriers can be estimated using the jellium model for metal nanoparticles and nanoshells [95, 96], but this approach cannot explain the material dependence of this process because the specific band structure of the metal is completely neglected. In order to capture the material properties, it is necessary to combine FDTD calculations for obtaining the plasmon modes with *ab initio* density functional theory (DFT) calculations to determine the initial energy distribution of generated hot carriers by decay of surface plasmon. **Figure 7b** shows the schematic of plasmonic hot-carrier generation [97].

The plasmon resonance frequency and dispersion can be obtained by evaluating the dynamic polarizability in the presence of the carrier-carrier Coulomb interaction. The dynamic polarizability in the random phase approximation (RPA) is given by

$$\chi^{\mathrm{RPA}}(\mathbf{q},\omega) = \frac{\chi^0(\mathbf{q},\omega)}{\varepsilon^{\mathrm{RPA}}(\mathbf{q},\omega)}, \tag{3}$$

where $\chi^0(\mathbf{q},\omega)$ is the noninteracting (zeroth-order) polarizability (single pair bubble) and $\varepsilon^{RPA}(\mathbf{q},\omega) = \varepsilon_m - v_c(\mathbf{q})\chi^0(\mathbf{q},\omega)$, with ε_m being the permittivity of the environment, and $v_c(\mathbf{q}) = e^2/2\varepsilon_0 q$ the Coulomb interaction between the carriers. The RPA method corresponds to the expansion of $1/\varepsilon^{\mathrm{RPA}}(\mathbf{q},\omega)$, leading to an infinite power series over the bubble diagrams.

If optical phonons are also considered, the effective dielectric function in the RPA expansion takes the form [98, 99]

$$\varepsilon^{\mathrm{RPA}}(\mathbf{q},\omega) = \varepsilon_m - v_c(\mathbf{q})\chi^0(\mathbf{q},\omega) - \varepsilon_m \sum_l v_{\mathrm{sph},l}(\mathbf{q},\omega)\chi^0(\mathbf{q},\omega) - \varepsilon_m v_{\mathrm{oph}}(\mathbf{q},\omega)\chi^0_{j,j}(\mathbf{q},\omega), \tag{4}$$

where ε_m is the average of dielectric constants of the environment. The second term represents the effective Coulomb interaction of electrons in graphene, and $v_c(\mathbf{q}) = e^2/2q\varepsilon_0$ is the direct Coulomb interaction. The third term is the effective dielectric function for different phonon modes (l) coming from electron-electron interaction mediated by substrate optical phonons, which couple to the electrons by means of the Fröhlich interaction, i.e.,

$$v_{\mathrm{sph},l}(\mathbf{q},\omega) = |M_{\mathrm{sph}}|^2 G_l^0(\omega), \tag{5}$$

where $|M_{\mathrm{sph}}|^2$ is the scattering and G_l^0 is the free phonon Green's function. The last term of Eq. (4) corresponds to the optical phonon mediated electron-electron interaction.

$$v_{\text{oph}}(\mathbf{q}, \omega) = |M_{\text{oph}}|^2 G^0(\omega). \tag{6}$$

Here $|M_{\text{oph}}|^2$ defines the scattering matrix element and $G^o(\omega)$ is the free phonon Green's function. In Eq. (4), $\chi^0_{j,j}(\mathbf{q}, \omega)$ is the current-current correlation function. This description is very general and can be applied to any metallic system.

For the simplest case, when only the carrier-carrier Coulomb interaction is present, we can derive the dynamical plasmon dispersion relation following standard textbooks [100]. The first step is to calculate noninteracting dynamical polarizability.

$$\chi^0(\mathbf{q}, \omega) = \frac{1}{\Omega} \sum_{\mathbf{k}, s} \frac{n_{0\mathbf{k}+\mathbf{q}, s} - n_{0\mathbf{k}, s}}{\varepsilon_{\mathbf{k}+\mathbf{q}} - \varepsilon_{\mathbf{k}} - \hbar\omega - i\hbar\eta'} \tag{7}$$

where Ω is the volume of the sample, $n_{0\mathbf{k},s}$ is the equilibrium electron density, and $\varepsilon_{\mathbf{k}}$ is the single-particle energy dispersion of the electrons. Assuming $|\mathbf{q}| \ll k_F$, we can expand $\chi^0(\mathbf{q}, \omega)$ in $\mathbf{q}$ by taking advantage of the expansions $\varepsilon_{\mathbf{k}+\mathbf{q}} = \varepsilon_{\mathbf{k}} + \mathbf{q} \cdot \nabla_{\mathbf{k}} \varepsilon_{\mathbf{k}} + \cdots$ and $n_{0\mathbf{k}+\mathbf{q}} = n_{0\mathbf{k}} + \frac{\partial n_0}{\partial \varepsilon} \mathbf{q} \cdot \nabla_{\mathbf{k}} \varepsilon_{\mathbf{k}} + \cdots$. At zero temperature, we have $\frac{\partial n_0}{\partial \varepsilon_{\mathbf{k}}} = -\delta(\varepsilon_{\mathbf{k}} - \varepsilon_F)$ and the velocity can be approximated by the Fermi velocity, i.e., $\nabla_{\mathbf{k}} \varepsilon_{\mathbf{k}} = \hbar \mathbf{v}_{\mathbf{k}} \approx \hbar v_F$, leading to the approximation

$$\begin{aligned}
\chi^0(\mathbf{q}, \omega) &\approx -\frac{2}{(2\pi)^3} \int d^3k \frac{\mathbf{q} \cdot \mathbf{v}_F \delta(\varepsilon_{\mathbf{k}} - \varepsilon_F)}{\mathbf{q} \cdot \mathbf{v}_F - \omega - i\eta} \\
&= \frac{2}{(2\pi)^2} \int_{-1}^{1} d(\cos\theta) \int_0^{\infty} k^2 dk \frac{q v_F \cos\theta}{q v_F \cos\theta - \omega - i\eta} \delta(\varepsilon_{\mathbf{k}} - \varepsilon_F) \\
&= \frac{k_F^2}{2\pi^2 \hbar v_F} \left[\frac{2}{3} \left(\frac{q v_F}{\omega + i\eta} \right)^2 + \frac{2}{5} \left(\frac{q v_F}{\omega + i\eta} \right)^4 + \cdots \right] \\
&\approx \frac{k_F^3 q^2}{3\pi^2 m (\omega + i\eta)^2} \left[1 + \frac{3}{5} \frac{q^2 v_F^2}{(\omega + i\eta)^2} \right] = \frac{n_0 q^2}{m(\omega + i\eta)^2} \left[1 + \frac{3}{5} \frac{q^2 v_F^2}{(\omega + i\eta)^2} \right].
\end{aligned} \tag{8}$$

Using the equation $k_F = (3\pi^2 n_0)^{1/3} \Leftrightarrow n_0 = \frac{k_F^3}{3\pi^2}$ for 3D metals, one obtains

$$\chi^0(\mathbf{q}, \omega) \approx \frac{\rho_0 q^2}{m(\omega + i\eta)^2} \left[1 + \frac{3}{5} \frac{q^2 v_F^2}{(\omega + i\eta)^2} \right] = \frac{\rho_0 q^2}{m(\omega + i\eta)^2} R(q, \omega)^2 \tag{9}$$

with

$$R(q, \omega)^2 = \left[1 + \frac{3}{5} \frac{q^2 v_F^2}{(\omega + i\eta)^2} \right]. \tag{10}$$

Consequently, the dynamical polarizability in the RPA and long wavelength regime is

$$\chi^{\mathrm{RPA}}(\mathbf{q},\omega)=\frac{\chi^0(\mathbf{q},\omega)}{\varepsilon(\mathbf{q},\omega)}=\frac{\frac{\rho_0 q^2}{m(\omega+i\eta)^2}R(q,\omega)^2}{1-\frac{4\pi e^2\rho_0}{m(\omega+i\eta)^2}R(q,\omega)^2}=\frac{\rho_0 q^2 R(q,\omega)^2}{m(\omega+i\eta)^2-4\pi e^2\rho_0 R(q,\omega)^2}$$
$$=\frac{\rho_0 q^2 R(q,\omega)}{2m\omega_p}\left\{\frac{1}{\omega+i\eta-\omega_p R(q,\omega)}-\frac{1}{\omega+i\eta+\omega_p R(q,\omega)}\right\}, \tag{11}$$

where the quantum plasma frequency of a 3D metal in the free electron approximation is given by the Pines and Bohm equation [101, 102]

$$\omega_p=\sqrt{\frac{4\pi n_0 e^2}{m}}, \tag{12}$$

with n_0 being the equilibrium electron density and m the electron mass. Using $\lim_{\eta\to 0_+}\frac{1}{z-i\eta}=\mathcal{P}\left(\frac{1}{z}\right)+i\pi\delta(z)$, it follows that

$$\mathrm{Im}[\chi^{\mathrm{RPA}}(\mathbf{q},\omega)]\approx\frac{\pi\rho_0 q^2 R(q,\omega)}{2m\omega_p}\left\{\delta\Big(\omega-\omega_p R(q,\omega_p)\Big)-\delta\Big(\omega+\omega_p R(q,\omega_p)\Big)\right\}, \tag{13}$$

which yields the bulk plasmon dispersion relation for 3D metals, i.e.,

$$\omega_q=\omega_p R(q,\omega_p)=\omega_p\left[1+\frac{3}{10}\frac{q^2 v_F^2}{(\omega_p+i\eta)^2}+\cdots\right]. \tag{14}$$

The same result can be obtained by solving $\varepsilon^{\mathrm{RPA}}(\mathbf{q},\omega)=0$. Note that the slope of the parabolic dependence at $q=k_F$ is

$$\frac{3v_F^2 k_F}{5\omega_p}=\frac{3v_F^2 k_F}{5\frac{4\pi e^2 n_0}{m}}=\frac{9\pi m v_F^2 k_F}{20e^2 k_F^3}=\frac{9\pi m v_F^2}{20e^2\left(\frac{m v_F}{\hbar}\right)^2}=\frac{9\pi\hbar^2}{20e^2 m}=0.67\frac{\mathrm{m}}{\mathrm{s}}\ll\frac{c}{\sqrt{\varepsilon_{\mathrm{d}}}}. \tag{15}$$

This difference in slopes between the 3D bulk plasmon and the photon-like surface plasmon polariton is clearly visible in **Figure 8** (see below).

It is well known that in 3D metals Landau damping occurs when the plasmon resonance energy enters the electron-hole continuum, which is determined by the condition

$$\hbar\omega=\varepsilon_{\mathbf{k}+\mathbf{q}}-\varepsilon_{\mathbf{k}}=\frac{\hbar^2(\mathbf{k}+\mathbf{q})^2-\hbar^2k^2}{2m}=\frac{\hbar^2q^2+2\hbar^2\mathbf{k}\cdot\mathbf{q}}{2m}=\frac{\hbar^2q^2+2\hbar^2kq\cos\theta}{2m}$$
$$=\frac{\hbar^2 q(q+2k\cos\theta)}{2m}\geq 0. \tag{16}$$

Since $-1\leq\cos\theta\leq 1$ and $0\leq k\leq k_F$, the electron-hole continuum is given by the gray shaded area in **Figure 8**. The plasmon can decay into an electron-hole pair when the plasmon dispersion

http://dx.doi.org/10.5772/67689

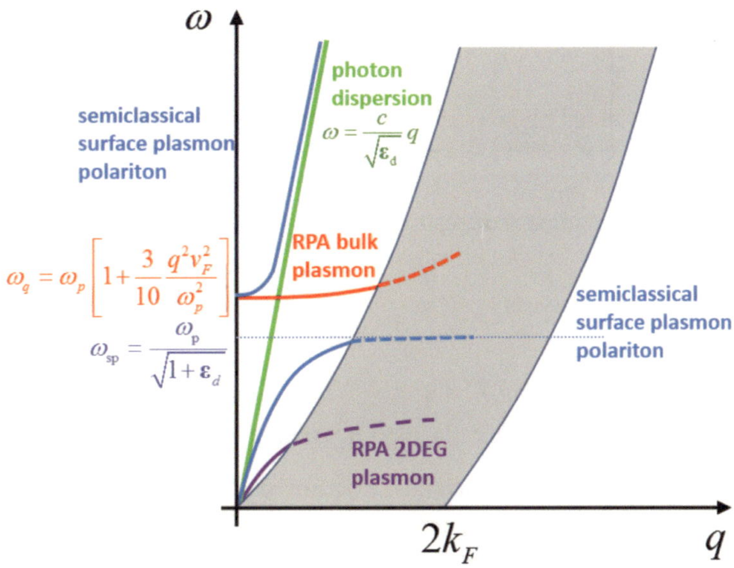

Figure 8. Dispersion curves of the RPA bulk plasmon resonance (red), the semi-classical surface plasmon polaritons (blue), and the RPA 2DEG plasmons. The shaded area marks the electron-hole continuum. When the bulk plasmon resonance enters the shaded area, the plasmon decays into electron-hole pairs which is called Landau damping.

curve enters the electron-hole continuum limit. This decay corresponds to intraband Landau damping.

The presence of a planar boundary for a 3D metal adds a new mode known as surface plasmon, which propagates at the metal-dielectric interface. Since the electron charge density of a metal leaks outside the interface into the dielectric in the order of $1/k_F \approx 1$ Å, a macroscopic description based on Maxwell's equations is sufficient to understand the surface plasmon qualitatively. Taking the boundary conditions into account, the dispersion relation of the surface plasmon is determined by Ritchie and Eldridge [103]

$$q_x = \frac{\omega}{c}\sqrt{\frac{\varepsilon_m \varepsilon_d}{\varepsilon_m + \varepsilon_d}} = \frac{\omega}{c}\sqrt{\frac{\omega^2 - \omega_p^2}{\left(1 + \frac{1}{\varepsilon_d}\right)\omega^2 - \frac{\omega_p^2}{\varepsilon_d}}} \tag{17}$$

in the Drude model, where $\varepsilon_m(\omega) = 1 - \omega_p^2/\omega^2$. This biquadratic equation can be solved analytically, yielding

$$\omega_{sp}(q_x) = \pm\frac{1}{\sqrt{2\varepsilon_d}}\sqrt{c^2 q_x^2(1+\varepsilon_d) + \varepsilon_d\omega_p^2 \pm \sqrt{c^4 q_x^4 (1+\varepsilon_d)^2 + 2c^2 q_x^2(-1+\varepsilon_d)\varepsilon_d\omega_p^2 + \varepsilon_d^2\omega_p^4}}. \tag{18}$$

Since $\omega_{sp}(q_x) \geq 0$, there are only two physical solutions, which are drawn in **Figure 8**. While the upper branch corresponds to the photon-like plasmon polariton, the lower branch represents the plasmon-like surface plasmon polariton (SPP) with the following asymptotics:

$$\omega_{\mathrm{sp}}(q_x) = \begin{cases} \dfrac{c}{\sqrt{\varepsilon_{\mathrm{d}}}} q_x \left[1 - \dfrac{1}{2} \left(\dfrac{c}{\omega_p} \right)^2 q_x^2 + \dots \right], & q_x \ll \dfrac{\omega_p}{c} \\ \dfrac{\omega_p}{\sqrt{1+\varepsilon_{\mathrm{d}}}} \left[1 - \dfrac{1}{2} \left(\dfrac{\omega_p}{c} \right)^2 \left(\dfrac{\varepsilon_{\mathrm{d}}}{1+\varepsilon_{\mathrm{d}}} \right)^2 q_x^{-2} + \dots \right], & q_x \gg \dfrac{\omega_p}{c} \end{cases} \tag{19}$$

which in the nonretarded regime reduces to Ritchie's equation [2],

$$\omega_{\mathrm{sp}} = \frac{\omega_p}{\sqrt{1+\varepsilon_d}} = \frac{\omega_p}{\sqrt{2}} = \sqrt{\frac{2\pi n_0 e^2}{m}} = \left(\frac{4e^2\sqrt{2m}}{3\pi\hbar^3} \right)^{1/2} \varepsilon_F{}^{3/4} \tag{20}$$

for wave vectors q in the range of $\omega_s/c \ll q \ll q_F$, where q_F is the Fermi wave vector [104]. Since $k_F = (3\pi^2 n_0)^{1/3}$ in 3D metals, the bulk plasmon frequency exhibits a $\varepsilon_F{}^{3/4} \propto n_0^{1/2}$ dependence.

To estimate the plasmon decay rate, the band structure of the materials should be calculated by means of DFT to find out the exact quasi-particle orbitals and energies. Different electronic structures can be used to calculate the band structure of noble metals. To estimate the decay rate, the electronic states and energies of the metals resulting from PBEsol + U DFT method are used in Eq. (2). To find the plasmon decay distribution, energy conservation is used to determine the regions of the first Brillouin zone, which are responsible for the electron-hole generation [97]. Since aluminum, which is a nearly free electron metal, has a band crossing close to the Fermi level at the W point, interband transitions from the valence band with energies $\hbar\omega$ from zero to the plasmon energy are allowed for creating a symmetric amount of both hot electrons and holes. There is an extra possibility for interband transitions at the K point, which mostly contribute to the generation of hot holes with energies >2 eV, giving rise to a modest asymmetry between hot electron and hole distributions. For copper, the interband transitions happen at the X, L and K, points, and for gold they occur at the X and L points [97]. In these two noble metals, since all the transitions originate from the "d" band, the hot hole energies are larger than 2 eV and consequently they have larger kinetic energies than the electrons. In metal-semiconductor interface, this asymmetry between hot hole and electron energy distribution plays a dominant role in the hot-carrier collection efficiency. For gold and copper, the holes have more kinetic energy than the electrons. This means that in a metal-semiconductor Schottky junction such as gold-n type gallium arsenide junctions, the hot electrons do not have enough energy to pass the Schottky barrier and therefore need to tunnel or get more thermal energy to overcome Schottky barrier. However, in gold-p type gallium arsenide junctions, the hot holes have sufficient kinetic energy to overcome the Schottky barrier. This means for copper and gold, the metal-p type semiconductor junction is more efficient for the collection of hot holes. But for aluminum and silver, the efficiency is almost the same for hot holes and electrons. According to the Fowler theory [96], the angular distribution of the hot-carriers should be isotropic. However, experiments show that none of the noble metals exhibit an isotropic distribution [97].

In nanoconfined structures, because of lack of translational symmetry, the crystal momentum q is not a good quantum number and there is no momentum conservation anymore. Therefore the transition creating an electron-hole pair is allowed to be nonvertical. This is the case for geometry-assisted intraband transition, where no phonon is required. A vector potential for one quantum of each symmetric and antisymmetric modes can be used in Fermi's golden rule to calculate the decay rate of an LSP to an electron-hole pair. By decreasing the thickness, the effect of confinement is to be more pronounced and more hot electrons can be generated by intraband transitions. This effect is more noticeable for the antisymmetric mode, because a smaller part of the light is located inside the gold film, which leads to reduction of interband transitions [97].

In metals, plasmons decay is not only in the ultraviolet and visible spectrum but also in the infrared and microwave regimes [105, 106]. Due to the conservation of momentum in infinite crystal lattices, the direct interband transition induced by plasmon decay is only possible for energies larger than the band gap energy. However, for energies below the visible spectrum, typically phonon-assisted and surface-assisted intraband electron-hole pair generations are able to bypass this selection rule [92].

The plasmon decay rate is related to the imaginary part of the dielectric tensor $\mathrm{Im}\,\overline{\varepsilon}\,(\omega)$ [92], i.e.,

$$\Gamma(\mathbf{r}) = \frac{1}{2\pi\hbar}\mathbf{E}^*(\mathbf{r})\cdot \mathrm{Im}\overline{\varepsilon}(\omega)\cdot\mathbf{E}(\mathbf{r}). \tag{21}$$

Let us consider a 3D semi-infinite metal slab extending in the negative z-direction with a single surface in the xy-plane at position $z = 0$. Using the electric field profile of a single quantum of the surface plasmon polariton with wave vector $\mathbf{k}$ and frequency ω and integration over the space gives the total nonradiative (nr) decay rate.

$$\Gamma_{\mathrm{nr}}(\mathbf{r}) = \frac{1}{2L(\omega)|\gamma(z<0)|}\boldsymbol{\lambda}^*\cdot \mathrm{Im}\,\overline{\varepsilon}\,(\omega)\cdot\boldsymbol{\lambda} \tag{22}$$

The decay rate of plasmon as a function of frequency can be calculated by substituting the experimental data for the complex dielectric function measured by ellipsometry. Within the random phase approximation (RPA), the nonradiative decay rate induced by direct interband transition is [97]

$$\Gamma_{\mathrm{nr(direct)}} = \frac{2\pi^2 e^2}{L(\omega)|\gamma(z<0)|{m_e}^2\omega^2}\int_{BZ}\frac{d\mathbf{q}}{(2\pi)^3}\sum\nolimits_{n'n}(f_{\mathbf{q}n}-f_{\mathbf{q}n'})\delta(\varepsilon_{\mathbf{q}n'}-\varepsilon_{\mathbf{q}n}-\hbar\omega)\times|\boldsymbol{\lambda}\cdot\langle\mathbf{p}\rangle^{\mathbf{q}}_{n'n}|^2 \tag{23}$$

where $\langle\mathbf{p}\rangle^{\mathbf{q}}_{n'n} = -\Psi^{\sigma*}_{\mathbf{q}n'}(\mathbf{r})i\hbar\nabla\Psi^{\sigma}_{\mathbf{q}n}(\mathbf{r})$ are momentum matrix elements between the quasi-particle orbitals $\Psi^{\sigma*}_{\mathbf{q}n'}(\mathbf{r})$ and $\Psi^{\sigma}_{\mathbf{q}n}(\mathbf{r})$. $f_{\mathbf{q}n}$ are the Fermi occupation functions of the quasi-particles, $\lambda \equiv \hat{k} - \hat{z}k/\gamma(z<0)$ is the polarization vector, and $\gamma(z)$ is the imaginary part of the wave vector in z-direction. Eq. (23) can also be obtained in first quantization and is based on the electric dipole approximation, which means the plasmon mode function slowly changes on the atomic scale. To include the effect of electron-electron and electron-phonon scattering on the decrease in the lifetime of carriers, the δ function is replaced by a Lorentzian with half width $\mathrm{Im}\,\Sigma_{\mathbf{q}n} + \mathrm{Im}\,\Sigma_{\mathbf{q}n'}$, where $\mathrm{Im}\,\Sigma_{\mathbf{q}n}$ is the total carrier lifetime due to the scatterings.

The contribution of the electron-phonon interaction to the plasmon decay through intraband transitions can be calculated in second-order perturbation theory [107, 108], giving

$$\Gamma_{\mathrm{nr(phonon)}} = \frac{2\pi^2 e^2}{L(\omega)|\gamma(z<0)|m_e{}^2\omega^2} \int\limits_{BZ} \frac{d\mathbf{q}' d\mathbf{q}}{(2\pi)^6} \sum\nolimits_{n'n\alpha\pm} (f_{\mathbf{q}n} - f_{\mathbf{q}'}n') \left(N_{\mathbf{q}'-\mathbf{q},\alpha} + \frac{1}{2} \mp \frac{1}{2} \right)$$

$$\times \delta(\varepsilon_{\mathbf{q}'}n' - \varepsilon_{\mathbf{q}n} - \hbar\omega \mp \hbar\omega_{\mathbf{q}'-\mathbf{q},\alpha}) \left| \boldsymbol{\lambda} \cdot \sum\nolimits_{n_1} \left(\frac{g^{\mathbf{q}'-\mathbf{q},\alpha}_{\mathbf{q}'n',\mathbf{q}n_1} \langle \mathbf{p} \rangle^{\mathbf{q}}_{n_1 n}}{\varepsilon_{\mathbf{q}n_1} - \varepsilon_{\mathbf{q}n} - \hbar\omega + i\eta} + \frac{\langle \mathbf{p} \rangle^{\mathbf{q}'}_{n'n_1} g^{\mathbf{q}'-\mathbf{q},\alpha}_{\mathbf{q}'n_1,\mathbf{q}n}}{\varepsilon_{\mathbf{q}'}n_1 - \varepsilon_{\mathbf{q}n} \mp \hbar\omega_{\mathbf{q}'} - \mathbf{q},\alpha + i\eta} \right) \right|^2, \tag{24}$$

where $\hbar\omega_{k,\alpha}$ is the energy of a phonon with wave vector $\mathbf{k}$ and polarization index α, $N_{\mathbf{k},\alpha}$ is the Bose occupation factor for the phonons, $g^{\mathbf{k},\alpha}_{\mathbf{q}'m',\mathbf{q}n}$ is the electron-phonon matrix element with $\mathbf{q}$ and n being the wave vector and band index of the electronic states, and the $\pm$ signs distinguish between absorption and emission of phonons.

Excitation of surface plasmon generates a strong field confinement on the surface with the exponential decay length ($|\gamma(z<0)|^{-1}$) inside the metal, which creates a Lorentzian distribution for the momentum of the plasmon in the z-direction. This momentum distribution leads to diagonal intraband transition and introduces a "surface-assisted" plasmon decay [92]

$$\mathrm{Im}\,\boldsymbol{\varepsilon}_{\mathrm{surface}}(\omega) = \frac{\omega_p^2}{\omega^3} \times \left(\frac{3}{4} |\gamma(z<0)| v_F \right) \frac{2k^2}{k^2 + |\gamma(z<0)|^2}, \tag{25}$$

where $\omega_p = \sqrt{4\pi n e^2/m_e}$ is the bulk plasma frequency of the metal, $v_F = (\hbar/m_e)(3\pi^2 n)^{3/2}$ is the Fermi velocity, and n is the bulk carrier density of the metal.

Numerical studies based on the free electron jellium model show that in nanostructures, due to the localization of electronic states and the nonconservation of the crystal momentum, intraband transitions are enhanced [95, 96, 109]. Using Fermi's golden rule together with the free electron eigen states and the dipole field profile, the nanoconfinement contribution is

$$\mathrm{Im}\boldsymbol{\varepsilon}_{\mathrm{sphere}}(\omega) = \frac{\omega_p^2}{\omega^3} \times \frac{6 v_F}{\pi^2 R}, \tag{26}$$

where R is the radius of the spherical nanoparticle.

In the Landau damping theory, the lowest-order processes consisting of direct, phonon-assisted, and surface/geometry-assisted electron-hole pair excitation contribute to the decay of plasmons [102, 110]. Higher-order processes leading to the excitation of many electron-hole pairs or many phonons are suppressed due to the phase-space factors at small energies [111]. Only at large energies, the higher-order processes become significant. Being completely

http://dx.doi.org/10.5772/67689

different from Landau damping, another source of plasmon damping is the resistive loss in the metal, which can be calculated by means of the linearized Boltzmann equation in the relaxation time approximation [92], giving

$$\mathrm{Im}\,\varepsilon_{\mathrm{resist}}(\omega) = \frac{4\pi\sigma_0}{\omega(1+\omega^2\tau^2)}, \tag{27}$$

where σ_0 is the zero frequency conductivity and τ is the momentum relaxation time, which can be derived from the carrier-carrier Coulomb and electron-phonon matrix elements.

4. Plasmons in a two-dimensional electron gas (2DEG)

According to Ritchie [2] and Stern [112], the dispersion relation for 2DEG plasmons is

$$\omega_{\mathrm{2DEG}} = \sqrt{\frac{n_s e^2}{2m^* \varepsilon_0 \bar{\varepsilon}(\mathbf{q},\omega)} q_x}, \tag{28}$$

where n_s is the surface carrier density. A typical dispersion curve is shown in **Figure 8**. In the case of a typical metal-oxide-semiconductor (MOS) system, the effective dielectric function is given by

$$\bar{\varepsilon}(\mathbf{q},\omega) = \frac{1}{2}\left[\varepsilon_{\mathrm{SC}}(\omega) + \varepsilon_{\mathrm{OX}}(\omega)\coth(dq_x)\right], \tag{29}$$

where $\varepsilon_{\mathrm{SC}}(\omega)$ and $\varepsilon_{\mathrm{OX}}(\omega)$ are the dielectric constants of the semiconductor and oxide layers, respectively. The oxide layer has a thickness of d. For example, Heitmann [113] and Wilkinson [114] were able to observe this type of 2DEG plasmons in AlGaAs-GaAs heterostructures. This result shows clearly that the dispersion curve can be tuned by changing the thickness of the oxide layer and by altering the gate voltage. Similarly, the plasmon dispersion can be tuned in atomically thin 2D materials. Since the dispersion relation in Eq. (28) exhibits a $\sqrt{q_x}$ dependence for a 2DEG, it is plausible to assume that atomically thin metallic 2D materials follow a similar dependence, which is indeed the case, as we discuss below in Sections 6 and 7. Since in a 2DEG

$$n_s = \frac{1}{\Omega}\sum_{|\mathbf{k}|\leq k_F,\, s} 1 = 2\int \frac{d^2k}{(2\pi)^2} 1 = 2\frac{\pi k_F^2}{(2\pi)^2} = \frac{k_F^2}{2\pi} = \frac{m^*\varepsilon_F}{\pi\hbar^2}, \tag{30}$$

the 2DEG plasmon frequency exhibits a $\sqrt{n_s} \propto \sqrt{\varepsilon_F}$ dependence. Note that the electron-hole continuum in a 2DEG is determined by a similar formula to Eq. (16). Therefore the gray shaded area in **Figure 8** has similar forms for 3DEGs and 2DEGs.

5. Static geometry of metallic objects and environment

When the size of the nanoparticle is much smaller than the wavelength of the incident light, the particle exhibits a dipolar oscillation mode (Fröhlich mode). As the diameter of the nanoparticle is increased, the electrostatic limit is not a good approximation anymore and the multipolar oscillation modes start to appear. Excitation of these modes gives rise to the broadening of the resonance [115], as seen from **Figure 9b**. The LSP resonance redshifts with increasing diameter of the sphere, which is due to retardation effects [90]. Decreasing the size of nanoparticle less than mean free path of electrons moves the material band structure and dielectric function away from the bulk properties and increases the surface scattering that gives rise to broadening of the absorption spectrum [115], as shown in **Figure 9a**. The internal field enhancement of an illuminated spherical nanoparticle is

$$\frac{E_i(\omega)}{E_0} = \frac{3\varepsilon_m}{\varepsilon(\omega) + \varepsilon_m}, \tag{31}$$

where ε and ε_m are the dielectric functions of the nanoparticle and the medium, respectively. An ellipsoidal nanoparticle with a shape defined by $(x/a)^2 + (y/b)^2 + (z/c)^2 = 1$ supports an internal field enhancement given by

$$\frac{E_i(\omega)}{E_0} = \frac{\varepsilon_m}{\varepsilon_m + L(\varepsilon(\omega) - \varepsilon_m)}, \tag{32}$$

where L is the shape factor, whose value lies between 0 and 1. For the resonance frequency, the real part of the denominator is zero, which means the resonance occurs at frequencies when

$$\varepsilon(\omega) = \varepsilon_m\left(1 - \frac{1}{L}\right) \tag{33}$$

is satisfied. The shape factor for the axis parallel to the polarization of the incident light (e.g., the axis a) is given by

$$L = \frac{abc}{2}\int_0^\infty \frac{dq}{(a^2+q)f(q)}, \tag{34}$$

where

$$f(q) = \left[(a^2+q)(b^2+q)(c^2+q)\right]^{1/2}. \tag{35}$$

For the core-shell spherical nanoparticles the resonance condition is given by

$$\varepsilon_c = -2\varepsilon_s\left[\frac{\varepsilon_s(1-f) + \varepsilon_h(2+f)}{\varepsilon_s(1+2f) + 2\varepsilon_h(1-f)}\right], \tag{36}$$

http://dx.doi.org/10.5772/67689

where $f = (r_c/r_s)^3$ is the core fill fraction and ε_c, ε_s, and ε_h are the core, shell, and host dielectric functions, respectively. The resonance frequency is tunable by modifying the shell thickness and the radius of the core. As shown in **Figure 10**, due to the hybridization of the plasmon modes, even in the electrostatic limit there are multiple resonance frequencies [116].

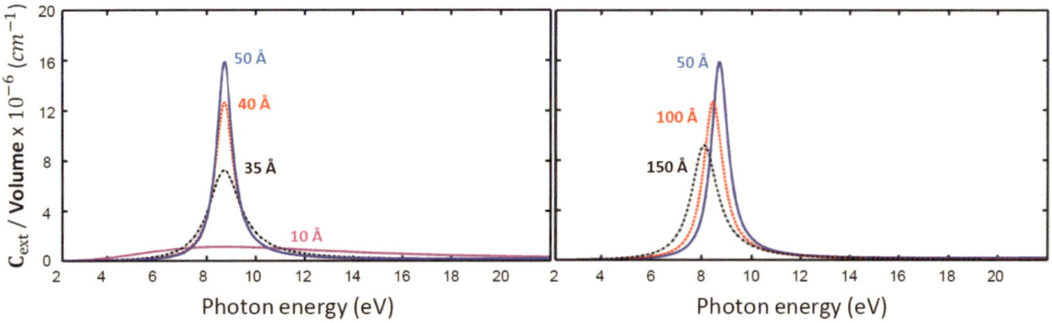

Figure 9. Calculated extinction coefficient per unit volume for a spherical Aluminum nanoparticle of various diameters. (a) For the case when the size of the nanoparticle becomes smaller than the mean free path of the electron, the linewidth becomes wider due to increased surface scattering. (b) For the case when the size of the nanoparticle becomes much larger than the mean free path of the electron, excitation of multipole modes leads to broadening of the linewidth and the retardation effect leads to a redshift of the LSP resonance peak.

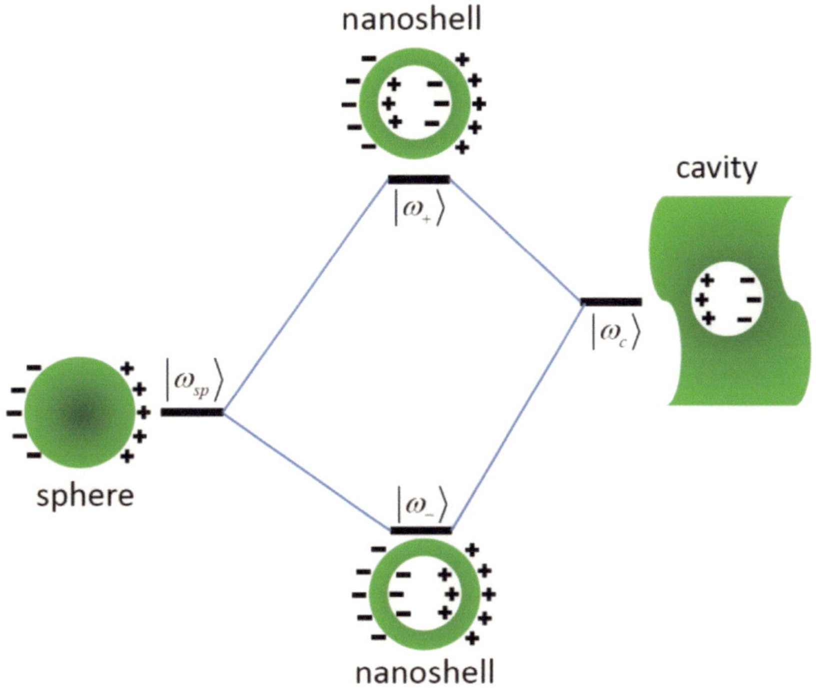

Figure 10. Energy-level diagram of plasmon hybridization in metallic core-shell nanoparticles. The hybridization results from the coupling between the cavity and sphere surface plasmons.

6. Plasmon theory for graphene

Graphene is a two-dimensional (2D) material comprised of a single layer of carbon atoms in a honeycomb lattice. It has unique electrical, optical, and mechanical properties due to its tunable band dispersion relation and atomic thickness. Because of its unique band structure graphene possesses a very high mobility and a fast carrier relaxation time [117–121], making it an attractive candidate for ultrafast electronics and optoelectronics. Exciting surface plasmons on graphene is a distinct technique to increase absorption with low damping rate. The surface plasmon couples the electromagnetic (EM) wave to the conductive medium, giving rise to direct absorption of light by monolayer graphene and providing the opportunity of electrical tunability of the plasmon resonance frequency, high degree of electric field confinement, and low plasmon damping rate [122–125]. The increased light-matter interaction results in an enhanced spontaneous emission rate close to the nanostructure edges [126, 127]. Recent experiments have achieved an absorption of 90% in the mid-IR range by connecting graphene with high carrier mobility to a silicon diffractive grating [128] and designing graphene nanoribbons [128, 129], nanodisks [130], and antidot array [125] theoretically. These high carrier mobilities can be achieved only for mechanically exfoliated graphene. Exciting plasmon on CVD-grown monolayer graphene with lower mobility than the mechanically exfoliated one reduces the absorbance to 19% and 28% for graphene nanoribbons [99, 127, 131] and nanodisks [132, 133], respectively. We show in experiments that the coupling of a patterned CVD-grown graphene sheet to an optical cavity amplifies the excited LSPs and enhances the light absorption to a current world record of 45% [134]. We also show that the theoretically achievable enhancement is 60% for a square lattice of holes [134].

The electric current of graphene in the interaction picture is given by

$$j_{xx}(\omega) = \sigma(\omega)E(\omega), \tag{37}$$

where $E(\omega)$ is the in-plane electric field and $\sigma(\omega)$ is the interband optical conductivity

$$\sigma_{\text{inter}}(\omega) = -\frac{ie^2}{\omega A}\sum_{\mathbf{k}} {v_x}^2 \frac{\rho_{vv} - \rho_{cc}}{\varepsilon_{c,\mathbf{k}} - \varepsilon_{v,\mathbf{k}} - \hbar(\omega + i\eta)} \tag{38}$$

with A being the cross-section area, v_x the velocity operator along x-direction, ρ_{vv} and ρ_{cc} the digonal elements of the density matrix, $\varepsilon_{c,\mathbf{k}}/\varepsilon_{v,\mathbf{k}}$ the conduction/valence band energy, and η an infinitesimal value. Using the energy dispersion, $\varepsilon_{n,\mathbf{k}} = nv_F\hbar|\mathbf{k}|$ with $n = +1/-1$ for conduction/ valence band and $v_F = 10^6 m/s$, and replacing the summation by the integration in Eq. (38) leads to

$$\begin{aligned}\sigma_{\text{inter}}(\omega) &= \frac{e^2 {v_f}^2}{2\omega}\frac{g}{4\pi}\int_0^\infty 2\pi k dk(\rho_{vv} - \rho_{cc})\delta(\varepsilon_{c,\mathbf{k}} - \varepsilon_{v,\mathbf{k}} - \hbar\omega) \\ &= \frac{e^2 {v_f}^2}{\omega}\int_0^\infty k dk(f(\varepsilon_v) - f(\varepsilon_c))\delta(\varepsilon_{c,\mathbf{k}} - \varepsilon_{v,\mathbf{k}} - \hbar\omega) \\ &= \frac{e^2}{4\hbar}\left[f\left(-\frac{\hbar\omega}{2}\right) - f\left(\frac{\hbar\omega}{2}\right)\right],\end{aligned} \tag{39}$$

Where $g = 4$ is the graphene degeneracy and ρ is replaced by the Fermi distribution function f. By inserting the explicit expression of the Fermi distribution, Eq. (39) is reduced to

$$\sigma_{\text{inter}}(\omega) = \frac{e^2}{4\hbar} \frac{\sinh\left(\frac{\hbar\omega}{2k_B T}\right)}{\cosh\left(\frac{\varepsilon_F}{2k_B T}\right) + \cosh\left(\frac{\hbar\omega}{2k_B T}\right)}, \tag{40}$$

where ε_F is the Fermi energy of graphene, k_B is the Boltzmann constant and T is the temperature. Eq. (40) is the graphene optical conductivity due to interband transition, which is valid in the visible light spectrum. This result can be derived from the Kubo formula too. When the energy of the incident light ($\hbar\omega$) and $k_B T$ are of the same order and ε_F is much larger than $k_B T$, the interband optical conductivity is constant, i.e.,

$$\sigma_{\text{inter}}(\omega) = \frac{e^2}{4\hbar}. \tag{41}$$

For infrared and THz radiation, the Fermi energy can be tuned to become much larger than the incident photon energy, and therefore due to Pauli blocking there are only intraband transitions. According ot the Boltzmann equation and under the relaxation time approximation, the carrier distribution in the presence of a constant electric field with x-polarization is given by

$$f(\mathbf{k}) = f^0(\mathbf{k}) - qE_x v_x(\mathbf{k})\tau(\mathbf{k}) \frac{\partial f^0(\mathbf{k})}{\partial \varepsilon}, \tag{42}$$

where $f^0(\mathbf{k})$ is the carrier distribution in the absence of the electric field, q is the carrier charge, and $\tau(\mathbf{k})$ is the relaxation time in the presence of electron-phonon interaction, electron-impurity interaction, and other scattering processes. By using Eq. (42), the electric current is given by

$$j_x = \frac{q}{A} \sum_{\mathbf{k}} f(\mathbf{k}) v_x(\mathbf{k}) = \frac{-q^2 E_x}{A} \sum_{\mathbf{k}} {v_x}^2(\mathbf{k}) \tau(\mathbf{k}) \frac{\partial f^0(\mathbf{k})}{\partial \varepsilon}. \tag{43}$$

In the absence of the external electric field, the net electric current is zero, so the summation of the first part of Eq. (43) in the electric current is zero. Since $\frac{\partial f^0(\mathbf{k})}{\partial \varepsilon}$ is so narrow around the Fermi surface, only the wave vectors near the Fermi energy contribute to the integration, and $\tau(\mathbf{k})$ is approximately constant. By replacing ${v_x}^2(\mathbf{k})$ with ${v_F}^2/2$, the intraband optical conductivity becomes

$$\sigma_{\text{intra}} = -\frac{q^2 \tau}{A} \frac{{v_F}^2}{2} \sum_{\mathbf{k}} \frac{\partial f^0(\mathbf{k})}{\partial \varepsilon} = -\frac{q^2 \tau {v_F}^2}{2\pi^2} \int_0^\infty dk 2\pi k \frac{\partial f^0(\mathbf{k})}{\partial \varepsilon}. \tag{44}$$

For each $\mathbf{k}$ there are two energy values in the conduction and the valence bands. Using the linear dispersion relation, the intraband optical conductivity is reduced to

$$\begin{aligned}\sigma_{\text{intra}} &= -\frac{q^2\tau}{\pi\hbar^2}\left(\int_0^{\infty}\varepsilon d\varepsilon\frac{\partial f^0(\varepsilon)}{\partial\varepsilon}+\int_0^{-\infty}\varepsilon d\varepsilon\frac{\partial f^0(\varepsilon)}{\partial\varepsilon}\right)\\ &= -\frac{q^2\tau}{\pi\hbar^2}\int_{-\infty}^{\infty}|\varepsilon|d\varepsilon\frac{\partial f^0(\varepsilon)}{\partial\varepsilon}\\ &= \frac{e^2\tau}{\pi\hbar^2}2k_BT\ln\left(2\cosh\left(\frac{\varepsilon_F}{2k_BT}\right)\right),\end{aligned} \tag{45}$$

where q is replaced by the electron charge $(-e)$. Assuming $\varepsilon_F \gg k_BT$, which is usually the case, the intraband optical conductivty simplifies to

$$\sigma_{\text{intra}} = \frac{e^2}{\pi\hbar^2}\varepsilon_F. \tag{46}$$

In the presense of an oscillating electric field, the relaxation time is a complex function $(\tau^{-1} \to \tau^{-1} - i\omega)$ and the intraband conductivity is given by

$$\sigma_{\text{intra}}(\omega) = \frac{e^2}{\pi\hbar^2}\frac{2k_BT}{\tau^{-1} - i\omega}\ln\left(2\cosh\left(\frac{\varepsilon_F}{2k_BT}\right)\right), \tag{47}$$

which in the case of $\varepsilon_F \gg k_BT$ is reduced to [118, 120]

$$\sigma_{\text{intra}}(\omega) = \frac{e^2}{\pi\hbar^2}\frac{\varepsilon_F}{\tau^{-1} - i\omega}, \tag{48}$$

where τ is determined by impurity scattering and electron-phonon interaction $\tau^{-1} = \tau_{\text{imp}}^{-1} + \tau_{\text{e-ph}}^{-1}$.

The dielectric function of graphene can be obtained via its AC conductivity by means of [135]

$$\boldsymbol{\varepsilon}(\omega) = 2.5 + i\frac{\sigma(\omega)}{\boldsymbol{\varepsilon}_0\omega d} \tag{49}$$

where $\boldsymbol{\varepsilon}_g = 2.5$ is the dielectric constant of graphite. Here d is the thickness of graphene.

The bandstructure of graphene is linear in the tight-binding approximation, as shown in **Figure 11b**. The dispersion relation for the TM mode in the geometry depicted in **Figure 11a**, which consists of graphene surrounded by dielectrics with constants $\boldsymbol{\varepsilon}_{r1}$ and $\boldsymbol{\varepsilon}_{r2}$, is given by [122]

$$\frac{\boldsymbol{\varepsilon}_{r1}}{\sqrt{q^2 - \frac{\boldsymbol{\varepsilon}_{r1}\omega^2}{c^2}}} + \frac{\boldsymbol{\varepsilon}_{r2}}{\sqrt{q^2 - \frac{\boldsymbol{\varepsilon}_{r2}\omega^2}{c^2}}} = -\frac{\sigma(q,\omega)i}{\omega\boldsymbol{\varepsilon}_0}, \tag{50}$$

where $\sigma(\omega,q)$ is the conductance of graphene. The semiclassical conductivity $\sigma(\omega)$ does not depend on q. In the nonretarded regime ($q = \omega/c$), the dispersion relation of the TM mode is reduced to [122]

$$q \approx \varepsilon_0 \frac{\varepsilon_{r1} + \varepsilon_{r2}}{2} \frac{2i\omega}{\sigma(q,\omega)}. \tag{51}$$

The larger the q (smaller plasmon wavelength), the higher is the transversal confinement of the TM mode due to excitation of the surface plasmon (SP) on graphene, which is similar to the case of regular metals. The main difference of SPs on metal and 2D materials is due to their band structures. While normal metals have parabolic bands, graphene has Dirac cones, which leads to a novel dispersion relation for the SPs on graphene, as shown in **Figure 11c**. The interactions between the substrate and graphene optical phonons and the graphene plasmons open a band gap in the graphene plasmon dispersion relation, as shown in **Figure 11d**, showing the plasmon loss function in the presence of plasmon-optical phonons interaction given by Eqs. (4)–(6) [134]. The electrons in 2D materials are bound in the normal direction, which leads to more confinement of the coupled electromagnetic wave due to plasmon excitation. In graphene, thermalization takes place on a time scale of 100 fs, while cooling is much slower and happens on a time scale of 10 ps, as seen from **Figure 11** [136, 137].

There are two different approaches to obtain the dispersion relation of the surface plasmon. In the semiclassical approximation, the Drude-like conductivity is used to obtain the plasmon dispersion relation [122, 138]

$$q(\omega) = \frac{\pi\hbar^2 \varepsilon_0(\varepsilon_{r1} + \varepsilon_{r2})}{e^2 \varepsilon_F}\left(1 + \frac{i}{\tau\omega}\right)\omega^2 \tag{52}$$

and the plasmon loss

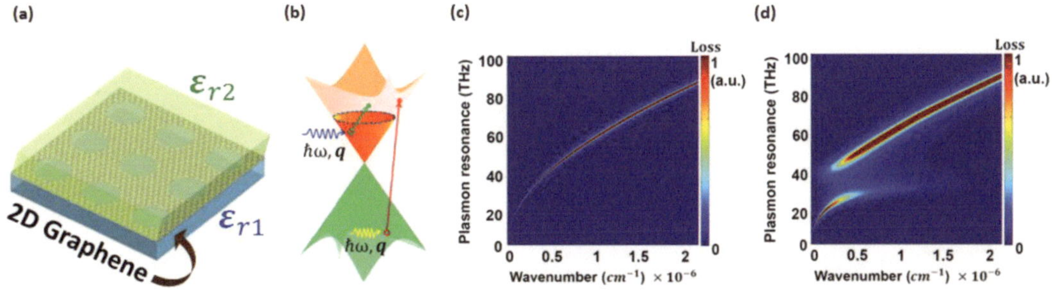

Figure 11. (a) Schematic of the patterned graphene layer surrounded by two media of dielectrics ε_{r1} and ε_{r1}. (b) The sketch of intraband (green arrow) and interband (red arrow) transitions after absorbing a photon on the Dirac cone. (c) The graphene plasmon loss function for the structure shown in part (a) without plasmon-optical phonon coupling. This function shows the graphene plasmon dispersion relation. (d) The graphene dispersion relation in presence of substrate/graphene optical phonons.

$$\frac{\text{Re}q}{\text{Im}q} = \omega\tau = \frac{2\pi c\tau}{\lambda_{\text{air}}}. \tag{53}$$

Another approach that works for both intraband and interband regimes is based on the selfconsistent linear response theory, known as random phase approximation (RPA) along with the relaxation time approximation defined by Mermin [139]. The dispersion relation of the plasmon can be obtained by solving

$$\varepsilon_{\text{RPA}}(q, \omega) = 0, \tag{54}$$

with the complex wave vector $q = q_1 + iq_2$. Considering the Coulomb interaction of electrons in graphene and the medium dielectric, first and second terms of Eq. (4), one obtains the plasmon dispersion relation

$$\frac{\varepsilon_{r1} + \varepsilon_{r1}}{2} + \frac{e^2}{2\varepsilon_0 q_1}\text{Re}\chi(q_1, \omega) = 0 \tag{55}$$

and the plasmon loss relation [122]

$$q_2 = \frac{\text{Im}\chi(q_1, \omega) + \frac{1}{\tau}\frac{\partial}{\partial\omega}\text{Re}\chi(q_1, \omega) + \frac{1}{\omega\tau}\text{Re}\chi(q_1, \omega)\left[1 - \frac{\chi(q_1, \omega)}{\chi(q_1, 0)}\right]}{\frac{1}{q_1}\text{Re}\chi(q_1, \omega) - \frac{\partial}{\partial q_1}\text{Re}\chi(q_1, \omega)}. \tag{56}$$

The plasmon losses (Req/Imq), wave localization ($\lambda_{\text{air}}/\lambda_p$), and the group velocity of the graphene plasmon ($v_g = \partial\omega/\partial q$) are calculated by means of the semiclassical approximation and RPA, as shown in **Figure 12** for the Fermi level of $\varepsilon_F = 0.135$ eV and relaxation time of $\tau = 1.35\times10^{-13}s$ [122]. Below the interband regime, the plasmon loss, the wave localization, and the group velocity calculated by RPA are in very good agreement with the semiclassical approach.

For 2D materials such as graphene, the amount of plasmon loss can be calculated by using the effective dielectric function of the material. The dynamical polarizability

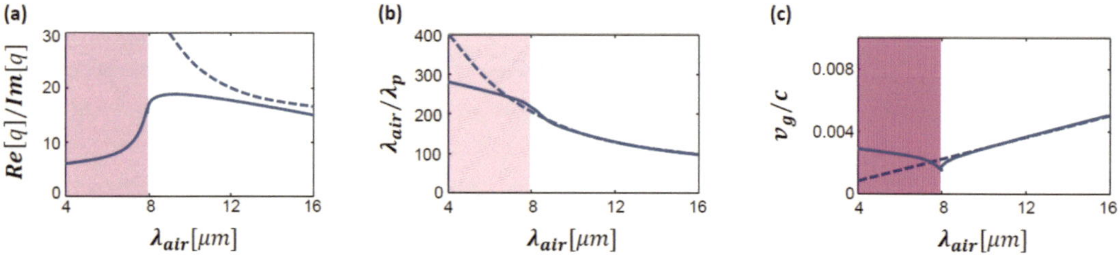

Figure 12. (a) Plasmon loss, (b) filed confinement, and (c) the group velocity of the graphene surface plasmon for $\varepsilon_F = 0.135$ eV and the carrier mobility of $\mu = 10000\ \text{cm}^2/\text{Vs}$. The solid and dashed line represent the RPA and semiclassical results, respectively, and the rose shaded area is the high-interband loss region.

$$\chi(\mathbf{q}, i\omega_n) = -\frac{1}{A}\int_0^\beta d\tau e^{i\omega_n \tau}\langle T\rho_q(\tau)\rho_{-q}(0)\rangle \tag{57}$$

determines several important quantities, such as the effective electron-electron interaction, plasmon spectra, phonon spectra, and Friedel oscillations. $\omega_n = \frac{2\pi n}{\beta}$ are the Matsubara frequencies, ρ_q is the density operator in q-space and A denotes the area. This quantity is calculated in the canonical ensemble for both of the sub-lattice density operators ($\rho = \rho_a + \rho_b$). The dynamical polarizability in zeroth-order in the electron-electron interaction in the long wavelength limit is [140]

$$\chi^0(\mathbf{q}, i\omega_n) = \frac{g_s g_v}{4\pi^2}\int d^2\mathbf{k} \sum_{s,\, s'=\pm} f^{ss'}(\mathbf{k}, \mathbf{q}) \frac{f(\varepsilon^s(\mathbf{k})) - f(\varepsilon^{s'}(|\mathbf{k}+\mathbf{q}|))}{\varepsilon^s(\mathbf{k}) - \varepsilon^{s'}(|\mathbf{k}+\mathbf{q}|) + i\hbar\omega_n}, \tag{58}$$

where $g_s = g_v = 2$ are the spin and valley degeneracy, $f(\varepsilon^s(\mathbf{k}))$ is the Fermi distribution function, and $\varepsilon^s(\mathbf{k}) = s\hbar v_F k - \varepsilon_F$ is the graphene energy. The band overlap of the wavefunctions

$$f^{ss'}(\mathbf{k}, \mathbf{q}) = \frac{1}{2}\left(1 + ss'\frac{k + q\cos\varphi}{|\mathbf{k}+\mathbf{q}|}\right) \tag{59}$$

is a specific property of graphene, where φ signifies the angle between $\mathbf{k}$ and $\mathbf{q}$. At zero temperature, the Fermi-Dirac distribution functions are step functions. In that case, using $\omega_n \to \omega + i\delta$, Eq. (58) is simplified to

$$\chi^{0\pm}_{T=0}(\mathbf{q}, \omega) = \frac{g_s g_v}{4\pi^2\hbar}\int d^2\mathbf{k} \sum_{\alpha=\pm} \frac{\alpha f^{\pm}(\mathbf{k}, \mathbf{q})}{\omega + \alpha v_F(k \mp |\mathbf{k}+\mathbf{q}|) + i\delta}, \tag{60}$$

where the + and – signs denote the intra and interband transitions, respectively. Integration over φ and k gives the retarded polarizability or charge-charge correlation function

$$\chi^0(\mathbf{q}, \omega) = \chi^0_{\varepsilon_F=0}(\mathbf{q}, \omega) + \Delta\chi^0(\mathbf{q}, \omega), \tag{61}$$

where

$$\chi^0_{\varepsilon_F=0}(\mathbf{q}, \omega) = -i\pi\frac{\mathfrak{I}(\mathbf{q}, \omega)}{\hbar^2 v_F^2}, \tag{62}$$

and

$$\Delta\chi^0(\mathbf{q}, \omega) = -\frac{g\varepsilon_F}{2\pi\hbar^2 v_F^2} + \frac{\mathfrak{I}(\mathbf{q}, \omega)}{\hbar^2 v_F^2}\left\{\begin{array}{l} \aleph\left(\frac{\hbar\omega + 2\varepsilon_F}{\hbar v_F q}\right) - \Theta\left(\frac{2\varepsilon_F - \hbar\omega}{\hbar v_F q} - 1\right) \times \left(\aleph\left(\frac{2\varepsilon_F - \hbar\omega}{\hbar v_F q}\right) - i\pi\right) \\ -\Theta\left(\frac{\hbar\omega - 2\varepsilon_F}{\hbar v_F q} + 1\right) \times \aleph\left(\frac{\hbar\omega - 2\varepsilon_F}{\hbar v_F q}\right) \end{array}\right\} \tag{63}$$

The two functions $\mathfrak{I}(\mathbf{q}, \omega)$ and $\aleph(x)$ are defined by

$$\Im(\mathbf{q},\omega) = \frac{g}{16\pi}\frac{\hbar v_F^2 q^2}{\sqrt{\omega^2 - v_F^2 q^2}} \tag{64}$$

and

$$\aleph(x) = x\sqrt{x^2-1} - \ln(x+\sqrt{x^2-1}), \tag{65}$$

Where $g = g_s g_v = 4$. For $\omega > qv_F$ and in the long wavelength limit $q \to 0$, $x = |(\hbar\omega \pm 2\varepsilon_F)/\hbar v_F q| \gg 1$, thus $x^2 - 1 \approx x^2$ and $\aleph(x) \approx x^2 - 2\ln(x)$. We derive here the dynamical polarizability in Eq. (63) in these regimes. The expansion of $\Im(\mathbf{q},\omega)$ gives

$$\Im(\mathbf{q},\omega) = \frac{g}{16\pi}\frac{\hbar v_F^2 q^2}{\omega}\left(1 - \frac{v_F^2 q^2}{\omega^2}\right)^{-1/2} \approx \frac{g}{16\pi}\frac{\hbar v_F^2 q^2}{\omega}\left(1 + \frac{v_F^2 q^2}{2\omega^2}\right). \tag{66}$$

In this condition and for intraband transitions $\hbar\omega < 2\varepsilon_F$ ($\hbar\omega < 2\mu$)

$$\aleph\left(\frac{\hbar\omega + 2\varepsilon_F}{\hbar v_F q}\right) - \aleph\left(\frac{2\varepsilon_F - \hbar\omega}{\hbar v_F q}\right) = \frac{8\hbar\omega\varepsilon_F}{\hbar^2 v_F^2 q^2} + 2\ln\left|\frac{2\varepsilon_F - \hbar\omega}{2\varepsilon_F + \hbar\omega}\right|. \tag{67}$$

As a result, $\Delta\chi^0(\mathbf{q},\omega)$ reduces to

$$\begin{aligned}\Delta\chi^0(\mathbf{q},\omega) &= -\frac{g\varepsilon_F}{2\pi\hbar^2 v_F^2} + \frac{\Im(\mathbf{q},\omega)}{\hbar^2 v_F^2}\left\{\frac{8\hbar\omega\varepsilon_F}{\hbar^2 v_F^2 q^2} + 2\ln\left|\frac{2\varepsilon_F - \hbar\omega}{2\varepsilon_F + \hbar\omega}\right| + i\pi\right\} \\ &= \frac{gq^2}{8\pi\hbar\omega}\left\{\frac{2\varepsilon_F}{\hbar\omega} + \frac{1}{2}\ln\left|\frac{2\varepsilon_F - \hbar\omega}{2\varepsilon_F + \hbar\omega}\right| + \frac{i\pi}{2}\right\}.\end{aligned} \tag{68}$$

If $2\varepsilon_F \gg \hbar\omega$, then

$$\Delta\chi^0(\mathbf{q},\omega) = \frac{gq^2}{8\pi\hbar\omega}\left\{\frac{2\varepsilon_F}{\hbar\omega} + \frac{i\pi}{2}\right\}. \tag{69}$$

By taking the decay rate $\omega \to \omega + i\tau^{-1}$ into account and inserting Eq. (62) into Eq. (61), the dynamical polarizability reduces to

$$\chi^0(\mathbf{q},\omega) \approx \frac{\varepsilon_F q^2}{\pi\hbar^2(\omega + i\tau^{-1})^2}. \tag{70}$$

In the presence of optical phonons, the effective dielectric function in the RPA regime is given by Eq. (4), which we restate here for convenience [98, 99]:

$$\boldsymbol{\varepsilon}^{\mathrm{RPA}}(\mathbf{q},\omega) = \boldsymbol{\varepsilon}_m - v_c(\mathbf{q})\chi^0(\mathbf{q},\omega) - \boldsymbol{\varepsilon}_m \sum_l v_{\mathrm{sph},l}(\mathbf{q},\omega)\chi^0(\mathbf{q},\omega) - \boldsymbol{\varepsilon}_m v_{\mathrm{oph}}(\mathbf{q},\omega)\chi^0_{j,j}(\mathbf{q},\omega), \tag{71}$$

where $\varepsilon_m = (\varepsilon_1 + \varepsilon_2)/2$ is the average of dielectric constants of graphene's environment. The collective oscillation modes of the electrons can be obtained by solving $\boldsymbol{\varepsilon}^{\mathrm{RPA}}(\mathbf{q},\omega) = 0$. The extinction function is identified by $Z = -\frac{\delta T}{T_0}$, or for the plasmonic structure coupled to an optical cavity $Z = -\frac{\delta R}{R_0}$, where $\delta R = R - R_0$ and R/R_0 is the reflectance with/without plasmon excitation, which corresponds to the enhanced absorbance at resonance frequencies [129, 131]

$$Z \propto -\,\mathrm{Im}\left(\frac{1}{\varepsilon^{\mathrm{RPA}}}\right). \tag{72}$$

The loss function represents the amount of energy dissipated by exciting the plasmon coupled to the substrate and graphene optical phonons. The collective oscillation modes of the electrons can be obtained by solving $\boldsymbol{\varepsilon}^{\mathrm{RPA}}(\mathbf{q}, \omega) = 0$. Considering the first two terms in Eq. (71) gives the plasmon dispersion relation of graphene

$$\hbar\omega_p(q) = \sqrt{\frac{g\alpha_G \varepsilon_F \hbar v_F q}{2\boldsymbol{\varepsilon}_0}}, \tag{73}$$

where the graphene fine structure constant is given by

$$\alpha_G \equiv \frac{e^2}{\boldsymbol{\varepsilon}\hbar v_F} = \frac{c}{\boldsymbol{\varepsilon} v_F}\alpha, \tag{74}$$

with $\alpha = e^2/\hbar c$ being the fine structure constant. In the case of graphene on a SiO_2 substrate, $\boldsymbol{\varepsilon}_{r1} = 1$ for air and $\boldsymbol{\varepsilon}_{r2} = 3.9$ for SiO_2, which yield $\alpha_G = 0.9$. For suspended graphene, $\boldsymbol{\varepsilon}_{r1} = \boldsymbol{\varepsilon}_{r2} = 1$ and therefore $\alpha_G = 2.2$. Thus, it is possible to tune α_G by altering the dielectric materials surrounding the graphene sheet [141].

It is also interesting to note that in contrast to the ${\varepsilon_F}^{3/4} \propto n_0^{1/2}$ dependence of the bulk plasmon resonance frequency in Eq. (20) and the ${\varepsilon_F}^{1/2} \propto n_s^{1/2}$ dependence of the 2DEG plasmon, the graphene plasmon exhibits a $\varepsilon_F^{1/2} \propto n_s^{1/4}$ dependence.

In order to show the results for $\varepsilon_F > 0$, we define two regimes for $\aleph(x)$, i.e.,

$$\begin{aligned} \aleph_>(x) &= x\sqrt{x^2 - 1} - \cosh^{-1}(x), \quad \text{for } x > 1 \\ \aleph_<(x) &= x\sqrt{1 - x^2} - \cosh^{-1}(x), \quad \text{for } x < 1 \end{aligned} \tag{75}$$

Using Eq. (60), the imaginary part of dynamical polarization is given by [140]

$$\mathrm{Im}\,\Delta\chi^0(\mathbf{q}, \omega) = \frac{\Im(\mathbf{q}, \omega)}{\hbar^2 v_F^2} \times \begin{cases} \aleph_>(\frac{2\varepsilon_F - \hbar\omega}{\hbar v_F q}) - \aleph_>(\frac{2\varepsilon_F + \hbar\omega}{\hbar v_F q}), \rightarrow 1A \\ \pi \rightarrow 1B \\ -\aleph_>(\frac{2\varepsilon_F + \hbar\omega}{\hbar v_F q}), \rightarrow 2A \\ -\aleph_<(\frac{\hbar\omega - 2\varepsilon_F}{\hbar v_F q}), \rightarrow 2B \\ 0, \rightarrow 3A \\ 0. \rightarrow 3B \end{cases} \tag{76}$$

and the real part is obtained by the Kramers-Kronig relation, yielding

$$\mathrm{Re}\,\Delta\chi^0(\mathbf{q},\omega) = -\frac{g\varepsilon_F}{2\pi\hbar^2 v_F^2} + \frac{\mho(\mathbf{q},\omega)}{\hbar^2 v_F^2} \times \begin{cases} \pi, \rightarrow 1A \\ -\aleph_>(\frac{2\varepsilon_F - \hbar\omega}{\hbar v_F q}) + \aleph_>(\frac{2\varepsilon_F + \hbar\omega}{\hbar v_F q}), \rightarrow 1B \\ -\aleph_<(\frac{\hbar\omega - 2\varepsilon_F}{\hbar v_F q}), \rightarrow 2A \\ \aleph_>(\frac{2\varepsilon_F + \hbar\omega}{\hbar v_F q}), \rightarrow 2B \\ -\aleph_<(\frac{\hbar\omega - 2\varepsilon_F}{\hbar v_F q}) + \aleph_<(\frac{2\varepsilon_F + \hbar\omega}{\hbar v_F q}), \rightarrow 3A \\ \aleph_>(\frac{2\varepsilon_F + \hbar\omega}{\hbar v_F q}) - \aleph_>(\frac{\hbar\omega - 2\varepsilon_F}{\hbar v_F q}). \rightarrow 3B \end{cases} \tag{77}$$

Different regions are shown in **Figure 13**. As we mentioned, the plasmon dispersion relation is determined by $\varepsilon(\mathbf{q},\omega_p - i\gamma) = 0$, where γ is the decay rate of the plasmon. For weak damping, the plasmon dispersion relation and damping are given by

$$\varepsilon_0 = v_c \mathrm{Re}\Delta\chi^0(\mathbf{q},\omega_p), \quad \gamma = \frac{\mathrm{Im}\Delta\chi^0(\mathbf{q},\omega_p)}{(\partial/\partial\omega)\mathrm{Re}\Delta\chi^0(\mathbf{q},\omega)\big|_{\omega_p}}. \tag{78}$$

The solution of the first part of Eq. (78) exists only for $\mathrm{Re}\Delta\chi^0(\mathbf{q},\omega_p) > 0$, which is valid only for finite graphene doping and $\omega > v_F q$. Interestingly, a plasmon does not decay if $\mathrm{Im}\Delta\chi^0(\mathbf{q},\omega_p) = 0$, which is the case in region 1B shown in **Figure 13**. For the finite doping ($\varepsilon_F > 0$), the acoustic phonon at long wavelength is inside the region 1A in **Figure 13**.

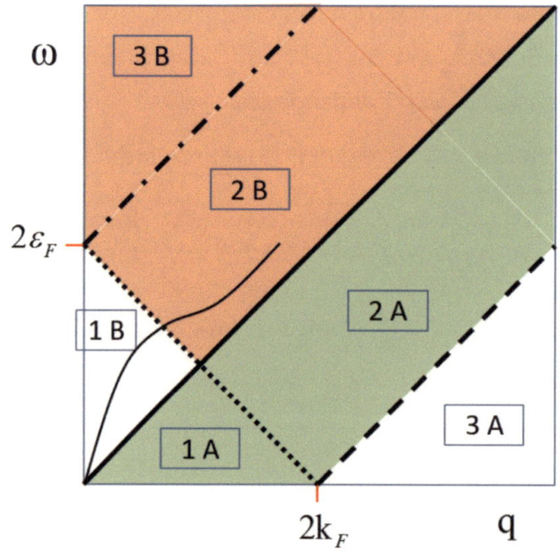

Figure 13. Different regions related to the dynamical polarization. The regions are separated by the straight line $\omega = v_F q$ (solid), $\hbar\omega = \hbar v_F q - 2\varepsilon_F$ (dashed), and $\hbar\omega = 2\varepsilon_F - \hbar v_F q$ (dotted). The red-shaded area depicts the region of interband Landau damping while the green-shaded area marks the region of intraband Landau damping. The radiative interband Landau damping originating from direct (vertical) transitions occurs at $q = 0$. The red-shaded area with $q \neq 0$ corresponds to the region of nonradiative interband Landau damping.

7. Surface plasmon resonances in 3D topological insulators

So far, we have discussed nanoplasmonics in 3D metals and graphene and their damping mechanisms. Now we focus on the RPA theory of nanoplasmonics in 3D topological insulator (TI) materials. In particular, we are going to identify the graphene plasmons as a limiting case of 3DTI plasmons in the case when the thickness of the 3DTI slab becomes atomically thin. Bismuth selenide (Bi_2Se_3) is the prime example of a 3DTI material that has a rhombohedral crystal structure and consists of five atomic layers arranged along the z-direction, known as quintuple layers (QLs). Two such QLs are coupled by a weak interaction, predominantly of the van der Waals type. At the vicinity of the Γ point in the Brillouin zone of Bi_2Se_3, the low energy of Dirac particles in the xy-plane can be described by the effective Hamiltonian [142, 143], $H_{TI} = \hbar v_f(\boldsymbol{\sigma}\times\mathbf{k}\cdot\hat{\mathbf{z}}) - E_F$, where $\mathbf{z}$ is the unit vector in a z-direction. In the case of graphene, it is the pseudospin, representing the two sub-lattice indices, which is locked to the momentum of the electron. The chiral nature of the electronic states with the spin locked in perpendicular direction to the momentum results in a unique type of collective excitations which are, indeed, accompanied by transverse spin fluctuations [142, 143].

Let us consider a slab of a 3DTI material of thickness d > 5 nm, which is large enough to suppress any overlap of the single-electron states between the top and the bottom layers. In contrast, long-range Coulomb interaction exists and couples the opposite surfaces as in 2D electron plasmas.

We consider a Hamiltonian that describes the properties of collective oscillations on the top and the bottom surfaces of a 3DTI material as [142]

$$H^{l/l'} = \hbar v_f \sum_{\alpha,\,\beta,\,k,\,l,\,l'} a^{+}_{k,\,l,\,\alpha}(\boldsymbol{\tau}\times\mathbf{k})\cdot\hat{z}a_{k,\,l,\,\beta} + \frac{1}{2S}\sum_{q,\,l,\,l'} v_{l,\,l'}(q)n_{q,\,l}n_{-q,\,l'}, \tag{79}$$

where $l(l')$ indicates the top (bottom) surface, α, β are spin indices, a is the annihilation operator, and S is the surface area. $n_{q,\,l} = \sum_{\mathbf{k},\,\alpha} a^{+}_{\mathbf{k}-\mathbf{q},\,l,\,\alpha}a_{\mathbf{k},\,l,\,\alpha}$ is the density operator, and $\boldsymbol{\tau} = \{\boldsymbol{\tau}_x,\ \boldsymbol{\tau}_y\}$ are the Pauli matrices. With a long-range Coulomb interaction within the RPA, the response function can be written as

$$\chi^{2x2}_{\mathrm{RPA}} = \frac{\chi^{2x2}_0(\mathbf{q},\omega)}{1 - v(\mathbf{q})\chi^{2x2}_0(\mathbf{q},\omega)}, \tag{80}$$

where $\chi^{2x2}_0(\mathbf{q},\omega) = \begin{pmatrix} \chi^{11} & 0 \\ 0 & \chi^{22} \end{pmatrix}$ is a 2×2 matrix. $v(\mathbf{q})$ is a 2×2 matrix whose diagonal and off diagonal elements are the two-dimensional Fourier transforms of Coulomb potentials and can be obtained by solving Poisson's equation [144]. Similar to the RPA dielectric function for graphene shown in Eq. (71), the equivalent equation for the RPA dielectric function in 3DTIs is

$$\varepsilon^{2x2}_{\mathrm{RPA}}(\mathbf{q},\omega) = 1 - v(\mathbf{q})\chi^{2x2}_0(\mathbf{q},\omega). \tag{81}$$

Compared to graphene, the main difference is that $v(\mathbf{q})$ is a 2×2 matrix accounting for the intrasurface and intersurface interactions. Here we write these expressions as

$$v^{11/22}(q) = \frac{4\pi e^2 \varepsilon_2 \left(\cosh(qd) + (\varepsilon_{3/1}/\varepsilon_2)\sinh(qd)\right)}{q\left(\varepsilon_2(\varepsilon_1 + \varepsilon_3)\cosh(qd) + (\varepsilon_1\varepsilon_3 + \varepsilon_2)\sinh(qd)\right)} \tag{82}$$

for the intrasurface interaction, and

$$v^{12}(q) = \frac{4\pi e^2 \varepsilon_2}{q\left(\varepsilon_2(\varepsilon_1 + \varepsilon_3)\cosh(qd) + (\varepsilon_1\varepsilon_3 + \varepsilon_2)\sinh(qd)\right)} \tag{83}$$

for the intersurface interaction. ε_i ($i = 1, 2, 3$) are the dielectric functions for the top, middle, and bottom layers, respectively. In the limit of equal dielectric constant, $\varepsilon_1 = \varepsilon_2 = \varepsilon_3 = \varepsilon$, $V_{11} = V_{22} \rightarrow 2\pi^2 e^2/\varepsilon q$, and $V_{12} = V_{21} \rightarrow (2\pi^2 e^2/\varepsilon q)e^{-qd}$. This shows that for a given wave vector the intrasurface Coulomb interaction depends on the dielectric constant of the materials, the intersurface Coulomb interaction not only depends on the dielectric constant, but is also modulated by the negative distance-dependent exponential factor. For a sufficiently thick slab, i.e., $qd \rightarrow \infty$, the top and the bottom surfaces are completely decoupled and therefore we obtain the limit of $V_{12} = V_{21} \rightarrow 0$, and thus we only have a nonzero intrasurface potential. This is the case that is equivalent to the Coulomb interaction in a sheet of graphene.

The response function χ provides important information about the collective states that are excited at small transferred momentum. The collective mode frequencies of the system can be obtained by solving $\mathrm{Det}[1 - v(\mathbf{q})\chi_0^{2x2}(\mathbf{q})] = 0$. In the region $v_f q < \omega < 2E_f/\hbar - v_f q$, collective modes of oscillations are undamped. Beyond that limit, such modes are not observed because the energy of the modes is transferred to the particle-hole excitations. In Section 6, we derived the dynamical polarizability for the case of a graphene sheet. For $v_f q < \omega$ with $\mathrm{Im}[\chi] = 0$, the linear response function can be written as [144]

$$\chi_{\mathrm{nn}}^{l/l'}(\mathbf{q}, \omega) = -\frac{E_F^{l/l'}}{2\pi\hbar^2 v_F{}^2} + \frac{1}{16\pi\hbar}\frac{q^2}{\sqrt{\omega^2 - (v_F q)^2}} \times \left[G\left(\frac{2E_F^{l/l'} + \hbar\omega}{\hbar v_F q}\right) - G\left(\frac{2E_F^{l/l'} - \hbar\omega}{\hbar v_F q}\right)\right], \tag{84}$$

where $G(x) = x\sqrt{x^2 - 1} - \cosh^{-1}(x)$. Using the relation for the relative permittivity $\varepsilon_r = 1 + \chi$, the linear response function given by Eq. (84) is obtained from Eq. (61) in the long wavelength limit. This charge-charge response function describes explicitly the dispersion of the collective modes exist on the two sides of a slab geometry of a 3DTI slab. Using Eq. (80), the pole of $\mathrm{Det}[1 - v(\mathbf{q})\chi_0^{2x2}(\mathbf{q})] = 0$ can be solved with the help of Eq. (35). The potential $v(q)$ is a 2×2 block diagonal (corresponding to spin and charge) for two surfaces. In the limit of an infinitely thick slab, we recover the intrasurface potential for $v(q)$ and the corresponding mode for graphene. The two modes in the long wavelength limit are: the symmetric (photon-like) mode and the antisymmetric (plasmon-like) mode. At a thickness below the Thomas-Fermi screening length, the symmetric mode is highly Landau damped. In **Figure 14**, we show the antisymmetric mode for a slab of thickness 120 nm at different Fermi energies of 0.3, 0.2, and 0.1 eV, taking into account the inter and intrasurface potentials given in Eqs. (82) and (83) in the $q \rightarrow 0$ limit of the RPA response function. This result can be compared with the experimental observation of the Dirac plasmon dispersion in Bi_2Se_3 [145].

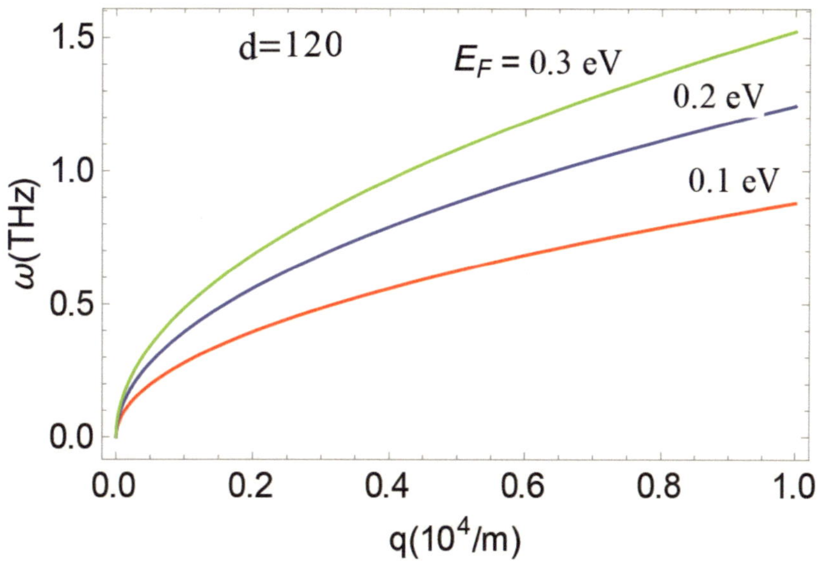

Figure 14. Plasmon dispersion in Bi_2Se_3 slab of thickness 120 nm at different Fermi energy with top and bottom dielectric constant of 1 and 9.3, respectively, taking full potential and $q \to 0$ limit. Only the antisymmetric mode is shown here.

In the limit of small thickness of the slab, when the condition $qd \to 0$ is satisfied, the antisymmetric photon-like mode behaves as

$$\omega_{\text{ant}}^2 = \frac{v_f^2 k_f \alpha_D}{\varepsilon_1 + \varepsilon_3}\left[\left(\frac{1+\xi}{2}\right)^{1/2} + \left(\frac{1-\xi}{2}\right)^{1/2}\right] q \tag{85}$$

and the symmetric plasmon-like mode as

$$\omega_{\text{sym}}^2 = \frac{v_f^2 k_f \alpha_D}{\sqrt{2}\varepsilon_2}\left[\frac{(1-\xi)^{1/2}(1+\xi)^{1/2}}{(1-\xi)^{1/2} + (1+\xi)^{1/2}}\right] dq^2, \tag{86}$$

where $\xi = (n_1 - n_2)/(n_1 + n_2)$ is the density polarization and $k_f = \sqrt{4\pi(n_1 + n_2)}$ is the total Fermi wave vector. $\alpha_D = e^2/\hbar v_f$ is the fine structure constant for the Dirac system. It is to be noted that indices 1 and 2 indicate two different surfaces. For equal Fermi energy at the top and the bottom surfaces, the density polarization reduces to zero, resulting in an equilibrium situation. When the thickness of the slab tends to zero, we recover only the antisymmetric mode, which is the mode obtained for a sheet of graphene. As the slab thickness goes to infinity, the two surfaces interact weakly and the intersurface potential falls rapidly as the thickness increases. In this case, we obtain that the two antisymmetric modes each correspond to a single sheet of graphene. Using a series expansion of the frequency, it is possible to obtain a more accurate solution for the symmetric mode, which is derived elsewhere [144]. The solution is given by

$$\omega_{\text{sym}} = \left(\sqrt{2}\varepsilon_2 + \alpha_D k_f d\right) v_f q / \left(\sqrt{2\varepsilon_2\left(\varepsilon_2 + \sqrt{2}\alpha_D k_f d\right)}\right) \tag{87}$$

For a size of the slab smaller than the mode wavelength, i.e., $d < \lambda_p$, the antisymmetric mode is independent of thickness, and the symmetric mode has a d^2 dependence. The plasmon dispersions for the limit of $qd \rightarrow 0$ for symmetric and antisymmetric modes with 10 and 20 nm of slab width at different dielectric environments are shown in **Figure 15a**. The electric field profiles along (E_x) and perpendicular (E_z) to the plasmon polarization direction are shown in **Figure 15b** for a thin slab of thickness 10 nm with $\varepsilon_1 = \varepsilon_3 = 3.8$ for an antisymmetric mode. We also show the Landau damping regions with a blue shade given by the line segments $\omega \leq v_F q$ and $\omega \geq 2E_F - v_F q$, in complete analogy to **Figure 13**. The distribution of excited "hot" carriers is limited to a narrow region around the Fermi level. In quasi-equilibrium, the hot carrier distribution can affect the plasmon lifetime.

The topological surface states are extended inside the bulk with a localization length given by $l_o = \hbar v_f / E_g$ where E_g is the bulk bandgap. l_o is equal to 1.5 nm (~2 quintuple layer (QL)) in Bi_2Se_3. Due to the quantum confinement of the carriers, two different conductivities arise: one due to the topological surface states that extend to 2 QL and the other one due to the 2DEG trapped below the surface, which extends up to 8 QL [146]. Theoretically, it should be possible to separate out the two different effects and write the response functions as $\chi = \chi^{TI} + \chi^{2DEG}$ for two independent channels. The contribution of the 2DEG to the plasmon dispersion can be calculated separately [147].

Surface plasmons in 3DTI surfaces do not consist only of charge density waves but are also accompanied by spin density waves due to inherent nature of spin-momentum locking. This can be qualitatively understood by calculating the surface current in terms of the spin and charge quantity by $\mathbf{J} = v_f \boldsymbol{\sigma} \times \hat{\mathbf{z}}$. The continuity equation $\partial n_q / \partial t = -\nabla \cdot \mathbf{J}$ shows that the charge density n_q oscillates in phase with the transverse spin $s^T = \hat{\mathbf{z}} \cdot (\hat{\mathbf{q}} \times \mathbf{s})$, and the amplitude of their ratio is given by $s^T / n_q = 1/\hbar v_f \left[E_F \alpha_D v_f / 2q\right]^{1/2}$, which can be much larger than unity in the

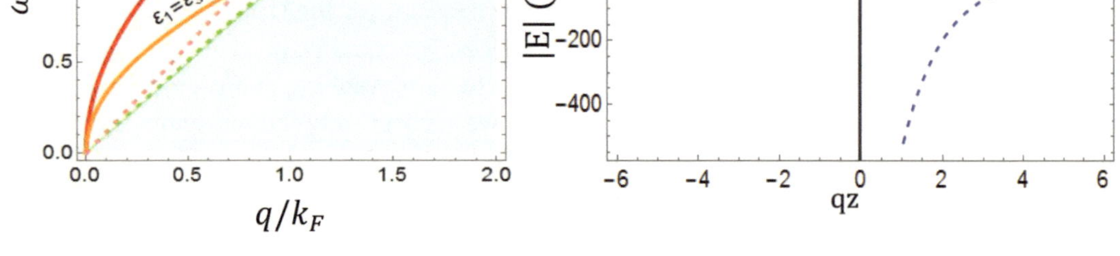

Figure 15. (a) Symmetric (dotted) and antisymmetric (solid) modes of plasmon dispersion in 3DTI in the limit of $qd \rightarrow 0$ for the slab thickness of 10 and 20 nm (Eqs. (7) and (9)). The Landau damping region is shown with the blue shade, determined by $\omega \leq v_F q$ and $\omega \geq 2E_F - v_F q$. (b) The electric field profile of surface plasmons along and perpendicular direction of polarization for antisymmetric mode (plasmon-like mode) in a slab of thickness 10 nm.

long wavelength limit [142]. To understand the electromagnetic response of spin and charge oscillations in a selfsustained mode, we write the induced potential $V_{\text{ind}} = v(\mathbf{q})\chi_0^{4x4}(\mathbf{q},\omega)V_{\text{ind}}$ under vanishing external perturbation [147]. One can solve again for the pole $\left(1 - v(\mathbf{q})\chi_0^{4x4}(\mathbf{q},\omega)\right)V_{\text{ind}} = 0$ in order to obtain the explicit expression for the induced potential. This time, both the spin and charge part of the response functions are included, i.e., the solutions include the explicit information of the spin and charge oscillations. The solutions to the induced potential proportional to

$$V^{\pm}{}_{\text{ind}} \propto \begin{pmatrix} 1 \\ 0 \\ y^{\pm} \\ 0 \end{pmatrix}, \tag{88}$$

where $y^{+} = 1$ and $y^{-} = -\frac{E_F^l}{E_F^{l'}}$. The induced charge density is calculated from the response function $\rho = \chi_0^{4x4}(\mathbf{q},\omega)V_{\text{ind}}$. Spins are inherently attached to the momentum as accounted by the complete response function given by Eq. (80). The symmetric and antisymmetric modes are purely spin-like and purely charge-like for a slab of 3DTI. This can be seen in the following expressions in the limit of $qd \to 0$ as

$$\begin{pmatrix} \varphi_1 \\ \varphi_2 \end{pmatrix} = \rho \begin{pmatrix} -\dfrac{1}{\sqrt{\dfrac{\alpha_D(E_F^l + E_F^{l'})}{\hbar v_F q(\varepsilon^l + \varepsilon^{l'})}}} \\ \dfrac{\dfrac{E_F^l}{E_F^{l'}}}{\sqrt{\dfrac{\alpha_D(E_F^l + E_F^{l'})}{\hbar v_F q(\varepsilon^l + \varepsilon^{l'})}\dfrac{E_F^l}{E_F^{l'}}}} \end{pmatrix} \tag{89}$$

for the antisymmetric mode, and

$$\begin{pmatrix} \varphi_1 \\ \varphi_2 \end{pmatrix} = \rho \begin{pmatrix} -\dfrac{1}{\sqrt{\dfrac{\alpha_D d(E_F^l \times E_F^{l'})}{\hbar v_F q \varepsilon_{TI}(E_F^l + E_F^{l'})}}} \\ -\dfrac{-1}{\sqrt{\dfrac{\alpha_D d(E_F^l \times E_F^{l'})}{\hbar v_F q \varepsilon_{TI}(E_F^l + E_F^{l'})}}} \end{pmatrix} \tag{90}$$

for the symmetric mode, where $\varphi_1 = \begin{pmatrix} \rho^l \\ s^l \end{pmatrix}$ and $\varphi_2 = \begin{pmatrix} \rho^{l'} \\ s^{l'} \end{pmatrix}$, and the indices l and l' can be chosen to represent the top and the bottom surfaces. Clearly, charges oscillate in phase and spins oscillate out of phase for antisymmetric mode, whereas spins oscillate in phase and charges oscillate out of phase for the symmetric mode on the top and the bottom surfaces.

It has been observed that the phonon modes in Bi_2Se_3 have frequencies of around 2 THz. Since the phonon modes overlap with the plasmon modes in energy space, there is a possibility of mode hybridization, similar to the case of graphene. In a micro-ring structure of Bi_2Se_3, there are bonding (lower) and antibonding (upper) plasmon modes [148]. The antibonding plasmon mode frequency is much larger than the phonon mode frequency, and the bonding plasmon mode overlaps with the phonon mode, resulting in hybridization in two branches with an interaction frequency of around 0.35 THz.

Recently, in Ref. [149], authors investigated the possibility of obtaining SPPs in Weyl semimetals. In Weyl semimetals such as TaAs, NbAs, $YbMnBi_2$, and $Eu_2Ir_2O_7$ [150–152] the valence and the conduction band touch in isolated points of the Brillouin zone close to the chemical potential, and their dispersion is described by an equation similar to the one for Dirac metals. A pair of Weyl nodes appears with opposite chiralities with a distance of **b** in the reciprocal space. Their topological properties are described by $S_\theta = \frac{e^2}{4\pi\hbar c}\int dt \int d^3r\, \theta\, \mathbf{E}.\mathbf{B}$, where $\theta = 2(\mathbf{br} - b_o t)$. S_θ is an important parameter that alters the electrodynamics of the bulk response to an applied electric field. In such a system, the authors found that the SPPs with $\mathbf{b} \neq 0$ have properties similar to those in ordinary metal SPPs in the presence of an external magnetic field, despite their unique origin. Therefore, the θ term that contributes to the dielectric tensor originating from the anomalous Hall displacement current induces an anomalous surface magnetoplasmon. Interestingly, for certain orientations of the surface of the Weyl semimetal, the authors predict the nonreciprocal dispersion of SPPs, i.e., the propagation of the SPP is unidirectional.

8. Conclusion

In this book chapter, we present descriptions of the plasmonic properties of metal nanostructures of different geometry, their size dependence, and applications in modern nanotechnology. We show dynamic control over the plasmon resonances where a shift up to 125 nm at a density of 10^{22} cm^{-3} can be achieved using a pump-probe technique. This provides the opportunity to utilize plasmonics in modern information processing devices. In addition to plasmonics in 3D metal nanostructures, we present a description of graphene and 3DTI plasmonics using classical and quantum perspectives. Using the RPA theory, we obtain symmetric and antisymmetric modes in a slab of 3DTI, which reduces to a graphene plasmonics in the limit of zero thickness of the slab. Surface plasmon (SP) damping mechanisms are interesting due to their potential applications for enhanced current density that comes from SPs nonradiative damping in nanostructures smaller than the skin depth. We present a quantum theory of SPs damping in metals and layered materials like graphene.

There are several potential applications of the nanoplasmonics in graphene and 3DTI. Graphene and 3DTI are potential candidates for nanospasers that utilize Dirac fermions, unlike the massive electrons or holes in the originally proposed spasing scheme by Bergman and Stockman in 2003 [153]. The spaser is a nanoplasmonic counterpart of a laser, where photons are not emitted. In Ref. [154], the authors have proposed a scheme of nanospasing using a sheet of graphene with an electrically pumped cascaded quantum well structure working as a gain medium. For the range of Fermi energy, it was shown that spasing could be potentially obtained for a plasmon relaxation time of several femtoseconds in the mid-IR range. A similar scheme with an optically pumped nanospaser for a slab 10 nm of 3DTI was proposed in Ref. [155]. It is advantageous to use a 3DTI, such as Bi_2S_3, as a nanospaser due to the possibility of using its bulk as a gain medium and the surface as a medium that supports SPPs. This configuration avoids the use of a separate gain medium to provide the feedback for the SPPs. Therefore, a 3DTI nanospaser can be truly nanoscopic and can be used for various applications in physics, chemistry, and biology.

Acknowledgements

We acknowledge the support provided by NSF grant CCF-1514089 and DARPA grant HR0011-16-1-0003.

Author details

Hari P. Paudel[1], Alireza Safaei[2] and Michael N. Leuenberger[2*]

*Address all correspondence to: michael.leuenberger@ucf.edu

1 Center for Nano-Optics, Georgia State University, Atlanta, Georgia, USA

2 NanoScience Technology Center, Department of Physics, and College of Optics and Photonics, University of Central Florida, Orlando, Florida, USA

References

[1] G. Mie, Annal of Physics, 25 (1908) 377.

[2] R.H. Ritchie, Plasma losses by fast electrons in thin films, Physical Review, 106 (1957) 874–881.

[3] J. Pendry, Applied physics—playing tricks with light, Science, 285 (1999) 1687–1688.

[4] B. Hecht, H. Bielefeldt, L. Novotny, Y. Inouye, D.W. Pohl, Local excitation, scattering, and interference of surface plasmons, Physical Review Letters, 77 (1996) 1889–1892.

[5] D.Y. Fedyanin, A.V. Krasavin, A.V. Arsenin, A.V. Zayats, Surface plasmon polariton amplification upon electrical injection in highly integrated plasmonic circuits, Nano Letters, 12 (2012) 2459–2463.

[6] X. Guo, Y.G. Ma, Y.P. Wang, L.M. Tong, Nanowire plasmonic waveguides, circuits and devices, Laser & Photonics Reviews, 7 (2013) 855–881.

[7] X. Guo, M. Qiu, J.M. Bao, B.J. Wiley, Q. Yang, X.N. Zhang, Y.G. Ma, H.K. Yu, L.M. Tong, Direct coupling of plasmonic and photonic nanowires for hybrid nanophotonic components and circuits, Nano Letters, 9 (2009) 4515–4519.

[8] R.W. Heeres, L.P. Kouwenhoven, V. Zwiller, Quantum interference in plasmonic circuits, Nature Nanotechnology, 8 (2013) 719–722.

[9] J.S. Huang, D.V. Voronine, P. Tuchscherer, T. Brixner, B. Hecht, Deterministic spatiotemporal control of optical fields in nanoantennas and plasmonic circuits, Physical Review B, 79 (2009) 195441.

[10] J.T. Kim, S.Y. Choi, Graphene-based plasmonic waveguides for photonic integrated circuits, Optics Express, 19 (2011) 24557–24562.

[11] S.E. Kocabas, G. Veronis, D.A.B. Miller, S.H. Fan, Transmission line and equivalent circuit models for plasmonic waveguide components, IEEE Journal of Selected Topics in Quantum Electronics, 14 (2008) 1462–1472.

[12] V.J. Sorger, R.F. Oulton, R.M. Ma, X. Zhang, Toward integrated plasmonic circuits, MRS Bulletin, 37 (2012) 728–738.

[13] H. Wei, H.X. Xu, Nanowire-based plasmonic waveguides and devices for integrated nanophotonic circuits, Nanophotonics, 1 (2012) 155–169.

[14] W.L. Barnes, A. Dereux, T.W. Ebbesen, Surface plasmon subwavelength optics, Nature, 424 (2003) 824–830.

[15] H.A. Atwater, The promise of plasmonics, Scientific American, 296 (2007) 56–63.

[16] K. Kneipp, Y. Wang, H. Kneipp, L.T. Perelman, I. Itzkan, R. Dasari, M.S. Feld, Single molecule detection using surface-enhanced Raman scattering (SERS), Physical Review Letters, 78 (1997) 1667–1670.

[17] S.M. Nie, S.R. Emery, Probing single molecules and single nanoparticles by surface-enhanced Raman scattering, Science, 275 (1997) 1102–1106.

[18] F. De Angelis, F. Gentile, F. Mecarini, G. Das, M. Moretti, P. Candeloro, M.L. Coluccio, G. Cojoc, A. Accardo, C. Liberale, R.P. Zaccaria, G. Perozziello, L. Tirinato, A. Toma, G. Cuda, R. Cingolani, E. Di Fabrizio, Breaking the diffusion limit with super-hydrophobic delivery of molecules to plasmonic nanofocusing SERS structures, Nature Photonics, 5 (2011) 683–688.

[19] S.J. Lee, Z.Q. Guan, H.X. Xu, M. Moskovits, Surface-enhanced Raman spectroscopy and nanogeometry: the plasmonic origin of SERS, Journal of Physical Chemistry C, 111 (2007) 17985–17988.

[20] W.H. Li, R. Zamani, P.R. Gil, B. Pelaz, M. Ibanez, D. Cadavid, A. Shavel, R.A. Alvarez-Puebla, W.J. Parak, J. Arbiol, A. Cabot, CuTe nanocrystals: shape and size control, plasmonic properties, and use as SERS probes and photothermal agents, Journal of the American Chemical Society, 135 (2013) 7098–7101.

[21] S.A. Maier, Plasmonic field enhancement and SERS in the effective mode volume picture, Optics Express, 14 (2006) 1957–1964.

[22] M.J. Mulvihill, X.Y. Ling, J. Henzie, P.D. Yang, Anisotropic etching of silver nanoparticles for plasmonic structures capable of single-particle SERS, Journal of the American Chemical Society, 132 (2010) 268–274.

[23] J.B. Song, J.J. Zhou, H.W. Duan, Self-assembled plasmonic vesicles of SERS-encoded amphiphilic gold nanoparticles for cancer cell targeting and traceable intracellular drug delivery, Journal of the American Chemical Society, 134 (2012) 13458–13469.

[24] R.W. Taylor, T.C. Lee, O.A. Scherman, R. Esteban, J. Aizpurua, F.M. Huang, J.J. Baumberg, S. Mahajan, Precise subnanometer plasmonic junctions for SERS within gold nanoparticle assemblies using cucurbit[n]uril "Glue," ACS Nano, 5 (2011) 3878–3887.

[25] J. Theiss, P. Pavaskar, P.M. Echternach, R.E. Muller, S.B. Cronin, Plasmonic nanoparticle arrays with nanometer separation for high-performance SERS substrates, Nano Letters, 10 (2010) 2749–2754.

[26] W. Xie, B. Walkenfort, S. Schlucker, Label-free SERS monitoring of chemical reactions catalyzed by small gold nanoparticles using 3d plasmonic superstructures, Journal of the American Chemical Society, 135 (2013) 1657–1660.

[27] J. Ye, F.F. Wen, H. Sobhani, J.B. Lassiter, P. Van Dorpe, P. Nordlander, N.J. Halas, Plasmonic nanoclusters: near field properties of the fano resonance interrogated with SERS, Nano Letters, 12 (2012) 1660–1667.

[28] D. Rodrigo, O. Limaj, D. Janner, D. Etezadi, F.J.G. de Abajo, V. Pruneri, H. Altug, Mid-infrared plasmonic biosensing with graphene, Science, 349 (2015) 165–168.

[29] J.N. Anker, W.P. Hall, O. Lyandres, N.C. Shah, J. Zhao, R.P. Van Duyne, Biosensing with plasmonic nanosensors, Nature Materials, 7 (2008) 442–453.

[30] A.E. Cetin, H. Altug, Fano resonant ring/disk plasmonic nanocavities on conducting substrates for advanced biosensing, ACS Nano, 6 (2012) 9989–9995.

[31] S.K. Dondapati, T.K. Sau, C. Hrelescu, T.A. Klar, F.D. Stefani, J. Feldmann, Label-free biosensing based on single gold nanostars as plasmonic transducers, ACS Nano, 4 (2010) 6318–6322.

[32] A.V. Kabashin, P. Evans, S. Pastkovsky, W. Hendren, G.A. Wurtz, R. Atkinson, R. Pollard, V.A. Podolskiy, A.V. Zayats, Plasmonic nanorod metamaterials for biosensing, Nature Materials, 8 (2009) 867–871.

[33] S.W. Lee, K.S. Lee, J. Ahn, J.J. Lee, M.G. Kim, Y.B. Shin, Highly sensitive biosensing using arrays of plasmonic au nanodisks realized by nanoimprint lithography, ACS Nano, 5 (2011) 897–904.

[34] J. McPhillips, A. Murphy, M.P. Jonsson, W.R. Hendren, R. Atkinson, F. Hook, A.V. Zayats, R.J. Pollard, High-performance biosensing using arrays of plasmonic nanotubes, ACS Nano, 4 (2010) 2210–2216.

[35] M.E. Stewart, N.H. Mack, V. Malyarchuk, J. Soares, T.W. Lee, S.K. Gray, R.G. Nuzzo, J.A. Rogers, Quantitative multispectral biosensing and 1D imaging using quasi-3D plasmonic crystals, Proceedings of the National Academy of Sciences of the United States of America, 103 (2006) 17143–17148.

[36] T. Vo-Dinh, H.N. Wang, J. Scaffidi, Plasmonic nanoprobes for SERS biosensing and bioimaging, Journal of Biophotonics, 3 (2010) 89–102.

[37] T.A. Larson, J. Bankson, J. Aaron, K. Sokolov, Hybrid plasmonic magnetic nanoparticles as molecular specific agents for MRI/optical imaging and photothermal therapy of cancer cells, Nanotechnology, 18 (2007) 325101.

[38] J.Z. Zhang, Biomedical applications of shape-controlled plasmonic nanostructures: a case study of hollow gold nanospheres for photothermal ablation therapy of cancer, Journal of Physical Chemistry Letters, 1 (2010) 686–695.

[39] B.V. Budaev, D.B. Bogy, On the lifetime of plasmonic transducers in heat assisted magnetic recording, Journal of Applied Physics, 112 (2012) 034512.

[40] B.C. Stipe, T.C. Strand, C.C. Poon, H. Balamane, T.D. Boone, J.A. Katine, J.L. Li, V. Rawat, H. Nemoto, A. Hirotsune, O. Hellwig, R. Ruiz, E. Dobisz, D.S. Kercher, N. Robertson, T.R. Albrecht, B.D. Terris, Magnetic recording at 1.5 Pb m(-2) using an integrated plasmonic antenna, Nature Photonics, 4 (2010) 484–488.

[41] N. Zhou, X.F. Xu, A.T. Hammack, B.C. Stipe, K.Z. Gao, W. Scholz, E.C. Gage, Plasmonic near-field transducer for heat-assisted magnetic recording, Nanophotonics, 3 (2014) 141–155.

[42] A. Yang, T.B. Hoang, M. Dridi, C. Deeb, M.H. Mikkelsen, G.C. Schatz, T.W. Odom, Real-time tunable lasing from plasmonic nanocavity arrays, Nature Communications, 6 (2015) 6939.

[43] M.T. Hill, M. Marell, E.S.P. Leong, B. Smalbrugge, Y.C. Zhu, M.H. Sun, P.J. van Veldhoven, E.J. Geluk, F. Karouta, Y.S. Oei, R. Notzel, C.Z. Ning, M.K. Smit, Lasing in metal-insulator-metal sub-wavelength plasmonic waveguides, Optics Express, 17 (2009) 11107–11112.

[44] Y. Hou, P. Renwick, B. Liu, J. Bai, T. Wang, Room temperature plasmonic lasing in a continuous wave operation mode from an InGaN/GaN single nanorod with a low threshold, Scientific Reports, 4 (2014) 5014.

[45] S.H. Kwon, J.H. Kang, C. Seassal, S.K. Kim, P. Regreny, Y.H. Lee, C.M. Lieber, H.G. Park, Subwavelength plasmonic lasing from a semiconductor nanodisk with silver nanopan cavity, Nano Letters, 10 (2010) 3679–3683.

[46] Y.J. Lu, C.Y. Wang, J. Kim, H.Y. Chen, M.Y. Lu, Y.C. Chen, W.H. Chang, L.J. Chen, M.I. Stockman, C.K. Shih, S. Gwo, All-color plasmonic nanolasers with ultralow thresholds: autotuning mechanism for single-mode lasing, Nano Letters, 14 (2014) 4381–4388.

[47] J. Pan, Z. Chen, J. Chen, P. Zhan, C.J. Tang, Z.L. Wang, Low-threshold plasmonic lasing based on high-Q dipole void mode in a metallic nanoshell, Optics Letters, 37 (2012) 1181–1183.

[48] V.V. Popov, O.V. Polischuk, A.R. Davoyan, V. Ryzhii, T. Otsuji, M.S. Shur, Plasmonic terahertz lasing in an array of graphene nanocavities, Physical Review B, 86 (2012) 195437.

[49] A.H. Schokker, A.F. Koenderink, Lasing at the band edges of plasmonic lattices, Physical Review B, 90 (2014) 155452.

[50] W. Zhou, M. Dridi, J.Y. Suh, C.H. Kim, D.T. Co, M.R. Wasielewski, G.C. Schatz, T.W. Odom, Lasing action in strongly coupled plasmonic nanocavity arrays, Nature Nanotechnology, 8 (2013) 506–511.

[51] J.A. Fan, C.H. Wu, K. Bao, J.M. Bao, R. Bardhan, N.J. Halas, V.N. Manoharan, P. Nordlander, G. Shvets, F. Capasso, Self-assembled plasmonic nanoparticle clusters, Science, 328 (2010) 1135–1138.

[52] J.B. Lassiter, H. Sobhani, J.A. Fan, J. Kundu, F. Capasso, P. Nordlander, N.J. Halas, Fano resonances in plasmonic nanoclusters: geometrical and chemical tunability, Nano Letters, 10 (2010) 3184–3189.

[53] B. Luk'yanchuk, N.I. Zheludev, S.A. Maier, N.J. Halas, P. Nordlander, H. Giessen, C.T. Chong, The Fano resonance in plasmonic nanostructures and metamaterials, Nature Materials, 9 (2010) 707–715.

[54] H. Wang, D.W. Brandl, F. Le, P. Nordlander, N.J. Halas, Nanorice: a hybrid plasmonic nanostructure, Nano Letters, 6 (2006) 827–832.

[55] H. Wang, D.W. Brandl, P. Nordlander, N.J. Halas, Plasmonic nanostructures: artificial molecules, Accounts of Chemical Research, 40 (2007) 53–62.

[56] Y.N. Xia, N.J. Halas, Shape-controlled synthesis and surface plasmonic properties of metallic nanostructures, MRS Bulletin, 30 (2005) 338–344.

[57] V. Amendola, O.M. Bakr, F. Stellacci, A study of the surface plasmon resonance of silver nanoparticles by the discrete dipole approximation method: effect of shape, size, structure, and assembly, Plasmonics, 5 (2010) 85–97.

[58] J. Bosbach, C. Hendrich, F. Stietz, T. Vartanyan, F. Trager, Ultrafast dephasing of surface plasmon excitation in silver nanoparticles: influence of particle size, shape, and chemical surrounding, Physical Review Letters, 89 (2002) 257404.

[59] K.S. Lee, M.A. El-Sayed, Gold and silver nanoparticles in sensing and imaging: sensitivity of plasmon response to size, shape, and metal composition, Journal of Physical Chemistry B, 110 (2006) 19220–19225.

[60] D.E. Mustafa, T.M. Yang, Z. Xuan, S.Z. Chen, H.Y. Tu, A.D. Zhang, Surface plasmon coupling effect of gold nanoparticles with different shape and size on conventional surface plasmon resonance signal, Plasmonics, 5 (2010) 221–231.

[61] C. Tabor, R. Murali, M. Mahmoud, M.A. El-Sayed, On the use of plasmonic nanoparticle pairs as a plasmon ruler: the dependence of the near-field dipole plasmon coupling on nanoparticle size and shape, Journal of Physical Chemistry A, 113 (2009) 1946–1953.

[62] J.H. Zhang, H.Y. Liu, Z.L. Wang, N.B. Ming, Shape-selective synthesis of gold nanoparticles with controlled sizes, shapes, and plasmon resonances, Advanced Functional Materials, 17 (2007) 3295–3303.

[63] P.J. Reece, Plasmonics—finer optical tweezers, Nature Photonics, 2 (2008) 333–334.

[64] B. Fluegel, A. Mascarenhas, D.W. Snoke, L.N. Pfeiffer, K. West, Plasmonic all-optical tunable wavelength shifter, Nature Photonics, 1 (2007) 701–703.

[65] A.N. Grigorenko, N.W. Roberts, M.R. Dickinson, Y. Zhang, Nanometric optical tweezers based on nanostructured substrates, Nature Photonics, 2 (2008) 365–370.

[66] M.L. Juan, M. Righini, R. Quidant, Plasmon nano-optical tweezers, Nature Photonics, 5 (2011) 349–356.

[67] R.D. Averitt, S.L. Westcott, N.J. Halas, Linear optical properties of gold nanoshells, Journal of the Optical Society of America. B-Optical Physics, 16 (1999) 1824–1832.

[68] E. Prodan, P. Nordlander, N.J. Halas, Electronic structure and optical properties of gold nanoshells, Nano Letters, 3 (2003) 1411–1415.

[69] J.R. Krenn, M. Salerno, N. Felidj, B. Lamprecht, G. Schider, A. Leitner, F.R. Aussenegg, J. C. Weeber, A. Dereux, J.P. Goudonnet, Light field propagation by metal micro- and nanostructures, Journal of Microscopy-Oxford, 202 (2001) 122–128.

[70] S.A. Maier, P.G. Kik, H.A. Atwater, S. Meltzer, E. Harel, B.E. Koel, A.A.G. Requicha, Local detection of electromagnetic energy transport below the diffraction limit in metal nanoparticle plasmon waveguides, Nature Materials, 2 (2003) 229–232.

[71] P. Nordlander, C. Oubre, E. Prodan, K. Li, M.I. Stockman, Plasmon hybridizaton in nanoparticle dimers, Nano Letters, 4 (2004) 899–903.

[72] D.E. Chang, A.S. Sorensen, P.R. Hemmer, M.D. Lukin, Quantum optics with surface plasmons, Physical Review Letters, 97 (2006) 4.

[73] D.E. Chang, A.S. Sorensen, E.A. Demler, M.D. Lukin, A single-photon transistor using nanoscale surface plasmons, Nature Physics, 3 (2007) 807–812.

[74] H.P. Paudel, M.N. Leuenberger, Light-controlled plasmon switching using hybrid metal-semiconductor nanostructures, Nano Letters, 12 (2012) 2690–2696.

[75] T.J. Echtermeyer, S. Milana, U. Sassi, A. Eiden, M. Wu, E. Lidorikis, A.C. Ferrari, Surface plasmon polariton graphene photodetectors, Nano Letters, 16 (2016) 8–20.

[76] G.X. Ni, L. Wang, M.D. Goldflam, M. Wagner, Z. Fei, A.S. McLeod, M.K. Liu, F. Keilmann, B. Ozyilmaz, A.H.C. Neto, J. Hone, M.M. Fogler, D.N. Basov, Ultrafast optical switching of infrared plasmon polaritons in high-mobility graphene, Nature Photonics, 10 (2016) 244.

[77] S. Chakraborty, O.P. Marshall, T.G. Folland, Y.J. Kim, A.N. Grigorenko, K.S. Novoselov, Gain modulation by graphene plasmons in aperiodic lattice lasers, Science, 351 (2016) 246–248.

[78] M.Z. Hasan, C.L. Kane, Colloquium: topological insulators, Reviews of Modern Physics, 82 (2010) 3045–3067.

[79] R. Kolesov, B. Grotz, G. Balasubramanian, R.J. Stohr, A.A.L. Nicolet, P.R. Hemmer, F. Jelezko, J. Wrachtrup, Wave-particle duality of single surface plasmon polaritons, Nature Physics, 5 (2009) 470–474.

[80] H.A. Atwater, A. Polman, Plasmonics for improved photovoltaic devices, Nature materials, 9 (2010) 205–213.

[81] J.A. Schuller, E.S. Barnard, W. Cai, Y.C. Jun, J.S. White, M.L. Brongersma, Plasmonics for extreme light concentration and manipulation, Nature materials, 9 (2010) 193–204.

[82] S.A. Maier, Plasmonic: Fundamentals and applications, Springer US, 2007.

[83] C. Sonnichsen, T. Franzl, T. Wilk, G. von Plessen, J. Feldmann, Plasmon resonances in large noble-metal clusters, New Journal of Physics, 4 (2002) 93.

[84] C. Sonnichsen, T. Franzl, T. Wilk, G. von Plessen, J. Feldmann, O. Wilson, P. Mulvaney, Drastic reduction of plasmon damping in gold nanorods, Physical Review Letters, 88 (2002) 077402.

[85] T.V. Shahbazyan, Landau damping of surface plasmons in metal nanostructures, Physical Review B, 94 (2016) 235431.

[86] S. Mukherjee, F. Libisch, N. Large, O. Neumann, L.V. Brown, J. Cheng, J.B. Lassiter, E.A. Carter, P. Nordlander, N.J. Halas, Hot electrons do the impossible: plasmon-induced dissociation of H2 on Au, Nano Letters, 13 (2013) 240–247.

[87] Y. Takahashi, T. Tatsuma, Solid state photovoltaic cells based on localized surface plasmon-induced charge separation, Applied Physics Letters, 99 (2011) 182110.

[88] F. Wang, N.A. Melosh, Plasmonic energy collection through hot carrier extraction, Nano Letters, 11 (2011) 5426–5430.

[89] N.V. Smith, Photoelectron energy spectra and the band structures of the noble metals, Physical Review B, 3 (1971) 1862–1878.

[90] V. Myroshnychenko, J. Rodriguez-Fernandez, I. Pastoriza-Santos, A.M. Funston, C. Novo, P. Mulvaney, L.M. Liz-Marzan, F.J.G. de Abajo, Modelling the optical response of gold nanoparticles, Chemical Society Reviews, 37 (2008) 1792–1805.

[91] H. Kuwata, H. Tamaru, K. Esumi, K. Miyano, Resonant light scattering from metal nanoparticles: practical analysis beyond Rayleigh approximation, Applied Physics Letters, 83 (2003) 4625.

[92] A.M. Brown, R. Sundararaman, P. Narang, W.A. Goddard, 3rd, H.A. Atwater, Nonradiative plasmon decay and hot carrier dynamics: effects of phonons, surfaces, and geometry, ACS Nano, 10 (2016) 957–966.

[93] Z. Fang, Y. Wang, Z. Liu, A. Schlather, P.M. Ajayan, F.H. Koppens, P. Nordlander, N.J. Halas, Plasmon-induced doping of graphene, ACS Nano, 6 (2012) 10222–10228.

[94] Y. Kang, S. Najmaei, Z. Liu, Y. Bao, Y. Wang, X. Zhu, N.J. Halas, P. Nordlander, P.M. Ajayan, J. Lou, Z. Fang, Plasmonic hot electron induced structural phase transition in a MoS2 monolayer, Advanced Materials, 26 (2014) 6467–6471.

[95] A. Manjavacas, J.G. Liu, V. Kulkarni, P. Nordlander, Plasmon-induced hot carriers in metallic nanoparticles, ACS Nano, 8 (2014) 7630–7638.

[96] A.O. Govorov, H. Zhang, Y.K. Gun'ko, Theory of photoinjection of hot plasmonic carriers from metal nanostructures into semiconductors and surface molecules, The Journal of Physical Chemistry C, 117 (2013) 16616–16631.

[97] R. Sundararaman, P. Narang, A.S. Jermyn, W.A. Goddard, 3rd, H.A. Atwater, Theoretical predictions for hot-carrier generation from surface plasmon decay, Nature Communications, 5 (2014) 5788.

[98] E.H. Hwang, R. Sensarma, S. Das Sarma, Plasmon-phonon coupling in graphene, Physical Review B, 82 (2010) 195406.

[99] H.G. Yan, T. Low, W.J. Zhu, Y.Q. Wu, M. Freitag, X.S. Li, F. Guinea, P. Avouris, F.N. Xia, Damping pathways of mid-infrared plasmons in graphene nanostructures, Nature Photonics, 7 (2013) 394–399.

[100] P. Pines, P. Nozieres, The Theory of Quantum Liquids, Benjamin, New York, W.A. Benjamin,1966.

[101] D. Pines, D. Bohm, A collective description of electron interactions: II. Collective **vs** individual particle aspects of the interactions, Physical Review, 85 (1952) 338–353.

[102] D. Bohm, D. Pines, A collective description of electron interactions: iii. coulomb interactions in a degenerate electron gas, Physical Review, 92 (1953) 609–625.

[103] R.H. Ritchie, H.B. Eldridge, Optical Emission from Irradiated Foils. I, Physical Review 126 (1962) 1935.

[104] R.H. Ritchie, Plasma Losses by Fast Electrons in Thin Films, Physical Review, 106 (1957) 874.

[105] A.D. Rakic, A.B. Djurisic, J.M. Elazar, M.L. Majewski, Optical properties of metallic films for vertical-cavity optoelectronic devices, Applied Optics, 37 (1998) 5271–5283.

[106] P.G. Etchegoin, E.C. Le Ru, M. Meyer, An analytic model for the optical properties of gold, The Journal of Chemical Physics, 125 (2006) 164705.

[107] E. Kioupakis, P. Rinke, A. Schleife, F. Bechstedt, C.G. Van de Walle, Free-carrier absorption in nitrides from first principles, Physical Review B, 81 (2010) 241201(R).

[108] J. Noffsinger, E. Kioupakis, C.G. Van de Walle, S.G. Louie, M.L. Cohen, Phonon-assisted optical absorption in silicon from first principles, Physical Review Letters, 108 (2012) 167402.

[109] H. Zhang, A.O. Govorov, Optical generation of hot plasmonic carriers in metal nanocrystals: the effects of shape and field enhancement, The Journal of Physical Chemistry C, 118 (2014) 7606–7614.

[110] J.M. Elson, R.H. Ritchie, Photon interactions at a rough metal surface, Physical Review B, 4 (1971) 4129–4138.

[111] J.B. Khurgin, How to deal with the loss in plasmonics and metamaterials, Nature Nanotechnology, 10 (2015) 2–6.

[112] F. Stern, Physical Review Letters, 18 (1967) 546.

[113] F. Stern, Polarizability of a Two-Dimensional Electron Gas, Physical Review Letters, 18 (1967) 546.

[114] R.J. Wilkinson, C.D. Ager, T. Duffield, H.P. Hughes, D.G. Hasko, H. Ahmed, J.E.F. Frost, D.C. Peacock, D.A. Ritchie, G.A.C. Jones, C.R. Whitehouse, N. Apsley, Plasmon excitation and self-coupling in a bi-periodically modulated 2-dimensional electron-gas, Journal of Applied Physics, 71 (1992) 6049–6061.

[115] D.R.H. Craih, F. Bohren, Absorption and Scattering of Light by Small Particles, John Wiley & Sons, Inc. , 1998.

[116] E. Prodan, C. Radloff, N.J. Halas, P. Nordlander, A hybridization model for the plasmon response of complex nanostructures, Science, 302 (2003) 419–422.

[117] F. Bonaccorso, Z. Sun, T. Hasan, A.C. Ferrari, Graphene photonics and optoelectronics, Nature Photonics, 4 (2010) 611–622.

[118] L.A. Falkovsky, Optical properties of doped graphene layers, Journal of Experimental and Theoretical Physics, 106 (2008) 575–580.

[119] L.A. Falkovsky, S.S. Pershoguba, Optical far-infrared properties of a graphene monolayer and multilayer, Physical Review B, 76 (2007) 153410.

[120] L.A. Falkovsky, A.A. Varlamov, Space-time dispersion of graphene conductivity, The European Physical Journal B, 56 (2007) 281–284.

[121] V. Singh, D. Joung, L. Zhai, S. Das, S.I. Khondaker, S. Seal, Graphene based materials: past, present and future, Progress in Materials Science, 56 (2011) 1178–1271.

[122] M. Jablan, H. Buljan, M. Soljacic, Plasmonics in graphene at infrared frequencies, Physical Review B, 80 (2009) 245435.

[123] F.H. Koppens, D.E. Chang, F.J. Garcia de Abajo, Graphene plasmonics: a platform for strong light-matter interactions, Nano Letters, 11 (2011) 3370–3377.

[124] Z. Fei, A.S. Rodin, G.O. Andreev, W. Bao, A.S. McLeod, M. Wagner, L.M. Zhang, Z. Zhao, M. Thiemens, G. Dominguez, M.M. Fogler, A.H. Castro Neto, C.N. Lau, F. Keilmann, D.N. Basov, Gate-tuning of graphene plasmons revealed by infrared nano-imaging, Nature, 487 (2012) 82–85.

[125] A.Y. Nikitin, F. Guinea, L. Martin-Moreno, Resonant plasmonic effects in periodic graphene antidot arrays, Applied Physics Letters, 101 (2012) 151119.

[126] S. Thongrattanasiri, F.J. Garcia de Abajo, Optical field enhancement by strong plasmon interaction in graphene nanostructures, Physcial Review Letters, 110 (2013) 187401.

[127] V.W. Brar, M.S. Jang, M. Sherrott, J.J. Lopez, H.A. Atwater, Highly confined tunable mid-infrared plasmonics in graphene nanoresonators, Nano Letters, 13 (2013) 2541–2547.

[128] W. Gao, J. Shu, C. Qiu, Q. Xu, Excitation of plasmonic waves in graphene by guided-mode resonances, ACS Nano, 6 (2012) 7806–7813.

[129] A.Y. Nikitin, F. Guinea, F.J. Garcia-Vidal, L. Martin-Moreno, Surface plasmon enhanced absorption and suppressed transmission in periodic arrays of graphene ribbons, Physical Review B, 85 (2012) 081405.

[130] S. Thongrattanasiri, F.H. Koppens, F.J. Garcia de Abajo, Complete optical absorption in periodically patterned graphene, Physical Review Letters, 108 (2012) 047401.

[131] L. Ju, B. Geng, J. Horng, C. Girit, M. Martin, Z. Hao, H.A. Bechtel, X. Liang, A. Zettl, Y.R. Shen, F. Wang, Graphene plasmonics for tunable terahertz metamaterials, Nature Nanotechnology, 6 (2011) 630–634.

[132] Z. Fang, Y. Wang, A.E. Schlather, Z. Liu, P.M. Ajayan, F.J. de Abajo, P. Nordlander, X. Zhu, N.J. Halas, Active tunable absorption enhancement with graphene nanodisk arrays, Nano Letters, 14 (2014) 299–304.

[133] Z. Fang, S. Thongrattanasiri, A. Schlather, Z. Liu, L. Ma, Y. Wang, P.M. Ajayan, P. Nordlander, N.J. Halas, F.J. Garcia de Abajo, Gated tunability and hybridization of localized plasmons in nanostructured graphene, ACS Nano, 7 (2013) 2388–2395.

[134] A. Safaei, S. Chandra, A. Vázquez-Guardado, J. Calderon, D. Franklin, L. Tetard, L. Zhai, M. N. Leuenberger, D. Chanda, Dynamically tunable extraordinary light absorption in monolayer graphene, Under Review.

[135] N.D.M. Neil W. Ashcroft, Solid state physics, Cengage Learning, 1976.

[136] J.C. Song, L.S. Levitov, Energy flows in graphene: hot carrier dynamics and cooling, Journal of physics. Condensed Matter: An Institute of Physics Journal, 27 (2015) 164201.

[137] M.L. Brongersma, N.J. Halas, P. Nordlander, Plasmon-induced hot carrier science and technology, Nature Nanotechnology, 10 (2015) 25–34.

[138] G.W. Hanson, Quasi-transverse electromagnetic modes supported by a graphene parallel-plate waveguide, Journal of Applied Physics, 104 (2008) 084314.

[139] N.D. Mermin, Lindhard dielectric function in the relaxation-time approximation, Physical Review B, 1 (1970) 2362–2363.

[140] B. Wunsch, T. Stauber, F. Sols, F. Guinea, Dynamical polarization of graphene at finite doping, New Journal of Physics, 8 (2006) 318.

[141] L.A. Ponomarenko, R. Yang, T.M. Mohiuddin, M.I. Katsnelson, K.S. Novoselov, S.V. Morozov, A.A. Zhukov, F. Schedin, E.W. Hill, A.K. Geim, Effect of a high-kappa environment on charge carrier mobility in graphene, Physical Review Letters, 102 (2009) 206603.

[142] V.A. Hari P. Paudel, Mark I. Stockman, Three-dimensional topological insulator based nanospaser.pdf, Physical Review B, 93 (2016) 155105.

[143] S. Raghu, S.B. Chung, X.-L. Qi, S.-C. Zhang, Collective Modes of a Helical Liquid, Physical Review Letters, 104 (2010) 116401.

[144] R.E.V. Profumo, R. Asgari, M. Polini, A.H. MacDonald, Double-layer graphene and topological insulator thin-film plasmons, Physical Review B, 85 (2012) 085443.

[145] P. Di Pietro, M. Ortolani, O. Limaj, A. Di Gaspare, V. Giliberti, F. Giorgianni, M. Brahlek, N. Bansal, N. Koirala, S. Oh, P. Calvani, S. Lupi, Observation of dirac plasmons in a topological insulator, Nature Nanotechnology, 8 (2013) 556–560.

[146] N. Bansal, Y.S. Kim, M. Brahlek, E. Edrey, S. Oh, Thickness-Independent Transport Channels in Topological Insulator Bi_2Se_3Thin Films, Physical Review Letters, 109 (2012) 116804.

[147] T. Stauber, G. Gómez-Santos, L. Brey, Spin-charge separation of plasmonic excitations in thin topological insulators, Physical Review B, 88 (2013) 205427.

[148] M. Autore, F. D'Apuzzo, A. Di Gaspare, V. Giliberti, O. Limaj, P. Roy, M. Brahlek, N. Koirala, S. Oh, F.J.G. de Abajo, S. Lupi, Plasmon-phonon interactions in topological insulator microrings, Advanced Optical Materials, 3 (2015) 1257–1263.

[149] J. Hofmann, S. Das Sarma, Surface plasmon polaritons in topological Weyl semimetals, Physical Review B, 93 (2016) 5.

[150] S.Y. Xu, I. Belopolski, N. Alidoust, M. Neupane, G. Bian, C.L. Zhang, R. Sankar, G.Q. Chang, Z.J. Yuan, C.C. Lee, S.M. Huang, H. Zheng, J. Ma, D.S. Sanchez, B.K. Wang, A. Bansil, F.C. Chou, P.P. Shibayev, H. Lin, S. Jia, M.Z. Hasan, Discovery of a weyl fermion semimetal and topological fermi arcs, Science, 349 (2015) 613–617.

[151] B.Q. Lv, H.M. Weng, B.B. Fu, X.P. Wang, H. Miao, J. Ma, P. Richard, X.C. Huang, L.X. Zhao, G.F. Chen, Z. Fang, X. Dai, T. Qian, H. Ding, Experimental Discovery of Weyl Semimetal TaAs, Phys. Rev. X, 5 (2015) 031013.

[152] A.B. Sushkov, J.B. Hofmann, G.S. Jenkins, J. Ishikawa, S. Nakatsuji, S. Das Sarma, H.D. Drew, Optical evidence for a Weyl semimetal state in pyrochlore $Eu_2Ir_2O_7$, Physical Review B, 92 (2015) 241108(R).

[153] D. Bergman, M. Stockman, Surface Plasmon Amplification by Stimulated Emission of Radiation: Quantum Generation of Coherent Surface Plasmons in Nanosystems, Physical Review Letters, 90 (2003) 027402.

[154] V. Apalkov, M.I. Stockman, Proposed graphene nanospaser, Light: Science & Applications, 3 (2014) e191.

[155] H.P. Paudel, V. Apalkov, M.I. Stockman, Three-dimensional topological insulator based nanospaser, Physical Review B, 93 (2016) 155105.

Chapter 4

Ultrafast Nonlinear Optical Effects of Metal Nanoparticles Composites

Kwang-Hyon Kim, Anton Husakou and
Joachim Herrmann

Additional information is available at the end of the chapter

http://dx.doi.org/10.5772/67412

Abstract

We present a theoretical method for the calculation of the transient nonlinearity in dielectric composites doped with metal nanoparticles and demonstrate some applications of this approach. First, we describe the theoretical basis of the linear and nonlinear properties of metal nanoparticles by using the time-domain discrete-dipole approximation. By using the two-temperature model for the description of the electron-electron and electron-lattice interaction, we derive an equation for the transient third-order nonlinear susceptibility. Based on this method and the effective medium approximation, we present numerical results for the nonlinear optical susceptibility for different nanocomposites media consisting of noble metal nanoparticles surrounded by a dielectric host. With increasing pump intensities, the plasmon resonance is shifted which leads to a saturation of the absorption. We present a theory of mode-locking of solid-state and semiconductor disk lasers using metal nanocomposites as saturable absorbers. Finally, we consider a novel slow-light device based on metal nanocomposites.

Keywords: metal nanoparticles, nanocomposite materials, nonlinear optical response, third-order nonlinear susceptibility, time-domain discrete-dipole approximation, effective medium approximation, saturation of absorption, mode-locking of lasers, slow light

1. Introduction

Metal nanoparticles (NPs) play a key role in the emerging field of nanooptics and plasmonics. The interaction between light and metal NPs is determined by charge-density oscillations on the surface of the particles, or so-called localized plasmon resonances. These elementary electronic excitations have been extensively studied, both in fundamental direction and with a view to applications (for reviews see e.g., Refs. [1–3]). Examples of the diverse fields of applications include localized plasmons, which allow greatly increased signal strengths in Raman spectroscopy and surface spectroscopy, significant enhancement of the emission rate of fluorescent molecules or quantum dots, increased efficiency and brightness of light-emitting devices (LEDs), as well as novel forms of fluorescence microscopy. The plasmonic properties of nanoparticles exhibit a highly sensitive dependence on its shape, and the plasmonic resonance is shifted in nonspherical particles, especially rods, spheroids, triangular prisms or bow-tie antennas to longer wavelengths [4].

The huge nonlinearity of metallic nanostructures after intense excitation with fs laser pulses is related to the response of intraband and interband nonlinear electronic processes to the electromagnetic field. Early theoretical studies of the Kerr-type nonlinear susceptibilities in composite media are reported in Refs. [5–7]. A review of main results can be found in Ref. [8]. More recently using the discrete-dipole approximation, linear and nonlinear optical characteristics of metal composites with nanoparticles with different sizes and shapes have been studied in Refs. [9–11], and the saturation of absorption in such materials in Ref. [12].

In Section 2 of this chapter, we present a method for the calculation of the linear and nonlinear properties of metal nanoparticles by using the time-domain discrete-dipole approximation and derive an equation for the transient nonlinear susceptibility based on the two-temperature model for the description of the electron-electron and electron-lattice interaction. In Section 3, we present numerical results for the ultrafast third-order nonlinear optical susceptibility for different nanocomposites media consisting of a noble metal nanoparticles surrounded by a dielectric host. Based on the predicted saturable absorption in Section 4, we describe ultrashort pulse generation by mode-locking of short-wavelength solid-state and semiconductor disk lasers using metal nanocomposites. In Section 5, we describe a mechanism for the realization of slow light phenomenon by using glasses doped with metal nanoparticles in a pump-probe regime near the plasmonic resonance due to the sensitive frequency dependence of the refractive index related with the saturable absorption [13].

2. Theoretical fundamentals

In this section, we introduce the basic theoretical model for ultrafast nonlinear optical responses of dielectric composites containing metal nanoparticles based on the time-domain discrete-dipole approximation and the two-temperature model.

2.1. Time-domain discrete-dipole approximation for the calculation of the ultrafast nonlinear optical responses of composites containing metal nanoparticles with different sizes and shapes

To evaluate the change of the dielectric function of metals, we should obtain the distribution of the amplitude of the enhanced field $A_{enh}(t)$ in metal NPs. The popular numerical method for calculating the transient field distribution in NPs is the finite-difference time-domain (FDTD) method. However, for the evaluation of the ultrafast nonlinear change of the dielectric function of a metal, we should know the amplitude $A_{enh}(t)$ which requires very significant

http://dx.doi.org/10.5772/67412

computational effort with FDTD. For the evaluation of $A_{\mathrm{enh}}(t)$, we use the time-domain discrete-dipole approximation (TDDDA) [10, 11]. Below, we briefly describe the basics of this numerical approach.

TDDDA is a numerical tool for evaluating the distribution of amplitude of real or enhanced field in metal NPs with arbitrary sizes and shapes. It is a time-domain version of conventional frequency-domain discrete-dipole approximation (DDA). TDDDA describes NPs as the collection of small dipoles. The amplitude $\mathbf{A}_m$ of the enhanced field at the m-th dipole is given by the following integral-differential equation:

$$\mathbf{A}_m = \left(1 - \frac{\varepsilon_{\mathrm{h}}^{1/2}}{c_0} r_{m\perp}\frac{\partial}{\partial t} + \frac{\varepsilon_{\mathrm{h}}}{2c_0^2} r_{m\perp}^2 \frac{\partial^2}{\partial t^2}\right)\mathbf{A}_m^{\mathrm{in}} e^{ik_0 r_{m\perp}} + \left\{-\frac{4\pi}{3} + \sum_{n\neq m}\left[G_{mn}^{(0)} + \mathrm{i}G_{mn}^{(1)}\frac{\partial}{\partial t} - G_{mn}^{(2)}\frac{\partial^2}{\partial t^2}\right]v\right\}\mathbf{Q}_n, \tag{1}$$

where ε_h is the permittivity of the host medium, c_0 is the light velocity in vacuum, $\mathbf{A}_m^{\mathrm{in}}$ and $\mathbf{Q}_m$ are the amplitudes of the incident field and polarization at the place of the m-th dipole, $r_{m\perp} = (\mathbf{k}\cdot\mathbf{r}_m)/k$ is the projection of $\mathbf{r}_m$ onto the direction of wavevector $\mathbf{k}$ of incident pulse, with $\mathbf{r}_m$ being the position vector at m-th dipole. In Eq. (1), $\mathbf{G}_{mn}^{(0)}$, $\mathbf{G}_{mn}^{(1)}$, and $\mathbf{G}_{mn}^{(2)}$ are tensors given by

$$\mathbf{G}_{mn}^{(0)} = \frac{\exp(ik_0R_{mn})}{R_{mn}}\left[k_0^2(\mathbf{1} - \hat{r}_{mn}\otimes\hat{r}_{mn}) + \frac{ik_0R_{mn} - 1}{R_{mn}^2}(\mathbf{1} - 3\hat{r}_{mn}\otimes\hat{r}_{mn})\right],$$

$$\mathbf{G}_{mn}^{(1)} = \frac{\varepsilon_{\mathrm{h}}^{1/2}}{c_0}\frac{\exp(ik_0R_{mn})}{R_{mn}}[k_0(2 + \mathrm{i}k_0R_{mn})(\mathbf{1} - \hat{r}_{mn}\otimes\hat{r}_{mn}) - k_0(\mathbf{1} - 3\hat{r}_{mn}\otimes\hat{r}_{mn})],$$

$$\mathbf{G}_{mn}^{(2)} = \frac{\varepsilon_{\mathrm{h}}}{2c_0^2}\frac{\exp(ik_0R_{mn})}{R_{mn}}[(-k_0^2R_{mn}^2 + 4\mathrm{i}k_0R_{mn} + 2)(\mathbf{1} - \hat{r}_{mn}\otimes\hat{r}_{mn}) - (\mathrm{i}k_0R_{mn} + 1)(\mathrm{i}k_0R_{mn} + 1)], \tag{2}$$

where $k_0 = \omega_0/c_0$ is the wave number at the central angular frequency ω_0 of incident light pulse, $R_{mn} = |\mathbf{r}_m - \mathbf{r}_n|$, $\hat{r}_{mn} = (\mathbf{r}_m - \mathbf{r}_n)/R_{mn}$. The amplitude of the polarization $\mathbf{Q}_m(t)$ in Eq. (1) is determined by $\mathbf{Q}_m = \beta\mathbf{A}_m + \mathbf{Q}_{m1}$, where $\beta = [(\varepsilon_\infty/\varepsilon_{\mathrm{h}}) - 1]/(4\pi)$, ε_∞ is the dielectric function for infinite frequency. In this formalism, we use the extended Drude model for the linear dielectric function of a metal given by $\varepsilon_{m0} = \varepsilon_\infty - \omega_{\mathrm{p}}^2/(\omega^2 + i\gamma\omega)$, where ω_{p} is the plasma frequency, and γ is the electron collision frequency. In the equation for $\mathbf{Q}_m$, $\mathbf{Q}_{m1}$ is determined by

$$\ddot{\mathbf{Q}}_{m1} + c\dot{\mathbf{Q}}_{m1} + b\mathbf{Q}_{m1} = a\mathbf{A}_m, \tag{3}$$

where $a = \omega_{\mathrm{p}}^2/(4\pi\varepsilon_{\mathrm{h}})$, $b = -(\omega_0^2 + \mathrm{i}\gamma\omega_0)$, and $c = \gamma - 2\mathrm{i}\omega_0$. The basic principle of TDDDA and the detailed derivations of the above equations can be found in Ref. [10].

2.2. Transient nonlinear Kerr-type optical nonlinearity

The transient nonlinear optical response of metal NPs is related with the electron and phonon temperatures. Upon illuminating metal nanoparticles, femtosecond laser pulses excite the nonequilibrium electron distribution and thermalization process follows through the interactions between electrons. The time scale of electron-electron interactions τ_{ee} is about a few hundreds femtoseconds. The electron system transfers its energy to the lattices through electron-phonon coupling, the time scale τ_{ep} of which is a few picoseconds. The thermalized part of electrons, in part, is cooled through heat exchange with lattices in the NPs. The above outlined processes can be described by the following phenomenological rate equations (see e.g., Refs. [14, 15]):

$$\begin{aligned} \frac{\partial N}{\partial t} &= -\frac{N}{\tau_{ee}} - \frac{N}{\tau_{ep}} + \delta P(t), \\ C_{e}\frac{\partial T_{e}}{\partial t} &= -G(T_{e} - T_{L}) + \frac{N}{\tau_{ee}}, \\ C_{L}\frac{\partial T_{L}}{\partial t} &= G(T_{e} - T_{L}) + \frac{N}{\tau_{ep}}, \end{aligned} \tag{4}$$

where N is the energy density stored in the nonthermalized part of the distribution; $P(t)$ is the absorbed laser power per unit volume; δ is a constant representing the contribution of light field to the generation of excited nonequilibrium electrons; C_e and C_L are the heat capacities, C_e and C_L are the temperatures of electrons and lattices, respectively; G is the cooling time constant.

The above rate equations describe only the temperatures of thermalized electrons and lattices. The relations between the optical responses of metal nanoparticles and these temperatures can be found by the following considerations. The change of transient transmittance ΔT, an experimentally observable nonlinear quantity, is a linear function of the dielectric function of NPs [14]:

$$\frac{\Delta T}{T} = \frac{\partial lnT}{\partial \varepsilon_1}\Delta\varepsilon_1 + \frac{\partial lnT}{\partial \varepsilon_2}\Delta\varepsilon_2, \tag{5}$$

where T is the transmittance of the composite, ε_1 and ε_2 are real and imaginary parts of the dielectric functions of metal, and Δ represents the change of the corresponding quantities. The change of dielectric functions depends on redistribution of the electron energy in the vicinity of the Fermi level. Such a change can be accurately described by a linear function of the electronic and lattice temperatures [14] in the case of weak pump energy and the above equation can be represented as

$$\frac{\Delta T}{T} = aT_{e}(t) + bT_{L}(t), \tag{6}$$

where a and b are constants.

From Eqs. (5) and (6), we can conclude that, in the considered case, the change of dielectric function of metal is proportional to the change of electronic and lattice temperatures. The empirical results show, however, that the influence of lattice temperature on the dielectric function change is negligible. From these facts, the change of the dielectric function of a metal $\Delta\varepsilon_m$ can be written [16]:

$$\Delta\varepsilon_{m}(t) = \eta\Delta T_{e}, \tag{7}$$

where η is a constant.

The disadvantage of this model is that it cannot directly account for the nonlinear optical parameters such as nonlinear refractive index and absorption. To solve this problem, below we relate the unknown parameters in Eqs. (4) and (7) with the nonlinear optical susceptibility of the metal which can be experimentally determined.

From the second and the third equations in Eq. (4), we obtain

$$\frac{\partial(T_e - T_L)}{\partial t} = -G\left(\frac{1}{C_e} + \frac{1}{C_L}\right)(T_e - T_L) + \left(\frac{1}{C_e\tau_{ee}} - \frac{1}{C_L\tau_{ep}}\right)N. \tag{8}$$

Introducing the cooling time $\tau_{ep} = C_eC_L/[G(C_e + C_L)]$ and taking into account that the increase of lattice temperature can be neglected compared to that of electrons and $T_e - T_L \approx T_e - T_0$, we obtain

$$\frac{\partial(T_e - T_0)}{\partial t} = -\frac{T_e - T_0}{\tau_{ep}} + \left(\frac{1}{C_e\tau_{ee}} - \frac{1}{C_L\tau_{ep}}\right)N, \tag{9}$$

where T_0 is the initial temperatures of electrons and lattices. On the other hand, integrating the first equation in Eq. (7) we have

$$N = \delta\int_{-\infty}^{t} P(t')exp\left(-\frac{t-t'}{\tau'_{ee}}\right)dt', \tag{10}$$

where $\tau'^{-1}_{ee} = \tau^{-1}_{ee} + \tau^{-1}_{ep}$. Combining Eqs. (9) and (10), we obtain

$$\frac{\partial(T_e - T_0)}{\partial t} = -\frac{T_e - T_0}{\tau_{ep}} + \delta\left(\frac{1}{C_e\tau_{ee}} - \frac{1}{C_L\tau_{ep}}\right)\int_{-\infty}^{t} P(t')exp\left(-\frac{t-t'}{\tau'_{ee}}\right)dt'. \tag{11}$$

From Eq. (7),

$$\frac{\partial\Delta\varepsilon_m}{\partial t} = \eta\frac{\partial(T_e - T_0)}{\partial t}. \tag{12}$$

By using Eqs. (11) and (12), we can write

$$\frac{\partial\Delta\varepsilon_m}{\partial t} = -\frac{\Delta\varepsilon_m}{\tau_{ep}} + \delta'\left(\frac{1}{C_e\tau_{ee}} - \frac{1}{C_L\tau_{ep}}\right)\int_{-\infty}^{t} P(t')exp\left(-\frac{t-t'}{\tau'_{ee}}\right)dt', \tag{13}$$

where ε_m is the dielectric function of metal as a function of time and incident intensity, and $\delta' = \eta\delta$. The absorbed energy density $P(t)$ must be proportional to the enhanced intensity in the metal NPs and therefore we find

$$\frac{\partial\Delta\varepsilon_m}{\partial t} = -\frac{\Delta\varepsilon_m}{\tau_{ep}} + \frac{\beta}{\tau_{ee}}\int_{-\infty}^{t} |A_{enh}(t')|^2 exp\left(-\frac{t-t'}{\tau'_{ee}}\right)dt', \tag{14}$$

where β is a constant proportional to $\delta'[C_e^{-1} - C_L^{-1}(\tau_{ee}/\tau_{ep})]$ and $A_{enh}(t)$ is the amplitude of enhanced field. To determine the constant β, we consider the steady-state nonlinear solution of

Eq. (14). For a steady-state pumping, Eq. (14) gives $\Delta\varepsilon_{\mathrm{m,\,static}} = \beta\tau_{\mathrm{ep}}|A_{\mathrm{enh}}|^2$ and we have $\beta = \chi_{\mathrm{m}}^{(3)}/\tau_{\mathrm{ep}}$, where $\chi_{\mathrm{m}}^{(3)}$ is the inherent degenerate third-order susceptibility of the metal. Consequently, we obtain [17]

$$\frac{\partial\Delta\varepsilon_{\mathrm{m}}}{\partial t} = -\frac{\Delta\varepsilon_{\mathrm{m}}}{\tau_{\mathrm{ep}}} + \frac{\chi_{\mathrm{m}}^{(3)}}{\tau_{\mathrm{ee}}\tau_{\mathrm{ep}}}\int_{-\infty}^{t} |A_{\mathrm{enh}}(t')|^2 exp\left(-\frac{t-t'}{\tau_{\mathrm{ee}}'}\right)dt', \tag{15}$$

Since the pure e-e scattering time τ_{ee} (typically around 100 fs) is much shorter than τ_{ep} (typically around or longer than 1 ps), in Eq. (15), τ_{ee}'can be replaced with τ_{ee}. The resultant dielectric function is determined by $\varepsilon_{\mathrm{m}}= \varepsilon_{\mathrm{m0}} + \Delta\ \varepsilon_{\mathrm{m}}(t)$, where $\varepsilon_{\mathrm{m0}}$ is the linear dielectric function of metal.

Equation (15) is the major governing equation for the ultrafast nonlinear optical responses of metal nanoparticles.

3. Numerical results for the ultrafast nonlinear response of metal NPs composites

3.1. The effective medium approximation

The transient distribution of the amplitude $\mathbf{A}_m$ obtained by using Eq. (1) is used for the calculations of the distribution of the nonlinear dielectric function $\varepsilon_n^{\mathrm{NL}}$ in metal nanoparticles, on each time step by using Eq. (15). The effective nonlinear dielectric function of the metal nanocomposite is calculated by using the effective medium approximation (EMA) $\varepsilon_{\mathrm{eff}}^{\mathrm{NL}} = \langle\varepsilon A\rangle/\langle A\rangle$ [9], where the average is done over the distribution of dipoles and A is, the amplitude of the enhanced field. In the calculations based on EMA, the dielectric function ε and the field amplitude A in the nanoparticle region are $\varepsilon_{\mathrm{m},n}^{\mathrm{NL}}$ (subscription m represents metal and n denotes the value at n-th dipole) and the enhanced field A_m calculated with the TDDDA. In the spatial areas outside of the nanoparticles, we take the permittivity of the host ε_{h} as ε and the amplitude of incident field as A. The volume of host medium is calculated using the filling factor f, which is the volume ratio of metal NPs to the volume of the host medium. After obtaining the effective nonlinear dielectric function of metal nanocomposite $\varepsilon_{\mathrm{eff}}^{\mathrm{NL}}$, one can calculate the nonlinear optical characteristics such as the nonlinear effective index and absorption coefficient. The effective refractive index of the composites is determined by $n_{\mathrm{eff}}^{\mathrm{NL}} = \mathrm{Re}\sqrt{\varepsilon_{\mathrm{eff}}^{\mathrm{NL}}}$. The nonlinear effective absorption coefficient of the composite is given by $\alpha_{\mathrm{eff}}^{\mathrm{NL}} = 2(c_0/\omega_0)\mathrm{Im}\sqrt{\varepsilon_{\mathrm{eff}}^{\mathrm{NL}}}$ and the transmittance by $T^{\mathrm{NL}} = \exp(-\alpha_{\mathrm{eff}}^{\mathrm{NL}} l)$, where l is the thickness of the composite film. Then, the relative change of transmittance $\Delta T/T$ is calculated by $(T^{\mathrm{NL}}-T^{\mathrm{L}})/T^{\mathrm{L}}$, where T^{L} is the linear transmittance.

For the special case of very small and spherical NPs, the amplitude of the enhanced field can be approximately calculated by $A_{\mathrm{enh}}(t) = x(t)A_0(t)$, where $x(t) = \varepsilon_{\mathrm{h}}/[\varepsilon_{\mathrm{m}}(t) + 2\varepsilon_{\mathrm{h}}]$ is the transient field enhancement factor and A_0 is the amplitude of the incident pump pulse. The nonlinearly modified $\varepsilon_m(t)$ is self-consistently calculated in combination with the above-mentioned enhanced amplitude $A_{enh}(t)$. The resultant effective dielectric function of the composite ε_{eff}^{NL} is calculated by using Maxwell-Garnett model.

3.2. Numerical results and discussions

In this section, we present the numerical results for the ultrafast nonlinear optical responses of metal NPs composites.

Figure 1 shows an example of the time-dependent nonlinear transmission of a metal nanocomposite. The figure shows $\Delta T/T$ of a 0.2-μm-thick Al_2O_3 film doped with Ag NPs. We consider NPs with diameter of 6.5 nm, and a filling factor of 1%. The value of pump fluence is assumed to be 100 μJ/cm^2. For the optimal fit with the experimental data [18], we use the e-e and cooling times of τ_{ee} = 100 fs and τ_{ep} = 0.70 ps, correspondingly. Though the pump pulse is in the sub-100 fs range, the transmittance responds slower, which is explained by the delayed thermalization of electrons. In both cases, the time necessary to reach a maximum change in the transmittance is roughly 200–250 fs longer than the e-e scattering time τ_{ee}. The exponential decay of the transmittance change depends on the cooling time τ_{ep} and also on the pump fluence. Here we assume for the two response times τ_{ee} = 100 fs and τ_{ep} = 1 ps.

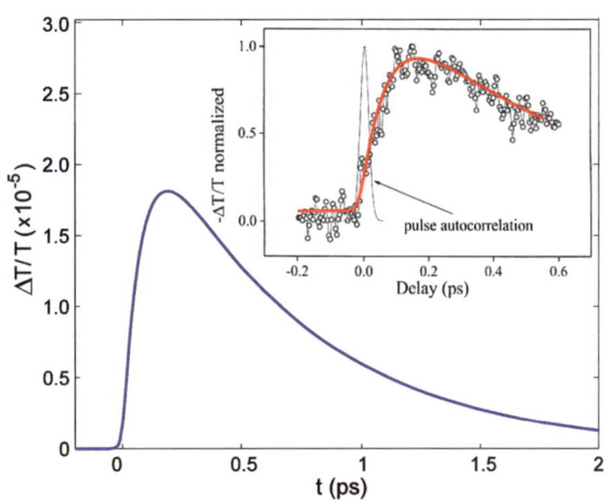

Figure 1. The transient nonlinear absorption of metal nanocomposites in an alumina film containing Ag NPs with filling factor of 1% and with particles diameter of 6.5 nm. The pump and probe have the central wavelength of 800 nm and 30 fs. We consider 0.25-μm-thick composite film. In the inset, the experimental (points, from Ref. [18]) and the theoretical (solid curve) results are compared.

Figure 2 presents the transient behavior of the effective refractive index (a) and the transmittance (b) of a 1-μm-thick silica glass layer doped with very small Ag nanospheres at pump light at 430 nm. In **Figure 2**, one can see that the relaxation of the transmittance is slower for a higher fluence of the incident light even for a fixed value of inherent cooling time of silver as large as 1 ps.

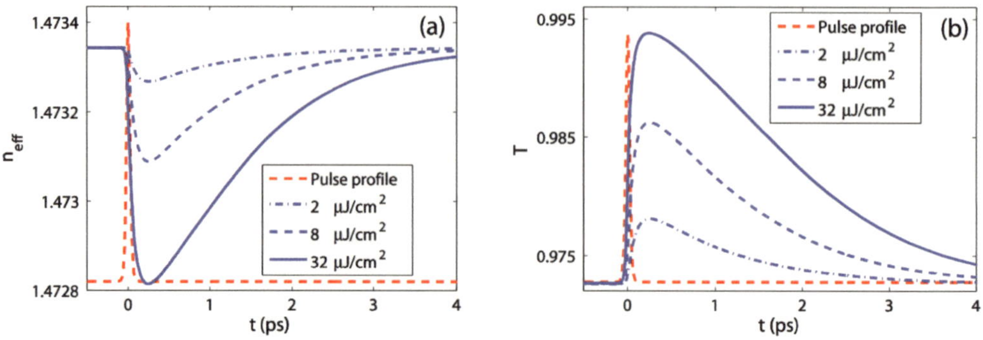

Figure 2. Refractive index (a) and transmittance (b) of silica glass doped with Ag NPs with diameters below 10 nm for different fluences of 50 fs pulses at 430 nm. Filling factor is 10^{-4} and thickness is considered to be 1 μm. The red-dashed curves in both panels represent the temporal profile of the pump pulse.

To investigate the dependence on the NP shape, we numerically investigate the ultrafast nonlinear responses of composites containing nonspherical metal nanoparticles, considering silica glass doped with silver nanorods. To study the dependency of the ultrafast nonlinear responses on the aspect ratio, we have taken nanorods with constant volumes. Note that the quantities presented in **Figures 3(a)** and **(b)** have been averaged over the possible polarization directions versus the orientations of nanorods. For the simulations, we have taken nanorods with different aspect ratio and geometry, as indicated in the figure. **Figure 3(a)** illustrates extinction spectra of silver nanorods with different aspect ratios, embedded in silica. As the figure shows, nanorod with a larger value of the aspect ratio (length divided by the diameter) exhibits surface plasmon resonance (SPR) at a longer wavelength. The main SPR peaks appear at 595, 680, and 800 nm for silver nanorods with aspect ratio of 1.68, 2.25, and 3.11, respectively. These peaks represent dipole plasmonic resonances for the case when the axes of nanorods are parallel to the polarization of incident light. The minor SPR peaks, which correspond to the dipole resonances for the polarization direction of incident wave perpendicular to the axes of nanorods, also appear

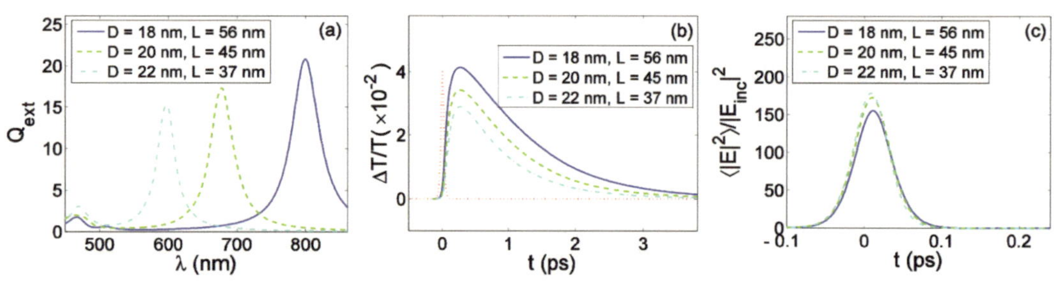

Figure 3. Ultrafast nonlinear response of silica glasses doped with silver nanorods with varying shape. The diameters D and the lengths L of nanorods are given in insets. The volumes of silver nanorods are kept the same for all the aspect ratios. Filling factor is 10^{-4} and the thickness of composite film is assumed to be 1 μm. (a) shows extinction spectra and (b) depicts the nonlinear change of relative transmittance $\Delta T/T$ for the different shapes of silver nanorods at the wavelengths corresponding to the main SPR peaks. The quantities shown in (a) and (b) are averaged over all the possible polarization directions relative the orientations of nanorods. Pulse duration and the fluence are taken to be 50 fs and 2 μJ/cm^2, respectively. In (b), red-dotted curve shows the temporal profile of incident pump pulse. (c) shows the space-averaged intensity enhancement when the axes of nanorods are parallel to the polarization of pump pulse.

(at around 470 nm). **Figure 3(b)** illustrates the nonlinear change of relative transmittance $\Delta T/T$ for the different aspect ratios of silver nanorods at the wavelengths corresponding to the main SPR peaks. The figure also shows that the maximum change of $\Delta T/T$ becomes larger for the nanorod with larger aspect ratio. We can expect that the stronger nonlinear response originates from a stronger intensity enhancement in the silver nanorod, which we can confirm from **Figure 3(c)**.

With the theoretical tool for calculation of the ultrafast nonlinear optical responses of metal NPs composites, below we theoretically demonstrate its possible use for the mode-locking and slow light devices by using these materials.

4. Ultrashort pulse generation by mode-locking of lasers using metal nanocomposites as saturable absorbers

In this section, we present the theory of mode-locking of visible solid-state [13] and semiconductor disk lasers [17] using metal nanocomposites as saturable absorbers.

4.1. Mode-locking of solid-state lasers

In this subsection, we present the general passive mode-locked operation of solid-state lasers taking Ho:YLF laser pumped with a diode laser as an example and a metal nanocomposite as saturable absorber.

The master equation for mode-locking of solid-state lasers can be written as follows [19]:

$$T_R \frac{\partial A(T,t)}{\partial T} = -iD\frac{\partial^2 A}{\partial t^2} + i\delta|A|^2 A + \left[g - l + D_{g,f}\frac{\partial^2 A}{\partial t^2} - q(T,t)\right]A(T,t), \tag{16}$$

where D is the second-order intracavity group delay dispersion (GDD), $D_{g,f} = g/\Omega_g^2 + 1/\Omega_f^2$ the gain and intracavity filter dispersion, Ω_g and Ω_f are the gain and filter bandwidth, respectively. The coefficient $\delta = (2\pi/\lambda_0 A_L)n_2 l_L$ represents the nonlinear effect by the Kerr nonlinearity in the gain crystal leading to self-phase modulation (SPM), where λ_0 is the central lasing wavelength, n_2 is the nonlinear index of laser active medium, A_L and l_L are the effective mode area and optical pass per round-trip. The gain coefficient is given by $g = g_0/[1 + E/(P_L T_R)]$, where g_0 is the small-signal gain, E is the pulse energy, P_L is the saturation power of laser material, l is the linear loss inherent to the passive resonator, and q is the loss by saturable absorber.

As an example, we investigate the mode-locking of a Ho:YLF laser operating at 545 nm and pumped at 445 nm which contains SCHOTT glass N-BAK 4 doped with Au NPs (with filling factor 2×10^{-3}). In such a case, the laser gain medium functions as a three-level system [20]. The Ho:YLF crystal has an emission linewidth of roughly 18 nm and a strong absorption peak around 450 nm which can be used for pumping with high power laser diodes. We have assumed the absorption cross sections at 445 and 545 nm to be 3×10^{-21} cm^2 and 5×10^{-21} cm^2, correspondingly, emission cross section at 545 nm to be 8×10^{-21} cm^2 [21], and the fluorescence lifetime of 110 μs [21]. The lasing transition corresponds to $^5S_2$–$^5I_8$ [20]. For the Ho^{3+} concentration of 1.2×10^{20} cm^{-3}, a beam area on the crystal with 1000 μm^2, a pump intensity 0.6 MW/cm^2 corresponding to a pump power of 6 W, a crystal length of 10 mm, and a resonator length of 1 m corresponding to a repetition rate of 0.15 GHz, we have the main laser parameters of g_0 = 0.22, P_L = 4.69 W, Ω_g = 57.08 THz, and δ = 3.94 MW^{-1}. We assumed the passive resonator loss to be l = 0.02. As the saturable absorber for passive mode-locking, we consider a metal nanocomposite given by a layer of SCHOTT glass (N-BAK 4) doped with small Au NPs. The plasmon resonance is found at 543 nm and we expect strong saturable absorption since the plasmon resonance peak is located near the central lasing wavelength. The saturable absorber loss is determined by $q(t) = -i(\omega/c)\sqrt{\varepsilon_{\rm eff}(t)}d$, where ω is central frequency, and d is the thickness of metal nanocomposite absorber. The cross section of beam on the metal nanocomposite absorber is taken to be 0.01 mm^2.

Figure 4 presents results for the passive mode-locking behavior of a Ho:YLF laser obtained by the solution of Eq. (16). Here, the GDD parameter has a value of −20 fs^2. As **Figure 4(a)** shows, the pulse duration rapidly decreases during the early stage of mode-locking and it is then stabilized.

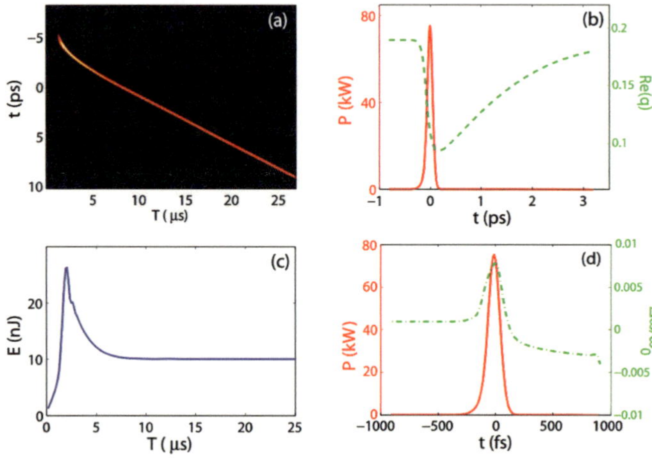

Figure 4. Mode-locking of Ho:YLF laser with SCHOTT glass N-BAK 4 doped with small Au NPs: (a) pulse evolution, (b) pulse profile and the dynamics of absorber loss, (c) evolution of pulse energy, and (d) intensity and frequency profile. D = −10 fs^2.

The numerical results show that the shortest pulse duration of about 100 fs is much smaller than the recovery time of 1 ps. In the early time of femtosecond pulse generation in solid-state lasers by the use of saturable absorbers for passive mode-locking, it was believed that the shortest pulse duration cannot be smaller than the recovery time of the absorber. However, due to the combined influence of spectral broadening by SPM and GVD, a slow response of the absorber with a recovery time larger than the pulse duration is sufficient for mode-locking [22, 23]. Pulse stabilization and shaping mechanism in this case can be explained by the fact that the action of the absorber steadily delays the pulse, similar as in solid-state lasers mode-locked by a semiconductor saturable absorber [24, 25]. Thus, the pulse moves slower than any noise structure behind the pulse maximum, and therefore, it fuses with the main pulse before it experiences high enough gain. This behavior is illustrated in **Figure 4(b)**.

Figure 5 shows results for the dependences of pulse duration and energy on the parameters of the laser.

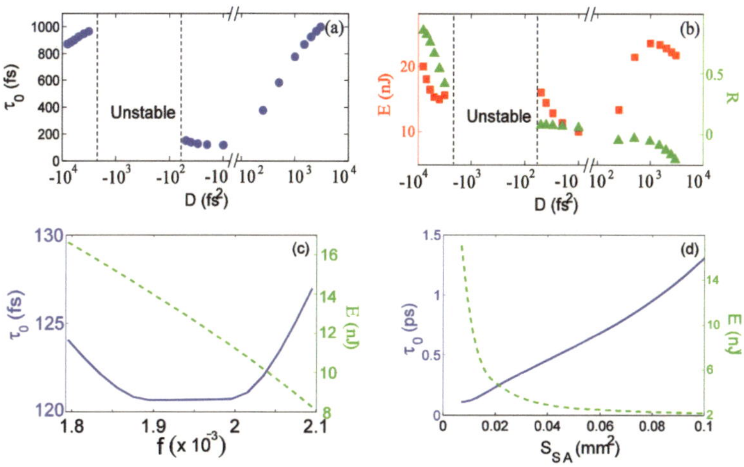

Figure 5. Dependence of pulse duration (a), energy [red squares in (b)], and the ratio R [green triangles in (b)] on the GDD parameter D. The beam area on the metal nanocomposite is 0.01 mm^2 and filling factor is 2×10^{-3}. (c) presents the dependences of pulse duration and pulse energy on filling factor. The beam area on the metal nanocomposite is 0.01 mm^2 and $D = -20$ fs^2. (d) depicts the dependencies of pulse duration and energy, and the ratio R on the beam area on the metal nanocomposite saturable absorber: $D = -20$ fs^2 and $f = 2 \times 10^{-3}$.

The resonator dispersion (the GDD parameter) is one of the key parameters affecting the mode-locking stability, pulse duration as well as energy. The simulations predict that the pulse evolution is unstable for the GDD parameter D from -55 fs^2 up to around -3200 fs^2 for the other parameters equal to those in the above subsections. The physical mechanism of the instability is the imbalance between the pulse delay, originating from the delayed absorber response in the metal nanocomposite, and pulse reshaping by the dispersion. In **Figure 5**, we show the dependence of the pulse duration and energy on the intracavity dispersion for a filling factor of 2×10^{-3}. As the dispersion increases, the pulse duration increases as well, independent of the sign of dispersion.

Figure 5(c) depicts the dependences of pulse duration and energy on the filling factor for $D = -20$ fs^2 and the beam area on the metal nanocomposite. For a small filling factor, we find a small maximum loss change leading to a larger pulse duration. For a filling factor smaller than 1.7×10^{-3}, mode-locking becomes unstable. For a filling factor larger than 2.1×10^{-3}, the total linear loss introduced by the saturable absorber is higher than the gain and the lasing itself becomes impossible.

In **Figure 5(d)**, pulse duration and energy are presented in dependence on the beam area for the metal nanocomposite absorber film with $D = -20$ fs^2 and $f = 2 \times 10^{-3}$. In the figure, the mode-locking process is stable for the beam area larger than 0.1 mm^2 or smaller than 6000 μm^2.

The pulse duration increases as the beam area increases, caused by the change of the reduced maximum. The possible duration range is from 110 fs to 1.31 ps. The shortest pulse duration of 100.7 fs is predicted for $D = 0$, beam area 6000 μm^2, and filling factor 2×10^{-3}.

Mode-locking of solid-state lasers by a slow absorber but shaped by the interplay between negative group-delay dispersion and SPM is often interpreted by the mechanism of soliton mode-locking, where the pulse is completely shaped by soliton formation and the saturable absorber only stabilizes the soliton against the growth of noise [23]. Whether such regime is indeed realized can be estimated by the ratio $R = -8ln(\sqrt{2}+1)D/(\delta E\tau_0)$, where τ_0 is the pulse duration full width at half maximum (FWHM). If the value of R is equal to 1, the pulse formation mechanism can be interpreted by soliton mode-locking. In **Figure 5(b)**, the ratio R is depicted by the green triangles. As can be seen in the left stability range, the factor R is in the range of 0.5, while in the right stability range, R is near zero. This means the predicted mode-locking behavior cannot be interpreted by the mechanism of soliton mode-locking, and in contrast to this interpretation, the action of the saturable absorber influences the pulse parameter. This is also confirmed by the calculated frequency chirp in **Figure 5(d)** which is not zero unlike that of a soliton. Note that the very specific regime of soliton mode-locking is not requested and femtosecond pulse generation in solid-state lasers by slow absorbers can be achieved under more general conditions [22].

4.2. Mode-locking of semiconductor lasers

Here, we investigate the possibility of femtosecond pulse generation by mode-locking of semiconductor disk lasers (SDLs) in visible range by using of metal nanocomposites as slow saturable absorbers. In contrast to the solid-state gain media, the semiconductors exhibit a dynamic gain in one round trip of the pulse in the resonator. This characteristic makes the process quite different from the mode-locked solid-state lasers but similar to one in passively mode-locked dye lasers. Mode-locking in this case rely on the interplay between a slow saturable absorber and slow gain saturation, which opens a net gain window in time, so that the pulse itself experiences gain per round trip and noise is discriminated outside this net gain window.

Passively mode-locked operation of the SDLs is described by the following master equation:

$$T_R \frac{\partial A(T,t)}{\partial T} = -iD\frac{\partial^2 A}{\partial t^2} + \left[(1-i\alpha)g - l + D_{g,f}\frac{\partial^2 A}{\partial t^2} - q(T,t)\right]A(T,t), \tag{17}$$

where α is the linewidth enhancement factor [26, 27]. In the above equation, the gain coefficient $g(t)$ is given by [26]

$$\frac{\partial g(t)}{\partial t} = -\frac{g(t)-g_0}{\tau_g} - g(t)\frac{|A(t)|^2}{E_g}, \tag{18}$$

where τ_g is the gain recovery time and E_g is the gain saturation energy dependent on the saturation fluence and the beam diameter on the gain medium. The temporal dynamics of the gain leads to a depletion of the leading edge of pulse which is absent in the case of solid-state lasers due to the quasi-static gain. On the other hand, the slow saturable nanocomposite absorber suppresses the trailing edge, and the combined action of both creates a net gain window with negative gain both on the leading and the trailing front of the generated pulse.

As an example, we consider a GaN-based SDL [27] operating with a central lasing wavelength of 420 nm. The small signal gain is chosen to be $g_0 = 2$, the gain linewidth $\Omega_g = 2\pi \times 20.39$ THz, the linewidth enhancement factor $\alpha = 2.8$ Lu et al. [27], the gain recovery time to be 1 ns [28], and the saturation energy for the gain medium to be 0.6 nJ. As a saturable absorber, we consider a silica glass film doped with Ag nanospheres with diameters smaller than 10 nm. The filling factor is taken to be 7×10^{-3}. In this case, the plasmon resonance is located at 414 nm and the composite exhibits strong saturated absorption near 420 nm.

In **Figure 6**, we present an example of passively mode-locked operation of this laser with a cavity dispersion of $D = 100$ fs^2 and a beam area on the metal nanocomposite of 0.002 mm^2. As can be seen in Figures 6(a) and (c) in this case, pulse shortening happens in the very beginning of the process up to about 20 ns, while the stabilization is achieved after about ~200 ns and is faster than in solid-state lasers (several μs). **Figure 6(c)** depicts the temporal behavior of gaun, saturable loss, and pulse power. **Figure 6(d)** shows both the intensity and frequency profile. The resultant pulse duration is about 83 fs and the pulse is positively chirped. This can be explained by the nonlinear index of the gain and absorber medium. The nonlinear refraction in the gain medium originates from the linewidth enhancement, due to the enhancement of the spontaneous emission into a lasing mode by the high cavity quality factor of the microcavity in the vertical cavity structure of the semiconductor disk gain medium.

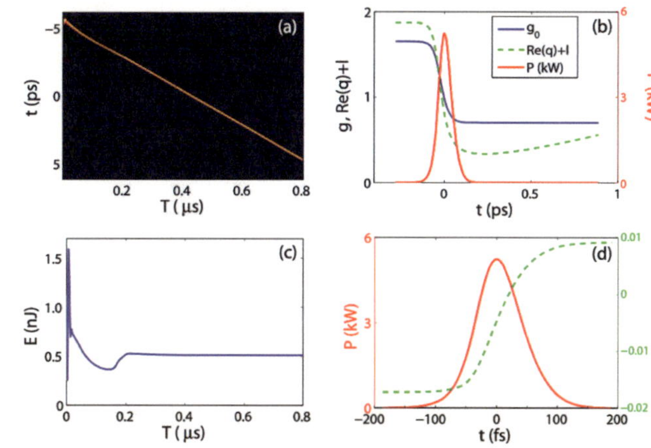

Figure 6. Evolution of pulse shape (a), power (b), duration (c), as well as frequency and power profile (d) in the GaN-based semiconductor disk laser (420 nm) with silica glass doped with Ag NPs. $D = 100$ fs^2.

Figures 7(a) and **(b)** show the dependences of pulse duration and energy on the cavity dispersion D. One can see that pulse shaping is unstable for the range of small dispersion (from $D = -150$ to 40 fs^2). The figure shows that the achievable shortest pulse duration is about 54.0 fs for negative dispersion of $D = -160$ fs^2 and about 55.3 fs for positive dispersion of $D = 50$ fs^2. For filling factor smaller than 6×10^{-3} due to the insufficient dynamic range of saturable loss, the mode-locked operation becomes unstable. For filling factor larger than 7×10^{-3}, the lasing itself is impossible due to the negative small signal net gain. For $f = 6 \times 10^{-3}$, $\tau_0 = 87.14$ fs and

$E = 0.69$ nJ, and for $f = 7 \times 10^{-3}$, $\tau_0 = 83.23$ fs and $E = 0.55$ nJ. These values are calculated for $D = -300$ fs^2.

In **Figures 7(c)** and **(d)**, we show the dependencies of pulse duration and energy on the beam area on the metal nanocomposite absorbers for both cases of positive and negative cavity dispersions, −300 and 300 fs^2.

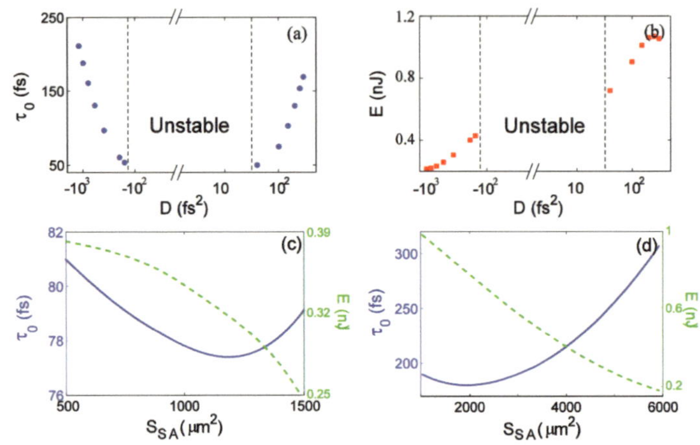

Figure 7. Dependencies of pulse duration (a) and energy (b) on the group delay dispersion D in the GaN-based semiconductor disk laser (420 nm) with silica composite doped with Ag NPs. Beam area on the saturable absorber composite is 1000 μm^2. Dependencies of pulse duration (c) and energy (d) on beam are on the silica glass doped with Ag NPs. In (c) $D = -300$ fs^2 and in (d) $D = 300$ fs^2.

In comparison with other well-established mode-locking elements such as semiconductor saturable absorber mirrors (SESAM), quantum dots, and carbon nanotubes, the operation range of metal nanocomposites as saturable absorber can be extended to much shorter wavelengths, down to the blue spectral range. Saturable absorbers in the visible spectral range can be realized by dyes dissolved in different solvents, but disadvantages such as maintaining a constant stream in the dye jet, temperature sensitivity, and long-term instability restrict their versatility. Different metal nanoparticle composites offer several advantages and allow the development of very compact and cheap mode-locking devices with tunable operation regions, extending from IR down to the blue spectral region. After the prediction of the possible application of metal nanocomposites for mode-locking of solid-state lasers [13] and semiconductor lasers [17], several groups successfully exploited nanoparticles for the generation of ultrashort pulses in mode-locked fiber lasers (see e.g. Refs. [29, 30], or solid-state lasers [31]).

5. Slow light by using metal nanoparticle composites

The possibility to reduce the group velocity of light significantly by various nonlinear effects has attracted much interest (see e.g., Refs. [32–35]). In this section, we show that metal nanocomposite materials can also significantly reduce the velocity of light. We assume that a metal nanocomposite is illuminated by a strong pump and a weak probe pulse with a frequency slightly different from that of pump which is in the range near the SPR [see **Figure 8(a)**]. For convenience, we consider probe pulse with ps durations which is much longer than e-e interactions time τ_{ee} (see Section 2). In this case, the nonlinear dielectric function change of metal can approximately be written as follows [see Eq. (15)]:

$$\Delta\varepsilon_{\mathrm{m}}(t) = \frac{\chi_{\mathrm{m}}^{(3)}}{\tau_{\mathrm{ep}}}\int_{-\infty}^{t} e^{-\frac{t-t'}{\tau_{\mathrm{ep}}}}|A^{\mathrm{enh}}(t')|^2 dt'. \tag{19}$$

As the slow light is a pulse propagating with a small group velocity, we have to study the propagation of a pulsed probe. Taking this into account, we write the amplitude of total enhanced field in metal NPs as

$$A^{\mathrm{enh}}(t) = A_{\mathrm{p}}^{\mathrm{enh}} + A_{\mathrm{pr}}^{\mathrm{enh}}(t)e^{-i\Omega t}, \tag{20}$$

where $\Omega = \omega_{\mathrm{pr}} - \omega_{\mathrm{p}}$ is the difference between probe ω_{pr} and pump frequencies ω_{p}. The above equation can be rewritten by using the frequency-domain probe amplitude $A_{\mathrm{pr}}^{\mathrm{enh}}(\Omega)$:

$$A^{\mathrm{enh}}(t) = A_{\mathrm{p}}^{\mathrm{enh}} + \int A_{\mathrm{pr}}^{\mathrm{enh}}(\Omega)e^{i\Omega t}d\Omega. \tag{21}$$

Assuming that the probe is much weaker than the pump, we obtain the dielectric functions of the metal at the pump and probe frequencies given by

$$\begin{aligned} \varepsilon_{\mathrm{m}}(\omega_{\mathrm{p}}) &= \varepsilon_{\mathrm{m0}} + \chi_{\mathrm{m}}^{(3)}|x_{\mathrm{p}}A_{\mathrm{p}}|^2, \\ \varepsilon_{\mathrm{m}}(\omega_{\mathrm{pr}}) &= \varepsilon_{\mathrm{m0}} + \chi_{\mathrm{m}}^{(3)}|x_{\mathrm{p}}A_{\mathrm{p}}|^2\left(1 + \frac{1}{1 + i\Omega\tau_{\mathrm{ep}}}\right), \end{aligned} \tag{22}$$

where $x_{\mathrm{p}} = A_{\mathrm{p}}^{\mathrm{enh}}/A_{\mathrm{p}}$ is the field enhancement factor for the pump depending on its intensity and is written by $x_{\mathrm{p}}(\omega_{\mathrm{p}}) = 3\varepsilon_{\mathrm{h}}/[\varepsilon_{\mathrm{m}}(\omega_{\mathrm{p}}) + 2\varepsilon_{\mathrm{h}}]$. We find the effective dielectric function for probe by using the Maxwell-Garnett formula. From the second equation in Eq. (22), we can expect that the effective dielectric function of metal nanocomposite has large dispersion in a narrow spectral range defined by τ_{ep}.

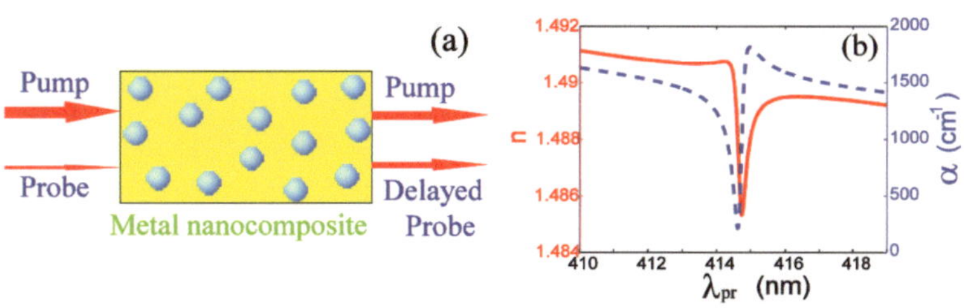

Figure 8. Scheme of slow-light device based on a metal-nanoparticle composite (a). The absorption coefficient (blue dashed line) and the effective index near the pump wavelength (red solid line) for the probe beam are shown in (b).

In **Figure 8(b)**, we show an example of the absorption spectrum and the effective refractive index for the probe in silica glass doped with Ag NPs. The steep increase of the effective index with the probe wavelength, as shown in the figure, leads to a large group index and strong slowing down of the probe wave.

For current applications, studying slow-light performance at telecommunication wavelengths is of particular importance. The plasmon resonance is shifted to a large extent for different shapes of the metallic NPs, with nanorods being particularly suitable to shift the plasmon resonance into the long wavelength range. Noble metals have extremely high inherent nonlinear susceptibilities at telecommunication wavelengths, e.g., for gold $\chi_{\mathrm{m}}^{(3)} = -1.5 \times 10^{-11}$ $\mathrm{m^2V^{-2}}$, which is 4 orders of magnitude higher than in the visible range [36]. Here, we solve the coupled propagation equations for the cw pump and the pulsed probe, to take into account the decrease of the pump intensity during the propagation due to the residual absorption and the change in the complex amplitude of probe, using the effective dielectric function calculated as explained above. The temporal profile of probe pulse is obtained by the Fourier transform.

Figure 9 shows typical incident and delayed pulses for a pump intensity of 6 MW/cm^2 at 1550 nm. We consider a TiO_2 film doped with gold nanorods with a diameter of 20 nm and a length of 66 nm with a surface plasmon resonance at 1540 nm. The propagation direction and polarization are the same for pump and probe beams. **Figure 9(a)** shows the evolution of probe pulse with propagation. Despite the saturation of the absorption for the pump, the output pump beam intensity is reduced down to 34 kW/cm^2 (not shown), which is about 200 times lower than that of input pump. The probe is sustained mainly by the coherent energy transfer from the pump and is attenuated by linear loss after the strong reduction of pump intensity. At a certain propagation length, the probe energy can become even larger than the incident energy. Here, gain peak for the probe arises already for quite low intensities. This can be seen in **Figure 9 (b)**: the probe peak intensity after 4 μm (green dash-dotted line) is 1.2 times larger than that of the incident pulse. For a propagation length of 5 μm, the fractional delay is roughly equal to 1.79, corresponding to a delay-bandwidth product of 2.68. As a central finding, the group velocity is 198 times smaller than the light velocity in vacuum. Because of the large inherent

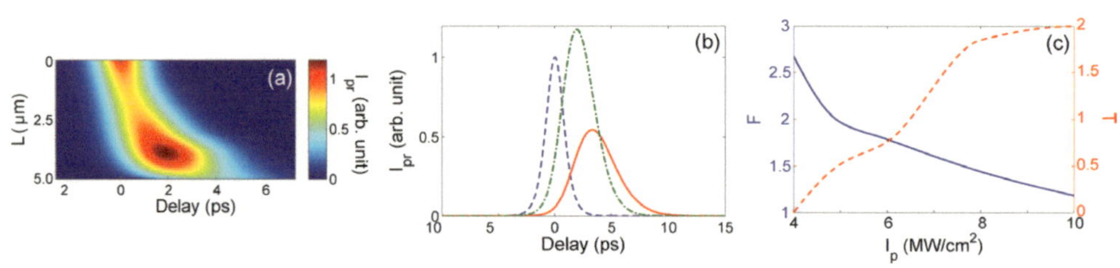

Figure 9. Slow light in TiO_2 film doped with Au nanorods, with a diameter of 20 nm and a length of 66 nm for pump intensity of 6 MW/cm^2 at 1550 nm. Filling factor is 2.5×10^{-2} and pulse duration is 1.85 ps. In (a) and (b), evolution of probe pulse and optical delay are shown, respectively. In (b), blue-dotted curve is the incident probe pulse, green dash-dotted and red solid curves are probe pulses for propagation lengths of 4 and 5 μm. In (c), we have shown the dependencies of fractional delay *F* (blue solid line), transmittance *T* (green-dotted line) on the pump intensity in TiO_2 film doped with Au nanorods at 1550 nm. The propagation length is 5 μm, the filling factor 5×10^{-2}, and the probe pulse duration 1.85 ps.

nonlinear susceptibility of gold in this wavelength range, the large fractional delay is achieved even with relatively low intensities of few MW/cm^2.

Figure 9(c) presents the dependence of fractional delay F (delay divided by the pulse duration of input probe) and transmittance T as functions of the pump intensity at 1550 nm. It can be seen that the fractional delay decreases as pump intensity increases, while the transmittance increases. In particular, for stronger pump, the transmittance for the probe can exceed unity due to amplification by an energy transfer from the pump. This feature constitutes a significant advantage compared to the conventional slow-light mechanism. **Figure 9(c)** shows that for a propagation length of 5 μm, a fractional delay in the range of 2 (corresponding to a delay-bandwidth product of 3) is predicted with a transmittance higher than 0.5.

Up to now, only a small number of experiments in different schemes have measured relative pulse delays exceeding these predictions. For comparison of results obtained with chip-compatible design, we refer to Ref. [32] with a fractional delay of 0.8 or to Ref. [33] with a value of 1.33. On the other hand, significantly larger fractional delays have been implemented by using cascaded microring resonators and photonic wire waveguides (but with relatively small probe transmission) [34] or by using double resonances of Cs atoms [35]. However, the numerical examples with a fractional delay of roughly two are not the physical limits of the method proposed here. Next, we will show that notably larger fractional delays can be realized using a modified setup.

The main drawback of the delay line in collinear configuration is due to the propagation length being limited by the attenuation of the pump. This can be circumvented using a transversely pumped setup as shown in **Figure 10(a)**, where the probe pulse is guided by the waveguide structure. The thickness of the film is 1 μm in this case, resulting in an attenuation of the pump intensity below 10%. Therefore, the pump intensity can be approximately considered to be constant. The other parameters are the same as for the case of **Figure 9** except the different pump intensity of 0.28 MW/cm^2. This value corresponds to a pump energy of 0.2 nJ for a 200 ps pump and a transverse waveguide width of 5 μm. Note that the polarization of pump and probe must be parallel due to the selective plasmon excitation for nonspherical NPs. **Figure 10(b)** shows the evolution of the probe pulse intensity normalized by the incident peak probe intensity. In this case, the peak intensity of the probe increases by a factor of 3.1 after propagation of 90 μm. For a more detailed presentation, in **Figure 10(c)**, we present the normalized probe pulses

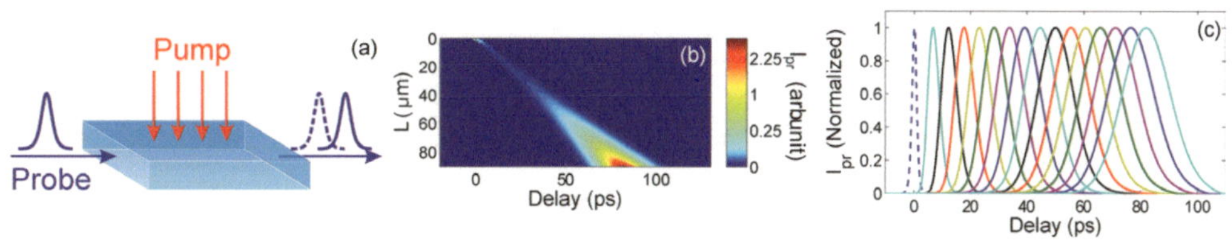

Figure 10. Delay for 1.85 ps probe pulse and pump intensity of 0.28 MW/cm^2 in TiO_2 film with a thickness of 1 μm in noncollinear configuration: (a) setup, (b) and (c) evolution of probe pulse. The remaining parameters are the same as in **Figure 9**. In (c), blue dashed curve is the incident pulse and solid curves are the delayed pulse profiles corresponding to propagation lengths L up to 90 μm with a step of 6 μm from the left to the right.

corresponding to propagation lengths up to 90 μm with a step of 6 μm. We predict a total fractional delay of about 43, with a delay-bandwidth product of 65 and a slowing down factor of 270. In this configuration, as the probe can be amplified, we have no principal limitation of the delay time; since with increasing propagation length, the delay time can be increased significantly which is only limited to the available pump source.

6. Conclusion

In this chapter, we presented a numerical method—the time-domain discrete-dipole approximation—for the calculation of the ultrafast nonlinear optical response of composites containing metal nanoparticles with different shapes. Based on the two-temperature model, an equation for the transient nonlinear response of metal nanocomposites is derived. This approach was applied for the study of the nonlinear refraction index and the transmittance of a thin fused silica glass layer doped with spherical nanoparticles and nanorods. The numerical results show that the total absorption coefficient exhibits strong saturation behavior near the plasmon resonance. The wavelength range of the saturated absorption can be significantly shifted in the whole ultraviolet, visible, and near-infrared spectral region by using metal NPs with nonspherical shapes. We have shown that the obtained results can be used for mode-locking of solid-state lasers and semiconductor lasers. In particular, in the short-wavelength range below 700 nm, useful saturable absorbers are still missing, but composites doped with metal NPs can enable the fabrication of new types of broadband mode-locking elements with a saturation intensity in the range of 10 MW/cm^2 and with an operation range from the near-IR up to the blue spectral range. As another example, we investigated theoretically a new slow-light mechanism based on composites doped with metal NPs. If two pulses with a frequency difference below the electron-phonon coupling rate propagate through such a composite, plasmon-induced oscillations of the nonlinear permittivity create a very high dispersion of the effective index for picoseconds-scale probe pulses, and correspondingly a strongly reduced group velocity. We have shown that utilizing these composites in a collinear setup reduced group velocities with a slow-down factor in the range of 200 and a fractional delay up to 2.5 for probe pulses with a few-ps duration can be implemented both at telecommunication wavelengths as well as in the optical range. Avoiding pump depletion, the relative delay can be significantly increased and a total fractional delay of 43 can be realized in a transversely pumped waveguide geometry.

Author details

Kwang-Hyon Kim[1], Anton Husakou[2] and Joachim Herrmann[2]*

*Address all correspondence to: jherrman@mbi-berlin.de

1 Institute of Lasers, State Academy of Sciences, Pyongyang, DPR Korea

2 Max Born Institute for Nonlinear Optics and Short Pulse Spectroscopy, Berlin, Germany

References

[1] Maier SA. Plasmonics: Fundamentals and Applications. Berlin: Springer; 2007.

[2] Zhang JZ, Noguez C. Plasmonic optical properties and applications of metal nanostructures. Plasmonics. 2008;**3**:127–150.

[3] Pelton M, Aizpurura J, Bryant G. Metal nanoparticle plasmonics. Laser Photon. Rev. 2008;**2**:136–159.

[4] Kelly KL, Coronado E, Zhao LL, Schatz GC. The optical properties of metal nanoparticles: the influence of size, shape, and dielectric environment. J. Phys. Chem. B. 2003;**107**:668–677.

[5] Sipe JE, Boyd RW. Nonlinear susceptibility of composite optical materials in Maxwell Garnett model. Phys. Rev. A. 1992;**46**:1614–1629.

[6] Hache F, Ricard D, Flytzanis C. Optical nonlinearities in small metal particle: surface-mediated resonance and quantum size effects. J. Opt. Soc. Am. B. 1986;**3**:1647–1655.

[7] Hache F, Ricard D, Flytzanis C. The optical Kerr effect in small metal particles and metal colloids: the case of gold. Appl. Phys. A. 1988;**47**:347–357.

[8] Palant B. Third-order nonlinear optical response of metal nanoparticles. In: Papadoulos MG et al, editor. Nonlinear Optical Properties of Matter. Berlin: Springer; 2006. pp. 461–508.

[9] Kim KH, Husakou A, Herrmann J. Linear and nonlinear optical characteristics of composites containing metal nanoparticles with different sizes and shapes. Opt. Express.2010; **18**:7488–7496.

[10] Kim KH, Yurkin MA. Time-domain discrete-dipole approximation for simulation of temporal response of plasmonic nanoparticles. Opt. Express.2015;**23**:15555–15564.

[11] Kim KH, Choe SH. Ultrafast nonlinear optical responses of dielectric composite materials containing metal nanoparticles with different sizes and shapes. Plasmonics (accepted). DOI: 10.1007/s11468-016-0335-x.

[12] Kim KH, Husakou A, Herrmann J. Saturable absorption in composites doped with metal nanoparticles. Opt. Express. 2010;**18**:21918–21925.

[13] Kim KH, Griebner U, Herrmann J. Theory of passive mode locking of solid-state lasers using metal nanocomposites as slow saturable absorbers. Opt. Lett. 2012;**37**:1490–1492.

[14] Tsen KT. Non-Equilibrium Dynamics of Semiconductors and Nanostructures. New York: CRC; 2005.

[15] Lysenko S, Jimenez J, Zhang G, Liu H. Nonlinear optical dynamics of glass-embedded silver nanoaprticles. J. Electron. Mater. 2006;**35**:1715–1721.

[16] Voisin C, Fatti ND, Christofilos D, Vallée F. Ultrafast electron dynamics and optical nonlinearities in metal nanoparticles. J. Phys. Chem. B. 2001;**105**:2264–2280.

[17] Kim KH, Griebner U, Herrmann J. Theory of passive mode-locking of semiconductor disk lasers in the blue spectral range by metal nanocomposites. Opt. Express. 2012;**20**:16174–16179.

[18] Bigot JY, Halté V, Merle JC, Daunois A. Electron dynamics in metallic nanoaprticles. Chem. Phys. 2000;**251**:181–203.

[19] Haus HA. Mode-locking of lasers. IEEE J. Sel. Top. Quantum Electron. 2000;**6**:1173–1185.

[20] Walsh BM, Barnes NP, Bartolo BD. Branching ratios, cross sections, and radiative lifetimes of rare earth ions in solids: application to Tm^{3+} and Ho^{3+} ions in LiYF4. J. Appl. Phys. 1998;**83**:2772–2787.

[21] Walsh BM, Barnes NP, Petros M, Yu J, Singh UN. Spectroscopy and modeling of solid state lanthanide lasers: application to trivalent Tm^{3+} and Ho^{3+} in $YLiF_4$ and $LuLiF_4$. J. Appl. Phys. 2004;**95**:3255–3271.

[22] Herrmann J, Müller M. Hidden mechanism for mode locking solid-state lasers by use of slow off-resonant saturable absorbers. Opt. Lett. 1995;**20**:994–996.

[23] Kaertner FX, Au JA der, Keller U. Soliton mode-locking with saturable absorbers, IEEE J. Sel. Top. Quantum Electron. 1996;**2**:540–556.

[24] Paschotta R, Keller U. Passive mode locking with slow saturable absorbers. Appl. Phys. B. 2001;**73**:653–662.

[25] Keller U. Ultrafast solid-state laser oscillators: a success story for the last 20 years with no end in sight. Appl. Phys. B. 2010;**100**:15–28.

[26] Saarinen EJ, Herda R, Okhotnikov OG . Dynamics of pulse formation in mode-locked semiconductor disk lasers. J. Opt. Soc. Am. B. 2007;**24**:2784–2790.

[27] Lu TC, Chu JT, Chu SW, Chen SW, Cheng BS, Kuo HC, Wang SC. Temperature dependent gain characteristics in GaN-based vertical-cavity surface-emitting lasers. Opt. Express. 2009;**17**:20149–20154.

[28] Smetanin IV, Vasil'ev PP, Boiko DL. Theory of the ultrafast mode-locked GaN lasers in a large-signal regime. Opt. Express. 2011;**19**:17114–17120.

[29] Kang Z, Xu Y, Zhang L, Jia Z, Liu L, Zhao D, et al. Passively mode-locking induced by gold nanorods in erbium-doped fiber lasers. Appl. Phys. Lett. 2013;**103**:041105.

[30] Wang XD, et al. Microfiber-based gold nanorods as saturable absorbers for femtosecond pulse generation in a fiber laser. Appl. Phys. Lett.2014;**105**:161074.

[31] Huang HT, et al. Gold nanorods as single and combined saturable absorbers for a high-energy q-switched Nd:YAG solid-sate laser. IEEE Photon. J.. 2015;**7**:4501210.

[32] Wu B, Hulbert JF, Lunt EJ, Hurd K, Hawkins AR, Schmidt H. Slow light on a chip via atomic quantum state control. Nature Photon. 2010;**4**:776–779.

[33] Okawachi Y, Foster MA, Sharping JE, Gaeta AL, Xu Q, Lipson M. All-optical slow-light on a photonic chip. Opt. Express. 2006;**14**:2317–2322.

[34] Xia F, Sekaric L, Vlasov Y. Ultracompact optical buffers on a silicon chip. Nature Photon. 2007;**1**:65–71.

[35] Camacho RM, Pack MV, Howell JC, Schweinsberg A, Boyd RW. Wide-bandwidth, tunable, multiple-pulse-width optical delays using slow light in Cesiumvapor. Phys. Rev. Lett. 2007;**98**:153601.

[36] Falcao-Filho EL, Barbosa-Silva R, Sobral-Filho RG, Brito-Silva AM, Galembeck A, Araujo CB. High-order nonlinearity of silica-gold nanoshells in chloroform at 1560 nm. Opt. Express. 2010;**18**:21636–21344.

Chapter 5

Understanding the Physical Behavior of Plasmonic Antennas Through Computational Electromagnetics

Xuezhi Zheng, Guy A.E. Vandenbosch and
Victor V. Moshchalkov

Additional information is available at the end of the chapter

http://dx.doi.org/10.5772/67589

Abstract

This chapter focuses on understanding the electromagnetic response of nanoscopic metallic antennas through a classical computational electromagnetic algorithm: volumetric method of moments (V-MoMs). Under the assumption that metals only respond to external electromagnetic disturbance locally, we rigorously formulate the light-nanoantenna interaction in terms of a volume integral equation (VIE) and solve the equation by using the method of moments algorithm. Modes of a nanoantenna, as the excitation independent solution to the volume integral equation (VIE), are introduced to resolve the antenna's complex optical spectrum. Group representation theory is then employed to reveal how the symmetry of a nanoantenna defines the modes' properties and determines the antenna's optical response. Through such a treatment, a set of tools that can systematically treat the interaction of light with a nanoantenna is developed, paving the road for future nanoantenna design.

Keywords: computational electromagnetics, nanoantennas, optical response, modal analysis, symmetries, group representation theory

1. Introduction

In the past decade, the grand leap in nanotechnology has granted us the very capability to reproduce every single object in the macroscopic world nanoscopically. Since classical antennas have been so optimized for controlling the low-frequency (normally below 1 THz) electromagnetic interaction, their topology naturally becomes the first option to be explored at nanoscale. The nanoscale version of classical antennas, which is always referred to as a "nanoantenna," has been proven to provide an effective route to couple photons in and out of nanoscale volumes. They are considered as excellent tools to probe and control light-matter interaction at the nanoscale (e.g., the interaction between light and molecules) and therefore have become an essential element in the discipline of nanophotonics [1, 2]. Till now, the concept of nanoantennas has been applied to many different fields [3–11] covering single-molecule detection, magnetic recording, bio-imaging, photochemistry, nanoscale signal processing, optical functional material, and so on.

The general physics of the interaction between electromagnetic fields and the collective oscillations of free electrons in a metal can be described by Maxwell's equations. This physical observation enables the reuse of traditional numerical techniques developed within the microwave community. On the one hand, well packaged in commercial solvers, differential equation-based techniques, the finite difference time domain (FDTD) method [12], and the finite element method (FEM) [13], to name a few, are widely employed by experimental physicists. On the other hand, integral equation-based techniques [14], where the light-nanoantenna interaction is mathematically depicted by surface integral equations (SIEs) or volume integral equations (VIEs), are also of essential importance. Since an integral equation involves a pre-calculated Green's function (analytically known) and only targets the scatterers, that is, the nanoantennas, it can provide an unbeatable accuracy, efficiency, and physical clarity compared to their differential equation counterpart.

To reach an integral equation formalism, one may start from the equivalence principle, match the electromagnetic field at the nanoantenna's boundary, and obtain surface integral equations. Another possibility is to use the volume equivalence principle. The nanoparticles are replaced by equivalent-induced electric polarization currents. With discretized boundaries or bodies, both the surface integral equation and the volume integral equation can be subsequently solved by the method of moments (MoMs) algorithm [15].

In this chapter, we solely focus on the volume integral equation formalism. A review of the surface integral equation formalism will be presented separately in the future. This chapter is organized as follows. In Section 2, under the local response material assumption, we formulate the interaction between light and a nanoantenna, which is of an arbitrary shape immersed in a generic environment composed of several dielectric regions, in terms of a volume integral equation. Then, the volume integral equation is solved by the method of moments algorithm. To validate our implementation and demonstrate its diversity, simulation results are contrasted with a wide range of experimental data.

In Section 3, the notions of eigenmodes and natural modes of a nanoantenna are presented. A mode is a fixed spatial distribution of some physical quantities (e.g., charge, current, electric field, magnetic field, etc.), and is only determined by the material, the geometry, and the environment of the structure. It is thus independent of the incident field. Two kinds of modes are introduced: eigenmodes and natural modes. While the eigenmodes are still frequency dependent, the natural modes are frequency independent. The latter represent the most fundamental properties of a structure. Different from a closed system (e.g., a cavity), the modes of an open system (i.e., a nanoantenna) are not necessarily orthogonal in an inner product sense. This implies a possible interference between the modes which is the very cause of the formation of an asymmetric line, that is, a Fano resonance, in a nanoantenna's scattering spectrum.

http://dx.doi.org/10.5772/67589

Section 3 indicates that the modes of a nanoantenna are not necessarily orthogonal. Nevertheless, the condition under which the modes are orthogonal or not orthogonal is not specified. In Section 4, we supply such a condition by using symmetry arguments. The relation between the symmetries of a nanoantenna and the mode orthogonality is rigorously put into the mathematical framework of the group representation theory. It is found that the modes that belong to different irreducible representations are orthogonal to each other in an inner product sense, and therefore no interference between these modes is allowed.

2. Volume integral equation formulation and volumetric method of moments algorithm for light–nanoantenna interaction

Assume a nanoantenna in space (see **Figure 1**) shone upon by an incident plane wave. This plane wave is generated by a current source which is very far from the scatterer. It oscillates with a time dependency $e^{j\omega t}$ with an angular frequency ω. A general field relation, which is the total field, is the sum of the scattered field and the incident field, must be satisfied at every point in the space,

$$\mathbf{E_{tot}}(\mathbf{r}, \omega) = \mathbf{E_{scat}}(\mathbf{r}, \omega) + \mathbf{E_{inc}}(\mathbf{r}, \omega), \tag{1}$$

Especially, at the nanoantenna the total electric field can be linked with the induced current which includes both the conduction currents due to the motion of free electrons and the displacement currents due to the motion of bound electrons through a dielectric function,

$$\mathbf{J_{ind}}(\mathbf{r}, \omega) = j\omega(\varepsilon(\mathbf{r}, \omega) - \varepsilon_0)\mathbf{E_{tot}}(\mathbf{r}, \omega). \tag{2}$$

In Eq. (2), ε_0 is the vacuum permittivity. However, in general it can be understood as the ambient background permittivity where the nanoantenna is positioned. $\varepsilon(\mathbf{r}, \omega)$ is a function of

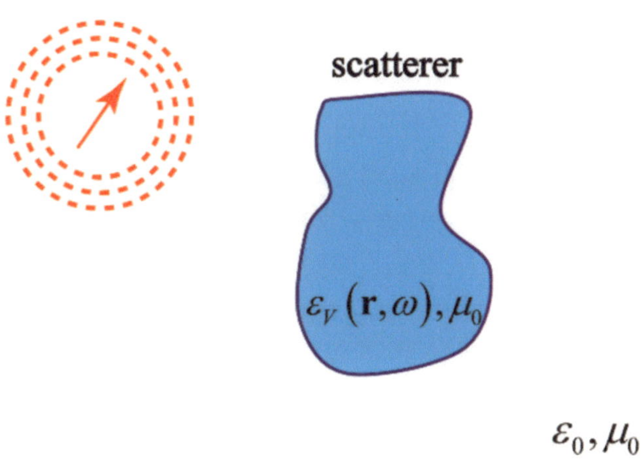

Figure 1. An abstract model for the light interaction with a nanoantenna (scatterer). In the figure, the scatterer is composed of a material characterized by electric and magnetic parameters, that is, $\varepsilon_V(\mathbf{r}, \omega)$ and μ_0. Here, the material is assumed to be non-magnetic. For the sake of conciseness, a homogeneous background is assumed and plotted but the background can be inhomogeneous. Lastly, the incident field is generated by a predefined current distribution.

observation point and frequency and represents the permittivity of the material constituting the nanoantenna. Here, the locality (i.e., the material property does not depend on the source point and hence no non-local behavior is assumed), the homogeneity (i.e., within the nanoantenna, the permittivity is a constant), and the isotropy (i.e., the dielectric function is a scalar) of the material are assumed. The relation stated in Eq. (2) is the so-called volume equivalence principle.

Further, it should be realized that the scattered field is induced by the induced current (as in Eq. (2)),

$$\mathbf{E_{scat}}(\mathbf{r}, \omega) = -j\omega\mu_0 \int_{V'} \overline{G}(\mathbf{r}, \mathbf{r}', \omega) \cdot \mathbf{J_{ind}}(\mathbf{r}', \omega) d\mathbf{r}'. \tag{3}$$

In Eq. (3), μ_0 is the vacuum permeability and, in the most general case, should be understood as the background permeability. $\overline{G}(\mathbf{r}, \mathbf{r}', \boldsymbol{\omega})$ is a Dyadic Green's function describing how an elementary-induced current $\mathbf{J_{ind}}(\mathbf{r}', \omega)$ at a source point $\mathbf{r}'$ generates an electric field at an observation point $\mathbf{r}$. The volume integral takes into account all the induced current spanning the nanoantenna and gives the scattered field at an observation point $\mathbf{r}$.

Focusing on the volume of the nanoantenna, substituting Eqs. (2) and (3) into Eq. (1) and putting all the terms containing the induced current on the left side of the equation, it is readily found that

$$\frac{\mathbf{J_{ind}}(\mathbf{r}, \omega)}{j\omega(\varepsilon(\mathbf{r}, \omega) - \varepsilon_0)} + j\omega\mu_0 \int_{V'} \overline{G}(\mathbf{r}, \mathbf{r}', \omega) \cdot \mathbf{J_{ind}}(\mathbf{r}', \omega) d\mathbf{r}' = \mathbf{E_{inc}}(\mathbf{r}, \omega). \tag{4}$$

Eq. (4) is the main equation in this chapter. Since the integral is taken with respect to the volume of the nanoantenna, in this manner, we formulate the interaction between light and a nanoscopic scatterer in terms of a volume integral equation. In short, Eq. (4) can be recast in an operator form,

$$Z(\mathbf{J_{ind}}(\mathbf{r}, \omega)) = \mathbf{E_{inc}}(\mathbf{r}, \omega). \tag{5}$$

The Z operator is named as the impedance operator since this operator links current density with electric field and has the same unit as an impedance. This operator consists of two parts,

$$Z = Z_{tot} + Z_{scat}. \tag{6}$$

The first term in Eq. (6) is called *the total field operator* that links the induced current with the total electric field,

$$\mathbf{E_{tot}}(\mathbf{r}, \omega) = Z_{tot}(\mathbf{J_{ind}}(\mathbf{r}, \omega)) = \frac{\mathbf{J_{ind}}(\mathbf{r}, \omega)}{j\omega(\varepsilon(\mathbf{r}, \omega) - \varepsilon_0)}. \tag{7}$$

The second term in Eq. (6) is called *the scattered field operator* that links the induced current with the scattered field,

$$Z_{\text{scat}}(\mathbf{J_{ind}}(\mathbf{r},\omega)) = j\omega\mu_0 \int_{V'} \overline{G}(\mathbf{r},\mathbf{r}',\omega)\cdot\mathbf{J_{ind}}(\mathbf{r}',\omega)\mathbf{dr}' \tag{8}$$

By inverting the impedance operator in Eq. (5), in principle the induced current $\mathbf{J_{ind}}(\mathbf{r},\omega)$ can be calculated. The scattered field in the space other than the region of the nanoantenna can be further calculated by using Eq. (3), and therefore the total field at any point in space can be recovered.

Since analytical solutions for the main equation (4) only exist for a few cases (e.g., the Mie solution in case of a sphere), in general the equation has to be solved numerically. To this end, a method of moments algorithm can be employed [15]. The algorithm starts with the discretization of the antenna with tetrahedral or hexahedral blocks (see **Figure 2**) and further assumes that the current density flowing from one tetrahedral (hexahedral) element to a neighboring one forms a certain basis function $\boldsymbol{\varphi}_i(\mathbf{r})$, for example, a rooftop shape [16, 17].

Approximating the induced current $\mathbf{J_{ind}}(\mathbf{r},\omega)$ by a weighted sum of the basis functions,

$$\mathbf{J_{ind}}(\mathbf{r},\omega) = \sum_{i=1}^{n} w_i\boldsymbol{\varphi}_i(\mathbf{r}), \tag{9}$$

and inserting Eq. (9) into Eq. (5), one equation with n unknowns is found

$$\sum_{i=1}^{n} Z\left(w_i\boldsymbol{\varphi}_i(\mathbf{r}')\right) = \mathbf{E_{inc}}(\mathbf{r},\omega) \tag{10}$$

In order to construct a system of N equations with N unknowns, a testing procedure is applied. Assuming that a test function is $\boldsymbol{\psi}_j(\mathbf{r})$, it is obtained that

$$\{z_{ji}\}\{w_i\} = \{e_j\}. \tag{11}$$

In Eq. (11), $\{z_{ji}\}$, $\{w_i\}$, and $\{e_j\}$ represent the impedance matrix, the vector containing the weighting coefficients as in Eq. (9), and the vector containing the projection of the incident field onto the jth basis function. $\{z_{ji}\}$ and $\{e_j\}$ are

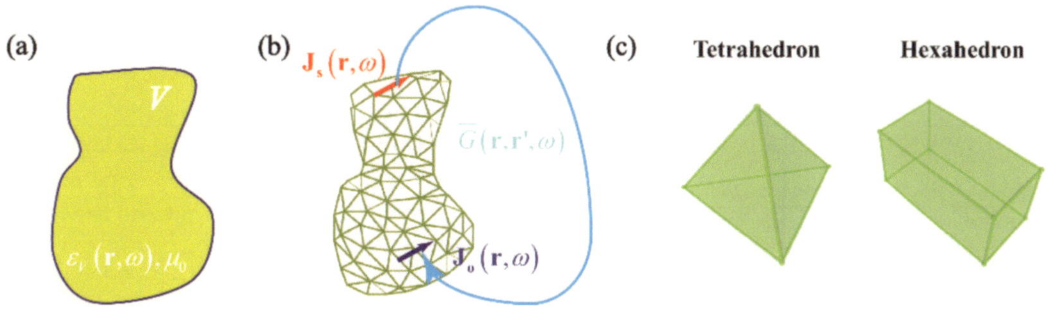

Figure 2. Continuum description (a), discretized description (b), and tetrahedral/hexahedral blocks (c) used to mesh the continuous volume.

$$z_{ji} = \langle \boldsymbol{\psi}_j(\mathbf{r})|Z|\boldsymbol{\varphi}_i(\mathbf{r})\rangle = \frac{\int_{V_j} \boldsymbol{\psi}_j(\mathbf{r})\cdot\boldsymbol{\varphi}_i(\mathbf{r})d\mathbf{r}}{j\omega(\varepsilon(\mathbf{r},\omega)-\varepsilon_0)} + \int\limits_{V_j} \boldsymbol{\psi}_j(\mathbf{r}) \int\limits_{V_i} \overline{G}(\mathbf{r},\mathbf{r}',\omega)\cdot\boldsymbol{\varphi}_i(\mathbf{r}')d\mathbf{r}', \tag{12}$$

$$e_j = \langle \boldsymbol{\psi}_j(\mathbf{r})|\mathbf{E_{inc}}(\mathbf{r},\omega)\rangle = \int\limits_{V_j} \boldsymbol{\psi}_j(\mathbf{r})\mathbf{E_{inc}}(\mathbf{r},\omega)dr. \tag{13}$$

Notice that in Eqs. (12) and (13), the integrations are with respect to the volume of the ith *basis function*, V_i, and the jth *test function*, V_j. By inverting the impedance matrix, the induced current can be reconstructed and the related physical quantities, for example, the near field and the far field of the nanoantenna, can be calculated.

In our method of moments implementation, Green's function for a planar stratified background is considered. Instead of using the Dyadic form of Green's function as in the above equations, a hybrid mixed-potential form is employed [18–22]. In the spirit of the Fourier transform, Green's function is analytically constructed in the Fourier space (the spectral domain) and then numerically transformed back to the real space (the spatial domain). In the latter step, all the singular behaviors in the Fourier space, for example, the branch point, the pole, the slowly decreasing asymptotic behavior when an observation point is close to a source point, are manually removed and analytically converted back to the real space so that the numerical integration is considerably facilitated.

Besides, if a hexahedral block (see Figure 2(c)) is used, a basis function that varies separately in the transverse direction and in the vertical direction is chosen. A transverse basis function reads,

$$\boldsymbol{\varphi}_i(\mathbf{r}) = \Delta_i(x,y)\boldsymbol{\Pi}_i(z). \tag{14}$$

In Eq. (14), the functions $\boldsymbol{\Delta}_i(x,y)$ and $\boldsymbol{\Pi}_i(z)$ are triangular and rectangular functions,

$$\Delta_i(x,y) = \begin{cases} \dfrac{\rho_i}{l_i}, & \rho_i = (x,y) \in S_i \\ 0, & \rho_i = (x,y) \notin S_i \end{cases} \tag{15}$$

$$\boldsymbol{\Pi}_i(z) = \begin{cases} 1, & z_i^- < z < z_i^+ \\ 0, & z_i^- > z \text{ or } z > z_i^+. \end{cases} \tag{16}$$

In Eq. (15), S_i denotes the base rectangle of the ith hexahedral block. ρ_i gives the distance of a point away from a given edge. l_i emphasizes the distance between the chosen edge and its opposite edge. In Eq. (16), the superscripts "+" and "−" delimit the upper limit and the lower limit of the block, respectively. A vertical basis function is

$$\boldsymbol{\varphi}_i(\mathbf{r}) = \boldsymbol{\Pi}_i(x,y)\Delta_i(z). \tag{17}$$

In Eq. (17), the triangular and rectangular functions are similarly defined as in Eqs. (15) and (16). Notice that Green's function for a general planar multilayer structure behaves differently in the transverse direction (which is described mathematically by a Bessel function and physically by a cylindrical wave) and in the vertical direction (which is described mathematically by an exponential function and physically by a plane wave). For example, a generic form of Green's function is

$$\int_0^{\infty} A_i(k_\rho) J_0(k_\rho \rho) e^{jk_{iz}|z-z'|} dk_\rho. \tag{18}$$

In Eq. (18), k_ρ is the transverse wave number and $k_{iz} = \sqrt{k_i^2 - k_\rho^2}$ is the corresponding vertical wave number in the ith layer. $A_i(k_\rho)$ represents the amplitude of a field component in the ith layer.

Based on the above facts, the sevenfold integral (including three from the inner volume, and three from the outer volume, and one from the inverse Fourier transform) in the second term of Eq. (12) can be reduced to a fivefold integral. Take the coupling between a horizontal test function and a horizontal basis function as example,

$$\begin{aligned} &\int_{V_j} \boldsymbol{\psi}_j(\mathbf{r}) \int_{V_i} \overline{G}(\mathbf{r}, \mathbf{r}', \omega) \cdot \boldsymbol{\varphi}_i(\mathbf{r}') d\mathbf{r}' = \\ &\int_{S_j} \Delta_j(x, y) \int_{S_i} \left\{ \int_0^{\infty} A_i(k_\rho) J_0(k_\rho \rho) \int_{z_j^-}^{z_j^+} \int_{z_i^-}^{z_i^+} e^{jk_{iz}|z-z'|} dz' dz\, dk_\rho \right\} \Delta_j(x, y) ds' ds \end{aligned} \tag{19}$$

In Eq. (19), the twofold integral with respect to the vertical direction can be performed analytically. In this way, an enormous amount of numerical efforts can be saved. Please notice that in order to cope with the recently increasing interest in nanoscale antenna arrays, the calculation

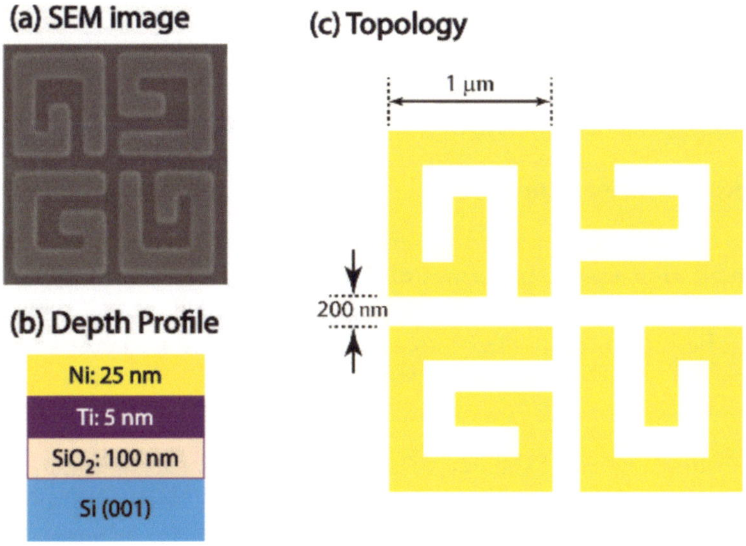

Figure 3. (a) The scanning electron microscope (SEM) image, (b) the depth profile, and (c) the geometry of a "G" shape structure with a period of 2400 nm. This figure is reproduced from [25].

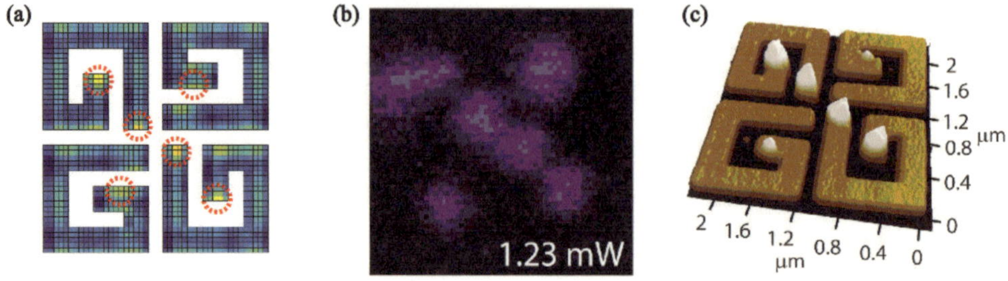

Figure 4. The optical response of the G-shape structure. (a) In the simulation, the G shape structure is excited by horizontally polarized normally incident light. The spatial distribution of the induced current intensity is plotted with the spatial positions where the current intensity is marked by red circles (in print: bold dashed lines). The plot is coded from blue to yellow to denote the strength of the induced current's intensity. (b) In the nonlinear experiment, the excitation condition is the same as in (a). The near field pattern is captured by a Second Harmonic Generation (SHG) imaging technique. The plot is coded from blue to white to denote the strength of the field intensity. (c) In the Atomic Force Microscopy (AFM) experiment, the bumps due to the local heat accumulation are demonstrated.

of periodic Green's functions has been implemented on the basis of a quasi three-dimensional (3-D) mixed potential integral equation formulation [23, 24].

The above MoM implementation is extremely efficient for the so-called extruded structures and can be immediately confirmed by experimental data. Take the Nickel (Ni)-based G shape structure (see **Figure 3**) as an example. In the simulation, the structure is illuminated by horizontally polarized light. As a result, the induced current is highly localized along the structure's diagonal (see **Figure 4(a)**). Remembering that the current is directly related with the electric field via Eq. (2), a corresponding intensive electric field concentration can be anticipated (see the second harmonic generation image of the structure's near field in **Figure 4(b)**). This electromagnetic response induces a local thermal effect, that is, heat accumulation (due to the large Ohmic loss in Ni). The generated heat can go beyond the melting point and finally decorates the optical response of the structure. Similar simulation—experiment comparison—can be found in [26–31].

3. Eigenmodes and natural modes of a nanoantenna

3.1. Eigenmodes of a nanoantenna

The impedance operator in Eq. (5) and accordingly the discretized version in Eq. (11) actually include all the electromagnetic properties of the light-nanoantenna interaction problem. As mentioned in the previous section, for a given incident electric field (e.g., a plane wave or the electromagnetic fields emitted by a fluorescent molecule), the induced polarization current in a nanoantenna can be calculated and the associated scattered fields can be readily derived. This analysis scheme indeed works for every nanoantenna.

Nevertheless, it should be noted that the solution is *incident field dependent*. In some cases, due to the complexity of a nanoantenna's geometry (e.g., the G-shape structure), an excessive amount of information can be read from a single simulation, for example, a very complex near-field pattern, multiple resonances in the extinction, and scattering spectra. In this sense, the *incident field-dependent* full solution may not be a good starting point for an analysis. To cope with this problem, in this section we pursue two incident field-independent solutions, that is, based on the eigenmodes and natural modes of a nanoantenna, and reveal their distinct roles in determining the antenna's response.

First, we investigate at the eigenvalue problem derived from Eq. (5) [32, 33],

$$Z(\mathbf{J_n}(\mathbf{r}, \omega)) = \lambda_n(\omega)\mathbf{J_n}(\mathbf{r}, \omega). \tag{20}$$

In Eq. (20), Z is the impedance operator defined in Eq. (5). For a specific angular frequency ω, $\mathbf{J_n}(\mathbf{r}, \omega)$ is the nth eigenmode with an eigenvalue $\lambda_n(\omega)$. Some extra attention should be paid to the terminology. Here, we define the eigenmodes of a nanoantenna based on the impedance operator Z. This definition is different from the term "eigenmode" (or "normal mode") in quantum optics. There, eigenmodes are defined based on a differential operator (e.g., the differential operator in the Helmholtz equation), while here eigenmodes are defined linked to the integral operator as in Eq. (20).

Under the assumption that the material constituting a nanoantenna is locally responding, homogeneous and isotropic, the total field operator Z_{tot} (defined in Eq. (7)) is a scalar. On this basis, we can distinguish two different contributions to the eigenvalue,

$$\lambda_n(\omega) = \lambda_{n,\,\text{tot}}(\omega) + \lambda_{n,\,\text{scat}}(\omega),\ \ \lambda_{n,\,\text{tot}}(\omega) = \frac{1}{j\omega(\varepsilon(\mathbf{r}, \omega) - \varepsilon_0)}. \tag{21}$$

Apparently, $\lambda_{n,\,\text{scat}}(\omega)$ is associated with the *scattering field* operator (defined in Eq. (8)) and hence is controlled by the geometry (as the volume integral in Eq. (8) is taken with respect to the volume of the nanoantenna) and the environment of the nanoantenna (as the involvement of the Dyadic Green's function in Eq. (8)). After removing the *total field* operator from both sides of Eq. (20), it is obtained that

$$Z_{\text{scat}}(\mathbf{J_n}(\mathbf{r}, \omega)) = \lambda_{n,\,\text{scat}}(\omega)\mathbf{J_n}(\mathbf{r}, \omega). \tag{22}$$

From Eq. (22), we can readily conclude that

When the material constituting a nanoantenna is local, homogeneous, and isotropic, the antenna's eigenmodes are indeed independent from the material of the nanoantenna. They only depend on the antenna's geometry and environment.

Due to the reciprocity of Green's function, Z_{scat} is a *complex symmetric* operator (i.e., the operator contains complex elements and its transpose is equal to itself $Z_{\text{scat}} = Z^T_{\text{scat}}$), which further implies that the eigenmodes are complex functions and orthogonal in a pseudo inner product sense [32],

$$\langle \mathbf{J_m}(\mathbf{r}, \omega) | \mathbf{J_n}(\mathbf{r}, \omega) \rangle = \int\limits_{V'} \mathbf{J_m}(\mathbf{r}, \omega) \cdot \mathbf{J_n}(\mathbf{r}, \omega) dr = \begin{cases} 1 & m = n \\ 0 & m \neq n \end{cases}. \tag{23}$$

By expanding the induced current in terms of the eigenmodes,

$$\mathbf{J_{ind}}(\mathbf{r},\omega)=\sum_{n=1}^{\infty}c_n\mathbf{J_n}(\mathbf{r},\omega), \tag{24}$$

and projecting Eq. (5) onto the eigenmodes in a pseudo inner product sense, another matrix equation can be received,

$$\{z_{mn}\}\{c_n\}=\{e_m\}. \tag{25}$$

In Eq. (25), $\{e_m\}$ is short for the projection of the incident field onto the mth eigenmode,

$$e_m=\int_{V'}\mathbf{J_m}(\mathbf{r},\omega)\cdot\mathbf{E_{inc}}(\mathbf{r},\omega)dr. \tag{26}$$

Eq. (25) is very similar to Eq. (11) in the sense that Eq. (11) uses *local* basis functions to approximate the induced current, while in Eq. (25) eigenmodes which can be deemed as *global* basis functions are employed. However, different from Eq. (11), which is in general a full matrix, due to the orthogonality property presented in Eq. (23), the impedance matrix $\{z_{mn}\}$ is a diagonal one. The weighting coefficients in Eq. (25) can be then evaluated as

$$c_n(\omega)=\frac{\int_{V'}\mathbf{J_n}(\mathbf{r},\omega)\cdot\mathbf{E_{inc}}(\mathbf{r},\omega)dr}{\lambda_n(\omega)}. \tag{27}$$

Thus, combining Eq. (24) with Eq. (27), it can be concluded that

> *The induced current is a weighted sum of eigenmodes. The contribution of an eigenmode is determined by two factors: on the one hand the fact whether the incident field can efficiently excite the eigenmode (the denominator of Eq. (27)); on the other hand the absolute value of the eigenvalue, that is, when the absolute value of the eigenvalue reaches its minimum, the corresponding weight reaches its maximum and hence the eigenmode reaches its resonance.*

Interestingly, it should be noted that we may use the concept of inner product as well in the construction of Eq. (25). In contrast to the definition of a pseudo inner product, which is

$$\int_{V}\mathbf{J_m}(\mathbf{r},\omega)\cdot\mathbf{J_n}(\mathbf{r},\omega)dr, \tag{28}$$

the inner product is defined as

$$\int_{V}\mathbf{J_m^*}(\mathbf{r},\omega)\cdot\mathbf{J_n}(\mathbf{r},\omega)dr. \tag{29}$$

Comparing Eq. (28) with Eq. (29) immediately reveals the key difference between the notion of the pseudo inner product and the notion of the inner product, namely the extra complex conjugate operation. In classical electromagnetism, the pseudo inner product is more associated with the reciprocal properties, while the inner product is emphasized in the definition of energy/power-related physical quantities, for example, the work done to a scatterer by an incident field.

In this spirit, we can reformulate Eqs. (24)–(27) while using the concept of inner product,

$$\left\{z'_{mn}\right\}\{c_n\} = \left\{e'_m\right\}. \tag{30}$$

$$z'_{mn} = \int_{V_j} \mathbf{J}^*_{\mathbf{m}}(\mathbf{r}, \omega) \int_{V_i} Z(\mathbf{J}_{\mathbf{n}}(\mathbf{r}', \omega))\mathbf{dr}'\mathbf{dr}, \tag{31}$$

$$e'_m = \int_{V'} \mathbf{J}^*_{\mathbf{m}}(\mathbf{r}, \omega) \cdot \mathbf{E}_{\mathbf{inc}}(\mathbf{r}, \omega)dr. \tag{32}$$

In Eq. (30), the impedance matrix and the vector containing the projection of the incident field on the mth eigenmode are primed to emphasize that the inner product is applied here. Notice that now the impedance matrix $\left\{z'_{mn}\right\}$ is not a diagonal matrix anymore. This is because the eigenmodes are only orthogonal in a pseudo inner product sense but are **not** orthogonal in an inner product sense.

Further, we can see the right-hand side of Eq. (20) as an "eigen" incident field,

$$\mathbf{E}_{\mathbf{inc,n}}(\mathbf{r}, \omega) = \lambda_n(\omega)\mathbf{J}_{\mathbf{n}}(\mathbf{r}, \omega). \tag{33}$$

Using Eq. (33), Eq. (31) becomes

$$z'_{mn} = \int_{V_j} \mathbf{J}^*_{\mathbf{m}}(\mathbf{r}, \omega) \cdot \mathbf{E}_{\mathbf{inc,n}}(\mathbf{r}, \omega)\mathbf{dr}. \tag{34}$$

Hence, physically z'_{mn} gives the work done by the nth"eigen" incident field on the mth eigenmode and quantifies the energetic coupling between different eigenmodes. We can give the following two definitions,

Definition 1: When $n = m$, $z'_{nn}(\omega)$ is defined as the self-impedance of the nth eigenmode.

Definition 2: When $n \neq m$, $z'_{mn}(\omega)$ is defined as the mutual impedance between the mth eigenmode and the nth eigenmode.

Quantitatively, by evaluating Eq. (34), a self-impedance and a mutual impedance are related with the eigenvalues,

$$z'_{\mathrm{mn}}(\omega) = d_{mn}(\omega)\lambda_n(\omega). \tag{35}$$

In Eq. (35), $d_{mn}(\omega)$ is the inner product of the mth and the nth eigenmodes and represents a unitless factor taking into account the fact that the eigenmodes are not orthogonal in the inner product sense, that is, not *energetically orthogonal*

$$d_{mn}(\omega) = \int_V \mathbf{J}^*_{\mathbf{m}}(\mathbf{r}, \omega) \cdot \mathbf{J}_{\mathbf{n}}(\mathbf{r}, \omega) dr. \tag{36}$$

Note that in Eqs. (30), (31), and (34), the factor $\frac{1}{2}$ has been systematically omitted.

3.2. Natural modes of a nanoantenna

As a special case of Eq. (5) where the eigenvalue is at zero, we can define the natural mode for a nanoantenna [34, 35]

$$Z(\mathbf{J}_{\boldsymbol{\alpha}}(r, \omega_\alpha)) = 0 \tag{37}$$

In Eq. (37), $\mathbf{J}_{\boldsymbol{\alpha}}(r, \omega_\alpha)$ is the nth natural mode with a natural frequency ω_α. Different from the eigenmodes discussed in the previous section, the natural modes are independent of frequency. Mathematically, the natural modes originate from the impedance operator's null space and mark the non-trivial solutions to Eq. (37). Physically, the natural modes (including the induced current and the associated field distribution) correspond to the resonant modes of an open resonator (in contrast to a closed resonator like a perfect electric conductor (PEC) cavity). Because of the radiation damping and the material loss in a nanoantenna, a natural frequency is a complex number (remember in a PEC cavity with no losses the natural frequency is real).

In our MoM algorithm, we can numerically solve for a natural mode by seeking for the (complex) frequency where the determinant of the impedance matrix is zero,

$$\det(\{z_{ji}(\omega_\alpha)\}) = 0, \quad \omega_\alpha \in \text{complex plane} \tag{38}$$

We are especially interested in how a natural mode and a natural frequency affect the optical response to an external perturbation at an angular frequency ω. The induced current is obtained by inverting the impedance operator,

$$\mathbf{J}_{\mathbf{ind}}(\mathbf{r}, \omega) = Z^{-1}(\mathbf{E}_{\mathbf{inc}}(\mathbf{r}, \omega)). \tag{39}$$

As in Eqs. (11), (25), and (30), the impedance operator can be recast for a specific set of basis functions. Eq. (39) can be further written as

$$\mathbf{J}_{\mathbf{ind}}(\mathbf{r}, \omega) = \frac{adj(Z)}{det(Z)} \mathbf{E}_{\mathbf{inc}}(\mathbf{r}, \omega). \tag{40}$$

adj(Z) and det(Z) are, respectively, the adjugate and the determinant of the impedance matrix (operator). In general, the determinant is a polynomial that can be decomposed into a product of $(\omega - \omega_\alpha)^i$ where i indicates the degree of degeneracy. In the discussion, it is assumed that the natural mode is not degenerate so that $i = 1$. The induced current can be rewritten as

$$\mathbf{J}_{\mathbf{ind}}(\mathbf{r}, \omega) = \sum_{\alpha=1}^{\infty} \frac{\mathbf{R}_{\boldsymbol{\alpha}}(\boldsymbol{\omega}_{\boldsymbol{\alpha}})}{(\omega - \omega_\alpha)} \mathbf{E}_{\mathbf{inc}}(\mathbf{r}, \omega). \tag{41}$$

R_α is the system residue matrix at a natural frequency. It can be shown that the residue matrix is a dyadic and can be decomposed as the outer product of a natural mode and its transpose,

$$\boldsymbol{R}_\alpha(\boldsymbol{\omega}_\alpha) = \lceil \mathbf{J}_{\boldsymbol{\alpha}}(\boldsymbol{r}, \omega_\alpha)\rangle\langle \mathbf{J}_{\boldsymbol{\alpha}}(\boldsymbol{r}, \omega_\alpha)\rfloor. \tag{42}$$

Substituting Eq. (42) into Eq. (41), the induced current can be expressed as

$$\mathbf{J}_{\mathbf{ind}}(\mathbf{r}, \omega) = \sum_{\alpha=1}^{\infty} \frac{\langle \mathbf{J}_{\boldsymbol{\alpha}}(\boldsymbol{r}, \omega_\alpha)|\mathbf{E}_{\mathbf{inc}}(\mathbf{r}, \omega)\rangle}{(\omega - \omega_\alpha)} \lceil \mathbf{J}_{\boldsymbol{\alpha}}(\boldsymbol{r}, \omega_\alpha)\rangle. \tag{43}$$

Based on Eq. (43), we can clearly define the weighting coefficient for a natural mode,

$$c_\alpha(\omega) = \frac{\langle \mathbf{J}_{\boldsymbol{\alpha}}(\boldsymbol{r}, \omega_\alpha)|\mathbf{E}_{\mathbf{inc}}(\mathbf{r}, \omega)\rangle}{(\omega - \omega_\alpha).} \tag{44}$$

As the natural frequency is a complex number,

$$\omega_\alpha = \omega_{\alpha r} + j\omega_{\alpha i}, \tag{45}$$

For an incident field with an angular frequency ω around $\omega_{\alpha r}$, we can approximate the coupling coefficient by

$$c_\alpha(\omega) \sim \frac{1}{\sqrt{(\omega - \omega_{\alpha r})^2 + \omega_{\alpha i}^2}}. \tag{46}$$

In the above approximation, the imaginary part of the natural frequency is assumed to be small. It is readily seen from Eq. (46) that when ω is around $\omega_{\alpha r}$, the weighting coefficient is maximized.

Moreover, consider an incoming electromagnetic pulse with a delta time dependence $\delta(t)$ and apply an Inverse Fourier Transform to Eq. (43) with respect to the frequency. In time domain, Eq. (44) reads,

$$\mathbf{J}_{\mathbf{ind}}(\mathbf{r}, t) = \sum_{\alpha=1}^{\infty} k_\alpha(\omega)\mathbf{J}_{\boldsymbol{\alpha}}(\boldsymbol{r}, \omega_\alpha)e^{j\omega_\alpha t}. \tag{47}$$

In Eq. (47), $k_\alpha(\omega)$ describes the coupling between a natural mode and an incident field. Due to the complex nature of the natural frequency, the natural mode oscillates at a frequency $\omega_{\alpha r}$ but decays with a rate of $\omega_{\alpha i}$. Since the decaying rate is linked with the dissipated power and the stored energy in a scatterer,

$$-2\omega_{\alpha i} = \frac{P_{\text{dissipated}}}{W_{\text{stored}}}. \tag{48}$$

Accordingly, the "life time" of a natural mode can be defined as $\frac{1}{-2\omega_{\alpha i}}$. In this way, a modal quality factor Q_{mod} can be defined for each natural mode,

$$Q_{\text{mod}} = \frac{\omega_{\alpha r} W_{\text{stored}}}{P_{\text{dissipated}}} = -\frac{\omega_{\alpha r}}{2\omega_{\alpha i}}. \quad (49)$$

Again consider the frequency domain and apply an incident field with an angular frequency ω around $\omega_{\alpha r}$. Assuming that the αth natural mode is dominant, the induced current is

$$\mathbf{J}_{\mathbf{ind}}(\mathbf{r}, \omega) \sim \frac{\langle \mathbf{J}_{\alpha}(r, \omega_{\alpha}) | \mathbf{E}_{\mathbf{inc}}(\mathbf{r}, \omega) \rangle}{(\omega - \omega_{\alpha})} \lceil \mathbf{J}_{\alpha}(r, \omega_{\alpha}) \rangle. \quad (50)$$

That is, the response near the resonance is largely determined by the excited natural mode. Since a fixed stored energy and dissipated power are associated with the natural mode, the modal quality factor Q_{mod} renders a good estimation for the quality factor Q_{res}at resonance.

3.3. Examples

In this section, we take real nanostructures to illustrate the above theoretical discussions on the eigenmodes and natural modes of a nanoantenna.

3.3.1. Eigenmodes of a nanobar

Assume a normal incident field polarized along the longest axis of the nanobar structure (**Figure 5**). The real part and the imaginary part of the self-impedance for the first three eigenmodes are demonstrated in **Figure 6(a)** and **(b)**. Note that the coupling coefficient is inversely proportional to the eigenvalue as in Eq. (27) and the eigenvalue is closely related with the self-impedance as in Eq. (35). When the absolute value of the self-impedance reaches its minimum, the coupling coefficient may maximize. Comparing **Figure 6(a)** and **(b)** with **Figure 6(c)**, this is indeed the case for the first mode (the L1 mode) and the third mode (the L2 mode). For the L2 mode, the coupling (the integral in the numerator of Eq. (27)) between the incident light and the mode pattern (see the top surface charge distribution for the first three

Figure 5. The topology, the depth profile, and the SEM image of the nanobar ($L = 370$ nm, $W = 70$ nm, and $H = 50$ nm) are presented. The white bar represents 100 nm.

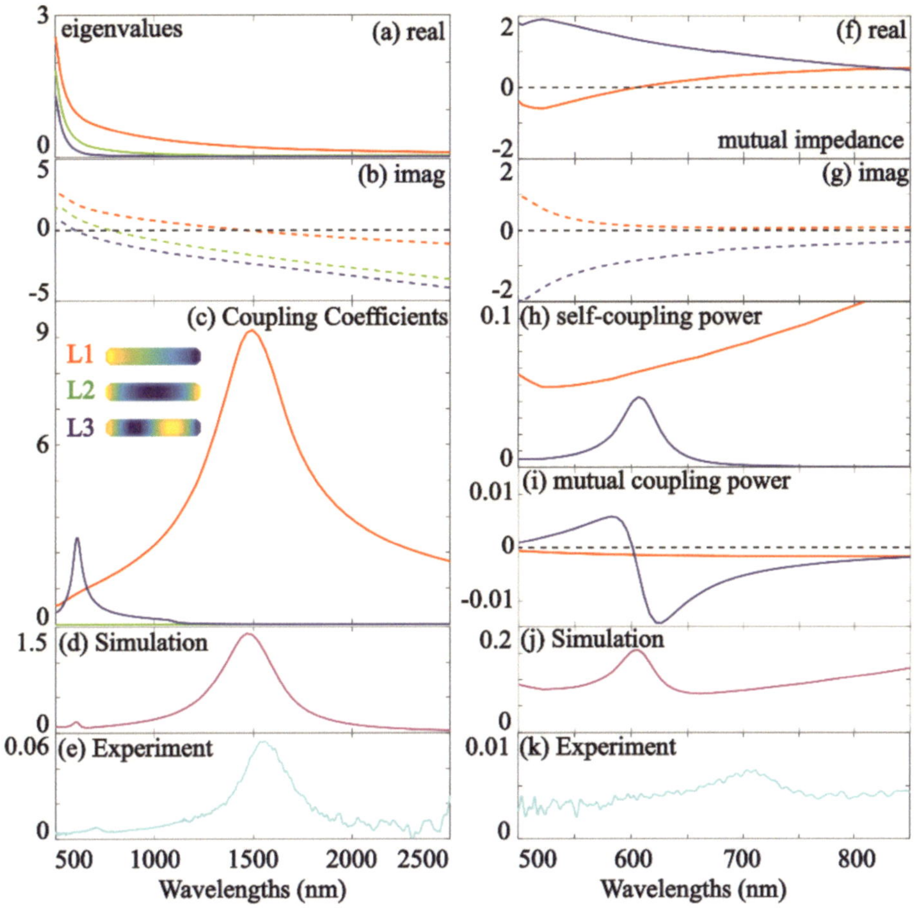

Figure 6. The left column illustrates the real part (a) and the imaginary part (b) of the self-impedance for the L1 – L3 mode, coupling coefficients (c) for the L1 mode and the L3 mode and extinction cross sections (d and e), while the right column demonstrates the real part (f) and the imaginary part (g) of the mutual impedance for the L1 mode and the L3 mode, the self-coupling power (h) of the L1 mode and the L3 mode, the mutual coupling power (i) between the L1 and L3 modes and the extinction cross sections (j and k). In the inset of (c), the top surface charge distribution for the first three modes is plotted. The colour there is coded from blue to yellow to mark the polarity of the surface charge (in print: grey scale). In (d), the V-MoM simulation is marked by the solid line. In (j), the total dissipated power is marked by the solid line.

modes in **Figure 6(c)**) should be taken into account. Since the L2 mode pattern is symmetric with respect to the short axis of the bar but the normal incident field is anti-symmetric, the coupling (the integral in the numerator of Eq. (27)) is zero, that is, the second mode does not contribute to the optical response at all. Further discussions on the symmetry can be found in Section 4. Lastly, comparing the numerically calculated (by the V-MoM algorithm) and experimentally obtained extinction cross sections with the coupling coefficients, one finds that although the coupling coefficients and the extinction cross sections follow the same trend, there is a shift. This shift is actually from the well-known near-field—far-field shift, which is a direct consequence of the radiation loss.

In the right column of **Figure 6**, the role of the mutual impedance in determining the antenna's optical response is studied. As discussed in Eq. (34), the mutual impedance describes the energetic coupling between different modes.

First, since the first mode and the third mode (i.e., the odd-order modes which are anti-symmetric with respect to the bar's short axis) have a different symmetry compared to the second mode (the even-order modes are symmetric with respect to the bar's short axis), the mutual coupling between the odd-order modes and the even-order modes is zero. Further, because the second mode is not excited by the incident field at all, the second mode does not contribute in any way to the antenna's response.

Second, the first mode and the third mode are energetically coupled (see the mutual impedance in **Figure 6(f)** and **(g)**). As a result, the excitation of one of these two modes would directly lead to the excitation of another. Taking this power transfer in between different modes into account breaks the symmetric line shape in the self-coupling power and generates the so-called asymmetric Fano line shape (see **Figure 6(h)–(k)**).

3.3.2. *Natural modes of nanopatches and nanodimer*

In this section, three nanostructures, that is, a gold (Au) nanopatch, a nickel (Ni) nanopatch, and a gold (Au) nanodimer, are considered. The nanopatches are fabricated on top of an SiO_2 (100 nm)/Si layer, while the nanodimer sits on an SiO_2 substrate (see **Figure 7**). In the numerical simulations, the same mesh ($5 \times 4 \times 1$) is used for the patch structures, while each monomer in the dimer is discretized by a $3 \times 2 \times 1$ mesh. For the sake of simplicity, the multilayer substrate is not explicitly taken into account but is modeled by a homogeneous surrounding environment with an effective refractive index ($n = 1.25$).

Essentially, two observations can be immediately made from the simulations and experiments shown in **Figure 8**:

First, natural modes are fundamental to the optical response of a nanoantenna. Depending on the incident field, some of the modes are excited, while some of the modes are forbidden. For the patch, due to the comparable dimensions along the long axis and the short axis, it is readily expected that the first two resonant modes are horizontal and vertical dipolar modes. Depending on the excitation, that is, horizontally (vertically) polarized normal incident light,

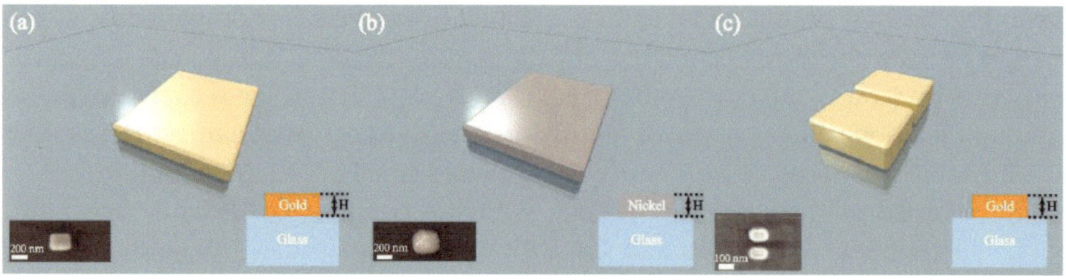

Figure 7. The design of an Au nanopatch (a), a nickel nanopatch (b) and a gold nanodimer (c). Both patches have the following dimensions: $L = 250$ nm, $W = 200$ nm, and $H = 25$ nm, while the dimensions of the dimer are $L = 135$ nm, $W = 100$ nm, $H = 50$ nm, and gap = 20 nm). This figure is reproduced from [35].

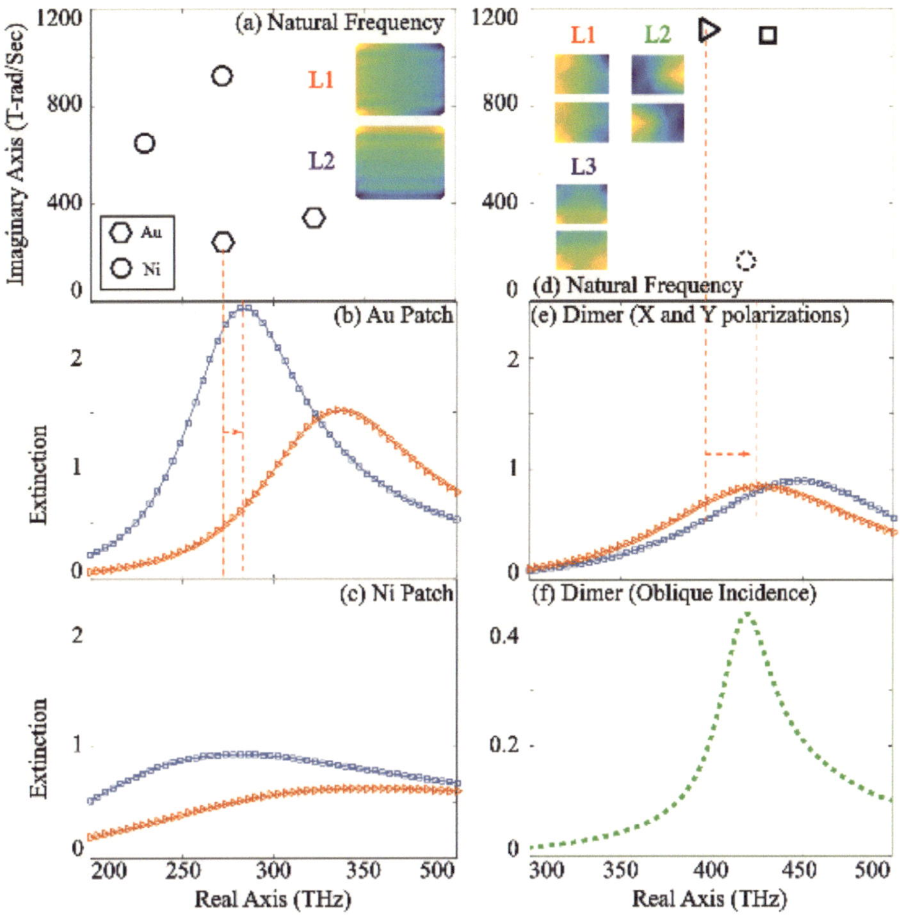

Figure 8. Natural frequencies, natural modes, and scattering/extinction spectra for the patches (the left column) and the dimer (the right column). (a) and (d) illustrate the natural frequencies and the natural modes for all the structures. In the inset of (a) and (d), the top surface charge distribution for the natural modes is illustrated. The colour is coded from blue to yellow to denote the polarity of the charge concentration (in print: grey scale). In (b-c) and (e-f), the scattering and extinction cross sections from simulations are demonstrated. In (b) and (c) of the left column, the red (in print: the line with triangle markers) and blue colour (in print: the line with square markers) respectively correspond to the response of the horizontal and the vertical modes while in (e) and (f) the right column, the red, green and blue colours (in print: the solid line with triangle markers, the solid line with square markers and the dotted line) represent the optical response from the L1, L2 and L3 modes. This figure is reproduced from [35].

the horizontal (vertical) dipolar mode is excited (see the left column in **Figure 8**). The same argument applies to the dimer structure (see **Figure 8**). Especially, the second mode (the L2 mode) can be only excited by an oblique incident field.

Second, natural frequencies control the resonances of a nanoantenna. From **Figure 8**, it is clear that the real part of a natural frequency gives a "baseline" for the resonance, while the imaginary part finally shifts the resonance to the "right" position. Another role that the imaginary part plays is that it is associated with the loss (both radiation and material dissipation) in the structure. A large imaginary part can "flatten" the resonance (e.g., the gold patch has less material loss than the nickel one, so it has a sharper resonance, or the L2 mode in the dimer has less radiation loss than the other two modes, resulting in a sharper resonance).

4. Symmetries

In this section, we introduce a set of symmetry arguments to determine under which circumstances the eigenmodes are orthogonal to each other in an inner product sense (as defined in Eq. (29)). Since a natural mode is just a special case of an eigenmode, the developed theory applies to the natural mode case as well. In the following, we shall concentrate solely on the symmetries and the eigenmodes of a nanoantenna.

To make our discussion more pedagogical, we again take the simple bar shape structure (see **Figure 9(a)**) as an example. The bar structure carries a C_2 symmetry group. There are two symmetry operations in this group: the identity operation E where no transformation is conducted, and a rotation of π about the z-axis, the C_2 symmetry operation.

Since symmetry operations are always applied based on coordinates, we should be able to find a corresponding set of matrices to "represent" these operations. Here, we especially focus on the matrices with the lowest dimensionalities, that is, the irreducible representations. Since the group under discussion is an abelian group, we have two irreducible representations and they are shown in **Figure 9(b)**.

Moreover, in contrast to these transformations operating on coordinates, we follow Wigner's conventions [36, 37] and define transformation operators which operate on **functions**,

$$P_R f(\boldsymbol{r}) = f\left(R^{-1}\boldsymbol{r}\right),\ P_R \boldsymbol{f}(\boldsymbol{r}) = R \cdot \boldsymbol{f}\left(R^{-1}\boldsymbol{r}\right). \tag{51}$$

In Eq. (51), this definition is illustrated for both scalar functions (such as charge, etc.) and vector functions (such as currents, electromagnetic fields, etc.). These transformation operators are commutative with the impedance operator defined in Eq. (5), that is, the sequence of applying the transformation operator and the impedance operator does not affect the final outcome of the calculation,

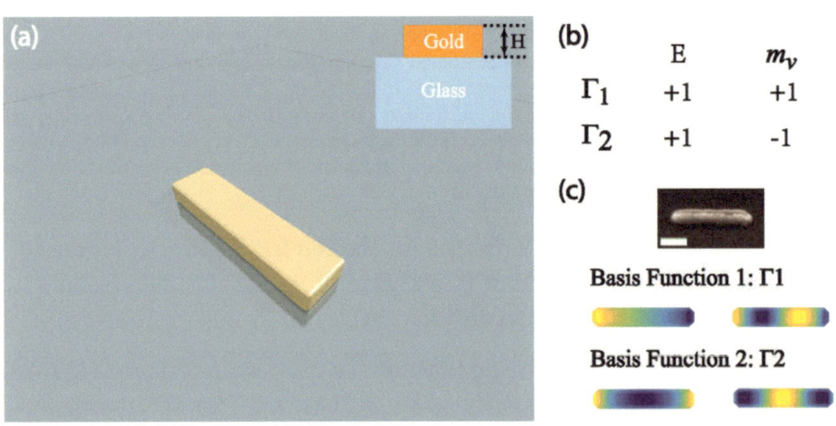

Figure 9. (a) The length (L), width (W) and height (H) of the nanobar are 370 nm, 70 nm and 50 nm. The bar sits on top of a SiO2 substrate characterized by a constant refractive index of 1.5. (b) The irreducible representation of the C2 group is shown. E and mv respectively represent the identity operation and the mirroring operation with respect to the short axis of the bar structure (marked by the red dash line in (c)). (c) The top surface charge pattern of the first four modes (L1-L4) are shown. The colours from blue to yellow are used to denote the charge polarity (in print: grey scale).

$$P_R Z(\mathbf{J_{ind}}(\mathbf{r}, \omega)) = Z(P_R \mathbf{J_{ind}}(\mathbf{r}, \omega)). \quad (52)$$

Combining the group's irreducible representations and its transformation operators, we can further construct a set of projection operators for the group under discussion,

$$\wp_j = \frac{l_j}{h} \sum_R \Gamma_j^*(R) P_R. \quad (53)$$

In Eq. (53), a projection operator is characterized by the subscript j which marks an irreducible representation. Here, j may run from one to two. l_j is the dimensionality of an irreducible representation. Since every irreducible representation has a dimensionality of one, l_j is equal to one. Then, the summation is carried out with respect to all the symmetry operations.

Based on the defined projection operator, we can categorize all the eigenmodes according to the irreducible representations (see an illustration of an eigenmode categorization in **Figure 9(c)**). It is well known that [36, 37] the functions (e.g., in the current case, the eigenmodes) that belong to different irreducible representations are orthogonal to each other in an inner product sense.

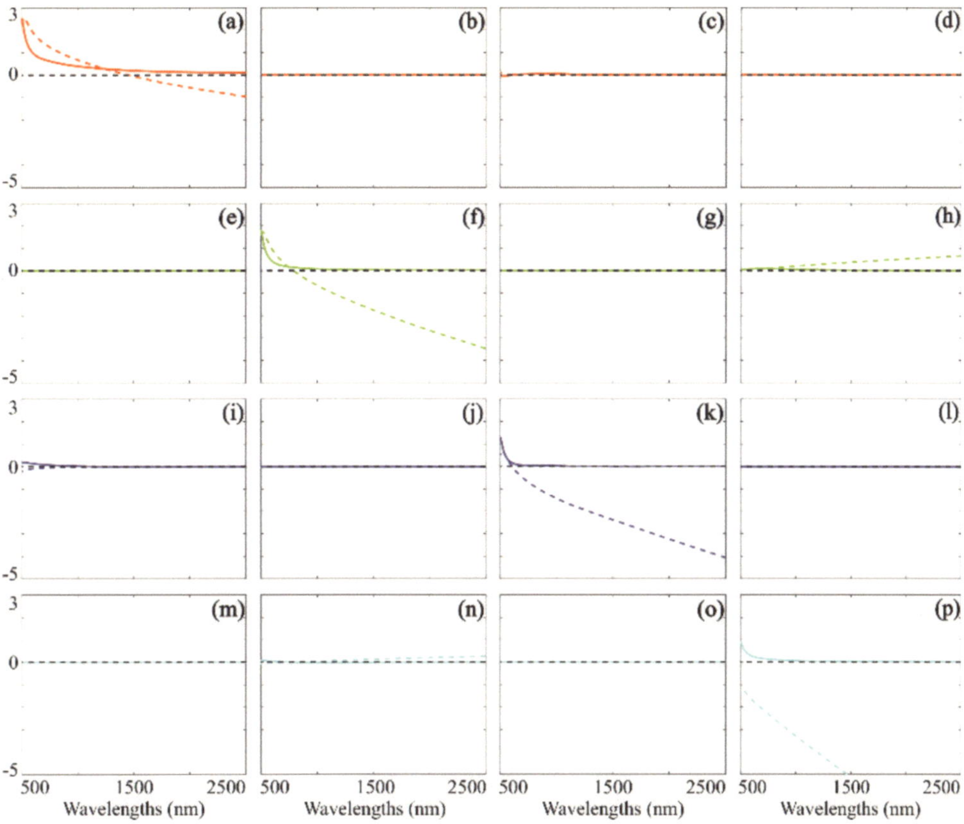

Figure 10. The mutual impedance (a) – (p) between the first four modes (see **Figure 9**) of the bar structure. This figure is organized as follows. A column represents a "source" mode that generates eigen-incident field as in Eq. (33) while a row represent an "observation" mode that the eigen-incident field is applied onto. Thus, the figure at the cross of the mth row and the nth column denotes the mutual impedance z'_{mn} as defined in Eq. (34). This figure is reproduced from [36].

This is the very condition that controls the modes' orthogonality (in an inner product sense). The above is equivalent to saying that, according to the irreducible representations, an eigenspace (in which one finds all the eigenmodes) can be split into several invariant subspaces (in which one finds all the eigenmodes belonging to an irreducible representation).

To demonstrate this condition, the energetic coupling among the first four eigenmodes for the bar structure is calculated with the V-MoM algorithm discussed in this chapter. Upon inspecting **Figure 10**, it can be immediately concluded that the even-order modes do not energetically couple with the odd-order modes. This is the same as we have seen in Section 3.3.1, where the vanishing energetic coupling is attributed to the integration of the product of an odd function and an even function. Now, this intuitive idea is formally incorporated in the framework of group theory. That is, since the even and the odd eigenmodes belong to different irreducible representations, they must be orthogonal in an inner product sense.

Following the same line of reasoning as in Section 3.3.1, an excitation is applied (see **Figure 11**). The incident field is polarized along the bar's long axis but the incident direction is oblique. As already seen in **Figure 6**, power transfer between modes of the same parity, that is, the even

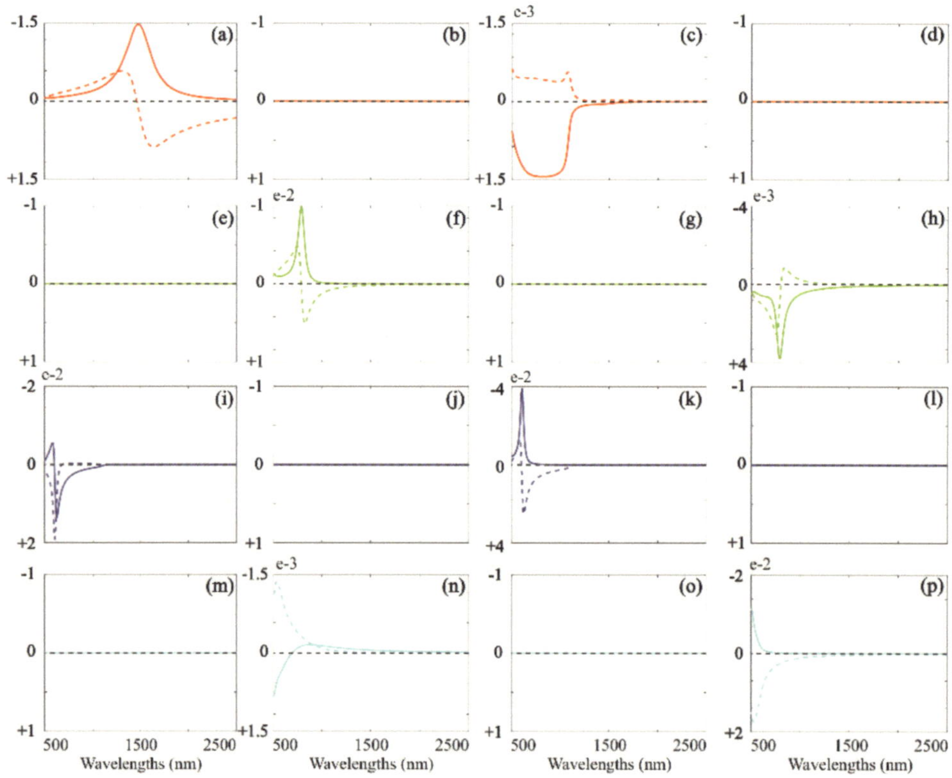

Figure 11. The power transfer (a) – (p) between the first four modes (see **Figure 9**) of the bar structure. This figure is organized similar to Figure 10. A column represents a "source" mode that generates eigen-incident field as in Eq. (33), while a row represent an "observation" mode that the eigen-incident field is applied onto. However, different from Figure 10, the coupling coefficients of the "source" and the "observation" modes are included. Thus, the figure at the cross of the mth row and the nth column denotes the power transfer p'_{mn} as defined in Eq. (54). This figure is reproduced from [36].

(odd)-order modes, is allowed and can be evaluated by including the effects of the different eigenmodes' coupling coefficients,

$$p'_{mn} = c^*_m(\omega)c_n(\omega)d_{mn}(\omega)\lambda_n(\omega). \tag{54}$$

Eq. (54) is a direct extension of Eq. (35). However, as stipulated by group theory, the power transfer between modes of different parity is forbidden. Lastly, as a comment, different from **Figure 6**, due to the obliquely incident field, the even-order modes are now excited, and hence the resonance and the energetic coupling between the second mode and the fourth mode is seen.

5. Conclusions

This chapter presents a complete set of numerical tools to tackle the nanoscale light-matter interaction problem. By comparing with experimental results, the V-MoM solver and its extensions are proven to be very efficient and accurate. The modal analysis scheme and the group theoretical-based analysis scheme are demonstrated to deliver a deep physical understanding of the physical system under study. Therefore, we believe that the developed integral equation-based solver may become an essential complement for commercial solvers which are mostly based on differential equations.

Author details

Xuezhi Zheng[1*], Guy A.E. Vandenbosch[1] and Victor V. Moshchalkov[2]

*Address all correspondence to: xuezhi.zheng@esat.kuleuven.be

1 Department of Electrical Engineering (ESAT-TELEMIC), KU Leuven, Belgium

2 Laboratory of Solid State Physics and Magnetism, KU Leuven, Belgium

References

[1] Novonty L, Hecht B. Principles of Nano-Optics. 2nd ed. Cambridge University Press (New York); 2012. P.578

[2] Gaponenko SV. Introduction to Nanophotonics. Cambridge University Press (New York); 2010. P.484

[3] Aslan K, Lakowicz JR, Geddes CD. Plasmon light scattering in biology and medicine: New sensing approaches, visions and perspectives. Current Opinion in Chemical Biology. 2005;**9**(5):538–544

[4] Hirsch LR, Stafford RJ, Bankson JA, Sershen SR, Rivera B, Price RE, et al. Nanoshell-mediated near-infrared thermal therapy of tumors under magnetic resonance guidance. Proceedings of the National Academy Of Sciences United States of America. 2003;**100** (23):13549–13554

[5] Liedberg B, Nylander C, Lunstrom I. Surface plasmon resonance for gas detection and biosensing. Sensors and Actuators. 1983;**4**:299–304

[6] Gray SK. Surface Plasmon—Enhanced spectroscopy and photochemistry. Plasmonics. 2007;**2**(3):143–146

[7] Schatz GC. Theoretical studies of surface enhanced Raman scattering. Accounts of Chemical Research. 1984;**17**(10):370–376

[8] Atwater HA, Polman A. Plasmonics for improved photovoltaic devices. Nature Materials. 2010;**9**:205–213

[9] Valev VK, Silhanek AV, De Clercq B, Gillijns W, Jeyaram Y, Zheng XZ, et al. U-shape switches for optical information processing at the nanoscale. Small. 2011;**7**(18):2573–2576

[10] Vercruysse D, Zheng XZ, Sonnerfraud Y, Verellen N, Di Martino G, Lager L, et al. Directional fluorescence emission by individual V-antennas explained by mode expansion. ACS Nano. 2014;**8**(8):8232–8241

[11] Tao D, Ruttiger C, Zheng X, Benz F, Ohadi H, Vandenbosch GAE, et al. Fast dynamic color switching in temperature-responsive plasmonic films. Advanced Optical Materials. 2016;**4**(6):877–882

[12] Lumerical. Available from: www.lumerical.com (2003-2017)

[13] COMSOL. Available from: www.comsol.com (2017)

[14] Chew WC. Waves and Fields in Inhomogeneous Media. Van Nostrand Reinhold (New York); 1990

[15] Harrington RF. Field Computation by Moment Methods. Wiley-IEEE Press (New York); 1993

[16] Rao SM, Wilton DR, Glisson AW. Electromagnetic scattering by surfaces of arbitrary shape. IEEE Transactions on Antennas and Propagation. 1982;**30**(3):409–418

[17] Schaubert DH, Wilton DR, Glisson AW. A tetrahedral modeling method for electromagnetic scattering by arbitrary shaped inhomogeneous dielectric bodies. IEEE Transactions on Antennas and Propagation. 1984;**32**(1):77–85

[18] Vandenbosh GAE, Van de Capelle AR. Mixed-potential integral expression formulation of the electric field in a stratified dielectric medium-application to the case of a probe current source. IEEE Transactions on Antennas and Propagation. 2002;**40**(7):806–817

[19] Demuynck FJ, Vandenbosch GAE, Van de Capelle AR. The expansion wave concept–Part I: Efficient calculation of spatial Green's functions in a stratified dielectric medium. IEEE Transactions on Antennas and Propagation. 1998;**46**(3):397–406

[20] Vandenbosch GAE, Demuynck FJ. The expansion wave concept—Part II: A new way to model mutual coupling in microstrip arrays. IEEE Transactions on Antennas and Propagation. 1998;**46**(3):407–413

[21] Vrancken M, Vandenbosch GAE. Hybrid dyadic-mixed-potential and combined spectral-space domain integral-equation analysis of quasi-3-D structures in stratified media. IEEE Transactions on Microwave Theory and Techniques. 2003;**51**(1):216–225

[22] Zheng X, Valev VK, Verellen N, Jeyaram Y, Silhanek AV, Metlushko V, et al. Volumetric method of moments and conceptual multi-level building blocks for nano topologies. IEEE Photonics Journal. 2012;**4**(1):267–282

[23] Zheng X, Valev VK, Volskiy V, Vandenbosch GAE, Moshchalkov VV. On the use of a hierarchical multi-level building block basis function scheme in periodic plasmonic structures. Applied Physics A. 2014;**115**(2):415–419

[24] Volskiy V, Vandenbosch GAE. Quasi 3D mixed potential integral equation formulation for periodic structures. IEEE Transactions on Antennas and Propagation. 2013;**61**(4): 2068–2076

[25] Valev VK, Silhanek AV, Jeyaram Y, Denkova D, De Clercq B, Petkov V, et al. Hotspot decorations map plasmonic patterns with the resolution of scanning probe techniques. Physics Review Letters. 2011;**106**:226803

[26] Valev VK, Baumberg JJ, De Clerq B, Braz N, Zheng XZ, et al. Nonlinear superchiral metasurfaces: Tuning chirality and disentangling non-reciprocity at the nanoscale. Advanced Materials. 2014;**26**(24):4074–4081

[27] Valev VK, De Clerq B, Biris CG, Zheng X, Vandendriessche S, Hojeij M, et al. Distributing the optical near field efficient field-enhancements in nanostructures. Advanced Materials. 2012;**24**(35):OP208–OP215

[28] Valev VK, Denkov D, Zheng XZ, Kuznetsov AI, Reinhardt C, Chichkov BN, et al. Plasmon enhanced sub-wavelength laser ablation: Plasmonic nanojets. Advanced Materials. 2012;**24**(10):OP29–OP35

[29] Valev VK, Zheng XZ, Biris CG, Silhanek AV, Volskiy V, De Clercq B, et al. The origin of second harmonic generation hotspots in chiral optical metamaterials (Invited). Optical Materials Express. 2011;**1**(1):36–45

[30] Valev VK, Clercq BD, Zheng XZ, Denkova D, Osley EJ, Vandendriessche S, et al. The role of chiral local field enhancements below the resolution limit of second harmonic generation microscopy. Optical Express. 2012;**20**(1):256–264

[31] Valev VK, Libaers W, Zywietz U, Zheng XZ, Centini M, Pfullmann N, et al. Nanostrip length dependence of plasmon induced material deformations. Optical Letters. 2013;**38** (13):2256–2258

[32] Hanson GW, Yakovlev AB. Operator Theory for Electromagnetics: An Introduction. Springer-Verlag (New York); 2013

[33] Zheng XZ, Verellen N, Volskiy V, Valev VK, Baumberg JJ, Vandenbsoch GAE, et al. Interacting plasmonic nanostructures beyond the quasi-static limit: A circuit model. Optical Express. 2013;**21**(25):31105–31118

[34] Zheng X, Volskiy V, Valev VK, Vandenbosch GAE, Moshchalkov VV. Line position and quality factor of plasmonic resonances beyond the quasi-static limit: A full-wave eigenmode analysis route. IEEE Journal of Selected Topics in Quantum Electronics. 2012;**19** (3):4600908

[35] Zheng X, Valev VK, Verellen N, Volskiy V, Herrmann LO, Van Dorpe Pol, et al. Implementation of the natural mode analysis for nanotopologies using a volumetric method of moments (V-MoM) algorithm. IEEE Photonics Journal. 2014;**6**(4):4801413

[36] Zheng X, Verellen N, Vercruysse D, Volskiy V, Van Dorpe P, Vandenbosch GAE, et al. On the use of group theory in understanding the optical response of a nanoantenna. IEEE Transactions on Antennas and Propagation. 2015;**53**(4):1589–1602

[37] Tinkham M. Group Theory and Quantum Mechanics. McGraw-Hill (New York) 1964

Chapter 6

Plasmonic Field Distribution of Homo- and Hetero Dimeric Ag and Au Nanoparticles

Nasrin Hooshmand

Additional information is available at the end of the chapter

http://dx.doi.org/10.5772/67411

Abstract

Silver (Ag) and gold (Au) nanoparticles are known to have very strong plasmonic fields among the other plasmonic metallic nanoparticles. When two Ag or Au nanoparticles are brought close together, hot spots (strong electromagnetic field) are formed between the particles, which can be exploited in imaging and sensing applications. In this chapter, we used the discrete dipole approximation (DDA) to investigate the interdimer separation dependence of the localized surface plasmon resonance (LSPR) of homo- and heterodimers of Ag and Au nanocubes (NCs) when the exciting incident light is polarized parallel to the dimer axis. It was found that as the interdimer separation changes, the plasmonic field distribution around the nanocubes' surface varied. The results from the homodimers showed that the primary plasmon band red-shifted in accordance with the universal scaling law and the hot spots geometry changed abruptly at small separations. The results simulated at very short distances showed that the hot spots formed in between the adjacent facets and away from the corners of these facets. However, at larger separations, it moved toward the adjacent corners. For heterodimers, unusual behavior was observed. It showed that the E-field resulting from excitation of the Ag-dominated plasmon resonance was significantly weaker than expected, and the red shift of the gold-dominated plasmon resonance did not follow the universal scaling law. It is likely that the silver plasmon mixes with the gold interband transition to form a hybrid resonance that produces weaker overall field intensity.

Keywords: plasmonics, DDA, coupling, nanoparticles, dimer

1. Introduction

Plasmonic nanoparticles made of Ag or Au received significant attention from researchers as it showed unique properties when they interact with the electromagnetic radiation, primarily in the visible region. The localized surface plasmon resonance (LSPR) is an example for one such property. This results from the resonant excitations of the collective oscillations of their conduction band electrons [1–4]. The LSPR is tunable and strongly dependent on the shape, size, composition, and relative dielectric function of the nanostructure [4–11]. The oscillation of the electrons results in a strong enhancement in the optical absorption, scattering and near-field intensities of noble metal nanoparticles allowing for their use in numerous applications such as biological imaging, selective photothermal therapy, surface enhanced Raman scattering (SERS), optical wave guiding, and biochemical sensing [12]. It is well known that as interparticle gaps are reduced, localized regions of intense electromagnetic fields known as "hot spots" [13, 14] will be formed, which are of interest in surface-enhanced Raman spectroscopy (SERS). This is primarily due to the spatial overlap of the individual plasmonic modes, which induces the formation of hybridized collective plasmonic modes. At nanoscale separations, the hot spots produced in these composite frameworks exhibit noteworthy enhancements in Raman scattering, fluorescence, infrared absorption; which has been useful for a variety of applications [15–17]. Substantial Raman enhancement can be generated, using larger aggregates experimentally [18].

The effect of coupling between the surface plasmons of adjacent particles, such as in nanoparticle aggregates, is very important and has received a great deal of attention in recent years. Coupling typically results in a shift in the LSPR wavelength and a great enhancement of the plasmonic electromagnetic field (E-field) [19, 20]. This is important in optical technologies such as chemical and biological imaging [21, 22], sensing [23–25], and therapeutics [26–29]. This interparticle plasmon coupling forms the basis of the intense enhancement of spectroscopic signals (e.g., SERS) from molecules adsorbed at nanoparticle junctions, providing the capability for single-molecule sensing and detection [30, 31]. The overall strength of coupling depends on the polarization direction of the exciting field and the distance between the nanoparticles. It has been observed that the LSPR shift due to coupling decreases exponentially with increasing interparticle distance [20, 32]. When normalized to particle size, the relationship holds for many nanoparticles with different sizes, shapes, metals, and media; this has been termed the universal scaling law [19, 33].

Ag and Au NCs with sharp corners are known to have very strong plasmonic fields concentrated at their corners. The study of the coupling between a pair of Ag NCs and the effect of rounding the cube corners on their coupling strength has recently been carried out [34]. This chapter will focus on the theoretical understanding of unique features of assembled nanocubes made of Au and/or Ag plasmonic nanoparticles in face-to-face (FF) orientation. Discrete dipole approximation (DDA) method has been used to study the distance dependence of the plasmonic field coupling between homodimers (Ag-Ag or Au-Au) and heterodimers (Ag-Au) of nanocubes with sharp corners. We examined the dependence of the interaction of particles with light at the interparticle separation gap. Finally, we looked at the LSPR extinction wavelength and the relative field intensity distribution and the hot spot formation (strong E-Field) between adjacent nanoparticles at small interparticle separations.

2. Research methods

Modeling and simulation play a key role in the advancement of nanoscience and nanotechnology. Among various numerical methods, DDA received significant attention as it can provide useful information on plasmonic phenomena. The DDA [35] is one of the most powerful theoretical techniques to model the optical properties of plasmonic nanoparticles of arbitrary geometry. This method was used to calculate near field interaction between of closely placed cubic Ag and Au nanoparticles (edge length = 42 nm) assembled in face-to-face orientation. The refractive index of Ag and Au NCs is assumed to be the same as that of the bulk metal [34]. The surrounding medium was set to be the water (n = 1.33). Simply, in DDA, the target (here Ag and/or Au NCs) represents as cubic arrays of several thousands of points acquire dipole moment in response to the local electric field located on a cubic lattice (with volume d^3). Details of the DDA method have been described elsewhere [35–37]. One of the important prospects of the LSPR to understand is the effect of the electromagnetic field distribution in determining sensing performance of the metal nanoparticles. This needs to go through the fundamental study of the plasmonic properties of nanoparticles. Using DDA [35, 37], we were able to calculate the plasmonic properties such as plasmonic field distribution of a pair of homo- and heterodimeric nanocubes made of Au and Ag at different separations upon exposure to resonant incident electromagnetic field. The Plasmonic field enhancement factor (in log-scale of $|E|^2/|E_0|^2$) was located on the surface of a dimer of cubes with the DDA technique at different excitation wavelengths.

3. Gold nanocube dimer

Plasmonic nanoparticles dimers are very important in this context due to large electromagnetic field formation between them when they are in close proximity and exposed to the incident light [32, 38–41]. The strong electromagnetic field (hot spot) only forms, if the incident light polarized along the dimer axis [42]. **Figure 1a** and **c** shows, as the separation distance between the cubes decreases, the intensity of the SPR bands enhances and red-shifts. The total extinction increases from 5 a.u. for a single particle (or a large separation) to 18 a.u. for 2 nm separation. The extinction band maximum red shifts from 588 to 642 nm, upon decreasing the separation gap between the cubes. These results follow the typical scaling law seen for other particles (**Figure 1b**).

To better understand the dipole coupling behavior between the two nanocubes, the E-field plasmonic enhancement was calculated for a single and the dimer of Au NCs at very short distance (2 nm) and larger distance (10 nm). For calculating the field distribution of single and dimer of Au NCs, the maximum plasmon band wavelength of interest was used to excite the single or dimeric nanoparticles. We excited the single Au NC at 585 nm, a dimer with 2 nm of separation at 641 nm, and a dimer with 10 nm separation at 613 nm (**Figure 2a–c**).

The E-field around the single nanocube showed the characteristic pattern of field concentrated at the corners. As the particles moved away (at 10 nm), it shows that the electromagnetic field became stronger near the adjacent corners compared to the adjacent facets. When the separation is reduced to 2 nm, the E-field strength moved away from the adjacent corners and in between the facing facets of the dimer.

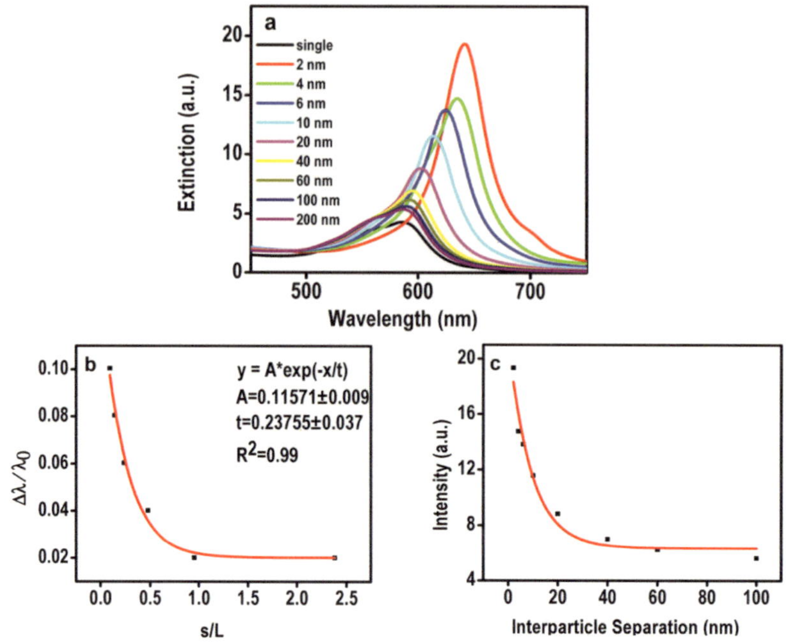

Figure 1. (a) Dependence of the extinction spectra on the separation distance of an Au NCs dimer in water. (b) Exponential behavior of fractional plasmon shift ($\Delta\lambda/\lambda_0$) as a function of the interparticle separation normalized by the length of the nanoparticle. (c) The exponential dependence of extinction maximum on the interparticle separation. This exponential dependence suggests that the extinction spectra of Au NC dimers follow the universal scaling law. Reprinted with permission from Ref. [41]. Copyright 2014 American Chemical Society.

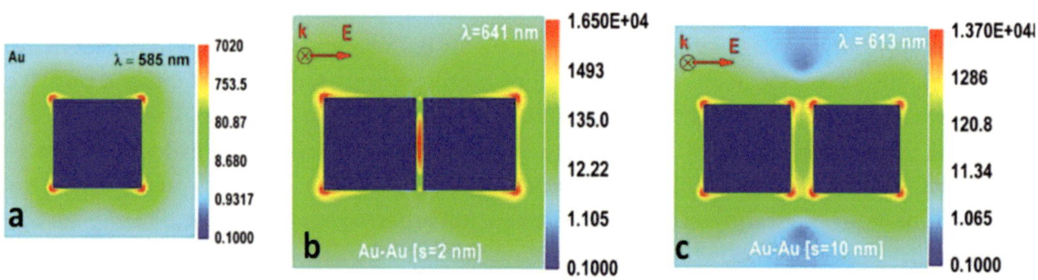

Figure 2. Plasmonic field enhancement for (a) single Au cube, (b and c) the dimer of Au-Au NCs with 2 nm of separation (excited at 641 nm) and 10 nm of separation (excited at 613 nm). As the separation distance increases to 10 nm, each cube in the dimer likely has the E-field distribution in character like which shown for the single cube. However, at 2 nm, the strongest field is located at the center of the adjacent faces. Reprinted with permission from Ref. [41]. Copyright 2014 American Chemical Society.

4. Silver nanocube dimer

In order to compare our results with a pair of Au NCs, we calculated extinction and field distribution for pairs of Ag NCs in the same condition. The extinction spectra for different separations of homodimer of Ag NCs are shown in **Figure 3a** and **c**.

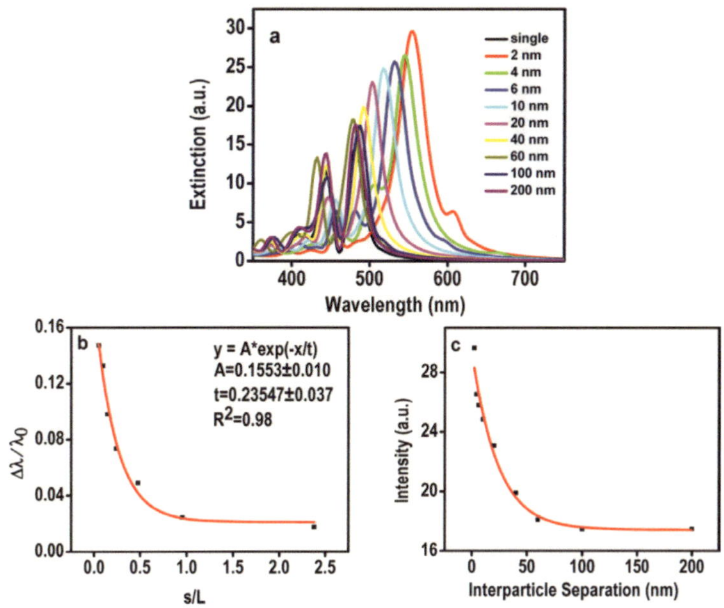

Figure 3. (a) DDA calculated extinction spectra at different distances, (b) exponential behavior of fractional plasmon shift ($\Delta\lambda/\lambda_0$) as a function of the interparticle separation normalized by the length of the nanoparticle and (c) the exponential decrease of the intensity of the strongest extinction peak as the interparticle separation increases. Reprinted with permission from Ref. [41]. Copyright 2014 American Chemical Society.

It shows that the extinction spectrum of a widely separated pair of cubes is nearly identical to that of the monomer. Contrary to the Au, it was found in the single and dimer Ag nanaoparticles that there are two strong bands at long wavelengths (probably dipolar) and two weak higher order polar bands at short wavelengths. As the gap between the cubes decreases, although the bands shift to longer wavelengths, the shorter wavelength bands show further red shift than the longer wavelength bands. While a strong doublet is observed for 60 nm of separation, for 20 nm separation, one band of the doublet becomes weaker and broader in the spectrum. It is possible that the higher order bands become stronger and the higher energy dipolar band is likely to be submerged under the strong higher order band.

As shown in **Figure 3b** with increasing the interparticle gap separation of the Ag NCs dimer, the magnitude of the red shift of the plasmon band decreases based on the universal scaling law [41]. **Figure 4a–c** shows the electromagnetic field distribution of the Ag NCs dimer and single Ag NC. Here, only the main plasmonic band for both single Ag NC (λ_{max} = 481 nm) and Ag NCs dimer at 2 nm (λ_{max} = 553 nm) and 10 nm (λ_{max} = 517 nm) was exited. As the separation distance increases, the E-field distribution becomes similar to the single Au NC.

In order to examine the hybridized dimer modes of the dimer of Ag nanocubes, the electromagnetic field was calculated at 2 and 10 nm separation distances. The results showed that at 10 nm of separation, the field strength is strongest at the corners (**Figure 4c**), whereas, at 2 nm separation, the strongest field moved toward the center of the adjacent faces (**Figure 4b**). Compared to the dimer of the Au NCs, in all cases, the electromagnetic fields for the Ag NCs were stronger, as expected.

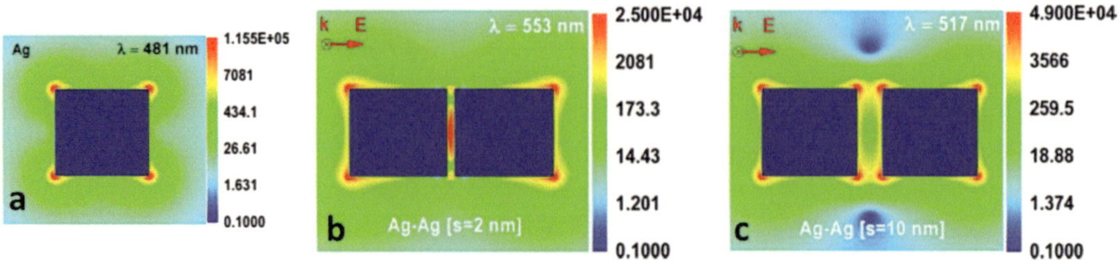

Figure 4. (a) Plasmonic electromagnetic field enhancement for a single Ag NC and the dimers of Ag-Ag NCs with (b) a separation distance of 2 nm (excited at 522 nm) and (c) separation distance of 10 nm (excited at 517 nm). As the separation distance decreases, the hot spots formation takes place in between the faces facing of the dimer and away from their corners. However, when the cubes move away from one another, the field distributions are mostly present around the corners of the faces facing facets of the dimer. Reprinted with permission from Ref. [41]. Copyright 2014 American Chemical Society.

5. Plasmon coupling of gold-silver heterodimeric nanocubes

The results for the Ag-Au hetrodimeric NCs showed unusual results and more exciting conclusions. We expect that the plasmonic modes of gold and silver will mix to form hybrid modes. Simply, if we consider Ψ_{Au} and Ψ_{Ag} as dipolar modes for gold and silver, respectively, we might think that the two hybrid modes would be $(\Psi_{Au} + \Psi_{Ag})$ and $(\Psi_{Au} - \Psi_{Ag})$ [41]. However, the hybrid modes will not have equal contribution of Ag and Au uncoupled modes because the energies of the uncoupled modes are so different (gold at 588 nm and silver at 482 nm). As discussed by Sheikholeslami et al. in their work on Au-Ag nanosphere heterodimers [43], because the interband transition of gold strongly overlaps energetically with the Ag NC's natural LSPR resonance, the LSPR resonance of the Ag NC may strongly couple to it. We showed at large dimer separations, the spectrum somewhat identical to the summation of the spectra for a single Au NC and an Ag NC, and there is no coupling between the Au and Ag cubes [41]. At short separating distance, mixing occurs between the heterodimer of Au and Ag cubes. The main band around 600 nm showed that significant enhancement in characteristic intensity corresponds to the Au-like plasmonic band. This band red shifts with decreasing separation and showed that the band does not follow the universal scaling law as we showed in the homodimer cubes.

In order to understand the interplay between the Au and Ag NCs, the E-fields were calculated at 2 and 10 nm with different excitation wavelengths. The field resulting from excitation of plasmon band at 2 nm (615 nm) and 10 nm (597 nm) is shown in **Figure 5a** and **b**. It shows that, at 10 nm of separation, the field is largely concentrated around the corners of the gold cube (on the right). In Ag NC dimer, at 2 nm of separation, the E-field enhances due to increased mixing-in of Ag dipolar mode. However, it is still weaker than the field around the Au NC. This hybrid mode can be represented as $\Psi_1 = a\Psi_{Au} + b\Psi_{Ag}$ where $a > b$. Just as with the homodimers, the E-field is strongest at the corners and between the facing facets (**Figure 6**).

It shows the field distributing of the heterodimer of Ag-Au nanocubes when exciting the dimer at 487 nm associates with 2 and 10 nm of separations, whereas the hybrid modes have

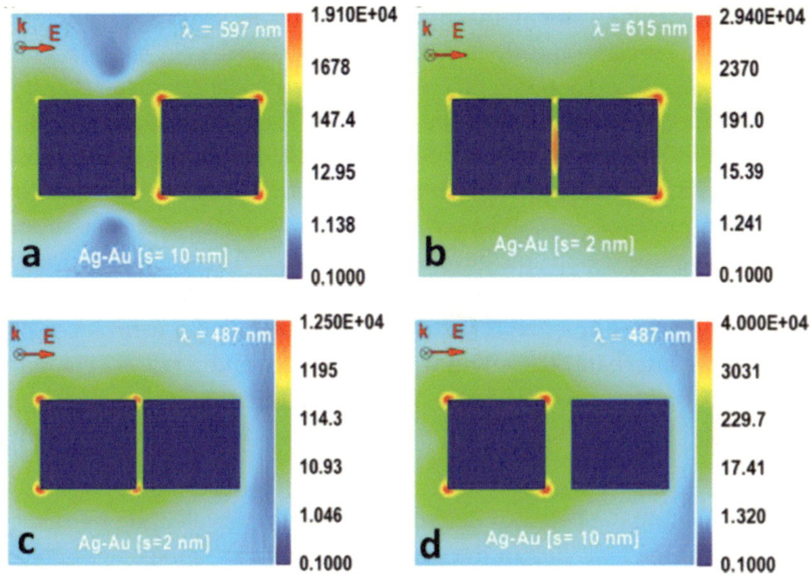

Figure 5. (a and d) Separating the cubes by 10 nm and exciting at 597 nm produces a field almost entirely located on the Au NC. However, exciting at 487 nm exhibited the same pattern as seen for separation distance of 2 nm. Reprinted with permission from Ref. [41]. Copyright 2014 American Chemical Society. (b and c) Plasmonic electromagnetic field enhancement for the dimers of Ag-Au when the particles are separated by 2 nm (exciting at 487 and 615 nm). It shows that the field is strongest around the corners of the silver cube at 487 nm and exciting at 615 nm, the most field strongest in the middle of facing facets as well as around the exterior corners of the gold cube.

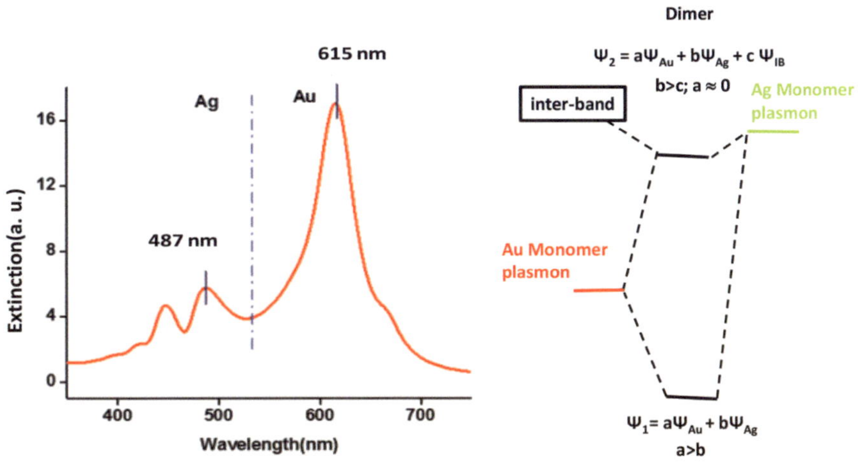

Figure 6. The spectrum of plasmonic coupling model for the heterodimer of Au-Ag NCs separated by 2 nm (left). It shows that there is significant mixing of the plasmonic modes of the two particles to form hybrid resonances at 2 nm. The two strongest bands at 615 and 487 nm associated with the corresponding field distribution in Figure 5b and c. A schematic represents that which modes mix to form the two strongest hybrid modes (right). Au interband transition can mix with Ag dipolar plasmonic resonance (around 487 nm), due to the similarity in energy. Adapted with permission from Ref. [41]. Copyright 2014 American Chemical Society.

more silver character. Herein, for both separations, we can see that the field concentrates around the Ag cubes' corners, and there is no field around the Au cube (**Figure 5c** and **d**). This suggests that this hybrid mode contains little gold dipole character. Compare to what was seen with all other dimeric pairs, as the intercube distance is reduced from 10 to 2 nm, the maximum field enhancement is reduced. This is expected due to the fact that silver's dipole mode hybridizes with gold's interband transition. Exciting the nanoparticle at the interband transition produces an incoherent excitation and little net E-field. Subsequently, a hybrid mode containing significant interband transition-character will produce a weaker overall E-field. We would also expect that there would be little to no E-field located on the Au NC, which was obvious. It is expected that this mode has the form $\Psi_2 = a\Psi_{Au} + b\Psi_{Ag} + c\Psi_{IB}$, where Ψ_{IB} is the interband transition mode and $b > c$ and $a \approx 0$. A schematic overview of the proposed mixing scheme for Au and Ag NCs is shown in **Figure 6** (on the right). The spectrum shows other modes that these high-energy resonances are likely included mostly of silver's higher order modes.

It concludes when exciting the heterodimer at gold plasmon wavelength, reducing the separation distance weakens the field due to the silver plasmon mode mixes with gold's interband transition, and since the interband transition does not show plasmonic behavior, it causes reduction in the overall field enhancement. As shown in **Figure 6** (on the left), the main band (the hybrid band) contains both Au and Ag components, and it is mostly gold-based resonance around 600 nm in character [41].

6. Conclusions

In this chapter, we summarize numerically the dependence of interparticle on the LSPR band shift and electromagnetic field distribution around face-to-face oriented Ag and/or Au NC homo- and heterodimers. To begin to study the plasmonic coupling behavior between cubes, the extinction spectrum was calculated for different separation distances. It was found that the separation distance between the cubes plays a significant role in the plasmonic shift in the extinction spectrum and corresponding E-field around the particles. Exciting homodimers of Au and Ag NCs parallel to the interparticle axis showed results consistent with expectations and follow the typical scaling law seen for other particles. This study is pointing toward the fact that in homodimers NC at very short separation of distance (i.e., 2 nm), the maximum field moved away from the corners of the cubes and became strong between the adjacent faces. Compared to homodimers, different behaviors are observed for the heterodimers. As the Au and Ag cubes in close proximity to each other, their plasmon modes mixed to form hybrid modes. As a result, Au dipolar mode weakly mixed with Ag higher energy dipolar mode. As a result, this hybrid mode enhanced the E-field primarily around the Au NC. Silver's plasmonic mode mixed with gold's interband transition, due to their similar energies. It was found that since the interband transition is an incoherent excitation, it results no E-field, and this hybrid mode had a weaker field enhancement compare to that of un-hybridized Ag mode.

http://dx.doi.org/10.5772/67411

Acknowledgements

The author would like to acknowledge the Georgia Institute of Technology for providing high performance computing resources and support. The author thanks B.T. Draine and P.J. Flatau for use of their DDA Cod DDSCAT 6.1. The financial support of NSF-DMR grant (1206637) is greatly appreciated.

Author details

Nasrin Hooshmand

Address all correspondence to: nhooshmand3@gatech.edu

Laser Dynamics Laboratory, School of Chemistry and Biochemistry, Georgia Institute of Technology, Atlanta, Georgia, United States

References

[1] Willets KA, Van Duyne RP. Localized surface plasmon resonance spectroscopy and sensing. Annual Review of Physical Chemistry. 2007;**58**:267-297

[2] Hao E, Schatz GC. Electromagnetic fields around silver nanoparticles and dimers. The Journal of Chemical Physics. 2004;**120**(1):357-366

[3] Byers CP, Zhang H, Swearer DF, Yorulmaz M, Hoener BS, Huang D, et al. From tunable core-shell nanoparticles to plasmonic drawbridges: Active control of nanoparticle optical properties. Science Advances. 2015;**1**(11)

[4] Link S, El-Sayed MA. Spectral properties and relaxation dynamics of surface plasmon electronic oscillations in gold and silver nanodots and nanorods. The Journal of Physical Chemistry B. 1999;**103**(40):8410-8426

[5] Anker JN, Hall WP, Lyandres O, Shah NC, Zhao J, Van Duyne RP. Biosensing with plasmonic nanosensors. Nature Materials. 2008;**7**(6):442-453

[6] El-Sayed MA. Some interesting properties of metals confined in time and nanometer space of different shapes. Accounts of Chemical Research. 2001;**34**(4):257-264

[7] Kelly KL, Coronado E, Zhao LL, Schatz GC. The optical properties of metal nanoparticles: The influence of size, shape, and dielectric environment. The Journal of Physical Chemistry B. 2002;**107**(3):668-677

[8] Knight MW, Wu YP, Lassiter JB, Nordlander P, Halas NJ. Substrates matter: Influence of an adjacent dielectric on an individual plasmonic nanoparticle. Nano Letters. 2009;**9** (5):2188-2192

[9] Murray WA, Auguié B, Barnes WL. Sensitivity of localized surface plasmon resonances to bulk and local changes in the optical environment. The Journal of Physical Chemistry C. 2009;**113**(13):5120-5125

[10] Sherry LJ, Chang S-H, Schatz GC, Van Duyne RP, Wiley BJ, Xia Y. Localized surface plasmon resonance spectroscopy of single silver nanocubes. Nano Letters. 2005;**5**(10):2034-2038

[11] Xu G, Chen Y, Tazawa M, Jin P. Surface plasmon resonance of silver nanoparticles on vanadium dioxide. The Journal of Physical Chemistry B. 2006;**110**(5):2051-2056

[12] Maier SA, Atwater HA. Plasmonics: Localization and guiding of electromagnetic energy in metal/dielectric structures. Journal of Applied Physics. 2005;**98**(1)

[13] Bordley JA, Hooshmand N, El-Sayed MA. The coupling between gold or silver nanocubes in their homo-dimers: A new coupling mechanism at short separation distances. Nano Letters. 2015;**15**(5):3391-3397

[14] Hooshmand N, Bordley JA, El-Sayed MA. Are hot spots between two plasmonic nanocubes of silver or gold formed between adjacent corners or adjacent facets? A DDA examination. The Journal of Physical Chemistry Letters. 2014;**5**(13):2229-2234

[15] Campion A, Kambhampati P. Surface-enhanced Raman scattering. Chemical Society Reviews. 1998;**27**(4):241-250

[16] Li Z, Shegai T, Haran G, Xu H. Multiple-particle nanoantennas for enormous enhancement and polarization control of light emission. ACS Nano. 2009;**3**(3):637-642

[17] Fraire JC, Pérez LA, Coronado EA. Cluster size effects in the surface-enhanced Raman scattering response of Ag and Au nanoparticle aggregates: Experimental and theoretical insight. The Journal of Physical Chemistry C. 2013;**117**(44):23090-23107

[18] Osawa M. Dynamic processes in electrochemical reactions studied by surface-enhanced infrared absorption spectroscopy (SEIRAS). Bulletin of the Chemical Society of Japan. 1997;**70**(12):2861-2880

[19] Jain PK, Huang W, El-Sayed MA. On the universal scaling behavior of the distance decay of plasmon coupling in metal nanoparticle pairs: A plasmon ruler equation. Nano Letters. 2007;**7**(7):2080-2088

[20] Rechberger W, Hohenau A, Leitner A, Krenn JR, Lamprecht B, Aussenegg FR. Optical properties of two interacting gold nanoparticles. Optics Communications. 2003;**220**(1-3):137-141

[21] El-Sayed IH, Huang X, El-Sayed MA. Surface plasmon resonance scattering and absorption of anti-EGFR antibody conjugated gold nanoparticles in cancer diagnostics: Applications in oral cancer. Nano Letters. 2005;**5**(5):829-834

[22] Sokolov K, Follen M, Aaron J, Pavlova I, Malpica A, Lotan R, et al. Real-time vital optical imaging of precancer using anti-epidermal growth factor receptor antibodies conjugated to gold nanoparticles. Cancer Research. 2003;**63**(9):1999-2004

[23] Alivisatos P. The use of nanocrystals in biological detection. Nature Biotechnology. 2004;**22**(1):47-52

[24] Haes AJ, Hall WP, Chang L, Klein WL, Van Duyne RP. A localized surface plasmon resonance biosensor: First steps toward an assay for Alzheimer's disease. Nano Letters. 2004;**4**(6):1029-1034

[25] Rosi NL, Mirkin CA. Nanostructures in biodiagnostics. Chemical Reviews. 2005;**105**(4): 1547-1562

[26] Hirsch LR, Stafford RJ, Bankson JA, Sershen SR, Rivera B, Price RE, et al. Nanoshell-mediated near-infrared thermal therapy of tumors under magnetic resonance guidance. Proceedings of the National Academy of Sciences of the United States of America. 2003;**100**(23):13549-13554

[27] Huang X, El-Sayed IH, Qian W, El-Sayed MA. Cancer cell imaging and photothermal therapy in the near-infrared region by using gold nanorods. Journal of the American Chemical Society. 2006;**128**(6):2115-2120

[28] Huang X, Jain PK, El-Sayed IH, El-Sayed MA. Determination of the minimum temperature required for selective photothermal destruction of cancer cells with the use of immunotargeted gold nanoparticles. Photochemistry and Photobiology. 2006;**82**(2):412-417

[29] Jain PK, El-Sayed IH, El-Sayed MA. Au nanoparticles target cancer. Nano Today. 2007;**2**(1):18-29

[30] Michaels AM, Jiang, Brus L. Ag nanocrystal junctions as the site for surface-enhanced raman scattering of single rhodamine 6G molecules. The Journal of Physical Chemistry B. 2000;**104**(50):11965-11971

[31] Nie S, Emory SR. Probing single molecules and single nanoparticles by surface-enhanced Raman scattering. Science. 1997;**275**(5303):1102-1106

[32] Su KH, Wei QH, Zhang X, Mock JJ, Smith DR, Schultz S. Interparticle coupling effects on plasmon resonances of nanogold particles. Nano Letters. 2003;**3**(8):1087-1090

[33] Reinhard BM, Siu M, Agarwal H, Alivisatos AP, Liphardt J. Calibration of dynamic molecular rulers based on plasmon coupling between gold nanoparticles. Nano Letters. 2005;**5**(11):2246-2252

[34] Grillet N, Manchon D, Bertorelle F, Bonnet C, Broyer M, Cottancin E, et al. Plasmon coupling in silver nanocube dimers: Resonance splitting induced by edge rounding. ACS Nano. 2011;**5**(12):9450-9462

[35] Draine BT, Flatau PJ. Discrete-dipole approximation for scattering calculations. Journal of the Optical Society of America. 1994;**11**(4):1491-1499

[36] Yang WH, Schatz GC, Vanduyne RP. Discrete dipole approximation for calculating extinction and Raman intensities for small particles with arbitrary shapes. The Journal of Chemical Physics. 1995;**103**(3):869-875

[37] Jensen TR, Schatz GC, Van Duyne RP. Nanosphere lithography: Surface plasmon resonance spectrum of a periodic array of silver nanoparticles by ultraviolet-visible extinction spectroscopy and electrodynamic modeling. The Journal of Physical Chemistry B. 1999;**103**(13):2394-2401

[38] Nordlander P, Oubre C, Prodan E, Li K, Stockman MI. Plasmon hybridization in nanoparticle dimers. Nano Letters. 2004;**4**(5):899-903

[39] Atay T, Song J-H, Nurmikko AV. Strongly interacting plasmon nanoparticle pairs: From dipole–dipole interaction to conductively coupled regime. Nano Letters. 2004;**4**(9): 1627-1631

[40] Marhaba S, Bachelier G, Bonnet C, Broyer M, Cottancin E, Grillet N, et al. Surface plasmon resonance of single gold nanodimers near the conductive contact limit. The Journal of Physical Chemistry C. 2009;**113**(11):4349-4356

[41] Hooshmand N, O'Neil D, Asiri AM, El-Sayed M. Spectroscopy of homo- and heterodimers of silver and gold nanocubes as a function of separation: A DDA simulation. The Journal of Physical Chemistry A. 2014;**118**(37):8338-8344

[42] Li G-C, Zhang Y-L, Lei DY. Hybrid plasmonic gap modes in metal film-coupled dimers and their physical origins revealed by polarization resolved dark field spectroscopy. Nanoscale. 2016;**8**(13):7119-7126

[43] Sheikholeslami S, Jun Y-w, Jain PK, Alivisatos AP. Coupling of optical resonances in a compositionally asymmetric plasmonic nanoparticle dimer. Nano Letters. 2010;**10**(7): 2655-2660

Section 2

Applications to Thermal and Infrared Domains

Chapter 7

Optical Absorption and Thermal Effects of Plasmonic Nanostructures

Jingzhi Wu and Yanhong Wang

Additional information is available at the end of the chapter

http://dx.doi.org/10.5772/67505

Abstract

With resonant light illumination, metallic nanostructures convert electromagnetic fields' energy into heat because of optical absorption associated with plasmonic resonance. The optical absorption triggers a heat generation process that involves not only the absorption of photon energy but also heat transfer from the nanostructures to the surrounding medium. In this chapter, we study enhanced optical absorption of plasmonic nanostructures. Moreover, thermal effects induced by optical absorption and heat transfer between nanostructures are analyzed.

Keywords: plasmonics, nanoparticle, thermal effect, absorption, optical trapping

1. Introduction

Optically excited nanostructures have large absorption cross-sections, which lead to efficient heat generation. Strength and localization of heat generation strongly depend on geometry and composition of plasmonic nanostructures [1, 2]. The regimes of optical heating of nanostructures can be collective or local. The collective heating is realized in a large and dense ensemble of nanoparticles (NPs) where heat fluxes from individual NPs add up, leading to high-temperature distribution [3]. The collective heating is typically realized in the regime of continuous wave (CW) illumination when the system has enough time to reach a non-equilibrium steady state with increased temperature [4, 5]. The local regime of heat generation can produce high temperature in confined volumes within or in the vicinity of a plasmonic nanostructure. The local and collective heating mechanisms depend on the composition and dimensions of nanostructures [6, 7]. Moreover, the temperature distribution and the efficiency of temperature generation depend on the shape of nanostructure [8, 9].

The theoretical description of plasmonic photothermal effect can be modelled as: the light scattering/absorption, the subsequent heat generation, and heat dissipation or transfer in the system of nanostructures [10–12]. The laser-induced heat generation and conduction in arbitrary plasmonic structures can be numerically modelled and solved, using methods as finite element method (FEM) [13], discrete dipole approximation (DDA) [14], boundary element method (BEM) [15], or finite-difference time-domain (FDTD) [16]. Experimental methods have been developed for observing light-induced heating effects in plasmonic systems and characterizing the temperature increase of nanostructures. In time domain, transient thermo-reflectance technique (TTR) was used to measure the pulsed-laser heating of a metal film by a pump-probe setup, with sub-picosecond resolution [17–19]. In a spatial domain, the development of scanning thermal microscope (SThM) enables thermal microscopic imaging with sub-500-nm resolution [20]. Other thermal microscopic imaging methods were also demonstrated, such as measuring the phosphor decay time [21], phase transition of organic molecules [22], or electric conductance of metal thin film [23].

In this chapter, the optical absorption of metallic nanostructures is simulated by finite element method. Enhanced optical absorption is demonstrated by designing periodic and random nanostructures. Broadband optical absorption is achieved based on refractory plasmonic nanostructures for high-temperature applications. Experiments show that optical absorption of 90% in the wavelength range from 200 to 1100 nm can be achieved using random nanostructures. The temperature distribution of NPs solution is measured using polarization anisotropy of fluorescence (FPA). The heat transfer between nanostructures in different surrounding media is analyzed using heat transfer theory. Thermal force induced by temperature gradient in the surrounding aqueous medium is also discussed. Nanoparticle trapping using both optical force and thermal force is demonstrated.

2. Optical absorption of plasmonic nanostructures

2.1. Plasmonic materials

Plasmonic materials are typically metals. Plasmonic resonance occurs when an electromagnetic radiation of a well-defined wavelength interacts with a metallic system: free electrons in the conduction band will oscillate with respect to fixed positive ions. When the exciting field is resonant with the electron oscillation frequency, a strong electromagnetic field develops near the nanostructure surface. The optical properties of noble metals can be described by a complex dielectric function ε_M that depends on the frequency of the incident light ω. There are two major contributions to ε_M: (1) the quasi-free movement of the conduction band electrons induced by the incident radiation and (2) the electronic interband transitions that may take place if the energy of photons exceeds the energy between energy bands. The dielectric function containing both contributions can be described by the Drude–Lorentz model [24]:

$$\varepsilon_M(\omega) = \varepsilon_r - \sum_j \frac{\omega_{Pj}^2}{\omega^2 + i\gamma_j\omega} + \sum_j \frac{\Delta\varepsilon_j\Omega_{Pj}^2}{\Omega_{Pj}^2 - \omega^2 - i\Gamma_j\omega} \tag{1}$$

where ε_r describes the optical response at high frequencies, which includes the background dielectric function of the ionic cores, γ is the characteristic collision rate responsible for

http://dx.doi.org/10.5772/67505

damping of electron oscillations that related to energy losses by heating and ω_p is the volume plasma frequency of the free electron system defined as: $\omega_P = \sqrt{ne^2/\varepsilon_0}$, where e is the elementary charge, ε_0 is the permittivity of free space and m is the electron mass. $\Delta\varepsilon_j$ is the contribution of interband transition to the dielectric function and Ω_p and γ are the plasma and damping frequencies for the bound electrons, respectively. The inclusion of more terms (j) in the sum provides a better fitting to the experimental data. The dielectric functions of commonly used metal materials gold (Au), silver (Ag) and one refractory material Tungsten (W) are shown in **Figure 1**.

Upon optical illumination, a nanostructure both absorbs and scatters light, which can be described by scattering, absorption and extinction cross-sections. Mie theory gives the exact analytical solution to the scattering and absorption of simple cases, for example, metallic spheres. Nevertheless, the scattering of particles with arbitrary shapes is beyond the ability of Mie solution. The calculation of the EM fields at all points in space when light interacts with a nanostructure of arbitrary shapes and dielectric properties is a complex problem that needs to be solved numerically. Several numerical methods, such as FDTD, DDA and FEM, are suitable for obtaining the scattering/absorption of nanostructures with arbitrary shapes. In this work, we use FDTD method for the numerical calculation of the scattering/absorption spectra of various plasmonic nanostructures.

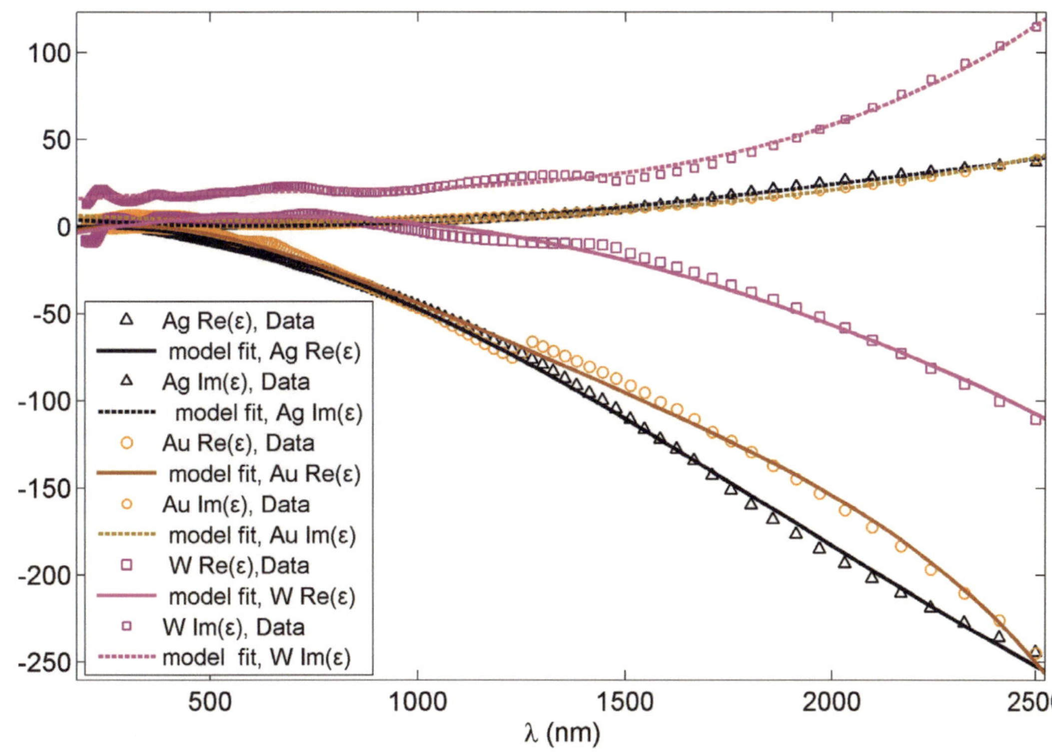

Figure 1. Dielectric function of metals.

Recent experimental findings show that the imaginary part of the dielectric function changes significantly with the increasing temperature [25], whereas the real part remains almost intact. The temperature-dependent deviations in the gold optical constants are quite significant. A temperature dependent non-linear permittivity model combining several oscillators (j) with lifetime Γ_j, strength f_j and frequency ω_j has been proposed recently [26]. Its expression comprises two terms: the first one taking into account the conduction electrons and the influence of temperature and the second one expressing the interband absorptions:

$$\varepsilon_M(\omega) = 1 - \frac{\Omega_P(T)^2}{\omega^2 - i\eta\omega\tilde{\Gamma}(T)} + \sum_j \frac{f_j\omega_P^2}{\omega_j^2 - \omega^2 + i\Gamma_j\omega} \quad (2)$$

$$\Omega_P(T) = \frac{\Omega_P}{\sqrt{1 + 3\zeta(T - T_0)}} \quad (3)$$

$$\tilde{\Gamma}(T) = \Gamma_0 + \Gamma(T) - \Gamma(T_0) \quad (4)$$

where $\Gamma(T)$ is the total damping coefficient at temperature T, γ_0 is the interband damping coefficient. $\Omega_P = \sqrt{f_D\omega_P}$ is the modified plasma frequency including the oscillator strength f_0. ζ is the thermal expansion coefficient and η is a dimensionless parameter that can be tuned to take into account defects-induced damping changes.

Temperature change can strongly modify the field enhancement and absorption characteristics of plasmonic devices. A deep understanding of the temperature dependence of the optical and thermal phenomena in plasmonic structures has a great importance in a variety of research areas, including plasmonics and near-field irradiative heat transfer. The increase in the imaginary part at elevated temperatures significantly reduces the surface plasmon polariton propagation length and the quality factor of the localized surface plasmon resonance. Even though the optical properties of bulk metals at elevated temperatures have been studied, the temperature-dependent data for plasmonic materials is still largely missing. Further experiments and analytical models need to be developed for accurate modelling of high-temperature plasmonic devices operating in a steady or dynamic state.

2.2. Broadband absorption

Enhancing light absorption of plasmonic materials has evoked great interest of researchers in many fields of thermophotovoltaics and solar thermal applications. One of the approaches to improve the absorbing performance is to utilize micro/nanostructures on material surface to produce unique optical characteristics that facilitate a route to achieving spectrum selectivity [27].

Applications in the high-temperature fields strongly demand for improving optical performances of refractory materials. The absorption wavelength of an ideal absorber is from 400 to 1500 nm for maximum absorption of the solar spectrum. This absorption spectrum selectivity is required to prevent re-radiation of the absorbed energy. It is possible to limit re-radiation by heating the material above 1200°C to match the narrow bandgap of photovoltaic cells [28]. In

http://dx.doi.org/10.5772/67505

order to obtain desired optical properties, different micro/nanostructures on refractory materials supporting plasmonic resonance have been proposed.

2.2.1. *Periodic plasmonic structures*

First, we study the absorption enhancement and spectrum selectivity of refractory material (Tungsten, W) with periodic structures comprising cylinder array. The incident light is a plane wave with wavelength range from 400 to 2500 nm. The material property is modelled to fit experimental data from the literature using a Drude-Lorentz model.

From **Figure 2**, we can see that the absorption enhancement can be achieved by adjusting the structure parameters of nanostructures in visible and near-infrared wavelength range. However, absorption enhancement is not obvious at a longer wavelength (> 2000 nm) because plasmonic resonance cannot be excited.

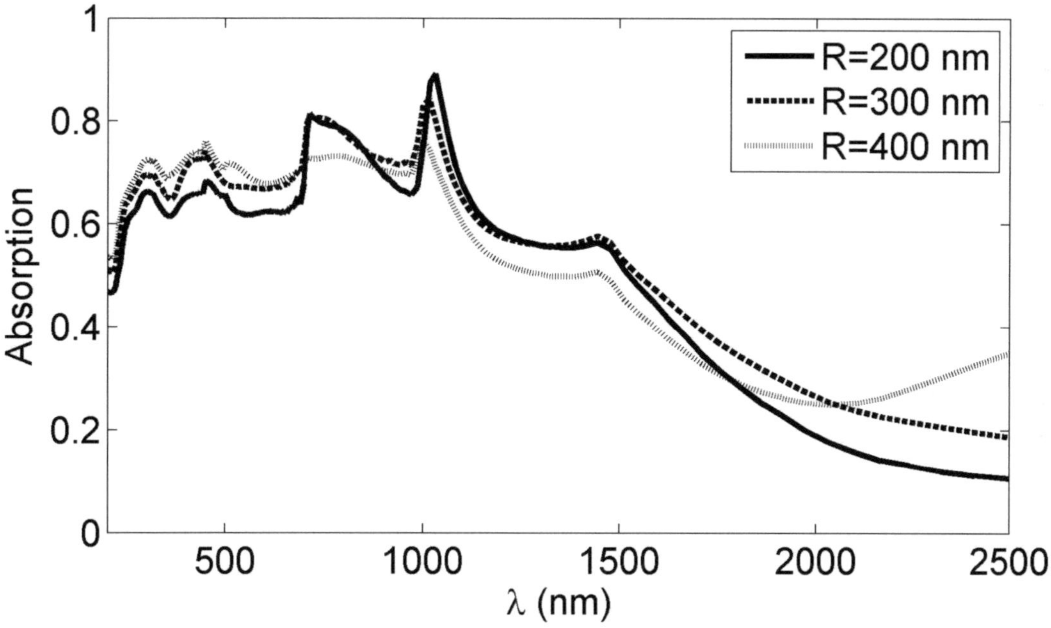

Figure 2. Absorption of the periodic plasmonic cylinders with different radii R, with P = 1000 nm and H = 500 nm. Inset is the model of periodic array of plasmonic cylinders with period P, height H and radius R.

Figure 3 shows that the absorption peak wavelength is dependent on the structure period. However, the absorption is gradually decreasing for the larger period because of weak plasmonic coupling between periodic structures. It is difficult to excite plasmonic resonance modes beyond the wavelength of 1200 nm. Therefore, there is no resonance peak in the mid-infrared range for achieving the spectrum selectivity absorption to match the narrow bandgap of photovoltaic cells.

The absorption of periodic structures with different height is shown in **Figure 4**. It shows that the absorption peak wavelength is strongly affected by the height of surface structure. Increasing the

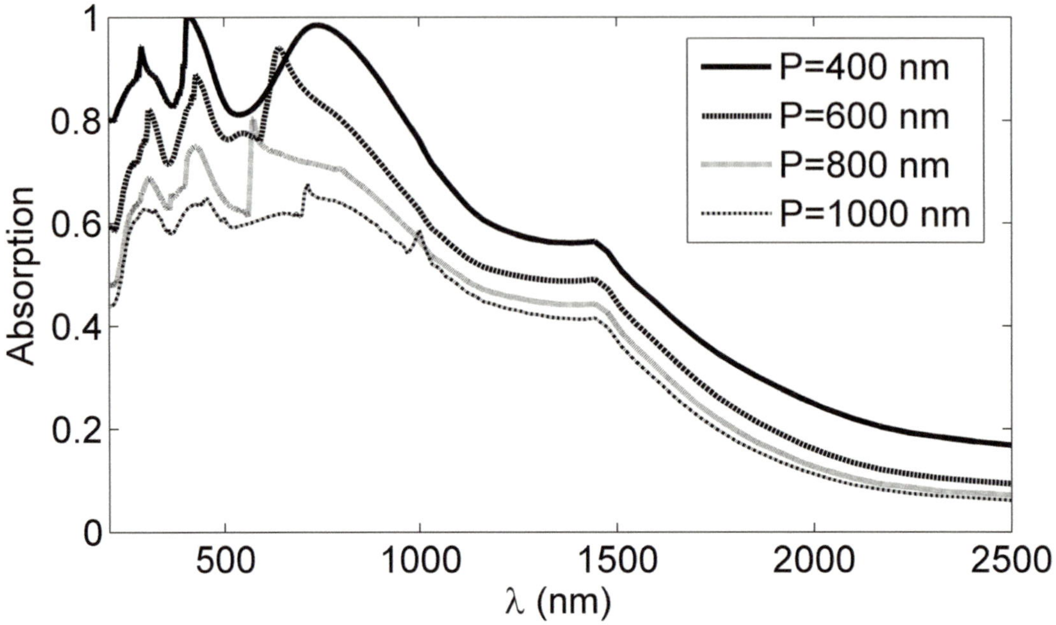

Figure 3. Absorption of the periodic structures with different period at R = 100 nm and H = 500 nm.

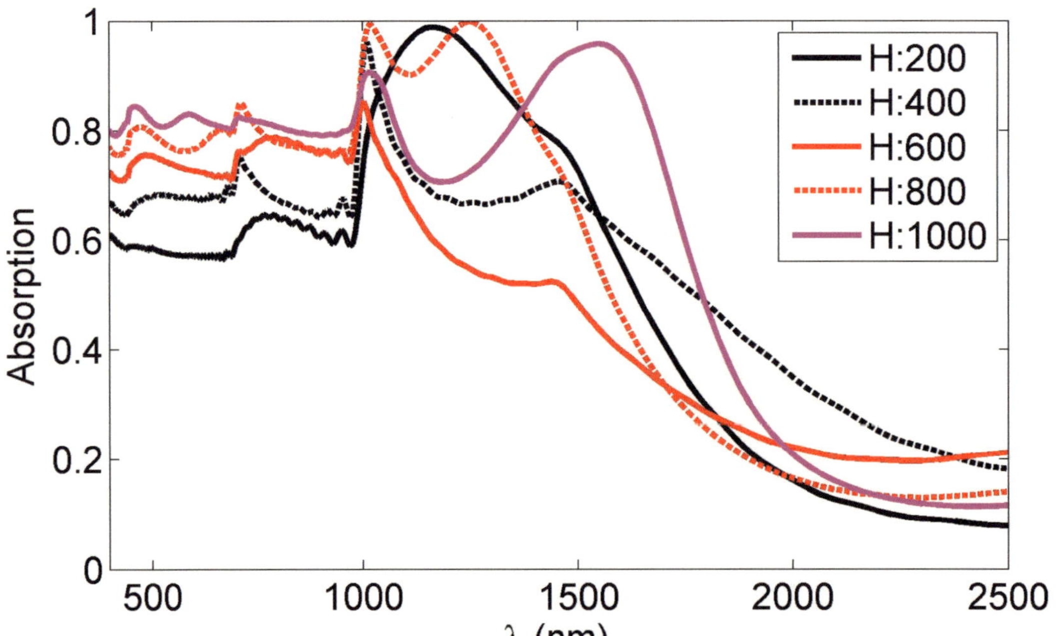

Figure 4. Absorption of the periodic structures with different heights at R = 300 nm and P = 1000 nm.

height of structures can enhance absorption efficiency as new resonant modes are supported. For instance, the absorption is greatly enhanced at wavelength 1500 nm for the structure with H = 1000 nm. Varying the height of structure can also shift the peak absorption wavelength, for example, the peak absorption is shifted to 1200 nm for the structure with H = 800 nm.

2.2.2. Random plasmonic structures

While optical absorption of periodic structures can be precisely controlled, random structures offer flexibility to design and fabricate. Random structures can be described by their height distribution and correlation function. There is an important statistical height parameter to be often used, root mean square (RMS) R_q, for evaluating the amplitude of surface roughness. The correlation length usually describes lateral dimensions and is another spatial parameter which is used for the determination of peak density corresponding to lateral dimensions [29].

The random structure is characterized by a specified RMS (R_q) and correlation length (L_c), which are related to the correlation function by [30]:

$$\langle H(\vec{r})H(\vec{r}+\vec{\delta})\rangle = {R_q}^2 \exp\left(-\left(\frac{\delta}{L_c}\right)^2\right) \tag{5}$$

where δ is the sampling resolution of the surface.

$${R_q}^2 = \frac{1}{L}\int_0^L \left(z(x)^2\right)dx \tag{6}$$

Here, L is the sample length of the profile and z(x) is the profile heights measured from a reference line. For convenience, the reference line is defined as x-axis along the lateral surface.

For the following FDTD simulations, the random surface roughness is generated by creating a matrix of uniform random numbers in k-space [31]. First, generating a square two-dimensional random rough surface F(x,y) with N×N surface points. The surface has a Gaussian height distribution and exponential auto covariance functions (in both x and y). L_S is the length of the surface side. Assuming $L_{cx} = L_{cy} = L_c$ are the correlation lengths in x and y, a Gaussian filter is applied to this matrix to remove high-frequency components and then a Fourier transform is used to transform the matrix back to real space.

Figure 5 shows the absorption of random structures with different RMS values. The correlation length (Lc) of each structure is set as 0.5 μm. We can see that absorption efficiencies are above 80% in the wavelength range of 400 to 1500 nm when RMS values are larger than 0.3 μm. For longer wavelength range > 1500 nm, the absorption efficiency is strongly dependent upon the value of RMS. Strong resonance peaks can be excited near 2 μm for the structure with RMS > 0.3 μm. The resonance peak (marked by red circles in **Figure 5**) obviously shifts toward longer wavelength for bigger RMS values. The overall absorption efficiency at off-resonance peak wavelength is sharply decreased. This property is beneficial for engineering spectrum selectivity in the mid-infrared wavelength range by adjusting the RMS parameter.

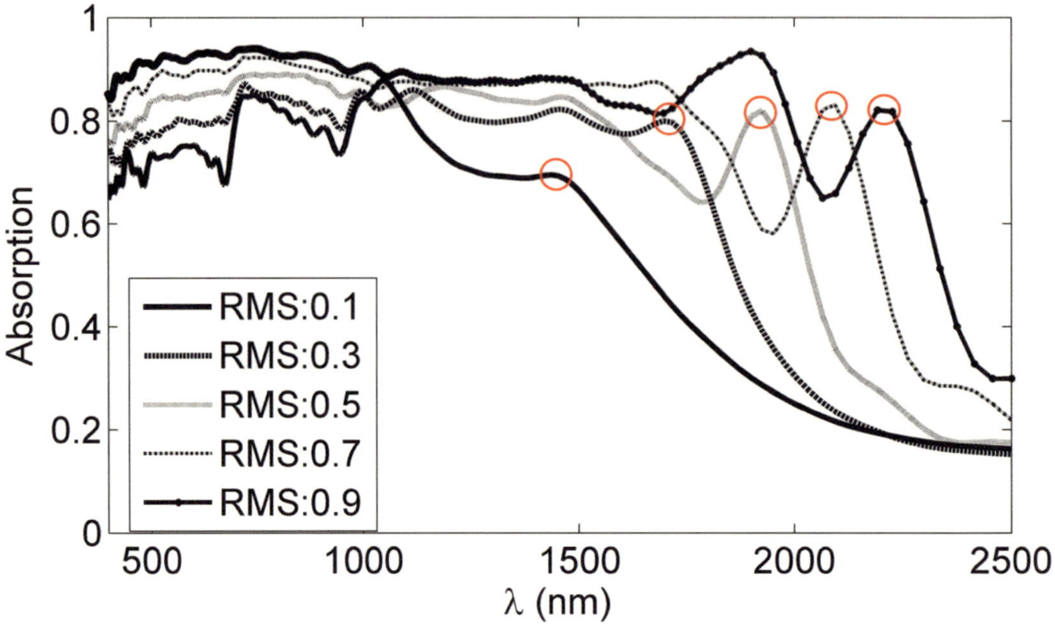

Figure 5. Absorption of random structures with various RMS values at Lc = 0.5 μm.

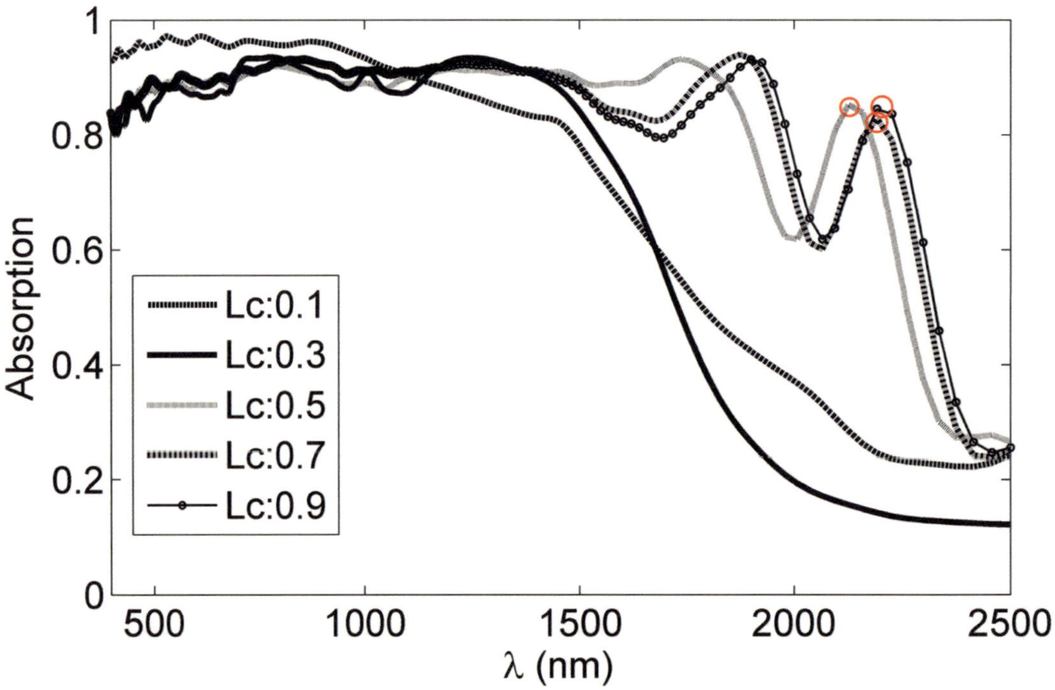

Figure 6. Absorption of random structures with different Lc values at RMS = 0.7 μm.

http://dx.doi.org/10.5772/67505

The optical absorption of structures with a different correlation length Lc and same RMS (0.7 μm) is plotted in **Figure 6**. It can be seen that the absorption efficiency is over 80% in the wavelength range from 400 to 1500 nm. However, the value of Lc has a much stronger effect on the absorption in the wavelength range from 1.5 to 2.5 μm. New resonance peaks with strong absorption are excited at Lc is equal to 0.5, 0.7 and 0.9 μm, but the resonance peaks are not obviously shifted with different Lc values (such as the positions marked by red circles in **Figure 6**). The absorption enhancement spectrum can be gradually broadened to 2 μm when Lc is 0.7 μm. Therefore, absorption enhancement and spectrum selectivity of refractory materials can be obtained by controlling RMS and Lc of the random structures.

To demonstrate the absorption enhancement, random nanostructures have been fabricated using femtosecond (fs) laser on the tungsten substrate (**Figure 7**). The optical absorption spectrum measured by a spectrophotometer shows that the absorption efficiency of nanostructures with RMS of 0.8 μm is greater than 90 % in the wavelength range from 200 to 1100 nm. However, the sample with surface structure RMS of 0.08 μm (polished surface) has much lower absorption (less than 70 % for $\lambda > 600$ nm). It indicates that the optical absorption of the refractory material depends on the size of microstructures on the surface.

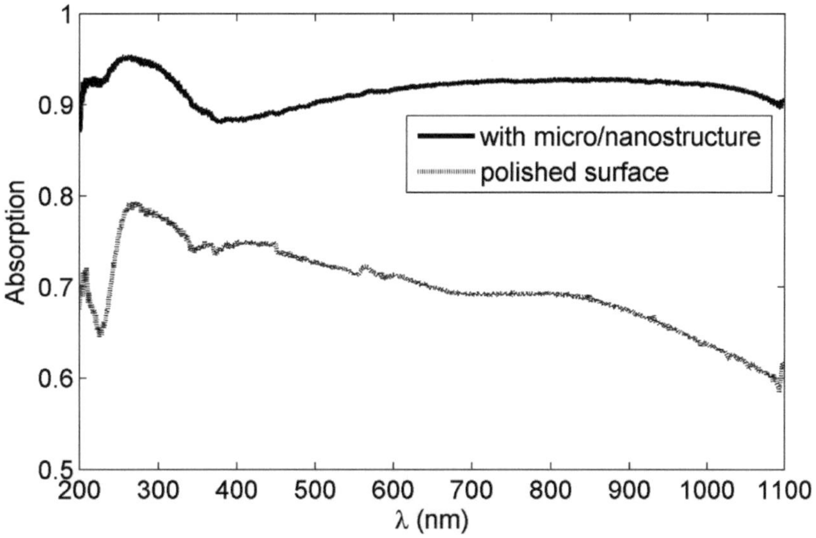

Figure 7. Absorption spectrum of laser-processed tungsten samples with RMS of 0.8 μm and polished surface with RMS of 0.08 μm. Inset shows the sample with polished surface (left) and the sample with micro/nanostructure (right).

3. Thermal effects of plasmonic resonance

The loss of metals, which is manifested by the imaginary part of permittivity $\mathrm{Im}(\varepsilon_M)$, implies that resistive heat will be generated while illuminated by EM fields. A large portion of the energy of incident photons is transferred to the energy of the collective oscillation of excited electrons within the nanostructure. Plasmonic nanostructures show strong scattering and absorption of light at specific wavelength in visible and near-infrared regions because of plasmonic resonances. Associated with the absorption of optical radiation is the generation of heat, which originates from non-radiative decay due to electron–electron and electron–photon interactions. Based on the Joule heating effect, the heat power volume density q_p can be written as

$$q_P = \frac{1}{2}(\mathbf{J}^*\mathbf{E} + \mathbf{J}\mathbf{E}^*) = \varepsilon_0 \omega \mathrm{Im}(\varepsilon_M)|\mathbf{E}|^2 \tag{7}$$

where **E** is the electric field in the nanostructure and **J** is the electric current density.

3.1. Localized temperature field

The resulting temperature distribution around a nanostructure is described by the heat transfer equation, a differential equation that must be solved numerically [32]:

$$\rho(\tilde{r})C_T(\tilde{r})\frac{\partial T(\tilde{r},t)}{\partial t} = \nabla\kappa(\tilde{r})\nabla T(\tilde{r},t) - q_P(\tilde{r},t) \tag{8}$$

here $\tilde{r}$ and t are the spatial coordinates, and time $T(\tilde{r}, t)$ is the local temperature, ρ, C_T and κ are the mass density, specific heat and thermal conductivity, respectively.

The time evolution of the temperatures T_i of N objects is governed by the following energy equations (i = 1,…,N) [33]:

$$\rho_i C_i V_i \frac{dT_i(t)}{dt} = Q_i^{abs}(t, T_1, \cdots, T_N, T_b) \tag{9}$$

here, the left-hand side is the time variation of the internal energy of object i, with ρ_i, C_i and V_i representing its mass density, heat capacity and volume, respectively.

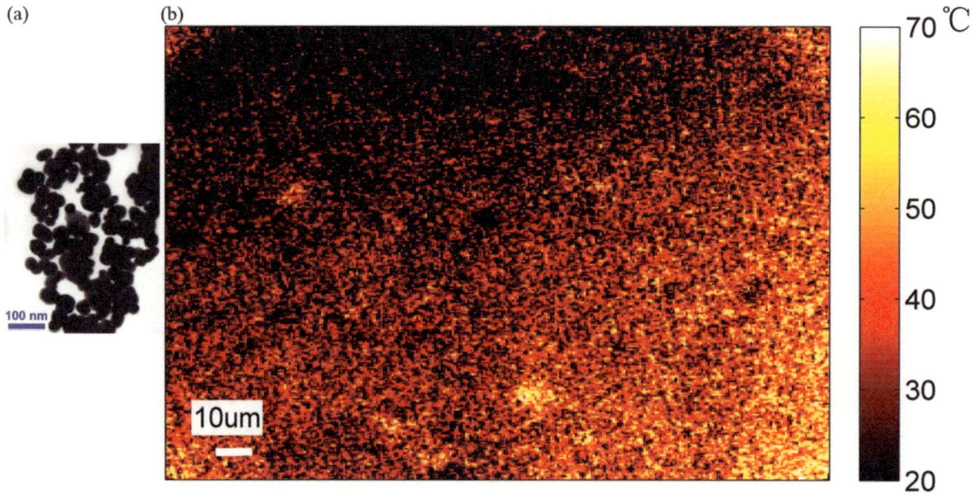

Figure 8. (a) The SEM image of Ag particles. (b) The temperature map of Ag particles in solution measured by FPA method.

Figure 8 shows the temperature increase of silver nanoparticles solution measured using fluorescence polarization anisotropy (FPA) method [34]. The FPA is directly related to rotational diffusion induced by molecular Brownian dynamics. In general, a population of fluorophores illuminated by linearly polarized incident light re-emits partially polarized fluorescence due to the random orientation of the molecules. By measuring the polarization anisotropy of fluorescence, the temperature field around the fluorophore molecules can be obtained.

3.2. Heat transfer between nanoparticles

Considering a collection of N objects located at positions $\mathrm{r_i}$ and maintained at different temperatures T_i with $i = 1, \cdots, N$, in a thermal bath with temperature fixed at $\mathrm{T_b}$. Suppose there is no phase and mass change of materials and the size of these objects is small enough compared with the thermal wavelength $\lambda_T = \hbar c/k_B T$ (T is the temperature, c is the speed of light in vacuum, k_B is Boltzmann's constant and $\hbar$ is Plank's reduced constant), so that all individual objects can be modelled as radiating dipoles. According to fluctuation-dissipation theorem, the net heat transfer on particle i as a sum of exchanges with the other particles and with the thermal bath [35] is

$$Q_i^{abs}(t, T_1, \cdots, T_N, T_b) = \left\langle \frac{d\mathrm{p}_i(t)}{dt} \cdot \mathrm{E}(r_i, t) \right\rangle \tag{10}$$

where the dipole moment $\mathrm{p}_i(t)$ is composed of its fluctuating and induced part, the local field $\mathbf{E}$ is the sum of incident part $\mathbf{E}^{\mathrm{b}}$ without scatters and its induced part $\mathbf{E}^{\mathrm{ind}}$ omitting the frequency dependence. The net heat transfer can be rewritten as [36]:

$$Q_i^{abs}(t, T_1, \cdots, T_N, T_b) = \int_0^{+\infty} \frac{d\omega}{2\pi} \hbar\omega \left[\sum_{j \neq i} \frac{4\chi_i \chi_j}{|\alpha_i|^2} n_{ji}(\omega) \mathrm{Tr}(\mathrm{T}_{ij}^{-1} \mathrm{T}_{ij}^{-1*}) + \frac{4\chi_i k^2}{|\alpha_i|^2} n_{bi}(\omega) \sum_{jk} \alpha_j \alpha_k^* \mathrm{Tr}\left(\mathrm{T}_{ij}^{-1} \mathrm{Im}(\mathrm{G}_{jk}^0) \mathrm{T}_{ki}^{-1*}\right) \right] \tag{11}$$

In the above equation, $\chi_j = \mathrm{Im}(\alpha_j) - \frac{k^3}{6\pi}|\alpha_j|^2$, $n_{ij}(\omega) = n(\omega, T_i) - n(\omega, T_j)$, where $n(\omega, T) = 1/[\exp(\hbar\omega/k_B T) - 1]$ is the Bose-Einstein distribution. $\mathrm{T}_{ij} = \delta_{ij}\mathrm{I} - (1 - \delta_{ij})k^2 \alpha_i \mathrm{G}_{ij}^0$, G_{ij}^0 is the dyadic Green tensor in free space

$$\mathrm{G}_{ij}^0 = \frac{\exp(ik\rho)}{4\pi\rho} \left[\left(1 + \frac{ik\rho - 1}{k^2\rho^2}\right) \mathrm{I} + \frac{3(1 - ik\rho) - k^2\rho^2}{k^2\rho^2} \hat{\rho} \otimes \hat{\rho} \right] \tag{12}$$

here, $k = \omega/c$, $\hat{\mathrm{r}} = \mathrm{r}/r$, $\boldsymbol{\rho} = \mathbf{r}' - \mathbf{r}$, and $\rho = |\boldsymbol{\rho}|$.

The power dissipated inside the particle i at a given frequency ω by the fluctuating field $\mathrm{E_{ij}}$ generated by the particle j can be calculated from the work of the fluctuating EM field on the charge carriers as:

$$P_{j \to i} = 3 \int_0^{\infty} \frac{d\omega}{2\pi} \Theta(\omega, T_j) \Gamma_{i,j}(\omega) \tag{13}$$

where $\Gamma_{i,j}(\omega)$ is the transmission coefficient (TC)

$$\Gamma_{i,j}(\omega) = \frac{4}{3}\frac{\omega^4}{c^4}\mathrm{Im}(\alpha_i)\mathrm{Im}(\alpha_j)\mathrm{Tr}[\mathrm{G}^{ij}\mathrm{G}^{ij*}] \tag{14}$$

Here α_i is the particle's polarizability. G^{ij} is the dyadic Green tensor.

The heat flux (HF) is rewritten as

$$P_{j\to i} = 3\left(\frac{\pi^2 k_B^2 T}{3h}\right)\overline{\Gamma}_{i,j}\Delta T \tag{15}$$

$\pi^2 k_B^2 T/3h$ is universal quantum of thermal conductance. $\overline{\Gamma}_{i,j}$ is the mean transmission coefficient. $\overline{\Gamma}_{i,j} = \int dx f(x)\Gamma_{i,j}(x) / \int_0^\infty f(x)dx$, where $f(x) = x^2 e^x/(e^x - 1)^2$ is a function reminiscent of the mean energy of a harmonic oscillator and $\int_0^\infty f(x)dx = \pi^2/3$.

Figure 9 shows the interparticle heat flux (HF) in a three-particle system (SiC particles) between particles 1 and 2 with particle 3 in the middle [35]. The HF is normalized by the HF between particles 1 and 2 without particle 3 in the same thermal conditions, for $R_2 = R_1$. It shows that the HF mediated by the presence of the third particle can be significantly larger than the value for two isolated dipoles. The enhancement of heat transfer can be over three orders of magnitude for particles separated by distance over 200 nm. The heat transfer between nanoparticles is strongly varied by the interactions with a third nanoparticle.

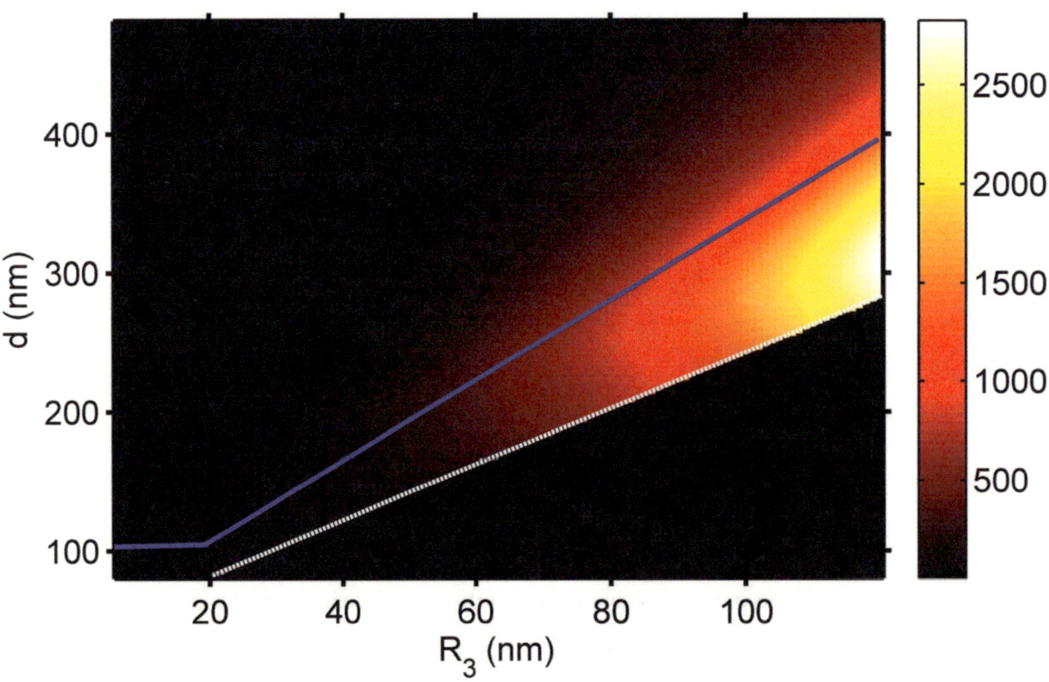

Figure 9. HF normalized to two-particle case with R2= R1. Normalized heat flux (HF) exchanged between particles 1 and 2 separated by different distances and particle 3 with various radii is located at the center in between. The white dashed line is where three nanoparticles are touching. The blue line is distance d–R1–R2–2R3 = max (R1,R2,R3) [35].

http://dx.doi.org/10.5772/67505

3.3. Thermal driving force

When colloidal particles in solutions are exposed to a temperature gradient, they are subjected to thermophoretic forces, which drive them toward one side of the gradient. The direction of particle movement depends on the ambient temperature and the details of particle solvent interactions. Despite much data available, the physical mechanisms of thermophoresis and the thermal driving force in liquids are not well understood. Particle motion in non-uniform gases is well described by kinetic theory. However, surface forces are essential in aqueous solutions; therefore, the steady state has to be characterized in terms of the mechanical equilibrium of hydrodynamic stress and surface forces.

The probability density distribution $p(\mathrm{r})$ for finding the particle at a certain position r is given by the Smoluchowski equation [37]:

$$\frac{\partial p(\mathrm{r})}{\partial t} = \nabla \cdot \left[\frac{-p}{f_c}F + D\nabla p\right] \tag{16}$$

Here f_c is the friction constant and F is the thermal force acting on the particle due to the present of the temperature gradient. D is the Brownian diffusion coefficient. For stationary solution, the thermal force F is written as the following by introducing the definition of the Soret coefficient $S_T = D_T/D$:

$$F = -\frac{S_T}{\beta}\nabla T \tag{17}$$

Here D_T is the thermal diffusion coefficient or thermophoretic mobility, and $\beta = 1/k_B T$, k_B is Boltzmann's constant.

Strong temperature gradients can be induced around plasmonic nanostructures as a result of light energy absorption and heat dissipation to surrounding medium. Recently, the thermal

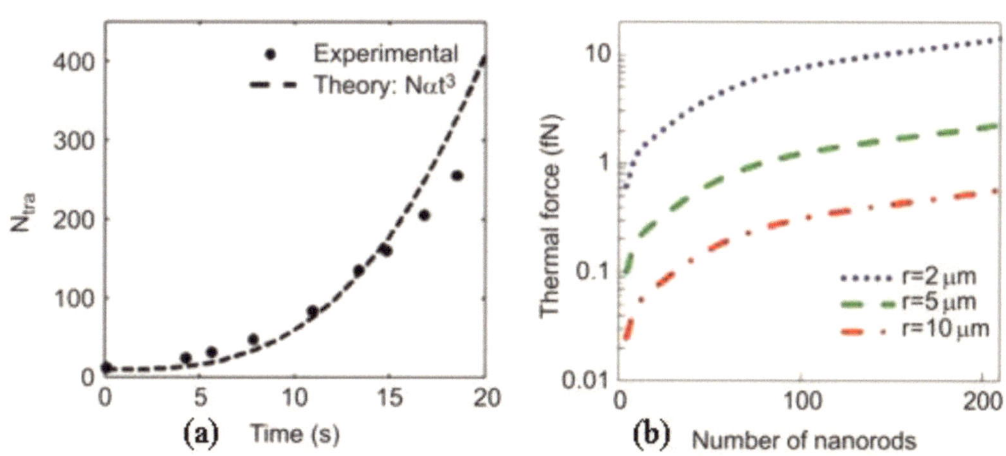

Figure 10. (a) Number of nanorods trapped in the focus as a function of the trapping time. (b) Opto-thermal force as a function of the number of nanorods trapped at different radii in the focal plane [38].

driving force induced by plasmonic resonances has been demonstrated in gold nanorods' solutions [38]. Experiments show that nanorods from a position up to 4–5 times the radius of the Airy spot of the laser beam can be attracted to the focal region. **Figure 10(a)** shows the dependence of the number of gold nanorods trapped in the focus as a function of trapping time. The trapped nanorods absorb a large amount of the energy from the trapping beam, which is converted to heating in the medium around the focal region and then to temperature increase in the surrounding medium. This opto-thermal process becomes significantly more pronounced when the number of trapped gold nanorods becomes large. The thermal attracting force arising from the temperature gradient can be large enough to overcome the Brownian motion of nanorods even at a distance several micrometres away from the focal spot, as shown in **Figure 10(b)**.

4. Conclusions

Resonant plasmonic structures allow for control of fundamental optical processes such as absorption and emission. By structuring metal surfaces on the subwavelength scale, plasmonic resonances can be designed to produce high-optical absorption in the broadband of wavelength. The absorbed optical energy is converted to heat, which can lead to significant local heating in metallic nanostructures and induce temperature gradient in the surrounding medium. The heat transfer between nanostructures is significantly enhanced by near field interaction. The thermal effects of plasmonic nanostructures can be used for energy conversion, optical trapping and thermal management.

Acknowledgements

This work was supported by the National Natural Science Foundation of China (11304289) and Shanxi Scholarship Council of China (2015-076).

Author details

Jingzhi Wu* and Yanhong Wang

*Address all correspondence to: jzwu@live.nuc.edu.cn

Laboratory of Nanophotonics, School of Information and Communications, North University of China, Taiyuan, Shanxi, China

References

[1] Halas NJ, Lal S, Chang WS, Link S, Nordlander P. Plasmons in Strongly Coupled Metallic Nanostructures. Chem Rev. 2011;111(6):3913–61.

[2] Wang Y-H, Wu J, Wang G, Yang F. Resonance Modulation and Optical Force of Nanostructures with Au Slabs Array. Optik Int J Light Electron Optics. 2016;127(5):2969–72.

[3] Kralik T, Hanzelka P, Zobac M, Musilova V, Fort T, Horak M. Strong Near-Field Enhancement of Radiative Heat Transfer Between Metallic Surfaces. Phys Rev Lett. 2012;109 (22):224302.

[4] Li Z, Nan J, Zhang X, Ye S, Shen H, Wang S, et al. Modulate the Morphology and Spectroscopic Property of Gold Nanoparticle Arrays by Polymer-Assisted Thermal Treatment. J Phys Chem C. 2015;119(21):11839–45.

[5] Wang H, Prasad Sivan V, Mitchell A, Rosengarten G, Phelan P, Wang L. Highly Efficient Selective Metamaterial Absorber for High-Temperature Solar Thermal Energy Harvesting. Sol Energy Mater Sol Cells. 2015;137:235–42.

[6] Li W, Zhang L, Zhou J, Wu H. Well-Designed Metal Nanostructured Arrays for Label-Free Plasmonic Biosensing. J Mater Chem C. 2015;3(25):6479–92. doi:10.1039/C5TC00553A.

[7] Fang Y, Jiao Y, Xiong K, Ogier R, Yang Z, Gao S, et al. Plasmon Enhanced Internal Photoemission in Antenna-Spacer-Mirror Based Au/TiO_2 Nanostructures. Nano Lett. 2015;15(6):4059–65.

[8] Braun M, Wurger A, Cichos F. Trapping of Single Nano-objects in Dynamic Temperature Fields. Phys Chem Chem Phys. 2014;16(32):17343.

[9] Wu JZ, Gan XS. Three Dimensional Nanoparticle Trapping Enhanced by Surface Plasmon Resonance. Opt Express. 2010;18(26):27619–26.

[10] Wu Y, Zhou L, Du X, Yang Y. Optical and Thermal Radiative Properties of Plasmonic Nanofluids Containing Core–Shell Composite Nanoparticles for Efficient Photothermal Conversion. Int J Heat Mass Transf. 2015;82:545–54.

[11] Coppens ZJ, Li W, Walker DG, Valentine JG. Probing and Controlling Photothermal Heat Generation in Plasmonic Nanostructures. Nano Lett. 2013;13(3):1023–8.

[12] Baffou G, Quidant R. Thermo-plasmonics: Using Metallic Nanostructures as Nano-sources of Heat. Laser Photon Rev. 2013;7(2):171–87.

[13] Chen Y, Nielsen TR, Gregersen N, Lodahl P, Mørk J. Finite-Element Modeling of Spontaneous Emission of a Quantum Emitter at Nanoscale Proximity to Plasmonic Waveguides. Phys Rev B. 2010;81(12):125431.

[14] Edalatpour S, Francoeur M. The Thermal Discrete Dipole Approximation (T-DDA) for Near-Field Radiative Heat Transfer Simulations in Three-Dimensional Arbitrary Geometries. J Quant Spectrosc Radiat Transf. 2014;133:364–73.

[15] Hohenester U. Quantum Corrected Model for Plasmonic Nanoparticles: A Boundary Element Method Implementation. Phys Rev B. 2015;91(20):205436.

[16] Lin W-C, Lin W-C, Tsai C-L, Lin K-P. Finite-Difference Time-Domain Simulation of Localized Surface Plasmon Resonance Adsorption by Gold Nanoparticles. In: Goh J, Lim CT,

editors. 7th WACBE World Congress on Bioengineering 2015: 6th to 8th July, 2015, Singapore. Cham: Springer International Publishing; 2015. pp. 138–41.

[17] Komarov PL, Burzo MG, Kaytaz G, Raad PE. Transient Thermo-reflectance Measurements of the Thermal Conductivity and Interface Resistance of Metallized Natural and Isotopically Pure Silicon. Microelectron J. 2003;34(12):1115–8.

[18] Smolin SY, Scafetta MD, Guglietta GW, Baxter JB, May SJ. Ultrafast Transient Reflectance of Epitaxial Semiconducting Perovskite Thin Films. Appl Phys Lett. 2014;105(2):22103.

[19] Suyuan B, Zhenan T, Zhengxing H, Jun Y, Jiaqi W. Thermal Conductivity Measurement of Submicron-Thick Aluminium Oxide Thin Films by a Transient Thermo-reflectance Technique. Chin Phys Lett. 2008;25(2):593–6.

[20] Wielgoszewski G, Gotszalk T. Chapter Four: Scanning Thermal Microscopy (SThM): How to Map Temperature and Thermal Properties at the Nanoscale. Adv Imag Elect Phys. 2015;190:177–221.

[21] Abou Nada F, Knappe C, Aldén M, Richter M. Improved Measurement Precision in Decay Time-Based Phosphor Thermometry. Appl Phys B. 2016;122(6):170.

[22] Bilde M, Barsanti K, Booth M, Cappa CD, Donahue NM, Emanuelsson EU, et al. Saturation Vapor Pressures and Transition Enthalpies of Low-Volatility Organic Molecules of Atmospheric Relevance: From Dicarboxylic Acids to Complex Mixtures. Chem Rev. 2015;115(10):4115–56.

[23] Zakaria I, Azmi WH, Mohamed WANW, Mamat R, Najafi G. Experimental Investigation of Thermal Conductivity and Electrical Conductivity of Al_2O_3 Nanofluid in Water: Ethylene Glycol Mixture for Proton Exchange Membrane Fuel Cell Application. Int Commun Heat Mass Transf. 2015;61:61–8.

[24] Reddy H, Guler U, Kildishev AV, Boltasseva A, Shalaev VM. Temperature-Dependent Optical Properties of Gold Thin Films. Opt Mater Express. 2016;6(9):2776–802.

[25] Shen P-T, Sivan Y, Lin C-W, Liu H-L, Chang C-W, Chu S-W. Temperature- and Roughness- Dependent Permittivity of Annealed/Unannealed Gold Films. Opt Express. 2016;24 (17):19254–63.

[26] Alabastri A, Tuccio S, Giugni A, Toma A, Liberale C, Das G, et al. Molding of Plasmonic Resonances in Metallic Nanostructures: Dependence of the Non-linear Electric Permittivity on System Size and Temperature. Materials. 2013;6(11):4879.

[27] Khodasevych IE, Wang L, Mitchell A, Rosengarten G. Micro- and Nanostructured Surfaces for Selective Solar Absorption. Adv Opt Mater. 2015;3(7):852–81.

[28] Cao F, McEnaney K, Chen G, Ren Z. A Review of Cermet-Based Spectrally Selective Solar Absorbers. Energy Environ Sci. 2014;7(5):1615–27. doi:10.1039/C3EE43825B.

[29] Mahdieh M, Badomi E. Determination of Roughness and Correlation Length of Dielectric Surfaces in Nano/Micro Scales Using Kirchhoff Approximation Method. J Opt. 2015;44 (3):240–248.

http://dx.doi.org/10.5772/67505

[30] Wang Y, Wu J. Broadband Absorption Enhancement of Refractory Plasmonic Material with Random Structure. Plasmonics. 2016:1–6.

[31] Orbanz P, Roy DM. Bayesian Models of Graphs, Arrays and Other Exchangeable Random Structures. IEEE Trans Pattern Anal Mach Intell. 2015;37(2):437–61.

[32] Boles YCAM. Thermodynamics: An Engineering Approach. 5th ed. Boston: McGraw-Hill;2006.

[33] Messina R, Tschikin M, Biehs S-A, Ben-Abdallah P. Fluctuation-Electrodynamic Theory and Dynamics of Heat Transfer in Systems of Multiple Dipoles. Phys Rev B. 2013;88 (10):104307.

[34] Baffou G, Kreuzer MP, Kulzer F, Quidant R. Temperature Mapping Near Plasmonic Nanostructures Using Fluorescence Polarization Anisotropy. Opt Express. 2009;17 (5):3291–8.

[35] Wang Y, Wu J. Radiative Heat Transfer Between Nanoparticles Enhanced by Intermediate Particle. AIP Adv. 2016;6(2):025104.

[36] Latella I, Pérez-Madrid A, Rubi JM, Biehs S-A, Ben-Abdallah P. Heat Engine Driven by Photon Tunneling in Many-Body Systems. Phys Rev Appl. 2015;4(1):011001.

[37] Wu J, Wang Y. Plasmonic Nanoparticle Trapping with Inhomogeneous Temperature Fields. Photon J IEEE. 2016;8(1):1–7.

[38] Gu M, Bao H, Gan X, Stokes N, Wu J. Tweezing and Manipulating Micro- and Nanoparticles by Optical Nonlinear Endoscopy. Light Sci Appl. 2014;3:e126.

Chapter 8

Infrared Solar Thermal-Shielding Applications Based on Oxide Semiconductor Plasmonics

Hiroaki Matsui and Hitoshi Tabata

Additional information is available at the end of the chapter

http://dx.doi.org/10.5772/67588

Abstract

This chapter describes plasmonic responses in In_2O_3:Sn nanoparticles (ITO NPs) and their assembled ITO NP sheets in the infrared (IR) range. ITO NPs clearly provide resonance peaks related to local surface plasmon resonances (LSPRs) in the near-IR range, which are dependent on electron density in the NPs. In particular, electron-impurity scattering plays an important role in determining carrier-dependent plasmon damping, which is needed for the design of plasmonic materials based on ITO. ITO NPs are mainly dominated by light absorption. However, a high light reflection is observed in the near- and mid-IR range when using assembled NP sheets. This phenomenon is due to the fact that the introduction of surface modifications to the NPs can facilitate the production of electric-field (*E*-field) coupling between the NPs. The three-dimensional (3*D*) *E*-field coupling allows for resonant splitting of plasmon excitations to the quadrupole and dipole modes, thereby obtaining selective high reflections in the IR range. The high reflective performances from the assembled NP sheets were attributed to the plasmon interactions at the internanoparticle gaps. This work provides important insights for harnessing IR optical responses based on plasmonic technology toward the fabrications of IR solar thermal-shielding applications.

Keywords: oxide semiconductor, surface plasmon, infrared and energy-saving

1. Introduction

Plasmonic nanomaterials based on transparent oxide semiconductors (TOSs, such as In_2O_3, ZnO and SnO_2) have received much attention as new optical phenomena with potential applications. In particular, oxide semiconductors with metallic conductivity by doping with intrinsic and/or extrinsic impurities show surface plasmon resonances (SPRs) in the infrared (IR) range [1–5]. Unlike noble metals (silver and gold), SPRs can be controlled by tuning the physical characters of a material [6–8], which provides new possibilities for optical manipulation of light. Studies of nanoplasmonics based on TOSs can break new ground in the areas of oxide semiconductors. A characteristic property cleared by these studies is that the optical nature of TOSs shows a low-loss plasmonic material even up to near-IR wavelengths because of IR transparency outside the reststrahlen band. The band structures on TOSs are simply composed of *s*- and *p*-orbitals [9], indicating no inter band transition-related *d*-orbitals such as those exhibited by the noble metals [10]. TOS materials with metallic conductivity have mainly applied to transparent electrodes [11, 12]. Therefore, SPRs on TOSs provide new insight for alternative plasmonic applications in the IR range.

SPR excitations on TOSs have been reported on different structures such as nanorods and nanowires [13–15]. In particular, nanoparticles (NPs) of In_2O_3:Sn and ZnO:Al produce localized surface plasmon resonances (LSPRs), which are strongly generated when confining the collective excitations of carriers into NPs [16, 17]. This makes use of localization of large electric fields in the vicinity of NP surfaces. Thus far, the majority of investigations concerning LSPRs have demonstrated on the noble metal NPs, which have been tailored for use in optical areas as diverse as waveguides and biochemical sensing [18–20]. Recently, In_2O_3:Sn NPs have launched as nanoplamonic materials. The careful choice of impurity dopants can show clear LSPR peaks in the near-IR range. The assembled films of In_2O_3:Sn NPs have shown optical enhancements of near-IR luminescence and absorption [21, 22]. These behaviours make use of the electric-fields (*E*-fields) excited on the NP film surfaces [23].

Assembled films of the noble metals have been utilized in surface-enhanced Raman and fluorescence spectroscopies, which are based on high *E*-fields derived from plasmon coupling between the NPs. When metal NPs are located near one another, coupling LSPR induces in a gap between NPs [24]. The strength of LSPR enhances with the magnitude due to interparticle coupling. This optical phenomenon has been utilized in enhanced light emissions, for example, hybrid layers of silver NPs and InGaN/GaN quantum wells [25]. However, assembled films of metal NPs have been limited in the visible range. The use of TOSs extends to longer wavelengths in the IR range.

Plasmonic properties of TOS materials have attracted attention for thermal-shielding applications in order to solar and radiant heat in the near- and mid-IR range, respectively [26]. To date, the composites and films of oxide semiconductor NPs have been studied regard to transmission and extinction spectra in the IR range because optical properties are dominated by absorbance [27–30]. The present thermal-shielding applications have strongly been desired to cut IR radiation not by absorption but through reflection. However, no previous paper has reported reflective performance on doped oxide semiconductor NPs. In addition, plasmonic applications exhibiting a thermal-shielding ability have not been previously studied in detail. The purpose of this chapter is to apply the plasmonic properties for satisfying recent industry demands for a material with thermal-shielding ability. These social requirements include the fabrication of flexible sheets with high heat-ray reflections, as well as visible and microwave transmissions. We use assembled NPs of In_2O_3:Sn as a concrete example. Plasmonic responses are dependent on electronic structure. For example, In_2O_3, ZnO and WO_3 have similar band structures with conduction and valence levels consisting of *s*- and *p*-orbitals. This indicates that the plasmonic properties of these materials can be well manipulated through same optical mechanism.

This chapter is organized as follows. In Section 2, we give a description of structural and optical properties of In_2O_3:Sn (ITO) NPs in the IR range from the viewpoint of local structural analyses. In Section 3, we focus plasmonic responses of ITO NPs from theoretical and experimental approaches, which is not as readily available in the noble metal NPs. To investigate mechanism of plasmonic excitations in ITO, NPs is valuable information for oxide-based plasmonics. Section 4 is devoted to discussion of thermal-shielding based on assembled films of ITO NPs for industrial applications. Above all, we describe plasmonic responses related to the 3*D* E-field coupling along the out-of-plane and in-plane directions. This has a key factor in producing selective high reflections in the IR Range, which provides important insights for harnessing IR optical properties towards the fabrications of solar-thermal shielding. In Section 5, we shortly provide electromagnetic (EM) responses of assembled ITO NP films in the microwave region. Finally, some concluding remakes are given in Section 6.

2. Fabrications and structures of ITO NPs

2.1. Fabrications of ITO NPs

In_2O_3:Sn nanoparticles (ITO NPs) were fabricated using a metal organic decomposition method. Various initial ratios of the metal precursor complexes of $(C_9H_{22}CO_2)_3In$ and $(C_9H_{22}CO_2)_4Sn$ were prepared as starting materials. Indium and tin carboxylates were heated with a chemical ratio of 95:5 in a flask supported by a mantle heater to 350°C. The temperature was maintained for 4 hours, and then the mixture was cooled to room temperature around 30°C. The obtain solutions produced a pale blue suspension, to which excess ethanol was introduced to cause precipitation. Centrifugation and repeated washing processes were carried out several times using ethanol, producing dried powders of ITO NPs with a pale blue colour. As a final step, the NP samples were dispersed in a nonpolar solvent of toluene. A zeta-potential measurement revealed that the NPs showed non-aggregated states in the solvent due to an electrostatic repulsion. The Sn concentration in the NPs in this chapter was measured by X-ray florescence spectroscopy.

2.2. Structural properties of ITO NPs

X-ray diffraction (XRD) measurements clarified that the NPs showed broad peaks characteristic of a colloidal sample with a crystal structure [**Figure 1(a)**] [31]. The *a*-axis length increased from 10.087 to 10.152 Å as a consequence of incorporations of Sn atoms in the host. Furthermore, local structures of the NPs were evaluated by transmission electron microscopy (TEM) [inset of **Figure 1(a)**]. The electron diffraction (ED) analyses revealed that the lattice interval (d_{222}) along the [222] direction at the centre region (d_{222} = 0.301 nm) was close to that at the edge region in the NP (d_{222} = 0.302 nm) [inset of **Figure 1(b)**]. In addition, the experimental results of energy-disperse X-ray microscopy showed that the Sn concentration [Sn] at the centre region,

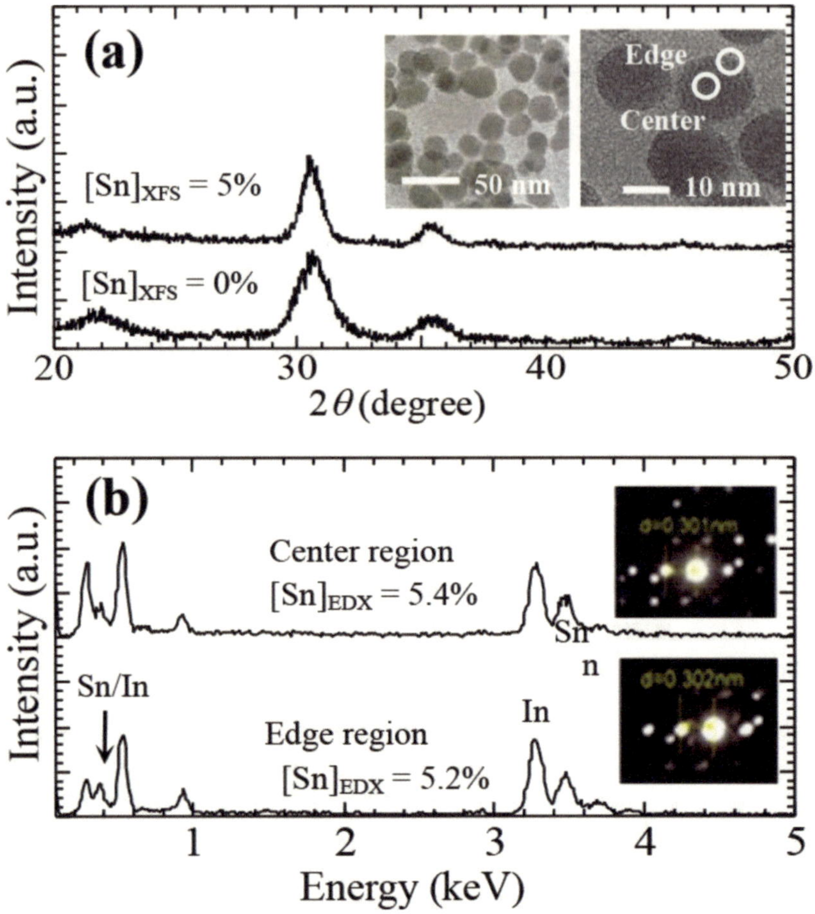

Figure 1. (a) XRD patterns of ITO NPs with Sn contents of 0 and 5%. Inset shows low-resolution (left) and high-resolution (right) TEM images of an ITO NP with a Sn content of 5%. (b) EDX spectra at the centre and edge regions in the NP. Inset represents ED patterns of the (222) plane at the centre and edge (Figure 1 of Ref. [22]). Copyright 2014 by the American Institute of Physics.

[Sn] = 5.4%, was similar to that at the edge region, [Sn] = 5.2%, which indicated that Sn atoms in the NP were spatially homogeneous [**Figure 1(b)**]. The NP diameter was around 20 nm, as confirmed using TEM and dynamic light-scattering methods. However, crystalline sizes derived from Scherer's equation from the (222) diffraction peak of XRD patterns were calculated as 1.58 and 2.20 nm for un-doped and doped NPs, respectively. Broadenings of the line-widths of the XRD patterns are attributed to structural imperfections such as defects and strains [22].

We further investigated structural properties by scanning-TEM (STEM) combined with electron-energy loss spectroscopy (EELS) [32]. A STEM-EELS technique can easily detect plasmonic response in a single NP. High-angle annular dark field (HAADF) images in **Figure 2(a)–(c)** cleared that Indium and Sn atoms in the NP were distributed homogeneously, which were consisted with the results of XRD. The EELS spectra at an edge and vacuum

http://dx.doi.org/10.5772/67588

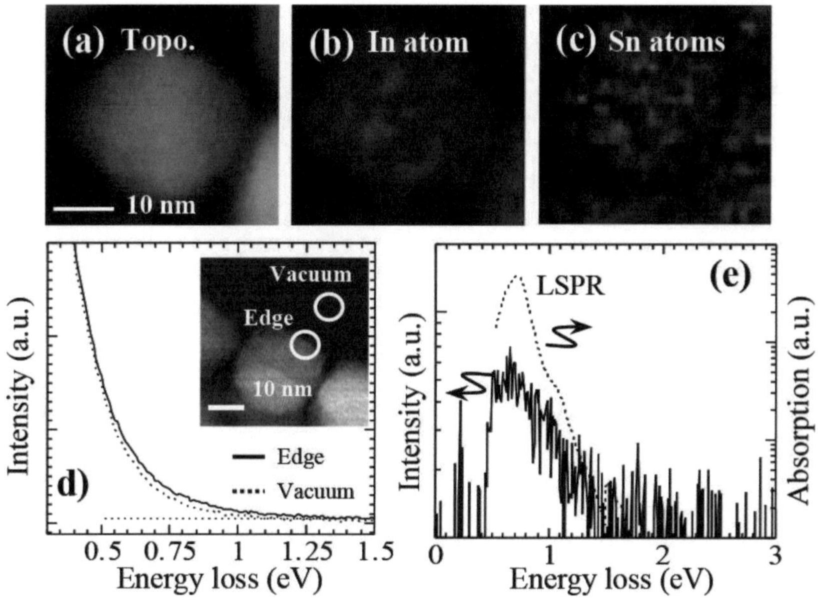

Figure 2. (a) Aberration-corrected STEM image, and HAADF images of In (b) and Sn atoms (c) in the ITO NP with a Sn concentration of 5%. (d) EELS spectra taken in the electron Probe position at vacuum (dotted line) and edge (solid line) regions, and a STEM-acquired particle image (inset). (e) Differential EELS spectrum (dotted line) and optical spectrum of ITO NP taken in the near-IR (solid line) (Figure 3 of Ref. [33]). Copyright 2014 by the American Institute of Physics.

region on the STEM-acquired particle image showed a slight spectral difference in energy-loss regions from 1.0 to 0.5 eV [**Figure 2(d)**]. In **Figure 2(e)**, a differential EELS spectrum had a maximum peak at 0.7 eV that was similar to the optical absorption in the near-IR, which was direct evidence of a LSP excitation on the NP surface as consequence of spatially homogeneous doping of Sn atoms in the NP.

3. Localized surface plasmons in ITO NPs

3.1. Theoretical calculations of optical properties

The absorption and scattering cross sections of a single ITO NP with a diameter (R) of 20 nm were theoretically estimated according to Mie theory on the basis of dielectric constants of a NP and a surrounding medium. A numerically analytical solution to Maxwell's equations describes the extinction (σ_{sca}) and scattering (σ_{scat}) of light by a perfect spherical particle structure as follows [34].

$$\sigma_{sca} = \frac{2\rho}{|k|^2} \sum_{L=1}^{\infty} (2L+1)\left(|a_L|^2 + |b_L|^2\right) \quad (1)$$

$$\sigma_{ext} = \frac{2\rho}{|k|^2} \sum_{L=1}^{\infty} (2L+1)\left[Re(a_L + b_L)\right] \quad (2)$$

where k is the incoming wave vector and L comprises integers representing the dipole, quadrupole, and high multipoles. a_L and b_L are represented by the following parameters, composed of the Riccati-Bessel equations of φ_L and δ_L:

$$a_L = \frac{m\,\varphi_L(mx)\,\varphi'_L(x) - \varphi'_L(mx)\,\varphi_L(x)}{m\,\varphi_L(mx)\,\delta'_L(x) - \varphi'_L(mx)\,\delta_L(x)} \tag{3}$$

$$b_L = \frac{\varphi_L(mx)\,\varphi'_L(x) - m\,\varphi'_L(mx)\,\varphi_L(x)}{\varphi_L(mx)\,\delta'_L(x) - m\,\varphi'_L(mx)\,\delta_L(x)} \tag{4}$$

where, $m = (n_R + in_I)/n_m$ is the complex refractive index of the metal, and n_m is the refractive index of the surrounding medium. In addition, $x = k_m r$ (r: is the radius of the particle). $k_m = 2\pi/\lambda_m$ is defined as the wavenumber in the medium. λ_m is the wavelength in the medium. **Figure 3(a)** shows absorption and scattering cross sections of an ITO NP with an R value of 20 nm. A peak position of LSP was located at around 1.8 μm. A spectral line-shape was a symmetric feature. A value of σ_{abs} was remarkably larger by three orders of magnitudes compared to σ_{scat}. The ratio of $\sigma_{abs}/\sigma_{scat}$ was not less than one for the particle diameter range above 200 nm [**Figure 3(b)**]. Therefore, it is considered that an ITO NP is dominated by light absorption in the near-IR region, indicating that it is not capable of light scattering an incident light in the near-IR range. Herein, we note that Full Mie theory in Eqs. (1)–(4) only has two functions of a dielectric constant and a particle diameter. The optical factors such as surface and radiation damping are excluded on the calculated spectra.

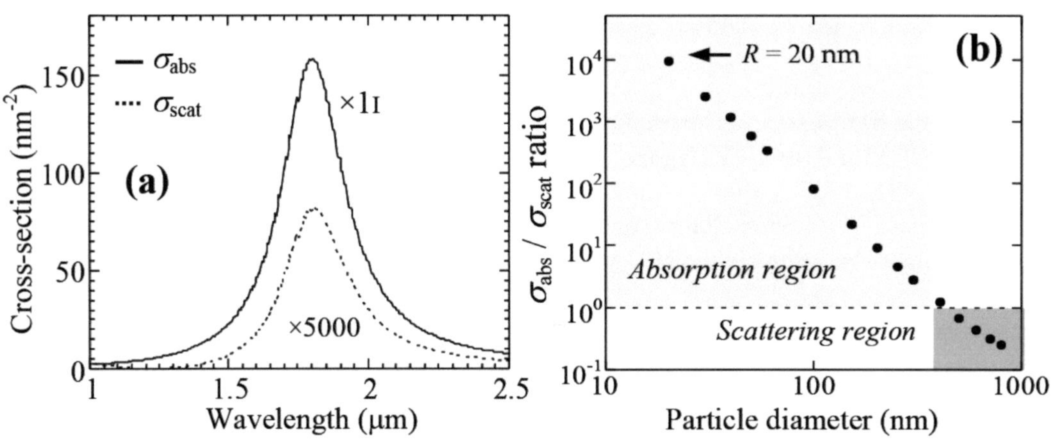

Figure 3. (a) Light absorption (σ_{abs}) and scattering (σ_{scat}) cross-sections of an ITO NP with an R value of 20 nm. (b) A value of $\sigma_{abs}/\sigma_{scat}$ ratio as a function of particle diameter.

3.2. Experimental optical properties

An optical absorption of an ITO NP solution (Sn concentration of 5%) was typically examined [**Figure 4(a)**]. Optical measurements in the IR-range were made at room temperature using a FT-IR spectrometer equipped with a liquid cooled HgCdTe (MCT) detector. An ITO NP solution was confined in an IR-transparent CaF_2 holder with an optical thickness of 25 μm, showing that a single absorption peak was located at 1.86 μm because of LSP excitations of

http://dx.doi.org/10.5772/67588

ITO NPs. This result was close to the theoretical data. We studied optical quality in plasmon resonances of ITO NPs using Mie theory as follows.

The absorption spectrum was fitted using the classical Mie theory with plasmon damping γ because of ionized impurity scattering derived from Sn impurities in the NPs. The theoretical fitting of an optical absorption (σ) to the experimental data in the quasi-static limit was employed [34]:

$$\sigma = 4\pi k R^3 Im\left\{ \frac{\varepsilon_p(\omega) - \varepsilon_H}{\varepsilon_p(\omega) + 2\varepsilon_H} \right\} \tag{5}$$

where, $k = 2\,\pi(\varepsilon H)^{1/2}\,\omega/c$ with c representing the speed of light, $\varepsilon_p(\omega)$ is the particle dielectric function, R is the particle radius, and ε_H indicates the host dielectric constants of toluene. The effective dielectric function (ε_{eff}) was used to obtain real nanoparticle dispersion by the Maxwell-Garnett effective medium approximation [35]:

$$\frac{\varepsilon_{eff}(\omega) - \varepsilon_H}{\varepsilon_{eff}(\omega) + 2\varepsilon_H} = f_v \frac{\varepsilon_p(\omega) - \varepsilon_H}{\varepsilon_p(\omega) + 2\varepsilon_H} \tag{6}$$

where, f_v is the particle volume fraction in the order of 10^{-5}. $\varepsilon_p(\omega)$ used the free-electron Drude term with a damping constant (γ) because ITO consists of free electrons due to the absence of interband transitions:

$$\varepsilon_p(\omega) = 1 - \frac{\omega_p^2}{\omega(\omega + i\gamma)} \tag{7}$$

The plasma frequency (ω_p) is defined to $\omega_p = n_e/\varepsilon_\infty \varepsilon_0 m^*$, where ε_∞ is the high-energy dielectric constant, and m^* is the effective electron mass. The fitted absorption was used with parameter values of $\varepsilon_H = 2.23$, $m^* = 0.3m_0$ and $\varepsilon_\infty = 3.8$ to determine $\varepsilon_p(\omega)$. In **Figure 4(a)**, the experimental absorption spectrum of the ITO NP solution could not be fitted to Mie theory based on Eqs. (1)–(3).

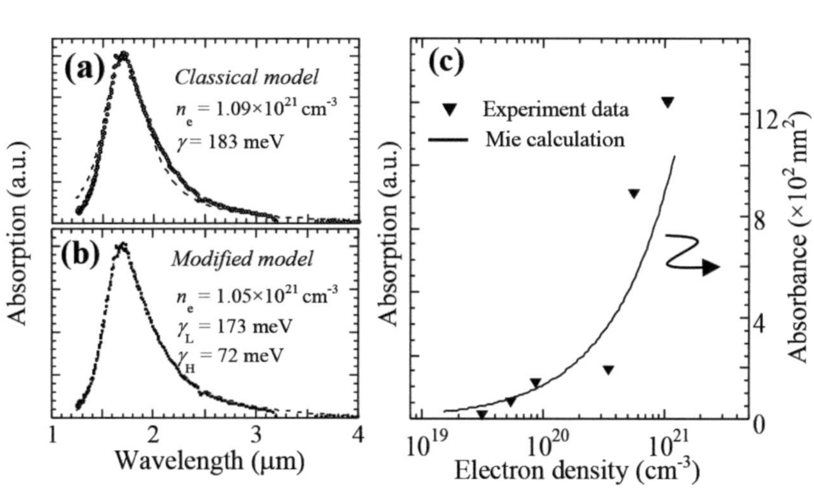

Figure 4. Experimental absorption spectrum of an ITO NP solution (Sn concentration of 5%) fitted to the classical (a) and modified Mie theory (b). Experimental and calculated spectra were indicated by open circles and dotted lines. (c) Absorption intensity of solution with ITO NPs with different electron density (black dots). Comparative data (black line) was extracted from the calculated absorption.

In general, plasmon damping γ of metallic NPs has been defined as a constant value. That is, γ is independent on frequency. However, plasmon damping of oxide semiconductors have been strongly dependent on frequency due to presence of electron-impurity scattering in a host. This is an interesting feature of the plasmonic response in doped oxide semiconductor NPs. Elucidation of the carrier-dependent damping mechanism is required for the design of plasmonic materials on TOSs. Therefore, frequency-dependent $\gamma(\omega)$ is introduced into the Drude term [36]. This is treated as the modified Mie theory for applying to doped oxide semiconductors.

$$\gamma(\omega) = f(\omega)\,\gamma_L + [\,1 - f(\omega)\,]\gamma_H\left(\frac{\omega}{\gamma_H}\right) \tag{8}$$

$$f(\omega) = \frac{1}{1 + exp\left(\frac{\omega - \gamma_X}{\sigma}\right)} \tag{9}$$

Above all, γ_L has information concerning electron-impurity scattering, resulting in an asymmetric spectral feature followed by spectral broadening in the long wavelength region [**Figure 4(b)**]. That is, it is considered that ionized impurity scattering attributed to Sn impurities in the NPs provided the asymmetric LSP response. In addition, strength in plasmon absorption remarkably increased with Sn concentration, as supported from experimental and theoretical aspects [**Figure 4(c)**]. LSPRs of ITO NPs were observed in electron density region above 3×10^{19} cm^{-3}, which was close to Mott critical density ($N_c = 9 \times 10^{18}$ cm^{-3}) of ITO. Above the N_c of ITO, the Fermi energy is determined by the highest occupied states in the conduction band. That is, LSP excitations are required to realize metallic states in the NPs, and became very strong at high electron densities above 10^{21} cm^{-3}.

4. Fabrications and structures of ITO NP sheets

4.1. Fabrications

Assembled NP sheets were deposited on IR transparent CaF_2 substrates by a spin-coating technique. Thick NP sheets were fabricated by way of multiple overglaze of a thin NP film obtained by a NP concentration of 0.2% in toluene. The spin-coating conditions were carried out using the following processes: 800 rpm (5 seconds) → 1200 rpm (10 seconds) → 2400 rpm (30 seconds) → 800 rpm (10 seconds). Fabricated sheet samples were heat-treated at above 150°C in air in order to evaporate the solvent. NP sheets with various thicknesses were obtained by repetition of the above coating sequences.

4.2. Structural evaluations

The assembled sheets of the NPs were evaluated by small-angle X-ray scattering (SAXS), providing an interesting insight into the scattering vector (q) of the SAXS intensity. A maximum SAXS peak (q_{max}) includes structural information about spatial ordering of nanoparticles estimated using $l = 2\,\pi/q_{max}$ with a spatial period (l). The SAXS pattern showed a maximum peak at $q = 0.33$ nm^{-1} followed with weak interferences [**Figure 5(a)**]. This resulted in an l value of 20 nm being close to the edge-to-edge between NPs, which was confirmed from in-plane surface scanning microscopy (SEM). A SEM image showed a close-packed structure because spin-coating produces

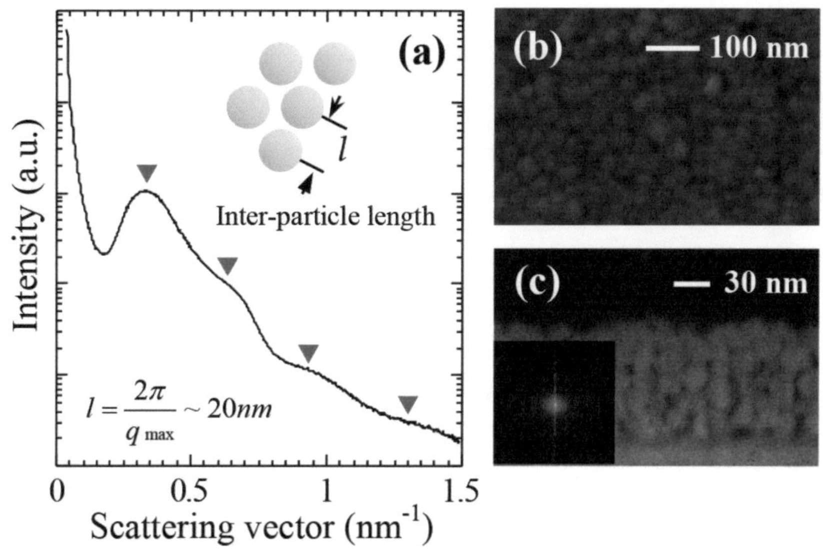

Figure 5. (a) SAXS pattern of a 96 nm-thick NP sheet. Inset indicates schematic picture to explain an inter-particle length (l). (b) In-plane surface (b) and cross-section SEM image (c) of 96 nm-thick NP sheets. Inset represents a FFT pattern of a cross-section SEM image (Figures 5 and 7 of Ref. [39]). Copyright 2014 by the American Chemical Society.

self-organizations of colloidal NPs into a hexagonally close-packed (HCP) structure based on shear and capillary forces on substrates [37, 38] [**Figure 5(b)**]. In addition, a cross-section SEM image also provided a close-packed structure [**Figure 5(c)**]. However, a particle-alignment is disordered packing feature between NPs, as indicated from a fast-Fourier transform (FFT) pattern of the SEM image [inset in **Figure 5(c)**]. This local structure is related to the broadened interferences of SAXS pattern, which gives influences to optical properties of the NP sheets.

5. Infrared optical responses of ITO NP sheets

5.1. Mono-layered NP sheets

The optical properties of a mono-layered ITO NP sheet are shown in **Figure 6**. Transmittance spectra exhibited a resonant peak at 2.64 μm, which revealed the red-shifted resonant wavelength because of a collective plasmon resonance (CPR) effect compared to those of NPs dispersed in toluene [**Figure 6(a)**] [40]. On the other hand, reflectance at the resonant wavelength was very small, indicating that the optical responses were mainly dominated by light absorption properties. Furthermore, the finite-difference time-domain (FDTD) simulation was carried out to evidence the experimental results. The modelled mono-layered NP sheet (R = 20 nm) has a HCP structure with an inter-particle length (l) of 2 nm along the X-Y (in-plane) direction. The modelled NP sheet was illuminated with light directed in the Z-direction from the air side. Periodic boundary conditions were applied to X and Y directions, and the bottom and top of the simulated domain in the Z-direction were analysed using perfectly matched layer (PML) boundary conditions. The refractive index (n) of the surface ligand between NPs was defined to 1.437. An ellipsometric measurement of an ITO film was conducted to obtain the complex dielectric constants within the visible-IR range. A surface ligand molecular of the NPs was a capric fatty acid in this chapter. A resonant peak at 2.45 μm was reproduced in transmittance and reflectance spectra, which was compared to the experimental results [**Figure 6(b)**]. The CPR effect was excited due to long-range coherences of *E*-field interactions between NPs, as supported from the two-dimensional (2D) image of the E-field distribution [inset of **Figure 6(b)**].

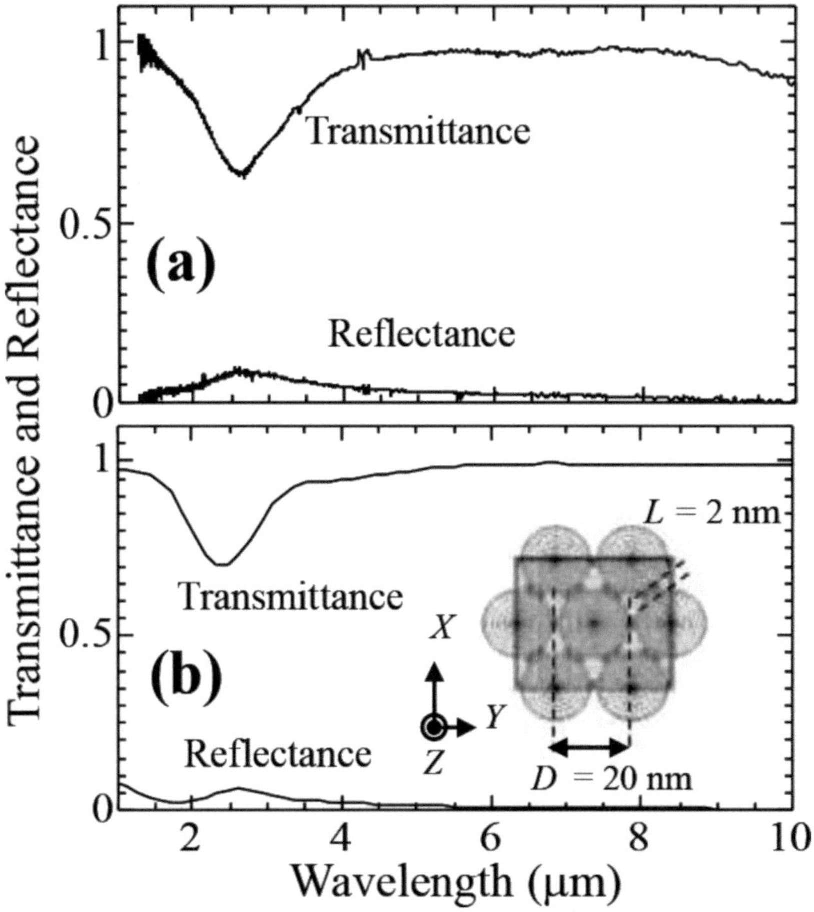

Figure 6. (a) Experimental and (b) simulated transmittance and reflectance of a mono layered NP sheet with a HCP structure. Inset in Figure 6(b) shows a model of a mono-layered NP sheet and an *E*-field distribution when an electric field of light is applied along the *X*-direction (Figure 4 of Ref. [39]). Copyright 2014 by the American Chemical Society.

5.2. Three-dimensional NP sheets

3D-stacked NP sheets showed a remarkable change in optical properties, which were clearly found on transmittance and reflectance spectra [**Figure 7(a)** and **(b)**]. Transmittance with a resonant wavelength at 2.20 μm decreased to a level close to zero with increasing sheet thickness. On the other hand, reflectance was enhanced at a close proximity of 0.6 in terms to the

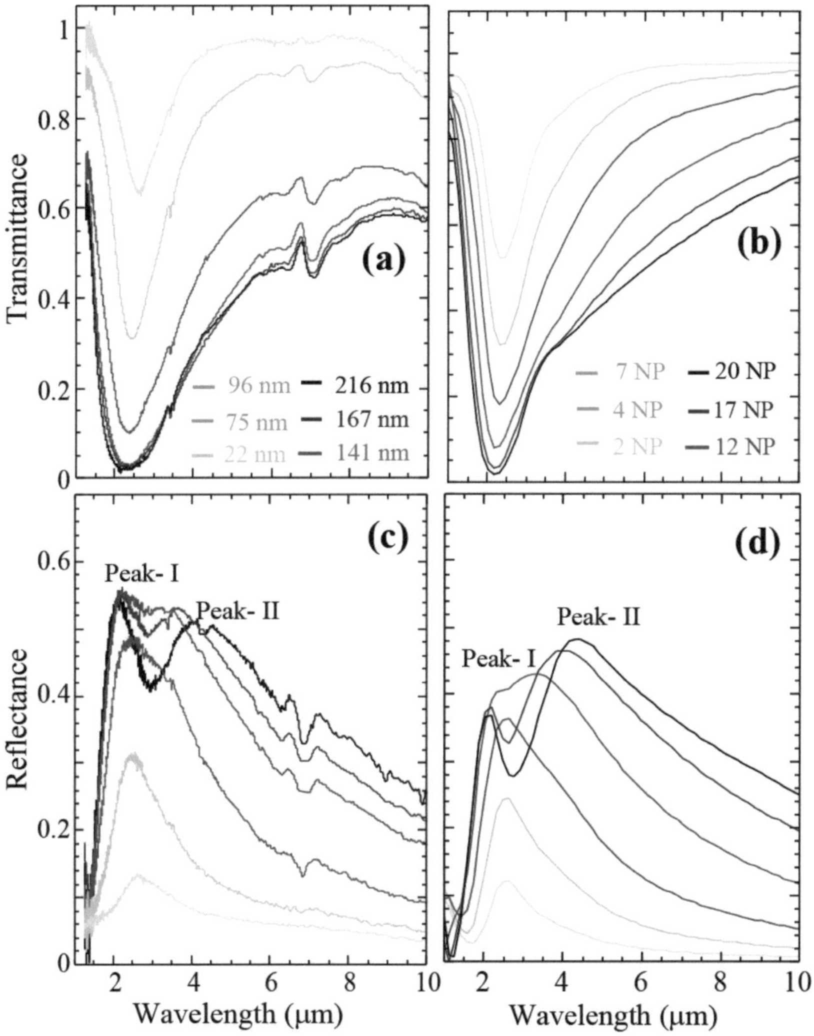

Figure 7. (a) Experimental and (b) simulated transmittance spectra of NP sheets with different thicknesses. (c) Experimental and (d) reflectance spectra of NP sheets with various NP layers. The modeled NP sheet was illuminated with light directed in the *Z*-direction from the air side. The direction of the *E*-field was perpendicular to the light and parallel to the *X*-direction (Figure 5 of Ref. [39]). Copyright 2014 by the American Chemical Society.

sheet thickness [**Figure 8(b)**]. The single peak of 22 nm-thick NP sheet was gradually separated into lower and higher wavelengths with the sheet thickness [**Figure 8(a)**]. We observed two types of resonant peaks (I and II) at 2.13 and 4.02 μm in the near- and mid-IR region on the 216 nm-thick NP sheet, respectively. The ratio of (*R*/*A*) of reflectance (*R*) and absorbance (*A*) increased quickly to a large value with increasing sheet thickness, which indicated that the assembled NP sheets showed reflectance-dominated optical properties.

FDTD calculations were conducted in order to clear the experimental results. From the cross-section SEM image, the modelled NP layers are based on a 3D HCP structure with an *l* value

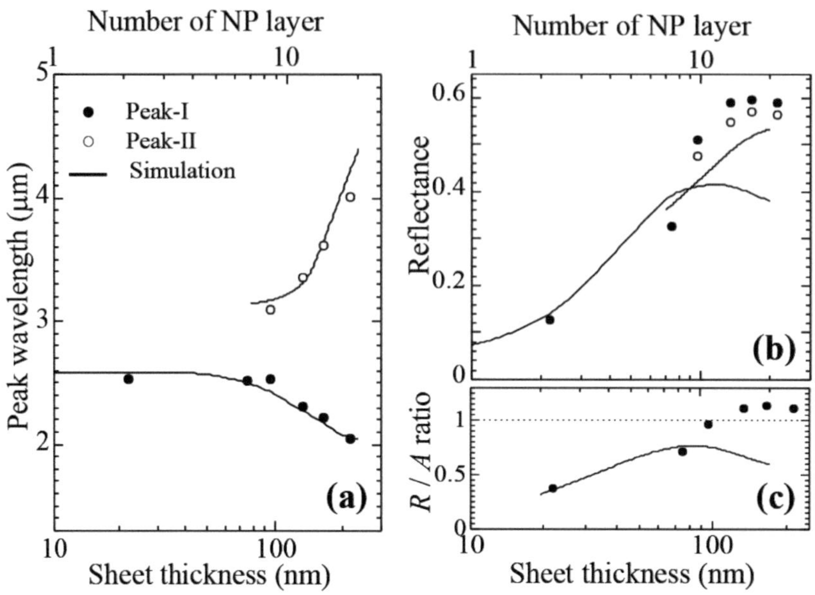

Figure 8. (a) Resonant wavelengths and (b) reflectance of peak-I and peak-II as a function of sheet thickness (bottom horizontal axis) and number of NP layers (upper horizontal axis). Black lines indicate calculated results of FDTD simulations. (c) Experimental and calculated ratios of *R*/*A* at peak position related to peak-I. *R* and *A* represent reflectance and absorbance, respectively (Figure 6 of Ref. [39]). Copyright 2014 by the American Chemical Society.

of 2 nm in the *Z*- (out-of-plane) direction. A layer structure along the in-plane (*X*-*Y*) direction employed the mono-layered NP layer in Section 5.1. The systematic change in the number of NP layers from 1 to 20 was capable of reproducing the experimental spectra [**Figure 7(c)** and **(d)**]. The employed model could describe the optical properties of the NP sheets. The increase in number of NP layers provided the resonant peaks in transmittance and peak separations in reflectance. These behaviours were similar to the experiment results [**Figure 8(a)** and **(b)**]. Herein, the reflectance for peak-I was smaller than that for peak-II in the case of calculations, resulting in a difference of *R*/*A* ratio between experimental and calculation aspects [**Figure 8(c)**].

5.3. Plasmon hybridization and reflectance mechanism

The resonant origins of reflectance of peak-I and peak-II were theoretically examined as a function of interparticle length between NPs. **Figure 9(a)** exhibits calculated reflectance of NP sheets with different *l* values on the basis of 20 NP layer model, revealing that reflectance monotonically enhanced with decreasing *l*. Peak-II showed a red-shift to longer wavelengths when decreasing *l* from 10 to 1 nm. In contrast, peak-I remained unchanged [**Figure 9(b)**]. These results suggest a difference in the origin of plasmon resonance between peak-I and peak-II. In general, the localized *E*-field from each metal NP usually overlaps when metal NPs are closely positioned, and plasmon coupling occurs [41]. In the plasmon hybridization process, the plasmon interaction can be categorized into bonding and anti-bonding levels. For example, the bonding level shows a red-shift of a resonant wavelength with decreasing interparticle length. On the other hand, there is a slight blue-shift a resonant peak is due to

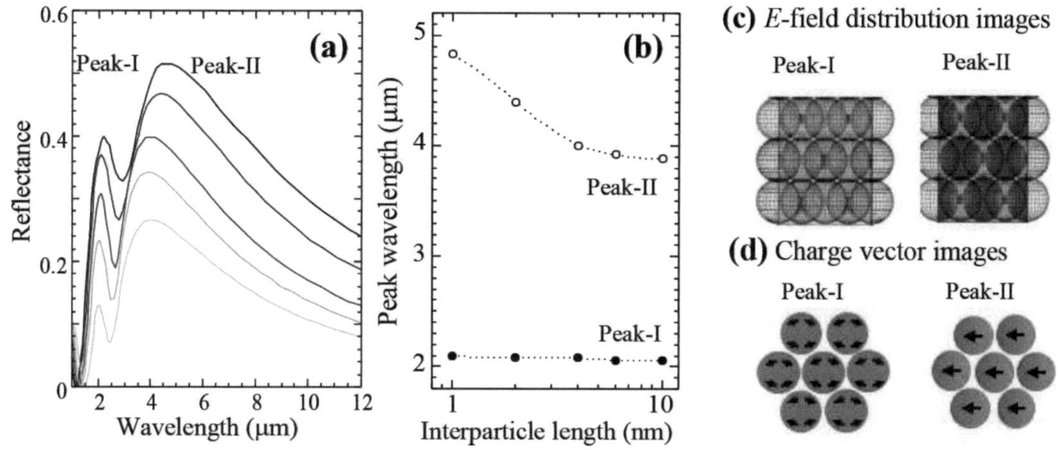

Figure 9. (a) Calculated reflectance of NP sheets with different interparticle lengths. (b) Resonant peaks of peak-I and peak-II as a function of interparticle length. (c) Images of *E*-field distributions. (d) Images of charge vectors at peak-I and peak-II (Figure 6 of Ref. [39]). Copyright 2014 by the American Chemical Society.

the anti-boding level. The peak shifts at peak-I and peak-II were similar to the anti-bonding and bonding states, respectively. Furthermore, *E*-field distributions and their charge vectors were investigated at peak-I and peak-II [**Figure 9(c)** and **(d)**]. We firstly focus on the mid-IR reflectance at peak-II. A resonant mode comprised of individual dipolar plasmons oscillating in-phase along the direction of incident polarization. The *E*-fields between the NPs were only localized along the in-plane *X*-direction. On the other hand, field analysis at peak-I exhibited that the dipolar plasmons in the NPs oscillate out-of-phase, providing a net dipole moment of nearly zero. Their *E*-fields were coupled with surrounding NPs along the out-of-plane and in-plane directions. The mode splitting of plasmon resonances was related to 3*D*-stacked assemblies of NPs. Accordingly, quadrupole and dipole modes were formed as peak-I and peak-II, respectively. These behaviours became pronounced with an increase in sheet thickness. We note that the differences in reflectance between experimental and simulation data could be considered in relation of a local structure and plasmon resonance. The NP sheets had a disordered structure of NPs from the SEM image. A dipole mode can be strongly observed in precise close-packed NP assemblies.

The character of *E*-field coupling in the NP sheets was further studied from polarized reflectance measurements. Two types of light polarized perpendicular (*s*-polarized) and parallel (*p*-polarized) to the plane of incidence were introduced the sample at an incident angle of 75°. **Figure 10** shows polarized reflectance spectra for *s*- and *p*-polarization configurations for a NP sheet. In a case of *s*-polarization, peak-I and peak-II were simultaneously obtained because the electric vector of the radiation produced electron oscillations in NPs parallel to the in-plane of the sample. On the other hand, peak-I only found for a *p*-polarization case. These optical properties concerning both polarizations are explained as follows.

The electric vector excites electron oscillations in NPs normal to the plane of the sample, and suppresses the *E*-field coupling along the in-plane direction. These results revealed that

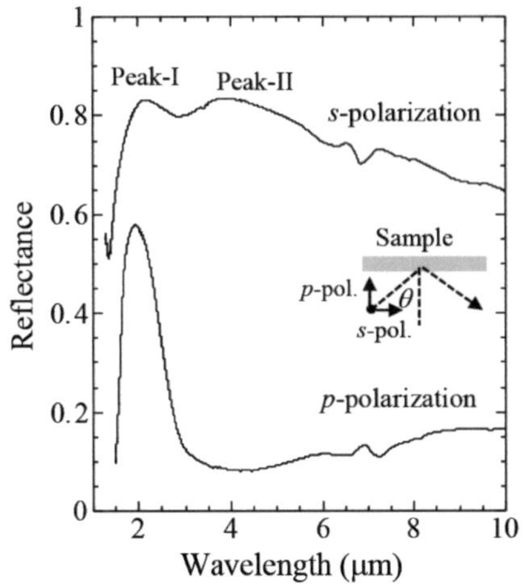

Figure 10. Reflectance spectra of a 216 nm-thick NP sheet for s- and p-polarized lights. Inset indicates the direction of the electric vector of incident light in addition to an *E*-field distribution under a *p*-polarization.

peak-II was excited by the field coupling along the in-plane directions. In contrast, the reflectance at peak-I was essential for *E*-field coupling along the out-of-plane direction. That is, the near- and mid-IR reflectance of the NP sheets was attributed to the 3*D* field coupling along the out-of-plane and in-plane directions. The film thickness-dependent plasmon splitting was attributed to the formation of field coupling along the out-of-plane direction, leading to the enhanced reflectance in the near-IR range.

5.4. Nanoparticle gap and reflectance of ITO NP sheets

The thermal behaviours of the NP samples were investigated by TG-DTA in an N_2 atmosphere with a heating rate of 10°C/min. The weight loss up to 250°C might be related to the loss of physically or chemically absorbed water. There was an obvious weight loss in the temperature range 270–320°C because of the generation of organic species confirmed by *m*/*z* peaks at 18 (H_2O) and 44 (CO_2, C_3H_8, C_2H_4O *etc.*) [**Figure 11(a)** and **(b)**]. The decomposed species were owing to thermal removal of the surface ligands composed of fatty acids on the NP surfaces. There phenomena are identified by FT-IR measurements. The relation between surface ligand molecules and optical properties in the NP sheets was cleared by the spectral changes after annealing at different temperatures (T_A = 150–550°C). **Figure 11(c)** shows temperature-dependent reflectance spectra taken in an inert atmosphere for a 216 nm-thick NP sheet, revealing remarkable spectral changes with increasing temperature. The two resonant peaks at 150°C were weakened gradually following the change in spectral shape with increasing temperature. Above all, the near-IR reflection at peak-I shifted to longer wavelengths at high temperatures above 300°C corresponding to the removal of the surface ligands. In addition, the annealing effects of NP sheets were checked by the electrical resistivity in the sheets. Electrical resistivity was in the order of 10^4Ω.cm below annealing temperatures below 250°C because the presence of the surfactant layers on NPs seriously impedes carrier transport in the NP sheets. The surface ligands act as interparticle insulating layers in NP networks [42].

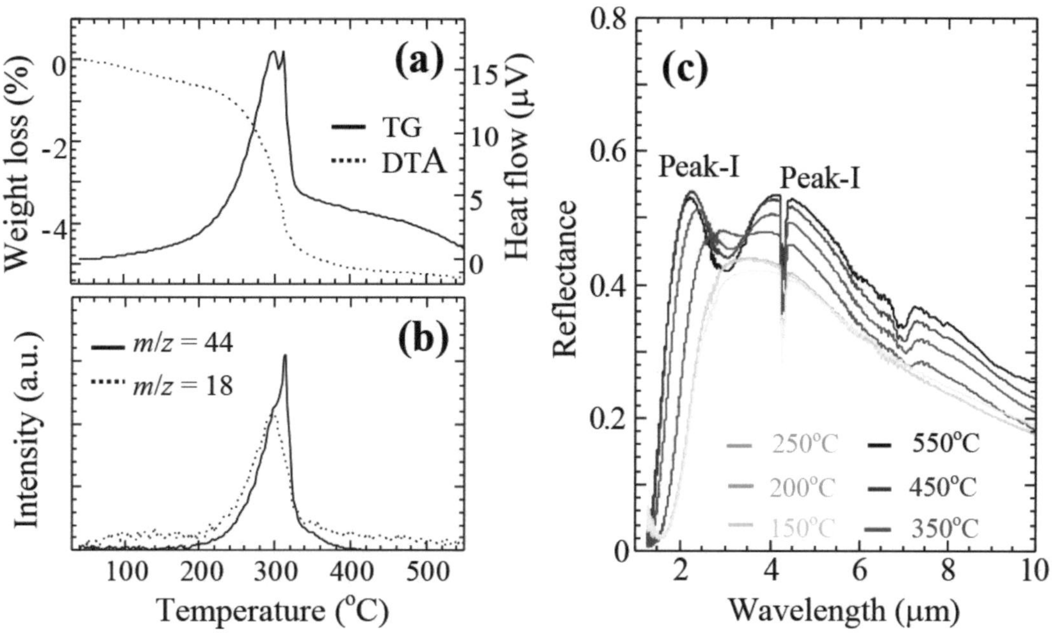

Figure 11. (a) TG-DTA curves of ITO NP samples in an N_2 atmosphere. (b) TOF-Mass spectroscopy combined with TG-DTA. *M/z* signals at 18 and 44 were detected in the range 27–550°C. (c) Dependence of reflectance spectra on annealing temperature for a 216 nm-thick NP sheet.

However, this was markedly reduced at high temperatures above 350°C due to the removal of the surface ligands. Therefore, the removal of surface ligand molecules from the NPs strongly affected the whole reflective phenomena, which also clarified the importance of interparticle length in obtaining a high reflectance performance.

5.5. Electromagnetic responses in microwave range

EM properties are shortly discussed on a NP sheet in the microwave range 0.5–40 GHz. This range is an important frequency range for telecommunications. Transparent solar-thermal shielding is effective techniques to prevent room heat in order to realize comfortable environment in vehicles. However, it is strongly required for vehicles to transmit EM waves in the microwave range through windows to carry out radio communications such as an Electronic Toll Collection System (ETC) and Information traffic system (ITS). Therefore, it is important to measure EM properties of NP sheets in addition to evaluate optical properties in the IR range.

250 nm-thick NP sheet with an A4 size was fabricated on a PET substrate (thickness: 0.2 mm) using a roll-coating method [inset of **Figure 12(a)**]. High reflectance with a close proximity of 0.6 was also obtained on a flexible PET substrate [**Figure 12(a)**]. The shielding effectiveness (SE) of the flexible NP sheet was almost zero, as different from that of a RF sputtered ITO film [**Figure 12(b)**]. The difference between the two materials related to electrical conductance (σ) in the sheets, which was in the order of 10^{-5} and 10^{-3} S/cm for the NP sheet and sputtered film, respectively. If the shielding material is thin, SE is mainly dominated by EM reflection as follows [43]:

$$SE = 20\log(\sqrt{2}\,\beta_0/2R) \tag{10}$$

where, β_0^2 is μ/ε_0, and R is the sheet resistivity (=1/σ). The significant obstruction of electron carrier transport between NPs produced low electrical conductance because of the presence of surface ligands on the NPs, and realized the high microwave transmissions.

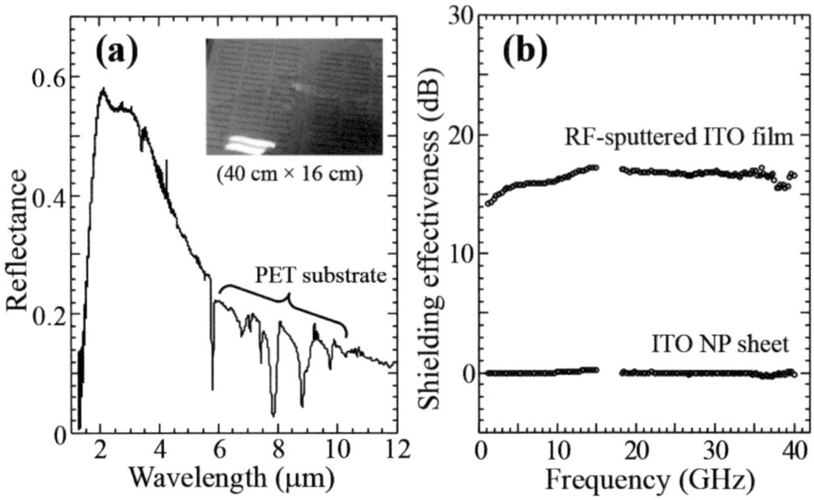

Figure 12. (a) Reflectance of a NP sheet on a PET film. Inset image represents a photograph of the fabricated sheet sample. (b) Shielding effectiveness (SB) in the microwave range 0.5–40 GHz for a NP sheet and RF-sputtered ITO film.

ITO NPs were used to obtain assembled NP sheets with small interparticle lengths by the presence of ligand molecules on the particle surfaces. This situation produced effective *E*-field coupling along the in-plane and out-of-plane directions. This caused the plasmon hybridization for the quadrupole and dipole modes, which played an important role in producing the high reflectance in the near- and mid-IR range. In addition, the *E*-field enhancements between NPs simultaneously caused a remarkable reduction of electrical contacts between the NPs, which contributed to the high microwave transmissions. The plasmonic control in assemblies of NPs represents promising potential for structural and optical designs used to fabricate a flexible thermal-shielding sheet with a reflection-type based on transparent oxide semiconductors. The knowledge gained in this study can be applied to NP sheets utilizing inexpensive ZnO and WO_3 [44, 45].

6. Summary

Crystallinity and local structures of oxide semiconductor NPs were conducted using ITO NPs by XRD and TEM measurements in Section 2. The plasmonic resonances of ITO NPs were clearly obtained in the near-IR range from the viewpoints of optical and EELS signals. In particular, electron-impurity scattering contributed towards plasmon damping as one of a factor that is absent in metal NPs on the basis of theoretical and experimental approaches, which was discussed in Section 3. In Sections 4 and 5, we described IR plasmonic applications in ITO NP sheets for solar-thermal shielding technology. Above all, the *E*-field coupling between NPs produced interesting plasmonic coupling because of the creation of narrow crevices in the interparticles. *3D* field interactions along the in-plane and out-of-plane directions caused high reflectance in the near- and mid-IR regions. Finally, ITO NP sheets could be extended to obtain large-size flexible films on PET substrates, which simultaneously showed microwave transmittance essential for telecommunications.

The above results provided important insights for basic science and practical applications based on plasmonic investigations based on oxide semiconductor NPs. However, plasmonic properties and applications are stand still-points at the present time. Hereafter, it will be needed to study plasmonic phenomena on oxide semiconductor NPs towards new concepts concerning optical manipulations in the IR range.

Acknowledgements

This research was supported in part by a grant-in-Aid from the JSPS Core-to-Core Program, A. Advanced Research Network, a grand from Toyota Physical and Chemical Research Institute, and for a grant-in-Aid for Exploratory Research and Scientific Research (B).

Author details

Hiroaki Matsui[1,2]* and Hitoshi Tabata[1,2]

*Address all correspondence to: hiroaki@ee.t.u-tokyo.ac.jp

1 Department of Bioengineering, the University of Tokyo, Japan

2 Department of Electric Engineering and Information Systems, the University of Tokyo, Japan

References

[1] C. Rhodes, M. Cerruti, A. Efremenko, M. Losego, D. E. Aspnes, J. -P. Maria and S. Franzen. Dependence of plasmon polaritons on the thickness of indium tin oxide thin films. Journal of Applied Physics. 2008;**103**:093108(1)–093108(8). doi:10.1063/1.2908862.

[2] H. Matsui, W. Badalawa, A. Ikehata and H. Tabata. Oxide surface plasmon resonance for a new sensing platform in the near-infrared range. Advanced Optical Materials. 2003;**1**:397–403. doi:10.1002/adom.201200075.

[3] W. Badalawa, H. Matsui, A. Ikehata and H. Tabata. Surface plasmon modes guided by Ga-doped ZnO layers bounded by different dielectrics. Applied Physics Letters. 2011;**99**:011913(1)–011913(3). doi:10.1063/1.3608313.

[4] E. Sachet, M. D. Losego, J. Guske, S. Franzen and J. -P. Maria. Mid-infrared surface plasmon resonance in zinc oxide semiconductor thin films. Applied Physics Letters. 2013;**102**:051111(1)–051111(4). doi:10.1063/1.4791700.

[5] H. Matsui, A. Ikehata and H. Tabata. Surface plasmon sensors on ZnO: Ga layer surfaces: Electric-field distributions and absorption-sensitive enhancements. Applied Physics Letters. 2015;**106**:019905(1)–019905(5). doi:10.1063/1.4905211.

[6] Y. Zhao, H. Pan, Y. Lou, X. Qiu, J. J. Zhu and C. Burda. Plasmonic $Cu_{2-x}S$ nanocrystals: Optical and structural properties of copper-deficient copper (I) sulfides. Journal of the American Chemistry Society. 2009;**131**(12):4253–4261. doi:10.1021/ja805655b.

[7] G. Garcia, R. Buonsanti, E. L. Runnerstrom, R. J. Mendelsberg, A. Llordes, A. Anders, T. J. Richardson and D. J. Milliron. Dynamically modulating the surface plasmon resonance of doped semiconductor nanocrystals. Nano Letters. 2011;**11**(10):4415–4420. doi:10.1021/nl202597n.

[8] G. V. Naik, V. M. Shalaev and A. Boltasseva. Alternative plasmonic materials: Beyond gold and silver. Advanced Materials. 2013;**25**(24):3264–3294. doi:10.1002/adma201205076.

[9] T. Yamada, H. Makino, N. Yamamoto and T. Yamamoto. Ingrain and grain boundary scattering effects on electron mobility of transparent conducting polycrystalline Ga-doped ZnO films. Journal of Applied Physics. 2010;**107**:123534(1)–123534(8). doi:10.1063/1.3447981.

[10] C. Sonnichesen, T. Franzl, T. Wilk, G. von Plessen and J. Feldmann. Plasmon resonances in large noble-metal clusters. New Journal of Physics. 2002;**4**:93(1)–93(8). doi:S1367-2630(02)39458-8.

[11] Z. C. Jin, I. Hamberg and C. G. Granqvist. Optical properties of sputter-deposited ZnO: Al thin films. Journal of Applied Physics. 1988;**64**:5117–5131. doi:10.1063/1.342419.

[12] N. Yamamoto, H. Makino, S. Osone, A. Ujihara, T. Ito, H. Hokari, T. Maruyama and T. Yamamoto. Development of Ga-doped ZnO transparent electrodes for liquid crystal display panels. Thin Solid Films. 2012;**520**(12):4131–4138. doi:10.1016/j/tsf.2011.04.067.

[13] S. Q. Li, P. Guo, B. Buchholz, W. Zhou, Y. Hua, T. W. Odom, J. B. Ketterson, L. E. Ocola, K. Sakoda and R. P. H. Chang. Plasmon-photonic mode coupling in indium-tin-oxide nanorod arrays. ACS Photonics. 2014;**1**(3):163–172. doi:10.1021/ph400038g.

[14] D. B. Tice, S. Q. Li, M. Taglazucchi, D. B. Buchholz, E. A. Weiss and R. P. H. Chang. Ultrafast modulation of the plasma frequency of vertically aligned indium tin oxide rods. Nano Letters. 2014;**14**(3):1120–1126. doi:10.1021/nl4028044.

[15] S. Wuestner, J. M. Hamm, A. Pusch, O. Hess. Plasmonic leaky-mode lasing in active semiconductor nanowires. Laser & Photonics Reviews. 2015;**9**(2):256–262. doi:10.1002/por.201400231.

[16] S. D. Lounis, E. L. Runnerstrom, A. Llordés and D. J. Milliron. Defect chemistry and plasmon physics of colloidal metal oxide nanocrystals. Journal of Physical Chemistry C. 2014;**5**:1564–1574. doi:10.1021/jz500440e.

[17] A. M. Schimpf, S. D. Lounis, E. L. Runnerstrom, D. J. Milliron and D. R. Gamelin. Redox chemistries and plasmon energies of photodoped In_2O_3 and Sn-doped In_2O_3 (ITO) nanocrystals. Journal of American Chemistry Society. 2014;**137**:518–524. doi:10.1021/a5116953.

[18] D. Solis Jr., A. Paul, J. Plson, L. S. Slaughter, P. Swanglap, W. S. Chang and S. Link. Turning the corner: Efficient energy transfer in bent plasmonic nanoparticle chain waveguides. Nano Letters. 2013;**13**(10):4779–4784. doi:10.1021/nl402358h.

[19] N. Tognalli, A. Fainstein, E. Calvo, C. Bonazzola, L. Pietrasanta, M. Campoy-Quiles and P. Etchegoin. SERS in PAH-Os and gold nanoparticle self-assembled multilayers. The Journal of Chemical Physics. 2005;**123**:044707(1)–044707(9). doi:10.1063/1.1954707.

[20] M. Fan, M. Thompson, M. L. Andreade and G. Brolo. Silver nanoparticles on a plastic platform for localized surface plasmon resonance biosensing. Analytical Chemistry. 2010;**82**(15):6350–6352. doi:10.1021/ac101495m.

[21] A. Furube, T. Yoshinaga, M. Kanehara, M. Eguchi and T. Teranishi. Electric-field enhancement inducing near-infrared two-photon absorption in an indium-tin oxide nanoparticle film. Angewandte Chemie. 2012;**51**(11):2640–2642. doi:10.1002/anie20117450.

[22] H. Matsui, W. Badalawa, T. Hasebe, S. Furuta, W. Nomura, T. Yatsui, M. Ohtsu and H. Tabata. Coupling of Er light emissions to plasmon modes on In_2O_3: Sn nanoparticle sheets in the near-infrared range. Applied Physics Letters. 2014;**105**(4):041903(1)–041903(5). doi:10.1063/1.4892004.

[23] K. Okamoto, B. Lin, K. Imazu, A. Yoshida, K. Toma, M. Toma and K. Tamada. Tning colors of silver nanoparticle sheets by multilayered crystalline structures on metal substrates. Plasmonics. 2012;**8**(2):581–590. doi:10.1007/s11468-012-9437-2.

[24] F. Le, D. W. Brandl, Y. A. Urzhumov, H. Wang, J. Kundu, N. J. Halas, J. Aizpurua and P. Nordlander. Metallic nanoparticle arrays: A common substrate for both surface-enhanced Raman scattering and surface-enhanced infrared absorption. ACS Nano. 2008;**2**(4):707–718. doi:10.1021/nn800047e.

[25] K. Okamoto, I. Niki, A. Scherer, Y. Narukawa. T. Mukai and Y. Kawakami. Surface plasmon enhanced spontaneous emission rate of InGaN/GaN quantum wells probed by time-resolved photoluminescence spectroscopy. Applied Physics Letters. 2005;**87**:071102(1)–071102(3). doi:10.1063/1.2010602.

[26] I. Hamberg and C. G. Granqvist. Evaporated Sn-doped In_2O_3 films: Basic optical properties and applications to energy-efficient windows. Journal of Applied Physics. 1986;**60**:R123–R157. doi:10.1063/1.337534.

[27] S. Y. Li, G. A. Nilsson and C. G. Granqvist. Plasmon-induced near-infrared electrochromism based on transparent conducting nanoparticles: Approximate performance limits. Applied Physics Letters. 2012;**101**:071903(1)–071903(3). doi:10.1063/1.4739792.

[28] K. Katagiri, R. Takabatake and K. Inumaru. Robust infrared-shielding coating films prepared using perhyropolysilazane and hydrophobized indium tin oxide nanoparticles

with tuned surface plasmon resonance. ACS Applied Materials & Interfaces. 2013;**5**(3):10240–10245. doi:10.1021/am403011t.

[29] X. Fang, C. L. Mak, J. Dai, K. Li, H. Ye and C. W. Leung. ITO/Au/ITO sandwich structure for near-infrared plasmonics. ACS Applied Materials & Interfaces. 2014;**6**(18):15743–15752. doi:10.1021/am5026165.

[30] P. Tao, A. Viswanath, L. S. Schadler, B. C. Benicewica and R. W. Siegel. Preparation and optical properties of indium tin oxide/epoxy nanocomposites with polyglycidyl methacrylate grafted nanoparticles. ACS Applied Materials & Interfaces. 2011;**3**(9):3638–3645. doi:10.1021/am200841n.

[31] Z. Sun, J. He, A. Kumbhar and J. Fang. Nonaqueous synthesis and photoluminescence of ITO nanoparticles. Langmuir. 2010;**26**(6):4246–4250. doi:10.1021/la903316b.

[32] H. Matsui, S. Furuta and H. Tabata. Role of electron carriers on local surface plasmon resonances in doped oxide semiconductor nanocrystals. Applied Physics Letters. 2014;**104**(12):211903(1)–211903(4). doi:10.1063/1.4880356.

[33] J. A. School, A. L. Koh and J. A. Dionne. Quantum plasmon resonances of individual metallic nanoparticles. Nature. 2012;**483**:421–427. doi:10.1038/nature10904.

[34] G. V. Hartland. Optical studies of dynamics in noble metal nanostructures. Chemistry Reviews. 2011;**111**(6):3858–3887. doi:10.1021/cr1002547.

[35] D. Ross and R. Aroca. Effective medium theories in surface enhanced infrared spectroscopy: The pentacene example. Journal of Chemical Physics. 2002;**117**:8095–8103. doi:10.1063/1.1502656.

[36] A. Pflug, V. Sitinger, F. Ruske, B. Szyszka and G. Dittmar. Optical characterization of aluminium-doped zinc oxide films by advanced dispersion theories. Thin Solid Films. 2004;**455–456**:201–206. doi:10.1016/j.tsf.2004.01.1006.

[37] S. M. Yang, S. G. Jang, D. G. Chio, S. Kim and H. K. Yu. Nanomachining by colloidal lithography. Small. 2006;**2**(4):458–475. doi:10.1002/smll.200500390.

[38] P. Jiang and M. J. McFarland. Large-scale fabrication of wafer-size colloidal crystals, macroporous polymers and nanocomposites by spin-coating. Journal of the American Chemistry of Society. 2004;**126**(42):13778–13786. doi:10.1021/ja0470923.

[39] H. Matsui, S. Furuta, T. Hasebe and H. Tabata. Plasmonic-field interactions at nanoparticle interfaces for infrared thermal-shielding applications based on transparent oxide semiconductors. ACS Applied Materials & Interfaces. 2016;**8**(18):11749–11759. doi:10.1021/acsami6b1202.

[40] C. Jiang, S. Markutsya and V. V. Tsukruk. Collective and individual plasmon resonances in nanoparticle films obtained by spin-assisted layer-by-layer assembly. Langmuir. 2003;**20**(3):882–890. doi:10.1021/la0355085.

[41] P. Nordlander, C. Oubre, E. Prodan, K. Li and M. I. Stockman. Plasmon hybridization in nanoparticle dimers. Nano Letters. 2004;**4**(5):899–903. doi:10.0121/nl049681c.

[42] J. Y. Kim and N. A. Kotov. Charge transport dilemma of solution-processed nanomaterials. Chemistry of Materials. 2014;**26**(1):134–152. doi:10.1021/cm402675k.

[43] C. A. Klein. A microwave shielding effectiveness of EC-coated dielectric slabs. IEEE Transactions of Microwave Theory Techniques. 1990;**38**:321–324.

[44] C. Guo, S. Yin, M. Yan, M. Kobayashi and T. Sato. Morphology-controlled synthesis of $W_{18}O_{49}$ nanostructures and their near-infrared absorption properties. Inorganic Chemistry. 2012;**51**(8):4763–4771. doi:10.1021/in300049j.

[45] H. Matsui, L. Y. Ho, T. Kanki, T. Tanaka, J. J. Delaunay and H. Tabata. Mid-infrared plasmonic resonances in two-dimensional VO_2 nanosquare arrays. Advanced Optical Materials. 2013;**3**:1759–1767.

Chapter 9

Surface Plasmonics and Its Applications in Infrared Sensing

Guiru Gu, Xuejun Lu, Thitikorn Kemsri and Yingjie Zhang

Additional information is available at the end of the chapter

http://dx.doi.org/10.5772/67410

Abstract

Surface plasmonic waves have been extensively researched due to their strong surface confinement. The strong surface confinement allows high absorption in an infrared (IR) detector with a thin active absorption region. The excitation of surface plasmonic resonance (SPR) depends on the metallic structures and the interface materials. This enables engineering of plasmonic-enhanced IR detector properties (e.g. detection wavelength, polarization and angular dependence) by properly designing the plasmonic structures. This chapter first gives a brief review of the surface plasmonic waves, followed by the description of SPR excitation in a metallic two-dimensional (2D) sub-wavelength hole array (2DSHA) structure. The applications of the 2DSHA SPR in IR detector enhancement are then presented with a discussion of the polarization and angular dependence.

Keywords: surface plasmonic waves, quantum dot, infrared detection

1. Introduction

Surface electromagnetic waves (also called surface plasmon polaritons, SPPs), predicted by Ritchie in 1957 [1], are coherent electron oscillations at a metal surface. The coherent electron oscillations generate electromagnetic (EM) waves that travel along the dielectric/metal interface [2]. Depending on the metal structures, certain resonant modes of the surface electromagnetic waves can be strongly absorbed (or excited) [2]. This is generally referred to as excitation of surface plasmonic resonance (SPR) modes. SPRs propagate along the metal-dielectric interface. The intensity profile exponentially decreases with the distance from metal-dielectric interface, that is, SPs are confined at the metal-dielectric interface. Such surface confinement effects have been used in Raman scattering, that is, surface-enhanced Raman scattering (SERS) [3–5] and spectroscopy [6, 7] and have been reported with thousands of times sensitivity enhancement [3, 4]. In addition to SERS, the SPR enhancement technology also provides a promising technique to concentrate EM energy on surface area and thus enables high quantum efficiency with a thin active absorption layer in a photodetector. Significant performance enhancements in a quantum dot infrared photodetector (QDIP) have been reported [8–15].

In addition, the SPR modes can also change EM field distribution and thus offers an effective technique for EM field engineering to achieve specific transmission and/or receiving patterns with polarization and detection spectrum engineering capability [11, 16]. In this chapter, a brief review of the fundamental concepts of SPRs will be first given and followed by the description of their applications in infrared detections. SPR-based EM field engineering will also be presented with the discussion of polarization and receiving angle control.

2. Review of surface plasmonic resonance

2.1. Plasma frequency, complex conductivity and the Drude model

Plasmonic waves are electromagnetic waves. They follow Maxwell equations, which can be expressed as follows in the phasor domain

$$\nabla \cdot \vec{D} = \rho, \tag{1}$$

$$\nabla \times \vec{E} = -j\omega\mu\vec{H}, \tag{2}$$

$$\nabla \cdot \vec{B} = \mathbf{0}, \tag{3}$$

$$\nabla \times \vec{H} = \vec{J} + j\omega\varepsilon\vec{E}, \tag{4}$$

where $\vec{J}$ is the current density, $\vec{D}$,$\vec{E}$, are the electric flux density and the electric field, respectively, and $\vec{B}$, $\vec{H}$ are the magnetic flux density and the magnetic field, respectively, ω is the angular frequency, and ε and μ are the permittivity and permeability of the material, respectively. The current density $\vec{J}$ is related to $\vec{E}$ by:

$$\vec{J} = \sigma\vec{E}, \tag{5}$$

where σ is the conductivity of the material. Eq. (4) can therefore be expressed as follows:

$$\nabla \times \vec{H} = j\omega\varepsilon\left(1 + \frac{\sigma}{j\omega\epsilon}\right)\vec{E}, \tag{6}$$

where $\varepsilon_c = \varepsilon\left(1 + \frac{\sigma}{j\omega\epsilon}\right)$ is the complex permittivity of the material. The Maxwell's current continuity equation is as follows:

$$\nabla \cdot \vec{J} = -e\frac{\partial}{\partial t}n, \tag{7}$$

where e is the charge of an electron, $e = 1.6\times10^{-19}(C)$, n is the density of electrons in metal.

The current density is related to the speed of the charge v by:

$$\vec{J} = e n_0 v, \tag{8}$$

where n_0 is the average electron density in metal. Combining Eqs. (**7**) and (**8**), one gets:

$$n_0 \nabla \cdot \vec{v} = -\frac{\partial}{\partial t} n, \tag{9}$$

Taking another partial derivative of t, one gets:

$$n_0 \nabla \cdot \frac{\partial}{\partial t} \vec{v} = -\frac{\partial^2}{\partial t^2} n, \tag{10}$$

Eq. (**10**) can be expressed as follows:

$$\frac{n_0}{m} \nabla \cdot m\vec{a} = -\frac{\partial^2}{\partial t^2} n, \tag{11}$$

where m is the mass of an electron. The $m\vec{a}$ term is the force on the electron, which is the electric force, $\vec{F} = e\vec{E}$. Eq. (**11**) can thus be changed to

$$\frac{e n_0}{m} \nabla \cdot \vec{E} = -\frac{\partial^2}{\partial t^2} n, \tag{12}$$

In phasor domain, Eq. (**12**) can be expressed as follows:

$$\frac{e n_0}{m} \nabla \cdot \vec{E} = \omega_p^2 n, \tag{13}$$

where ω_p is the plasma frequency. Combining Eqs. (**1**) and (**13**), one gets

$$\omega_p = \sqrt{\frac{e^2 n_0}{m\epsilon}} \tag{14}$$

Taking the partial derivative of t for Eq. (**8**), one gets

$$\frac{\partial}{\partial t} \vec{J} = \frac{e n_0}{m} m \frac{\partial}{\partial t} v, \tag{15}$$

The $m\frac{\partial}{\partial t}\vec{v}$ term can be replaced by the electric force, $\vec{F} = e\vec{E}$. Eq. (**15**) becomes

$$\frac{\partial}{\partial t} \vec{J} = \frac{e^2 n_0}{m} \vec{E}, \tag{16}$$

In phasor domain, Eq. (**16**) can be rewritten as follows:

$$j\omega \vec{J} = \frac{e^2 n_0}{m} \vec{E}, \tag{17}$$

Combining Eqs. (5) and (17), one gets:

$$\sigma = \frac{e^2 n_0}{j\omega m}, \tag{18}$$

The complex permittivity ε_c is thus given as follows:

$$\varepsilon_c = \varepsilon\left(1 + \frac{\sigma}{j\omega\epsilon}\right) = \varepsilon\left(1 - \frac{e^2 n_0}{\omega^2 m\epsilon}\right) = \varepsilon\left(1 - \frac{\omega_p^2}{\omega^2}\right) \tag{19}$$

The derivations and equations above are for the ideal lossless free-electron gas model. The loss term can be included in the model by introducing a relaxation rate γ, that is,

$$\varepsilon_c = \varepsilon\left(1 - \frac{\omega_p^2}{\omega^2 + j\omega\gamma}\right) \tag{20}$$

Eq. (20) is the complex permittivity given by the well-known Drude model [17].

2.2. Plasmonic waves at the dielectric and metal interface

Following Raether in Ref. [2], one can analyse plasmonic waves using a transmagnetic (TM) plane wave. **Figure 1** shows the incident of a TM plane EM wave with the *H*-field in the y-direction. Note that the phasor expressions of the H-fields follow the convention of $i\vec{k}\cdot\vec{R}$ indicating the wave travels in the $\vec{R}$ direction, whereas $-i\vec{k}\cdot\vec{R}$ indicating the wave travels in the $-\vec{R}$ direction. A negative sign needs to be included in Eq. (6) to use this convention.

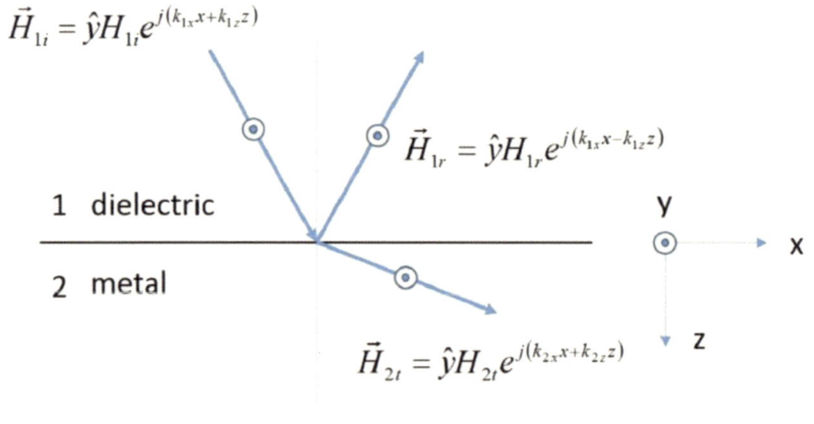

Figure 1. Scheme of TM wave incidence at the dielectric and metal interface.

Under the aforementioned convention, the E-field defined by Eq. (**6**) can be changed to

$$\vec{E} = -\frac{1}{j\omega\varepsilon_c}\nabla\times\vec{H} \tag{21}$$

From Eq. (**21**), the E_x can be expressed as follows:

$$\vec{E}_{1i} = -\hat{x}\frac{jk_{1z}}{j\omega\varepsilon_1}H_{1i}e^{j(k_{1x}x+k_{1z}z)} \tag{22}$$

$$\vec{E}_{1r} = \hat{x}\frac{jk_{1z}}{j\omega\varepsilon_1}H_{1r}e^{j(k_{1x}x-k_{1z}z)} \tag{23}$$

$$\vec{E}_{2t} = -\hat{x}\frac{jk_{2z}}{j\omega\varepsilon_{2c}}H_{2t}e^{j(k_{2x}x+k_{2z}z)} \tag{24}$$

where k_{1x}, k_{1z} are the x and z components of the propagation constant k_1, respectively. Similarly, k_{2x}, k_{2z} are the x and z components of the propagation constant k_2, respectively.

$$\boldsymbol{k}_{1x}^2 + \boldsymbol{k}_{1z}^2 = \boldsymbol{k}_1^2 = \omega^2\epsilon_1\mu_1 \tag{25}$$

$$\boldsymbol{k}_{2x}^2 + \boldsymbol{k}_{2z}^2 = \boldsymbol{k}_2^2 = \omega^2\epsilon_{2c}\mu_2 \tag{26}$$

where $\mu_1 \approx \mu_2 \approx \mu_0$ are the permeability of the material 1, 2 and vacuum, respectively. The boundary conditions are H_y and E_x continuous at the interface (i.e. $z = 0$). Applying the boundary conditions, one gets the following:

$$H_{1i}e^{jk_{1x}x} + H_{1r}e^{jk_{1x}x} = H_{2t}e^{jk_{2x}x} \tag{27}$$

$$\frac{k_{1z}}{\varepsilon_1}H_{1i}e^{jk_{1x}x} - \frac{k_{1z}}{\varepsilon_1}H_{1r}e^{jk_{1x}x} = \frac{k_{2z}}{\varepsilon_{2c}}H_{2t}e^{jk_{2x}x} \tag{28}$$

Since Eq. (**27**) and Eq. (**28**) are valid for any x, $k_{1x} = k_{2x}$.

When there is no reflection, that is, $H_{1r} = 0$, Eqs. (**27**) and (**28**) become

$$H_{1i} = H_{2t} \tag{29}$$

$$\frac{k_{1z}}{\varepsilon_1}H_{1i} = \frac{k_{2z}}{\varepsilon_{2c}}H_{2t} \tag{30}$$

Combining Eqs. (**29**) and (**30**), one gets the following:

$$k_{2z} = \varepsilon_{2c}\frac{k_{1z}}{\varepsilon_1} \tag{31}$$

Plugging Eq. (**31**) into Eq. (**26**), one gets the following:

$$k_{2x}^2 + k_{2z}^2 = \frac{\varepsilon_{2c}^2}{\varepsilon_1^2} k_{1z}^2 + k_{2x}^2 = \omega^2 \epsilon_{2c} \mu_0 \tag{32}$$

Combing Eq. (**32**) with Eq. (**25**), one can get k_{1x} as follows:

$$k_{1x}^2 = \omega^2 \mu_0 \frac{\epsilon_{2c} \epsilon_1}{(\varepsilon_1 + \varepsilon_{2c})} \tag{33}$$

Eq. (**33**) can be expressed using relative permittivity ε_{1r}, and ε_{2cr}.

$$k_{1x}^2 = \omega^2 \mu_0 \epsilon_0 \frac{\epsilon_{2cr} \epsilon_{1r}}{(\varepsilon_{1r} + \varepsilon_{2cr})} \tag{34}$$

where $\varepsilon_{1r} = \frac{\varepsilon_1}{\varepsilon_0}$, and $\varepsilon_{2cr} = \frac{\varepsilon_{2c}}{\varepsilon_0}$.

$$k_x = k_0 Re\left[\sqrt{\frac{\epsilon_{2cr} \epsilon_{1r}}{(\varepsilon_{1r} + \varepsilon_{2cr})}}\right] + jk_0 Im\left[\sqrt{\frac{\epsilon_{2cr} \epsilon_{1r}}{(\varepsilon_{1r} + \varepsilon_{2cr})}}\right] \tag{35}$$

k_{2z} can be obtained from Eq. (**34**) as follows:

$$k_{2z} = k_0 Re\left[\sqrt{\frac{\epsilon_{2cr} \epsilon_{2cr}}{(\varepsilon_{1r} + \varepsilon_{2cr})}}\right] + jk_0 Im\left[\sqrt{\frac{\epsilon_{2cr} \epsilon_{2cr}}{(\varepsilon_{1r} + \varepsilon_{2cr})}}\right] \tag{36}$$

where Re(.) and Im(.) are taking the real and the imaginary parts. The plasmonic wave propagates in x and z directions with loss described by the imaginary parts of the corresponding directions, that is, $k_{\mathrm{spp}} = k_x$. Since ε_{2cr} is generally much larger than ε_{1r}, Eq. (**36**) can be simplified as follows:

$$k_x = k_0 \sqrt{\epsilon_{1r}} \tag{37}$$

2.3. Plasmonic wave excitation in a metallic two-dimensional (2D) sub-wavelength hole array (2DSHA) structure

Surface plasmonic waves can be excited by numerous structures. The metallic 2DSHA structure [18] is one of the commonly used plasmonic structures for performance enhancement in infrared photodetectors [8, 10–12]. **Figure 2** shows a schematic structure with the light incident scheme of the metallic 2DSHA array structure. The 2DSHA structure is a square lattice with the period of p, and the diameter of the hole is d. The periodical structure provides a grating vector of $\Lambda = \frac{2\pi}{p}$. Surface plasmonic wave can be excited when the grating vector of Λ matches the k_{spp}, that is, $k_{spp} = \frac{2\pi}{p}$.

Figure 3 shows the cross section of the light incidence on the 2DSHA structure. The excited plasmonic waves and the scattered waves by the 2DSHA structure are indicated in the figure.

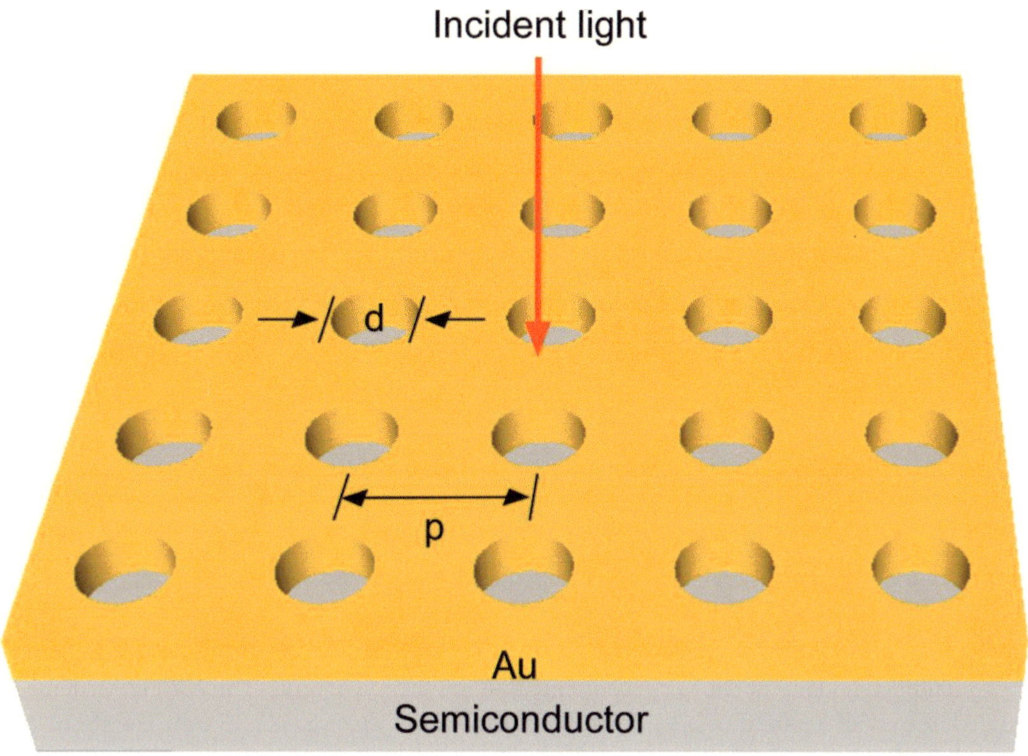

Figure 2. Schematic structure with the light incident scheme of the metallic 2DSHA array structure. The 2DSHA structure is square lattice with a period of p, and the diameter of the hole is d.

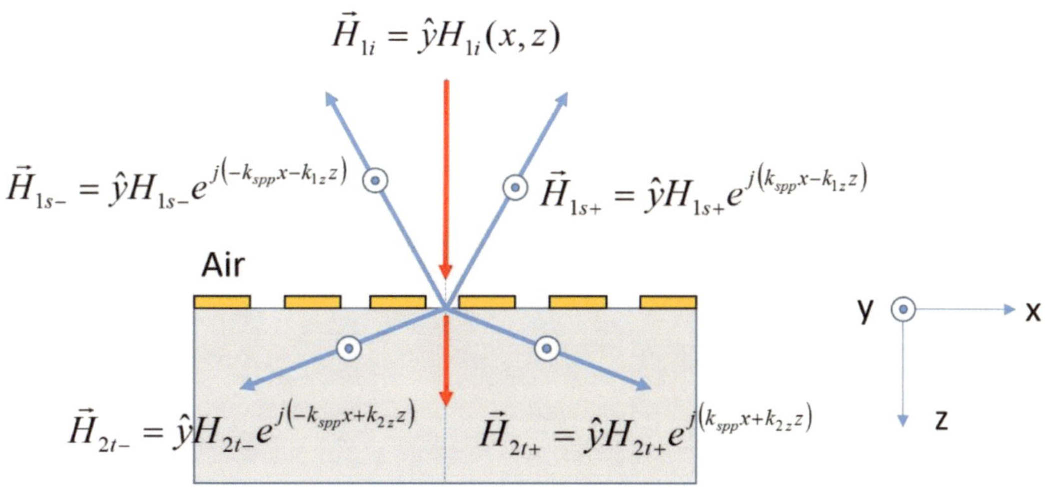

Figure 3. Schematic of the cross-section of the light incidence on the 2DSHA structure. The 2DSHA structure generates plasmonic waves and scattering waves.

From E-M field boundary conditions, the following relations hold:

$$\boldsymbol{H}_{1s-} = \boldsymbol{H}_{2t-} \tag{38}$$

$$\boldsymbol{H}_{1s+} = \boldsymbol{H}_{2t+} \tag{39}$$

$$\frac{k_{1z}}{\varepsilon_1} = -\frac{k_{2z}}{\varepsilon_{2c}} \tag{40}$$

The excitation efficiency η can be defined by the overlap integral of the incident filed with the plasmonic waves, that is,

$$\eta = \int \boldsymbol{H}(\boldsymbol{x}, \boldsymbol{z}) \boldsymbol{e}^{i(k_{spp}x + k_{2z}z)}|_{z=0} \boldsymbol{dx} \tag{41}$$

where $H(x, z)$ is the magnetic field of the incident wave modified by the 2DSHA, that is, the periodical 2DSHA structure will introduce periodical variation in $H(x, z)$ as well.

2.4. Near-field E-field distribution in the 2DSHA structure

The near-field E-field component can be simulated as well. **Figure 4** shows the top (x-y) and cross-section (x-z) views of the E_x component at different wavelengths. The 2DSHA structure is square lattice with a period of 2.3 μm, and the diameter of the hole is 1.15 μm. The simulation is performed using CST Microwave studio®. The incident light is linearly polarized surface normal plane wave with E-field magnitude of 1 V/m. The colour scale bar represents the magnitude of the E_x.

Figure 5 shows the top (x-y) and cross-section (x-z) views of the E_z component at different wavelengths. From **Figures 4** and **5**, both E_x and E_z strongly depend on the excitation wavelengths. E_z is high between 7.6 and 8.5 μm, whereas E_x is strong between 7.6 and 8.0 μm. E_z is primarily at the edge of the holes, whereas E_x is inside the holes. The period of the 2DSHA $p = 2.3$ μm, $k_{spp} = \frac{2\pi}{p}$.

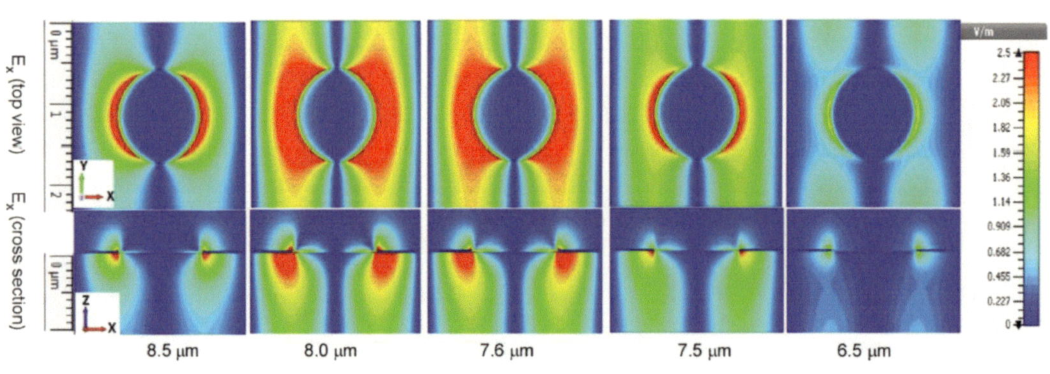

Figure 4. Top and cross-section views of the E_x component at different wavelengths simulated using CST Microwave studio®. The incident light is a linearly polarized surface normal plane wave with E-field magnitude of 1 V/m.

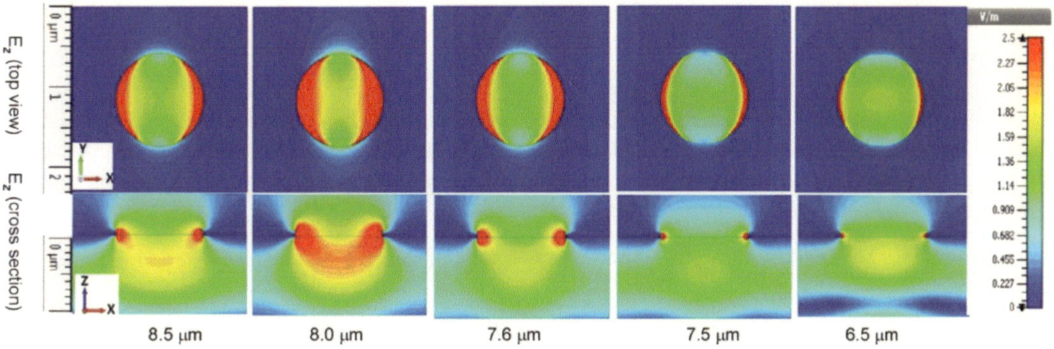

Figure 5. Top and cross-section views of the E_z component at different wavelengths simulated using CST Microwave studio®. The incident light and the simulation conditions are the same.

From Eq. (35), k_{spp} is

$$k_{spp} = \frac{2\pi}{\lambda_0} Re\left[\sqrt{\frac{\epsilon_{2cr}\epsilon_{1r}}{(\varepsilon_{1r} + \varepsilon_{2cr})}}\right] + j\frac{2\pi}{\lambda_0} Im\left[\sqrt{\frac{\epsilon_{2cr}\epsilon_{1r}}{(\varepsilon_{1r} + \varepsilon_{2cr})}}\right] \quad (42)$$

where λ_0 is the free space wavelength, ε_m and ε_d are the relative permittivity of the metal and GaAs, respectively. The dielectric constant of GaAs in the long-wave infrared region is 10.98 [19]. The dielectric constant of Au is calculated to be −1896 + *i*684 at 7.6 µm [1]. The calculated plasmonic resonant wavelength is λ_{sp} = 7.6 µm. At the resonant wavelength (i.e. $\lambda = 7.6$ µm), the E-fields are strongly confined near the surface. Such strong surface confinement of plasmonic waves enables high absorption using a thin absorption layer and thus leads to significant enhancement in quantum dot infrared photodetectors (QDIPs) [8, 10–13, 15]. In the following sections, we will first give a brief introduction of QDIPs, followed by the description of surface plasmonic enhancement in QDIPs.

3. QDIPs in mid-wave infrared (MWIR) and long-wave infrared (MWIR/LWIR) detection

QDIPs are based on intersubband transitions in self-assembled InAs quantum dots (QDs). The simplified band diagram and the schematic structures of a QD infrared photodetector (QDIP) are shown in **Figure 6(a)** and **6(b)**, respectively. The s, p, d, f, represent the energy levels of a QD with the wetting layers (WL). A typical QDIP consists of vertically-stacked InAs quantum dots layers with GaAs capping layers. The electrons are excited by the normal incident light and subsequently collected through the top electrode and generate photocurrent. This is a unipolar photodetector, where only conduction band is involved in the photodetection and photocurrent generation process.

The QDIP technology offers a promising technology in MWIR and LWIR photodetection due to the advantages provided by the three-dimensional (3D) quantum confinement of

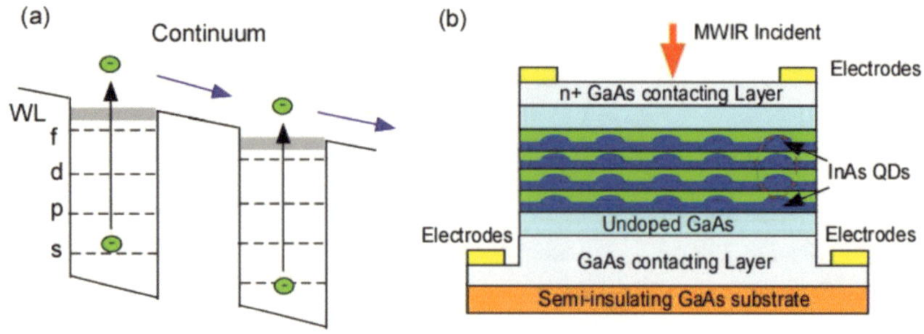

Figure 6. Schematic structure a unipolar QDIP with the intersubband transition: (a) simplified band diagram and (b) schematic of the QDIP. The s, p, d and f, represent the energy levels of a QD.

carriers—including intrinsic sensitivity to normal incident radiation [13], high photoconductive (PC) gain, high quantum efficiency [15] and photoresponsivity [16]. The normal incidence detection capability greatly simplifies the fabrication complexity for a large format (1K × 1K) FPA. The high photoconductive (PC) gain and high photoresponsivity provide a promising way for MWIR and LWIR sensing and detection.

Figure 7 shows the photodetection spectrum of a QDIP. It covers a broadband IR spectrum from 3 to 9 μm. The insert of **Figure 7** shows an atomic force microscopic (AFM) picture of the QDs by self-assembled epi-growth. The growth gives high-density QDs with uniform sizes. The InAs/

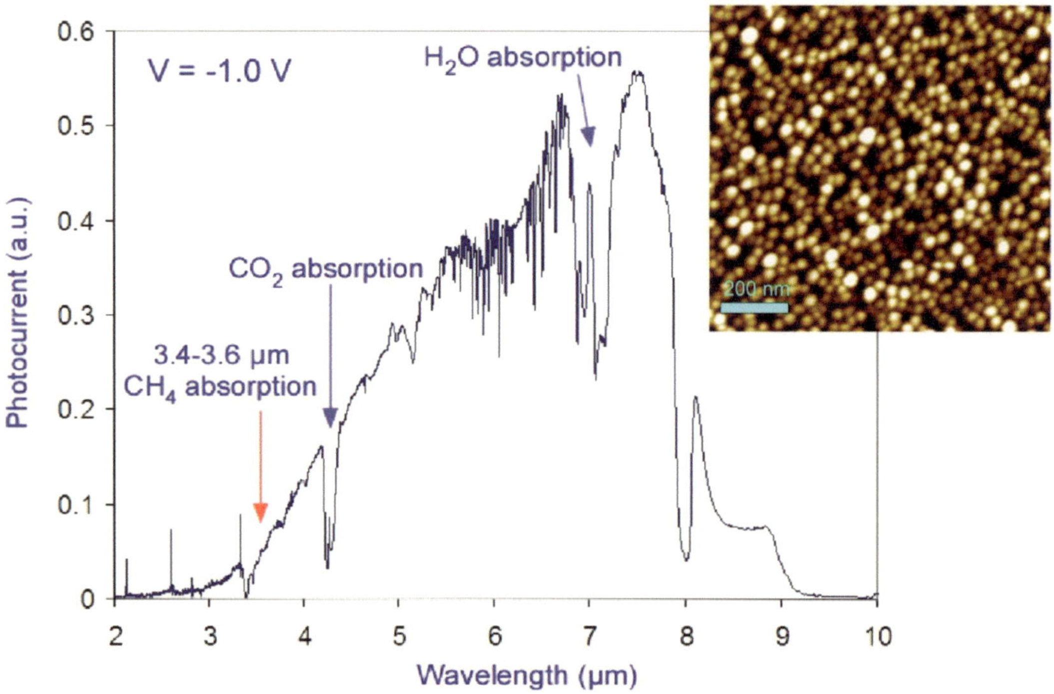

Figure 7. Photodetection spectrum of a QDIP. It shows a broad IR detection spectrum from 3 to 9 μm. The insert shows an AFM image of the QDs by self-assembled epi-growth.

http://dx.doi.org/10.5772/67410

GaAs QD material is a mature material system with low substrate cost, high material quality and large wafer growth capability. It can offer low-cost MWIR and LWIR photodetectors and focal plane arrays (FPAs) with simplified fabrication processes and high yield. Detailed QDIP performance, such as photoconductive gains, noise, photoresponsivity and photodetectivity can be found in literature [20–27].

4. 2DSHA plasmonic-enhanced QDIPs

4.1. Backside configured 2DSHA-enhanced QDIPs

Surface confinement of plasmonic waves enables high absorption using a thin absorption layer and thus leads to significant enhancement in QDIPs [8, 10–13, 15]. We have developed a backside-configured plasmonic enhancement technology by fabricating the 2DSHA plasmonic structures on top of a QDIP and illuminating the QDIP from the opposite side of the plasmonic structure (i. e. backside illumination) [12]. **Figure 8(a)** and **8(b)** show the schematic cross-section structures of the backside-configured plasmonic QDIP and the reference QDIP without the

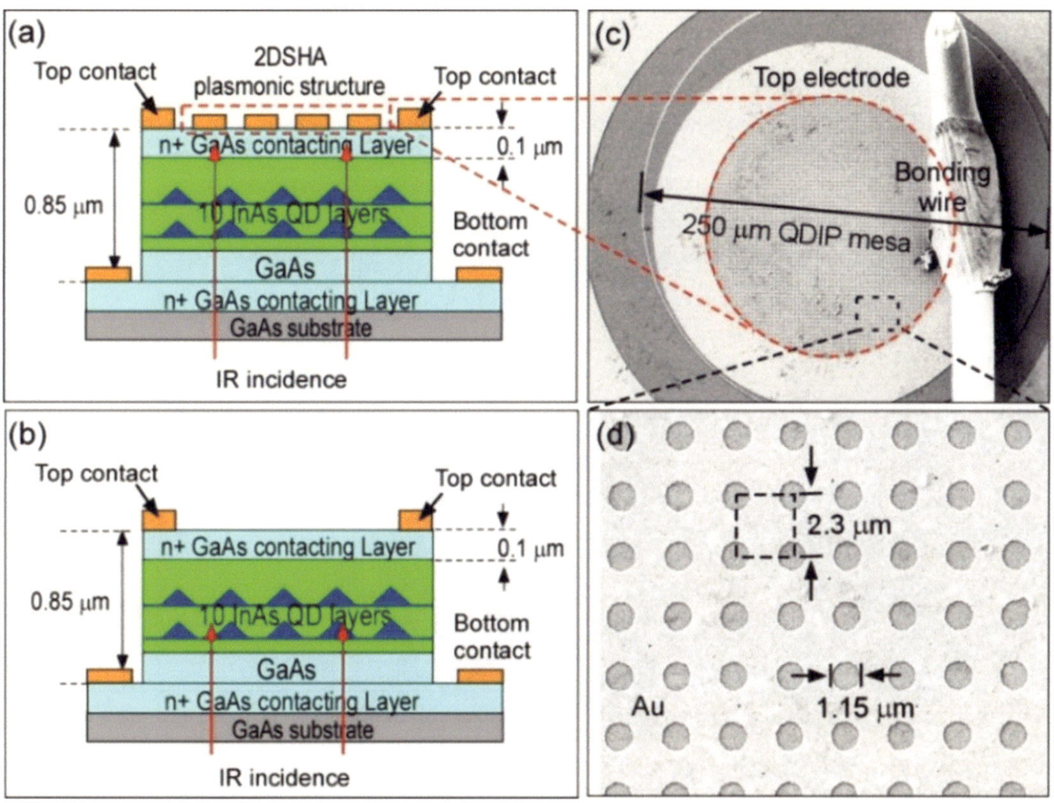

Figure 8. (a) Schematic cross-section structure of the backside-configured 2DSHA plasmonic structure on the QDIP. The IR incident light illuminates from the substrate side; (b) ref-QDIP without the plasmonic structure; (c) SEM image of the 2DSHA plasmonic structure on the QDIP and (d) close-up view of the plasmonic structure ().

plasmonic structure, respectively. **Figure 8(c)** and **8(d)** show scanning electron microscopic (SEM) images of the plasmonic structures on a QDIP and a close-up view of the plasmonic structure, respectively.

Figure 9 shows the measured photocurrent spectra of the backside-configured plasmonic QDIP compared with the top-side configured plasmonic QDIP and the reference QDIP. The backside-configured plasmonic QDIP clearly shows higher plasmonic enhancement than those of the top-side configured plasmonic QDIP and the reference QDIP.

Figure 10(a) shows the simulated E-field intensities of the top and backside-configured plasmonic structures, respectively. The backside-configured plasmonic structure can induce stronger E-field at the interface and therefore can provide larger plasmonic enhancement [12].

4.2. Wavelength-tuning and multispectral enhancement

From Eq. (**42**) and $k_{spp} = \frac{2\pi}{p}$, one gets:

$$\lambda_{sp} = \frac{p}{\sqrt{m^2 + n^2}} Re\left[\sqrt{\frac{\epsilon_{2cr}\epsilon_{1r}}{(\varepsilon_{1r} + \varepsilon_{2cr})}}\right] \tag{43}$$

where (m, n) are the orders of the grating vectors. Therefore, the plasmonic resonance wavelength can be tuned by changing the period of the 2DSHA plasmonic structure. The plasmon resonant

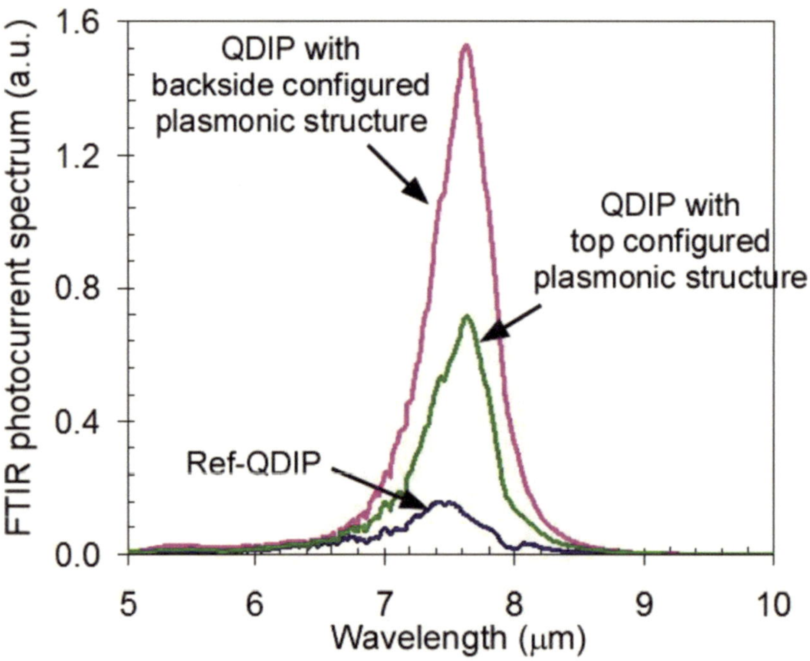

Figure 9. Measured photocurrent spectra of the backside-configured plasmonic QDIP compared with the top-side configured plasmonic QDIP and the reference QDIP. The backside-configured plasmonic QDIP clearly shows higher plasmonic enhancement.

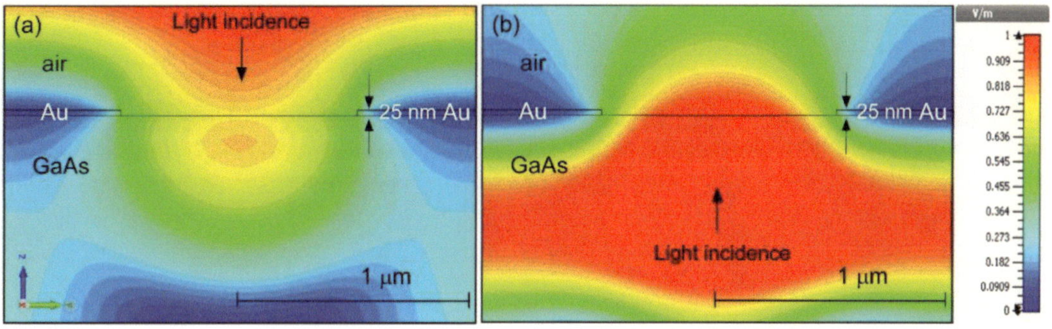

Figure 10. Simulated E-field intensities: (a) top-configured plasmonic structure and (b) backside-configured plasmonic structure. The backside-configured plasmonic structure can induce stronger E-field at the interface and therefore can provide larger plasmonic enhancement (© IOP Publishing. Reproduced with permission from Ref. [12]. All rights reserved).

peaks λ_{sp} of different *(m, n)* orders for various 2DSHA periods *p* are shown in **Figure 11** that exhibits the linear dependence of the resonant peak wavelength λ_{sp} with the period *p* of the 2DSHA for different *(m, n)* orders. As predicted by Eq. (**43**), good linearity is obtained for both the (1, 0)th or (0, 1)th and the (1, 1)th order plasmonic resonant peaks.

We define the photocurrent enhancement ratio *R* as follows:

$$R = \frac{I_{\text{plasmonic}}}{I_{\text{ref}}} \tag{44}$$

where $I_{\text{plasmonic}}$ is the photocurrent of the plasmonic QDIP and I_{ref} is the photocurrent of the reference QDIP without the 2DSHA structures.

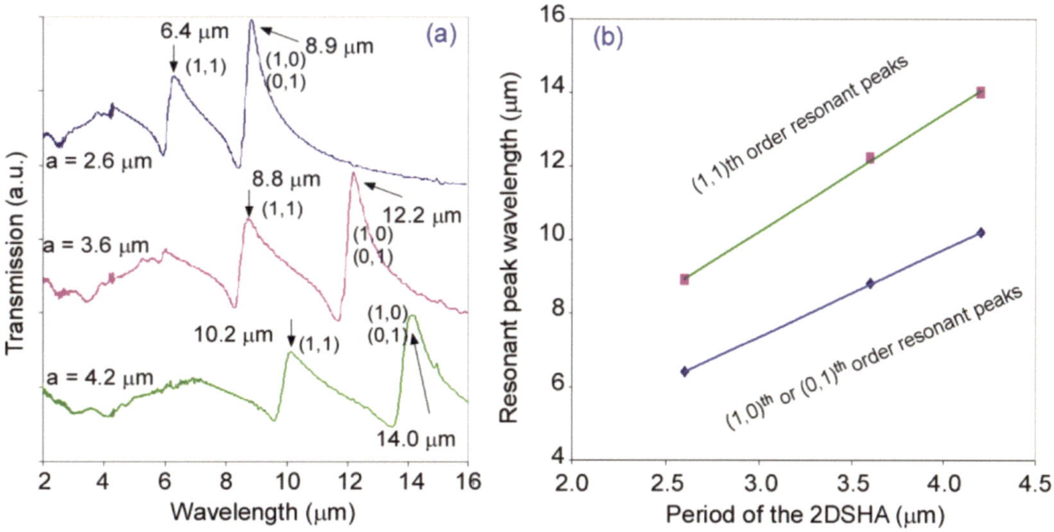

Figure 11. (a) Plasmonic resonant peaks λ_{sp} of different (*i, j*) orders for various 2DSHA periods; (b) linear dependence of the resonant peak wavelength λ_{sp} with the period of different (*m, n*) orders.

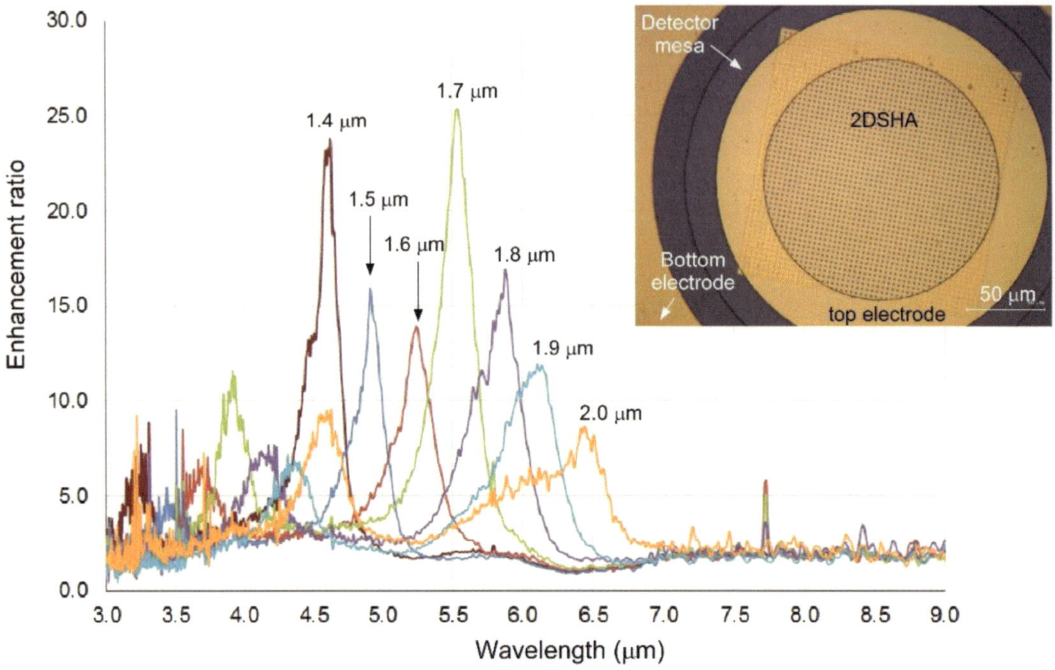

Figure 12. Photocurrent enhancement ratio spectra for plasmonic QDIPs with different periods varying from 1.4 to 2.0 μm. The insert is a microscopic picture of a QDIP with a 2DSHA plasmonic structure (Reproduced from Ref. [28] with the permission of AIP Publishing).

Figure 12 shows the photocurrent enhancement ratio spectra for all the plasmonic QDIPs with different periods from 1.4 to 2.0 μm. The periods of the 2DSHA plasmonic structures are marked on the curves. The insert shows a microscopic picture of a QDIP with a 2DSHA plasmonic structure. The IR incidence is from the backside of the QDIPs, that is, backside-configured 2DSHA plasmonic structures.

As shown in **Figure 12**, the enhancement wavelengths can be tuned by varying the periods of the 2DSHA plasmonic structures. This allows one to achieve multispectral QDIPs without fabricating an individual filter on each photodetector [28].

4.3. Polarization dependence and polarimetric detection

From Eq. (**43**), the plasmonic resonant wavelength depends on the period p. By changing the plasmonic lattice from a square lattice to a rectangular lattice with different periods in the x and y directions, one can engineer the resonant wavelengths at different polarizations.

Figure 13(a) shows an SEM picture of a rectangular lattice 2DSHA plasmonic structure with the periods of 2.6 and 3.0 μm in the x-direction and the y-direction, respectively. **Figure 13(b)–(d)** shows the transmission spectra of the un-polarized, 90° polarized (i.e. polarization in y-direction), and 0° polarized (i.e. polarization in x-direction), respectively. The resonant wavelengths correspond to the orders in the x- and y- directions are marked in the figures. The un-polarized light shows transmission in both (0, 1) and (1, 0) orders. However, the 0° polarized

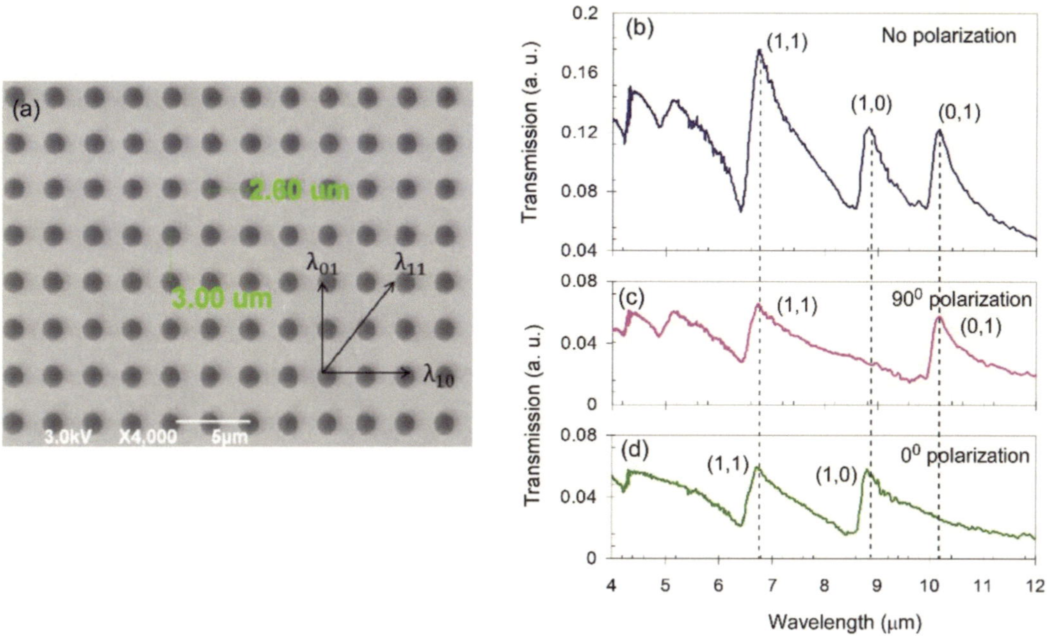

Figure 13. (a) SEM picture of rectangular lattice 2DSHA structure. The periods in x- and y- directions are 2.6 and 3.0 µm, respectively. (b), (c) and (d) transmission spectra at of the un-polarized, 90° polarized (i.e. polarization in y-direction), and 0° polarized (i.e. polarization in x-direction), respectively. The polarized light only shows transmission at their corresponding resonant wavelengths (© IOP Publishing. Reproduced with permission from Ref. [11]. All rights reserved).

light only shows transmission at the (1, 0) order resonant wavelength and the 90° polarized light only shows transmission at the (0, 1) order.

Figure 14 shows a microscopic picture of a QDIP with the rectangular lattice 2DSHA structure. The insert shows an SEM picture of the rectangular lattice 2DSHA structure.

Figure 15(a) and **(b)** shows the measured detection spectra of the plasmonic QDIP at 90° polarization and 0° polarization, respectively. The photo-response of a reference QDIP without the plasmonic structure is also shown in the figures for comparison. The plasmonic QDIP shows different detection wavelengths at different polarizations.

4.4. Angular dependence of the 2DSHA plasmonic enhancement

Figure 16 shows the schematic view of the top-illuminated 2DSHA plasmonic QDIP with an incident angle θ.

From Eq. **(41)**, the coupling efficiency η to the plasmonic waves depends on the overlap integral of $H(x,z)$ with the plasmonic wave $e^{i(k_{spp}x+k_{2z}z)}$. At the incident angle θ, the $H(x,z)$ not only periodically modulated by the 2DSHA structure, but also added additional phase term $e^{i(k_0 sin\theta x)}$, that is,

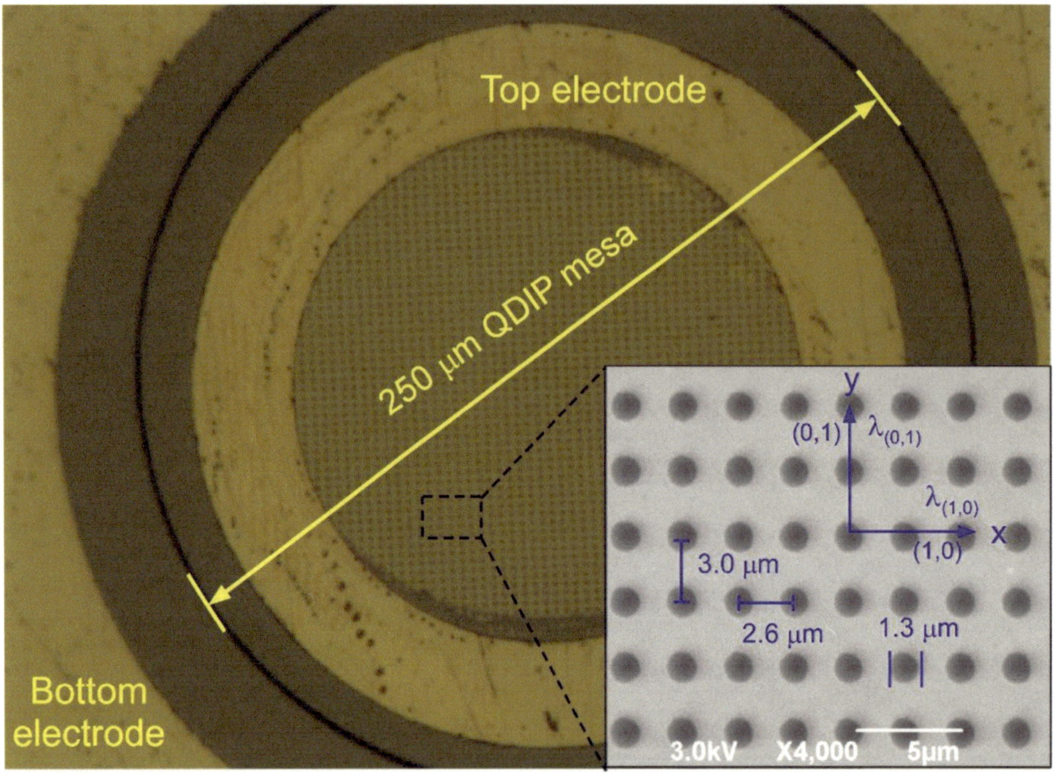

Figure 14. Microscopic picture of a QDIP with the rectangular lattice 2DSHA structure. Insert: SEM picture of the rectangular lattice 2DSHA structure (© IOP Publishing. Reproduced with permission from Ref. [11]. All rights reserved).

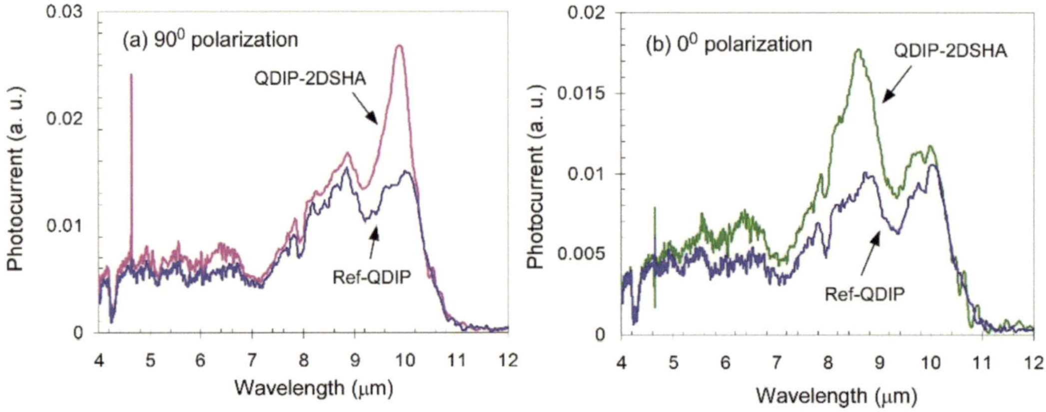

Figure 15. Measured detection spectra of the plasmonic QDIP at different polarizations: (a) 90° polarization and (b) 0° polarization. Different polarizations show different detection wavelengths (© IOP Publishing. Reproduced with permission from Ref. [11]. All rights reserved).

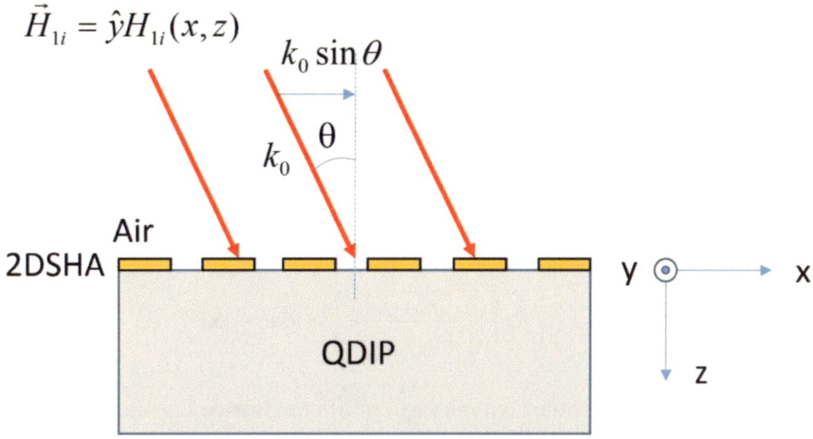

Figure 16. Schematic view of the top-illuminated 2DSHA plasmonic QDIP with an incident angle θ.

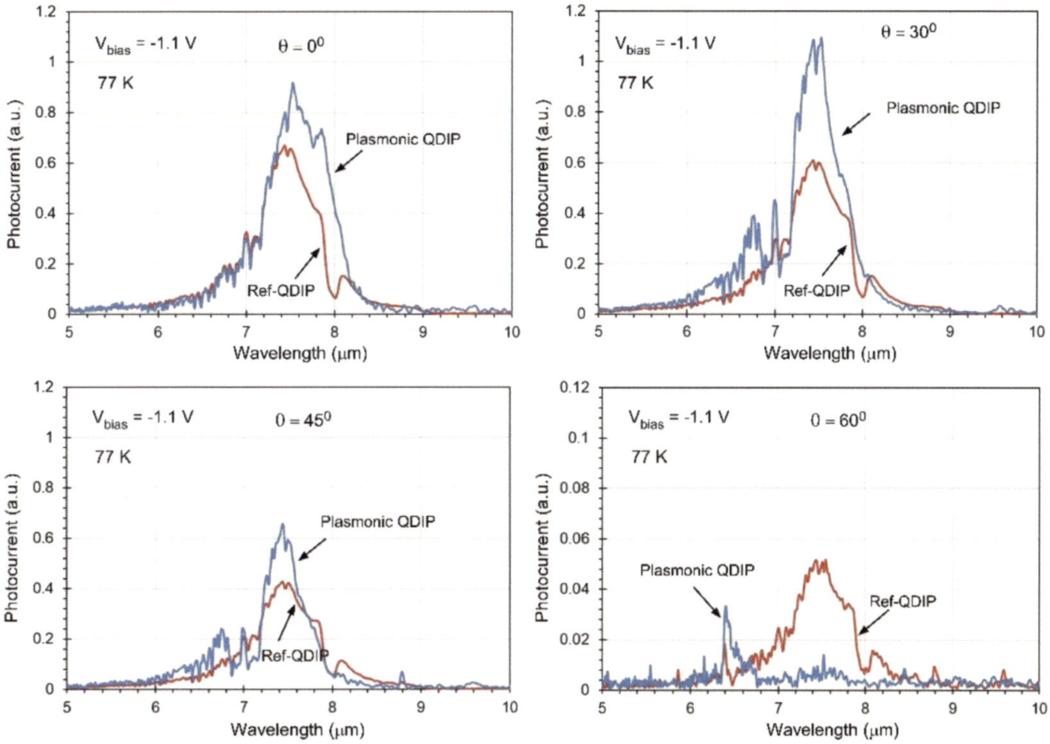

Figure 17. Measured photocurrent of the 2DSHA plasmonic QDIP at different incident angles of 0°, 30°, 45° and 60° compared with the reference QDIP at the same incident angles. The plasmonic resonant wavelength λ_0 shift to shorter wavelengths as the incident angle increases.

$$H(x,z) \propto e^{\pm i\Lambda x} e^{ik_0 sin\theta x} \tag{45}$$

From Eq. (**45**), the coupling condition can be expressed as follows:

$$k_0 sin\theta \pm \Lambda = k_{spp} \tag{46}$$

Taking the + sign, one can write Eq. (**46**) in terms of plasmonic resonant wavelength λ_0 as follows:

$$\lambda_0 = \left\{ Re\left[\sqrt{\frac{\epsilon_{2cr}\epsilon_{1r}}{(\varepsilon_{1r} + \varepsilon_{2cr})}} \right] - sin\theta \right\} \Lambda \tag{47}$$

Eq. (**47**) shows that the enhancement wavelength shifts to shorter wavelengths as the incident angle increases.

Figure 17 shows the measured photocurrent of the 2DSHA plasmonic QDIP at different incident angles of 0°, 30°, 45°, and 60° compared with the reference QDIP at the same incident angles. Blue shift of the plasmonic resonant wavelengths λ_0 is experimentally observed as the incident angle increases.

To quantitatively analyse the relationship between the plasmonic wavelength shift and the incident angle, we plot the plasmonic resonant wavelengths λ_0 versus *sin*θ. **Figure 18** shows the λ_0 versus *sin*θ. A linear relation is obtained, which agrees well with Eq. (**47**). The interception and the slope are 7.6 and −1.4 μm, respectively. The interception 7.6 μm matches the value

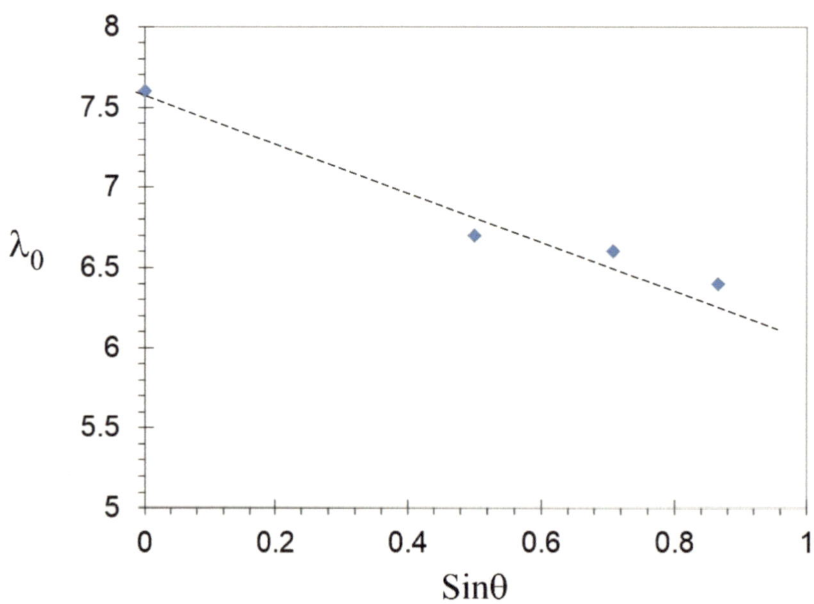

Figure 18. λ_0 versus *sin*θ plot. λ_0 varies linearly with *sin*θ. The interception and the slope are 7.6 and −1.4 μm, respectively.

$Re\left[\sqrt{\frac{\epsilon_{2cr}\epsilon_{1r}}{(\varepsilon_{1r}+\varepsilon_{2cr})}}\right]\Lambda$ in Eq. (**47**). However, the slope of -1.4 μm is smaller than $\Lambda = 2.3$ μm as predicted in Eq. (**47**). This may be due to the plasmonic scattering-induced phase distribution in $H(x, z)$ in addition to the $e^{i(k_0 sin\theta x)}$ term.

5. Conclusion

In conclusion, 2DSHA plasmonic structures can effectively induce plasmonic resonant waves. The resonant wavelength depends on the period of the plasmonic structure. The surface confinement effect allows one to improve the performance of QDIPs. The plasmonic enhancement also strongly depends on the polarization and the incident angle. The polarization and incident angle dependence allow polarimetric and polarization engineering in QDIP designs.

Author details

Guiru Gu[1]*, Xuejun Lu[2], Thitikorn Kemsri[2] and Yingjie Zhang[2]

*Address all correspondence to: ggu@stonehill.edu

1 Stonehill College, Easton, MA, USA

2 University of Massachusetts Lowell, Lowell, MA, USA

References

[1] R. H. Ritchie, "Plasma losses by fast electrons in thin films," *Physical Review*, vol. 106, pp. 874–881, 06/01/1957.

[2] H. Raether, *Surface Plasmons on Smooth Surfaces*. Springer, Berlin Heidelberg, 1988.

[3] K. Kneipp, Y. Wang, H. Kneipp, L. T. Perelman, I. Itzkan, R. R. Dasari, *et al.*, "Single molecule detection using surface-enhanced Raman scattering (SERS)," *Physical Review Letters*, vol. 78, pp. 1667–1670, 03/03/1997.

[4] S. Nie and S. R. Emory, "Probing single molecules and single nanoparticles by surface-enhanced Raman scattering," *Science*, vol. 275, pp. 1102–1106, 1997.

[5] K. Kneipp and H. Kneipp, "SERS signals at the anti stokes side of the excitation laser in extremely high local optical fields of silver and gold nanoclusters," *Faraday Discuss*, vol. 132, pp. 27–33; discussion 85–94, 2006.

[6] K. A. Willets and R. P. Van Duyne, "Localized surface plasmon resonance spectroscopy and sensing," *Annual Review of Physical Chemistry*, vol. 58, pp. 267–297, 2007.

[7] H. Xu, E. J. Bjerneld, M. Käll, and L. Börjesson, "Spectroscopy of single hemoglobin molecules by surface enhanced Raman scattering," *Physical Review Letters*, vol. 83, pp. 4357–4360, 11/22/1999.

[8] C.-C. Chang, Y. D. Sharma, Y.-S. Kim, J. A. Bur, R. V. Shenoi, S. Krishna, *et al.*, "A surface plasmon enhanced infrared photodetector based on InAs quantum dots," *Nano Letters*, vol. 10, pp. 1704–1709, 2010.

[9] P. Senanayake, C.-H. Hung, J. Shapiro, A. Lin, B. Liang, B. S. Williams, *et al.*, "Surface plasmon-enhanced nanopillar photodetectors," *Nano Letters*, vol. 11, pp. 5279–5283, 2011.

[10] P. Vasinajindakaw, J. Vaillancourt, G. Gu, R. Liu, Y. Ling, and X. Lu, "A Fano-type interference enhanced quantum dot infrared photodetector," *Applied Physics Letters*, vol. 98, pp. 211111–1 to 211111–3, 2011.

[11] P. Vasinajindakaw, J. Vaillancourt, G. Gu, and X. Lu, "Surface plasmonic enhanced polarimetric longwave infrared photodetection with band pass spectral filtering," *Semiconductor Science and Technology*, vol. 27, pp. 65005–65009, 2012. http://dx.doi.org/10.1088/0268-1242/27/6/065005.

[12] G. Gu, J. Vaillancourt, P. Vasinajindakaw, and X. Lu, "Backside-configured surface plasmonic structure with over 40 times photocurrent enhancement," *Semiconductor Science and Technology*, vol. 28, p. 105005, 2013. http://dx.doi.org/10.1088/0268-1242/28/10/105005.

[13] G. Gu, N. Mojaverian, J. Vaillancourt, and X. Lu, "Surface plasmonic resonance induced near-field vectors and their contribution to quantum dot infrared photodetector enhancement," *Journal of Physics D: Applied Physics*, vol. 47, p. 435106, 2014.

[14] G. Gu, J. Vaillancourt, and X. Lu, "Analysis of near-field components of a plasmonic optical antenna and their contribution to quantum dot infrared photodetector enhancement," *Optics Express*, vol. 22, pp. 24970–24976, 2014/10/20 2014.

[15] J. Vaillancourt, N. Mojaverian, and L. Xuejun, "A longwave infrared focal plane array enhanced by backside-configured plasmonic structures," *IEEE Photonics Technology Letters*, vol. 26, pp. 745–748, 2014.

[16] R. Liu, P. Vasinajindakaw, G. Gu, J. Vaillancourt, and X. Lu, "Optimizing light absorption in quantum dot infrared photodetectors by tuning surface confinement of surface plasmonic waves," *Journal of Physics D: Applied Physics*, vol. 46, p. 015102, 2013.

[17] P. Drude, "Electron theory of metals," *Annalen der Physik*, vol. 306, pp. 566–613, 1900.

[18] T. W. Ebbesen, H. Lezec, H. Ghaemi, T. Thio, and P. Wolff, "Extraordinary optical transmission through sub-wavelength hole arrays," *Nature*, vol. 391, pp. 667–669, 1998.

[19] *Handbook of Optical Constants of Solids*. Academic Press, San Diego London New York Boston Sydney Tokyo Toronto, 1998.

[20] D. Pan, E. Towe, and S. Kennerly, "Normal-incidence intersubband (In, Ga)As/GaAs quantum dot infrared photodetectors," *Applied Physics Letters*, vol. 73, pp. 1937–1939, 1998.

[21] H. C. Liu, M. Gao, J. McCaffrey, Z. R. Wasilewski, and S. Fafard, "Quantum dot infrared photodetectors," *Applied Physics Letters*, vol. 78, pp. 79–81, 2001.

[22] A. D. Stiff, S. Krishna, P. Bhattacharya, and S. Kennerly, "High-detectivity, normal-incidence, mid-infrared ($\lambda \sim 4 \mu m$)InAs/GaAs quantum-dot detector operating at 150 K," *Applied Physics Letters*, vol. 79, pp. 421–423, 2001.

[23] L. Jiang, S. S. Li, N.-T. Yeh, J.-I. Chyi, C. E. Ross, and K. S. Jones, "In0.6Ga0.4As/GaAs quantum-dot infrared photodetector with operating temperature up to 260 K," *Applied Physics Letters*, vol. 82, pp. 1986–1988, 2003.

[24] S. Krishna, S. Raghavan, G. v. Winckel, P. Rotella, A. Stintz, C. P. Morath, *et al.*, "Two color InAs/InGaAs dots-in-a-well detector with background-limited performance at 91 K," *Applied Physics Letters*, vol. 82, pp. 2574–2576, 2003.

[25] S. Chakrabarti, A. D. Stiff-Roberts, P. Bhattacharya, S. Gunapala, S. Bandara, S. B. Rafol, *et al.*, "High-temperature operation of InAs-GaAs quantum-dot infrared photodetectors with large responsivity and detectivity," *Photonics Technology Letters, IEEE*, vol. 16, pp. 1361–1363, 2004.

[26] J. C. Campbell and A. Madhukar, "Quantum-dot infrared photodetectors," *Proceedings of the IEEE*, vol. 95, p. 1815, 2007.

[27] X. Lu, J. Vaillancourt, and M. J. Meisner, "Temperature-dependent photoresponsivity and high-temperature (190K) operation of a quantum dot infrared photodetector," *Applied Physics Letters*, vol. 91, pp. 051115–1 to 051115–3, 2007.

[28] Y. Zhang, J. Vaillancourt, G. Gu, W. Guo, and X. Lu, "Quantum selection rule dependent plasmonic enhancement in quantum dot infrared photodetectors," *Journal of Applied Physics*, vol. 119, p. 193103, 2016. http://dx.doi.org/10.1063/1.4949511.

Section 3

Applications to Optoelectronics

Chapter 10

Equivalent Nanocircuit Theory and Its Applications

Qing Zhang, Jun You and Chengpu Liu

Additional information is available at the end of the chapter

http://dx.doi.org/10.5772/67681

Abstract

A new methodology termed as equivalent nanocircuit (EN) theory is briefly introduced, and its recent important progress in designing meta-material devices with peculiar characteristics in optical or infrared frequency domain is reviewed. Three representative EN-based designs of infrared window meta-materials, such as Butterworth filter, metal-insulator-metal absorber and design-simplified TCO-based super-flat absorber, are demonstrated. All these progresses clearly indicate that the EN theory provides an inspiring advancement on the way of designing more complicated meta-devices.

Keywords: meta-material, metatronics, nanocircuit theory, surface plasmon polariton

1. Introduction

The functionalities of plasmonic nanodevices are dependent on their artificial structure parameters that usually are smartly engineered. To introduce convenient and feasible methods based on new paradigms to guide, the structure design is eagerly demanded. In the radio or microwave frequency domains, lumped circuit elements, e.g. resistors (*R*), inductors (*L*) and capacitors (*C*), can be effectively and flexibly used as blocks for designing complicated micro-electronic devices, and the conventional circuit theory is widely and successfully adopted. One question naturally arises: can such a circuit theory, along with its accompanying mathematical machinery, be extended and applied to the design of plasmonic nanostructures working at the optical domain?

Within this context, in this chapter, a new methodology, 'equivalent nanocircuit (EN) theory' is briefly introduced, and its representative applications in designing plasmonic devices with peculiar characteristics in optical or infrared frequency domains are enumerated. First, to start with the short introduction of the basis of 'metatronics', the analogy between micro-electronic lumped circuit elements (*R*, *L*, *C*) and optical nanocircuit elements (nanoparticles, nanoantennas, nanogratings and optical meta-surfaces) is established. Second, the method describing how to use the above optical nanocircuit elements to construct a single-layer or multiple-layer complicated meta-material structure is proposed and thus the 'metatronics' concept moves forward to the multi-order circuit theory. Finally, three representative applications of multi-order EN theory to design infrared window meta-materials are demonstrated: (1) a synthesis procedure for designing a third-order Butterworth filter is proposed; (2) a metal-insulator-metal (MIM) ultra-broadband absorber is successfully designed in the infrared range; (3) with the transparent conductive oxides (TCOs) semiconductor materials as building blocks, a design-simplified broadband super-flat perfect infrared absorber is realized.

2. Metatronics and optical nanocircuit

2.1. 'Metatronics': meta-material-inspired optical circuitry

Optical meta-material bridges the gap between the conventional optics and the nanoworld, which gives rise to a diversity of surprising and profound effects fully appreciated and technologically explored in recent years [1–3]. The electromagnetic property of meta-materials is dependent on their specifically and smartly engineered artificial structures. To explore new design methodology is eagerly demanded. Historically, in electronics, basic functionalities are synthesized by 'lumped' circuit elements, such as resistors, inductors, capacitors and transistors, and more complicated operations can be realized by combining them into a complicated circuit in some specific serial or parallel ways. Now, great interests in pushing classic circuit operation to infrared or visible optical frequency range have boomed in order to achieve an optical analogy [4–7]. Generally, just simply reducing the basic unit size of circuit to the micrometre (μm) or nanometre (nm) level would be technically difficult to achieve the above goals.

The first challenge arises in the state-of-the-art nanoscale fabrication technique is its inevitable difficulty to achieve deeply sub-wavelength dimension becoming serious. The second challenge lies in material dispersion which is sometimes vital. As we know that metals such as gold, silver, aluminium and copper are highly conductive materials at RF and microwaves, commonly used in conventional circuits. However, at optical frequency, these metals would behave differently, and they exhibit plasmonic resonance instead of the usual conductivity, i.e. the coupling of optical signals with collective oscillation of conductive electrons at these metal surfaces is dominant where the real part of permittivity is negative. In other words, at optical wavelengths, the conduction current is no longer the dominant current circulating through lumped optical elements. Therefore, the traditional circuit theory and the corresponding methodology for micro-electronics lose their functionality in optical frequency domain.

In 2007, Engheta et al. have made an important breakthrough and first proposed the concept of 'metatronics' [6] which bridges the gap between low frequency circuitry design and high frequency nanodevice design. In their opinions, it is possible to realize the performance of lumped-circuit-like elements at optical frequency just by properly designing and suitably arranging plasmonic or/and non-plasmonic nanoparticles, as shown in **Figure 1**. The theoretical framework of 'metatronics' is very simple and is based on one Maxwell equation $\nabla \times H = J_c + (-i\omega D)$, and this equation has two terms on its right-side: one is the conduction

http://dx.doi.org/10.5772/67681

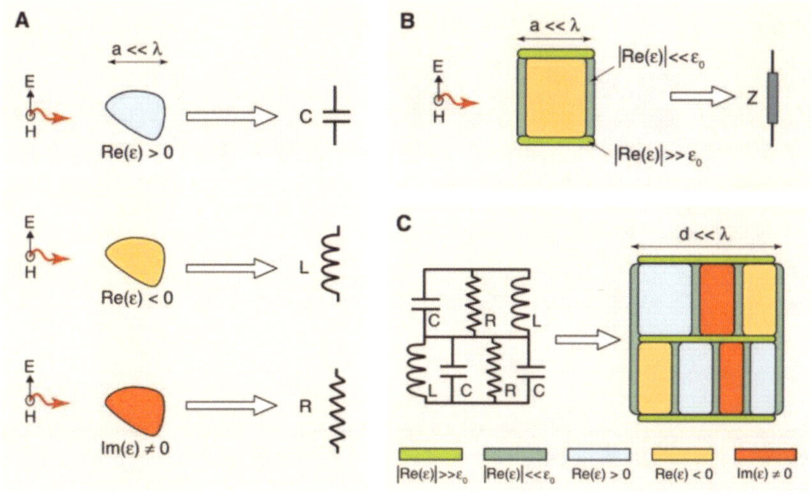

Figure 1. Analogy between sub-wavelength nanoparticles and conventional lumped nanocircuit elements at optical frequency domain. Here, one nanoparticle can take a different role when the sign of its complex permittivity ε is different. If combined in a specific way, different nanoparticles can possess some defined functionalities (adapted from Ref. [6], **Figure 1**).

current $J_c = \sigma E$ and the other is the displacement current $J_d = -i\omega D$. As mentioned above, the material dispersion plays a crucial role at optical wavelength, and the conduction current (originally dominant at low-frequency) is no longer dominant. Instead, the electric displacement current can be taken as the 'flowing optical current'. With the left of displacement current, the next and most important step is to re-build the formulas to calculate the 'lumped impedance' of nanoparticles as for an optical circuit.

2.2. Optical lumped elements

Based on the above 'metatronic' concept, the optical impedance (an intrinsic parameter) of nanoparticles is similar to the electrical counterpart ($Z=V/I$) [8], which is independent of the surrounding environments. As sketched in **Figure 2**, two isolated sub-wavelength nanospheres (radius R) with complex permittivity ε are immersed in homogenous space (with permittivity ε_0 being real) and illuminated uniformly by a vertical electric field E_0. Because of small enough size of nanoparticles (with respect to the central wavelength of illuminating field), the scattered electromagnetic (EM) fields in the vicinity of nanospheres and the total fields inside it can be obtained under suitable approximations by means of the well-known time-harmonic and quasi-static approach [9]. If integrating the flux of displacement current J_d across the sphere induced by the scattered electric field, one can get the 'total' displacement current as,

$$I_{imp} = I_{sph} + I_{fringe} = -i\omega(\varepsilon - \varepsilon_0)\pi R^2|E_0|, \tag{1}$$

$$I_{sph} = -i\omega\varepsilon\pi R^2(\varepsilon - \varepsilon_0)|E_0|/(\varepsilon - 2\varepsilon_0), \tag{2}$$

$$I_{fringe} = -i\omega\varepsilon_0 2\pi R^2(\varepsilon - \varepsilon_0)|E_0|/(\varepsilon + 2\varepsilon_0). \tag{3}$$

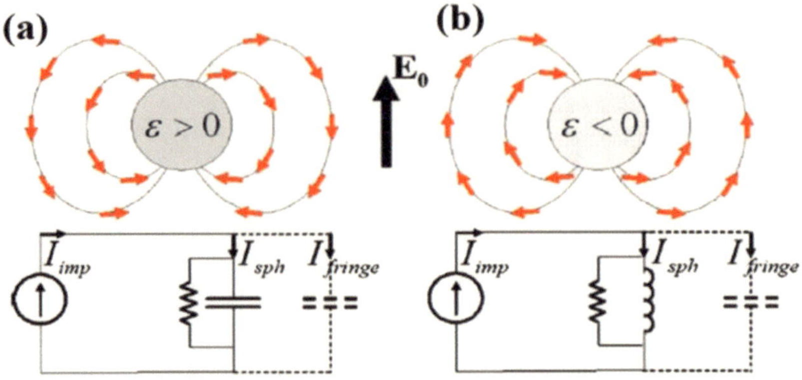

Figure 2. Optical nanocircuit models for (a) dielectric or (b) plasmonic nanoparticles. E_0 denotes the incident electric field, and thinner field lines together with the arrows represent the electric dipolar fringe field from nanospheres. (adapted from Ref. [8], **Figure 1**).

Here, as shown at the bottom of **Figure 2**, I_{imp} is the 'impressed' displacement current, I_{sph} is displacement current circulating in nanospheres and I_{fringe} is displacement current for the fringe (dipolar) field. Such currents are all related to the polarization charges at the surface of nanospheres induced by light source and all can be intuitively interpreted as branch currents at nodes in a parallel circuit.

The 'average' potential difference between the upper and lower hemi-spherical surfaces of the sphere is given by

$$\langle V\rangle_{\text{sph}} = \langle V\rangle_{\text{fringe}} = R(\varepsilon - \varepsilon_0)|E_0|/(\varepsilon - 2\varepsilon_0) \tag{4}$$

Thus, after having obtained the ratio between the potential difference (Eq. 4) and the effective current evaluated from Eq. (2) or (3), one can, respectively, get the equivalent impedance for the 'nanosphere' or the 'fringe' branch of the circuit as

$$Z_{\text{sph}} = (-i\omega\varepsilon\pi R)^{-1}, \quad Z_{\text{fringe}} = (-i\omega 2\pi R\varepsilon_0)^{-1}. \tag{5}$$

From Eq. (5), one can clearly see that the two parallel elements in the circuits (**Figure 2**) may behave differently, which is determined by the permittivity sign of the nanospheres as shown in **Figure 1**. For example, a non-metallic (or dielectric) sphere can be thought as a capacitor (because of ($\text{Re}(\varepsilon) > 0$), parallel with a resistor ($Im(\varepsilon) \neq 0$), indicating a loss) (left-bottom, **Figure 2**). In contrast, the impedance of the outside fringe is always capacitive, since its permittivity is positive. Therefore, here the equivalent circuit elements are expressed in terms of nanosphere parameters as,

$$C_{\text{sph}} = \pi R\text{Re}[\varepsilon], \quad G_{\text{sph}} = \pi\omega R\text{Im}[\varepsilon], \quad C_{\text{fringe}} = 2\pi R\varepsilon_0. \tag{6}$$

Instead, if such a sphere is made of plasmonic materials (e.g. Ag, Au), it behaves as an inductive element, since ($\text{Re}(\varepsilon) < 0$), and similarly the equivalent inductor for such a sphere becomes,

http://dx.doi.org/10.5772/67681

$$L_{\mathrm{sph}} = \left(-\omega^2 \pi R \mathrm{Re}[\varepsilon]\right)^{-1}. \tag{7}$$

Naturally, one can implement a more complicated circuit by a series and parallel operations among different kinds of such lumped particles. If two nanoparticles share a common interface and electric field is tangential to the common interface, a parallel operation between them would be made effective (left-column, **Figure 3**).

In contrast, if the electric displacement vector locally is normal to that of common interface, the displacement current J_{d} passing across one element through the common interface must flow inside the second element under the confinement of the continuity condition, and a series cascade of these two nanoelements is realized (right-column, **Figure 3**). This represents an important difference with respect to those in the conventional circuit theory, and in the latter, the lumped elements are completely isolated from the external world, and their interaction with the circuit is determined by the facts that how they pass through their terminals and how they are connected with the rest of the circuit. In contrast, in the former, i.e. the configuration presented here, since the fringing and external fields (**Figure 2**) play dominant roles in the interaction of nanoelements with their neighbouring particles and surrounding environment,

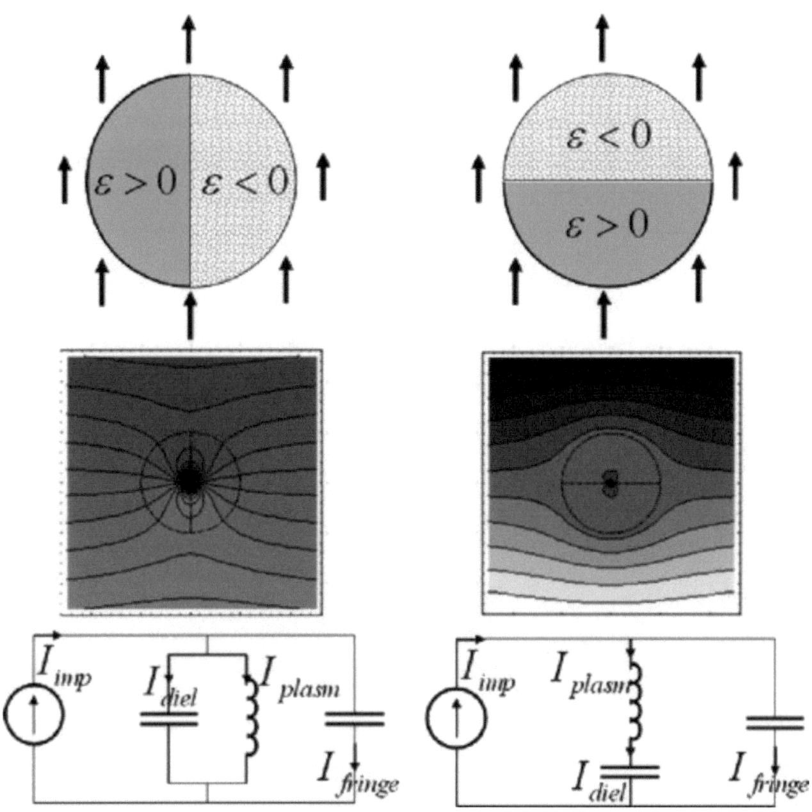

Figure 3. Two conjoined two-dimensional half-spheres with different signs of permittivity, illuminated by a uniform electric field. Parallel (left) and series (right) configurations are, respectively, shown (adapted from Ref. [8], **Figure 2**).

the originally same connection now may consists of both series and parallel interconnections which sensitively depend on the orientation of the external applied field. This point may be taken as an important and additional degree of freedom in optical nanocircuit design.

In the above, just nanosphere is taken as an example particle. In fact, except for this, the basic 'alphabets' of metatronics can also be gratings, cylinder pillars, rectangle bricks and other antennas, and they have been widely applied to various wavelength regimes,for example in a series of theoretical analyses and numerical simulations [4, 7, 10–12].

2.3. Metatronics-based applications

The recent experimental progresses [7, 13] have verified the validity and potentiality of the above optical nanocircuit paradigm introduced in Section 2.2, and also demonstrated the possibility of re-configuring the circuit responses just by changing the orientation and polarization of illuminating field to induce a specific feature not available in conventional electric circuits. No doubt that this metatronics concept provides us an effective and practical tool to design optical nanodevices, such as designing and tuning of optical nanoantennas [14, 15] and meta-surfaces [16].

One representative example is shown in **Figure 4** where a polarization-selective optical filter making use of a simple sub-wavelength grating is designed. Different incident illuminations and different effective optical connections between the nanoelements of this 'stereo-nanocircuit' are chosen to conveniently control the light transmittance, i.e. to make the circuit functions either band-stop or band-pass. This scheme may be exploited for parallel processing of multiple flows of information through a single nanostructure.

In detail, when electric field E is perpendicular to the nanorods (**Figure 4a**), the optical displacement current J_d 'flows' transversely across the nanorods and air gaps (in other words,

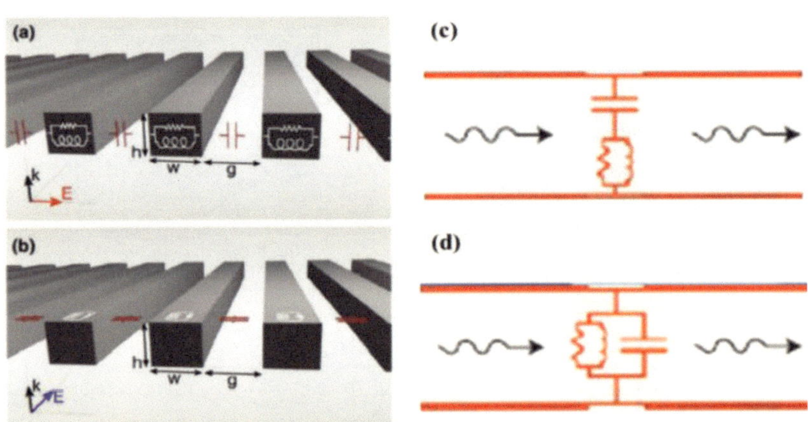

Figure 4. Sub-wavelength grating with parallel plasmonic nanorods working as a two-dimensional optical nanocircuit with a stereo-functionality. (a) The E field is perpendicular to the nanorods and nanoinductors (including nanoresistors) and nanocapacitors form a series configuration (c). In contrast, if the E field is parallel to nanorods (b), a parallel configuration (d) of lumped circuit elements is formed instead (adapted from [10], **Figure 1**).

the displacement current is in the plane of the arrays but perpendicular to the nanorods). Therefore, the final equivalent series impedance is written as,

$$\begin{aligned} Z_{\perp(\text{equivalent})} &= Z_{\perp(\text{nanorod})} + Z_{\perp(\text{airgap})}\,, \\ Z_{\perp(\text{nanorod})} &= iw/\omega h\varepsilon, \quad Z_{\perp(\text{airgap})} = ig/\omega h\varepsilon_{\text{air}}. \end{aligned} \tag{8}$$

In contrast, when E is parallel to the nanorods (**Figure 4b**), the optical displacement current J_d 'flows' along both the nanorods and air gaps. Therefore, each nanorod and its neighbouring air gap collectively have equivalent parallel impedance,

$$\begin{aligned} Z_{\|(equivalent)} &= Z_{\|(nanorod)} Z_{\|(airgap)} / (Z_{\|(nanorod)} + Z_{\|(airgap)}), \\ Z_{\|(nanorod)} &= i/\omega wh\varepsilon\,, \quad Z_{\|(airgap)} = i/\omega gh\varepsilon_{air} \end{aligned} \tag{9}$$

These equivalent impedances (series or parallel,in Eqs. (8) and (9)) finally determine the transmittance of the incident optical signal, and the latter is naturally derived as,

$$\begin{aligned} T_{\perp} &= |Z_{\perp(\text{equivalent})} / (Z_{\perp(\text{equivalent})} + [\eta_0 (w+g)/2])|^2, \\ T_{\|} &= |Z_{\|(\text{equivalent})} / (Z_{\|(\text{equivalent})} + [\eta_0 / (2(w+g))])|^2. \end{aligned} \tag{10}$$

To test this optical circuit approach, one can compare the results in **Figure 5** calculated from Eq. (10) with those experimentally measured (10) or exactly numerical results with the help of

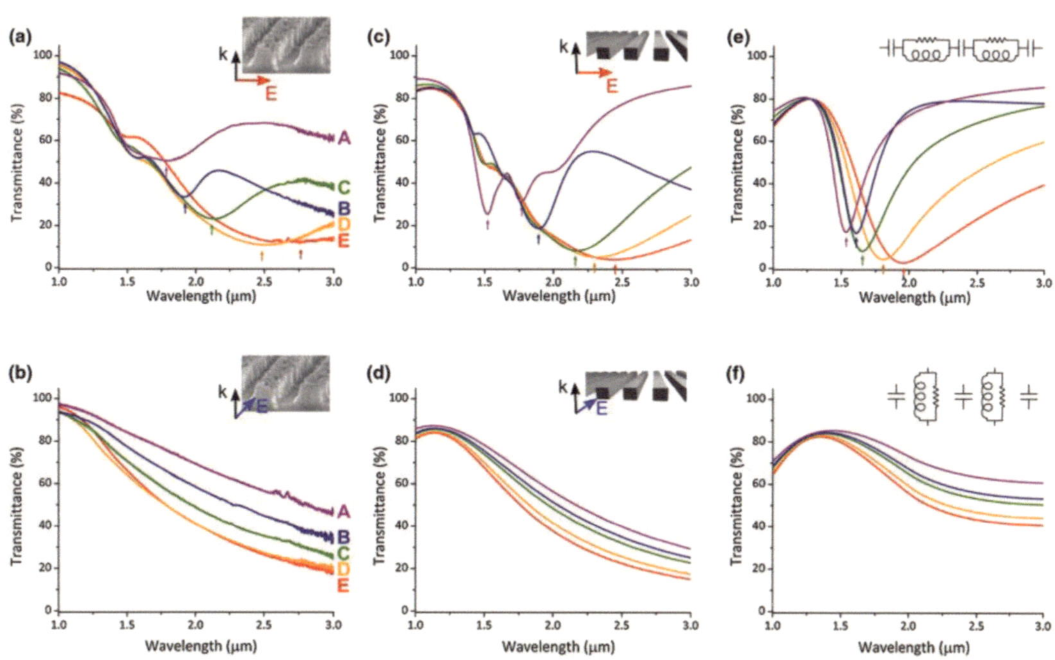

Figure 5. Transmittance spectra for five different samples from A to E. Experimentally measured data for (a) perpendicular or (b) parallel polarization of the incident wave impinging on the nanorod arrays. (c and d) Full-wave simulation results and (e and f) calculation results from nanocircuit theory for polarization perpendicular (upper) or parallel (lower) to nanorods (adapted from Ref. [10], **Figure 2**).

a commercial software. As one expects, the grating with a 'series' configuration behaves as a band-stop filter, however the same grating with a 'parallel' configuration as a band-pass filter instead.

Another representative example is related to the radio-frequency (RF) antennas which have been widely investigated and applied in wireless telecommunication system in the last century. The functionality of RF antenna is information revolution. It is usually used with a radio transmitter or radio receiver. In transmission, a radio transmitter supplies an electric current oscillating at radio frequency (i.e. a high frequency alternating current (AC)) to the antenna's terminals, and the antenna radiates the energy from the current as electromagnetic waves (radio waves). In analogy with their RF counterparts, optical antennas made of plasmonic nanoparticles are able to efficiently coupled localized sources or guided waves at the nanoscale level to far-field radiation, and in turn, to convert the impinging radiation from the far-field into sub-wavelength localized or guided fields [17]. As the counterpart of RF antenna, an optical nanoantenna exhibits novice and interesting characteristics because of their plasmonic nature.

In detail, as depicted in **Figure 6a**, a conventional linear RF antenna is loaded at its feeding gap with lumped circuit elements and changing the antenna input impedance allowance to operate at a given frequency or to achieve a good match for a specific feeding network. Analogously, an optical nanocircuit opens the same possibility for an optical nanoantenna (**Figure 6b**), and the complex optical input impedance can be interpreted as the parallel combination of the dipole intrinsic impedance Z_{dip} and the gap impedance Z_{gap} [11, 15]. The former impedance is assumed to be a fixed property of the nanoantenna geometry and surrounding environment, and the latter can be engineered to a large extent by loading the gap with different materials.

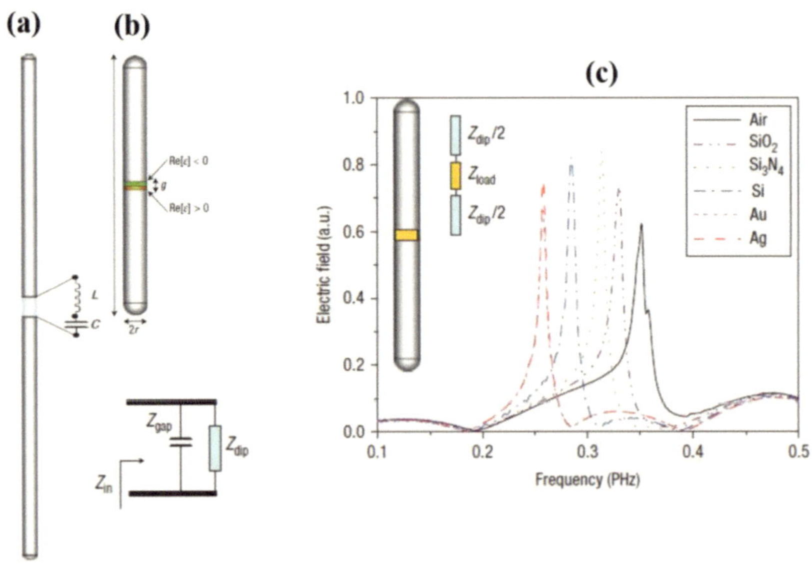

Figure 6. Optical nanoantennas. RF dipole antenna (a) loaded with lumped circuit elements at its feeding gap, and analogously, a plasmonic dipole nanoantenna (b) loaded with optical nanocircuits (adapted from Ref. [15], **Figure 1**). The low inset shows the circuit model of antenna input impedance. (c) Tuning of nanoantenna resonance by 'gap loading' with different realistic materials (adapted from Ref. [15], **Figure 3**).

http://dx.doi.org/10.5772/67681

Such as for a cylindrical gap with height of t, radius of a and excited by an incident electric field parallel to its axis, the gap impedance is given by,

$$Z_{\text{gap}} = \frac{it}{\omega \varepsilon \varepsilon_0 \pi a^2}. \tag{11}$$

Thus, by filling the gap with different materials (or their proper series or parallel combination), the impedance of the gap can be tailored to a large degree. As a result, one can tune the frequency response (**Figure 6c**) or radiation pattern easily.

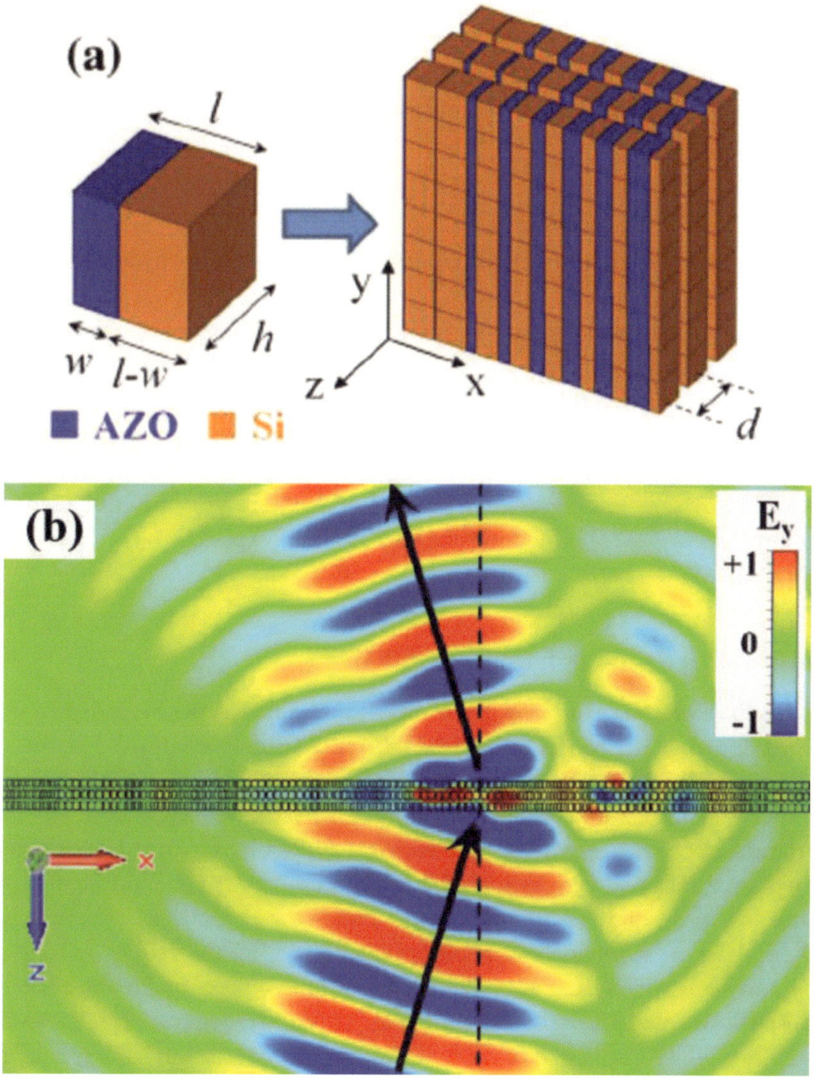

Figure 7. Meta-transmit-array used for full control of nanoscale optical transmission. (a) Basic nanocircuit building block (left) and resulting meta-screen (right) with transverse inhomogeneous profiles of surface impedance. (b) Full-wave simulation of meta-transmit-array for light deflection with high efficiency and minimized reflection (adapted from Ref. [16]).

In addition, except for the above sub-wavelength grating and non-antenna, meta-surface represents another important type [18], the planar counterparts of meta-materials that provide the unprecedented control of the amplitude, phase or polarization of light waves at the sub-wavelength (nano) scale. For example, these two-dimensional surfaces can alter the wave-front of incident light for a widespread application in beam shaping [19–24], polarizers [25, 26] and flat lenses [27–29]. As proposed in [16], the fundamental building blocks are paired plasmonic or dielectric nanorods collectively working as an inductor–capacitor nanocircuit (**Figure 7a**, left), whose impedance depends directly on the filling ratio of plasmonic and dielectric materials. By suitably alternating these nanocircuit blocks on the transverse direction, one can synthesize a meta-surface with the required inhomogeneous impedance profile. The configuration composed of a stack of three meta-surfaces (**Figure 7a**, right), can fully control the nanoscale optical transmission, while simultaneously minimizing the reflection (impedance-matching to free-space), allowing, for example light deflection with an almost ideal efficiency as shown in **Figure 7b**.

3. Equivalent nanocircuit theory for multi-layer meta-material design

In this Section, the 'design stack' is moved upward, from the 'physical layer' of optical lumped elements (pure nanoparticles, nanogratings, nanoantennas, etc.) to more complex functional devices, including: (1) infrared third-order Butterworth filters; (2) metal-insulator-metal (MIM) ultra-broadband absorbers; (3) simplified broadband super-flat perfect infrared absorbers only composed of single transparent conductive oxides (TCOs). Among these design procedures for multi-layer nanostructures in our research group, the suitability of the equivalent nanocircuit theory is confirmed once more which in turn enriches and expands the application of equivalent nanocircuit theory.

3.1. Infrared broadband third-order Butterworth filters

Frequency selective surfaces (FSSs) have been the subject of investigations by many researchers for decades. An FSS is a periodic structure usually composed of an assembly of identical elements arranged in one- or two-dimensional lattice. These structures are used in a variety of important applications ranging from microwave systems and antennas to radar and satellite communications. The simplest FSS device is a filter. By means of circuit elements (e.g. resistors (R), inductors (L) and capacitors (C)) FSSs can be effectively and flexibly designed to functionalize a low-pass, high-pass, band-pass or band-stop responses in the microwave or RF domain [30, 31]. The great interests in FSS's application in higher frequency range to achieve high-density and high-speed optical analogues [5, 10] have been pushed and the goals have been synthesized as realistic nanostructures at optical frequency, thanks to the optical nanocircuit theory which proves once more to be an essential design tool to construct optical FSSs or filters.

Different from those basic 'alphabets' mentioned in the previous section (nanogratings, nanoantennas, etc.), here, nanobricks are chosen as the building blocks. One reason is that they are widely used as the atoms of meta-material and plasmonic structure, and the second is that the brick's planar profile makes us easily calculate the equivalent impedance.

As a demonstration, one layer of periodic nanosquare array based on indium-tin-oxide (ITO) used for an infrared FSS filer is presented (**Figure 8**). The reason why ITO material is chosen for making up of FSS filter cells is just that, it possesses low electrical resistance and high transmittance in the visible range and widely used as an electrode for displays [33]. Especially, in the infrared spectral range, ITO material can demonstrate a metallic performance and this makes it to become a counterpart of noble metal. In addition, in the practical calculations, the permittivity ITO is usually modelled by the Drude dispersion relation.

When such a nanosquare array is illuminated vertically (along z-direction) by an optical signal from the bottom side with electric field E polarized parallel to the x-direction, the optical displacement current J_d 'flows' along both the unit cells and air gaps [7, 32], the nanosquare array first acts as a 'parallel' combination of lumped elements R, L and C_1, and then in series with a capacitance C_2, as shown in **Figure 8b**. Following the general capacitor impedance formula $Z_c = i/(\omega C)$ with the capacitance $C = \varepsilon a/b$, and a, b being the element's two-dimensions, the effective lumped impedance of the unit nanosquare cells can be written as,

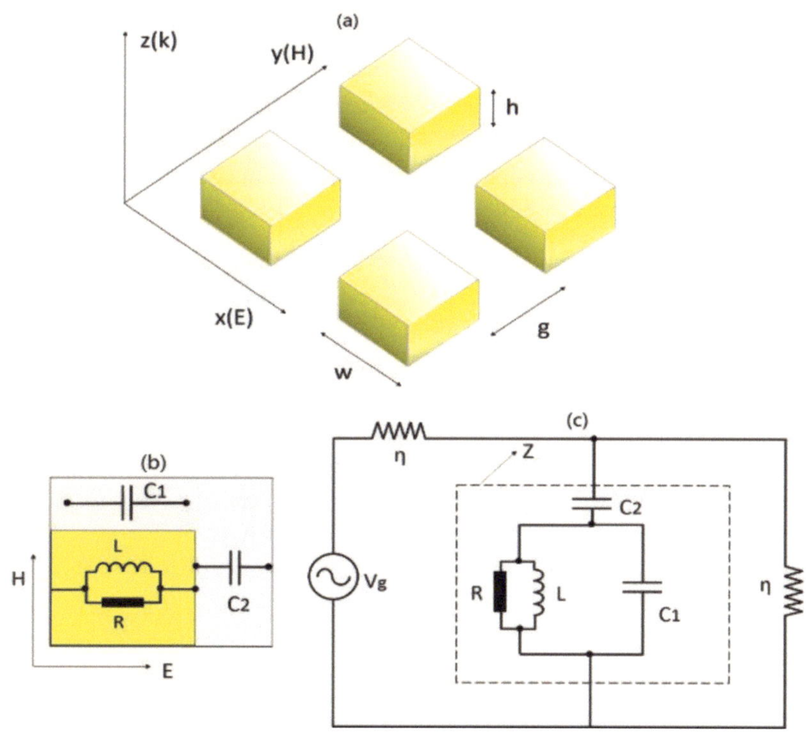

Figure 8. Schematic diagrams of (a) periodic nanosquare array (as for its unit cell, the width and height, as well as the gap distance between two adjacent unit cells are denoted as w, h and g), (b) equivalent lumped circuit elements and (c) whole equivalent circuit used for designing single-layer FSS filter (adapted from Ref. [32], **Figure 1**).

$$Z_{\mathrm{ITO}} = Z_{\mathrm{R}} \| Z_{\mathrm{L}} = iw/\omega h w \varepsilon_{\mathrm{ITO}},$$
$$Z_{C_1} = iw/\omega h g \varepsilon_{\mathrm{air}}, \quad Z_{C_2} = ig/\omega h (w+g) \varepsilon_{\mathrm{air}} \tag{12}$$

Thus, the input impedance of a single layer is given by,

$$Z = Z_{\mathrm{ITO}} \| Z_{C_1} + Z_{C_2}. \tag{13}$$

Subsequently, the equivalent circuit model of this whole FSS system (**Figure 8c**), where $\eta = \sqrt{\mu/\varepsilon}$ is the intrinsic impedance of surrounding medium. Within the equivalent circuit impedance theory [34], the reflectance and transmittance of the incident optical signal are naturally obtained as,

$$S_{11} = (Z\|\eta - \eta)/(Z\|\eta + \eta), \quad S_{21} = S_{11} + 1,$$
$$R = |S_{11}|^2, \quad T = |S_{21}|^2 = |2Z/(2Z+\eta)|^2. \tag{14}$$

Obviously, in order to achieve different optical responses, the impedances of nanocircuit elements can be changed, by tuning the structural size (w, h or g), the constituent material property ($\varepsilon_{\mathrm{ITO}}$), or the illumination light wavelength ($\lambda = 2\pi c/\omega$) and then the property of such a band-stop filter could be flexibly controlled. As a demonstration, samples from A to D with different widths of 100, 200, and 300 nm, but with fixed gap of 100 nm and height of 150 nm for simplicity are chosen.

The transmittances of the FSS filters composed of samples from A to D, shown in **Figure 9a**, are obtained through Eq. (14). In addition, one can employ a full-wave FDTD simulation to check the validity of such an equivalent nanocircuit theoretical model, and the corresponding transmittance spectra are shown in **Figure 9b** for comparison. The comparison result indicates that they are consistent well with each other.

As investigated in **Figure 9c**, the band-stop behaviour of the FSS filters is dependent on their geometric size. With the increment of w/g, the band-stop centre has a red-shift which can be well interpreted by the optical circuit theory [15]: the larger the width of the nanosquare unit cell, the bigger the inductor L induced between the two adjacent air gaps, hence leading to a higher resonant wavelength. In contrast, with the increment of w/g, the band-stop depth decreases and this behaviour is a result of the fact that the larger the width of nanosquare unit cell the stronger the guidance of light. From the electric field distribution for one representative sample at the resonance frequency, one can see that the unit cell structure works as an antenna and the incident light is localized at the regions between two adjacent metal unit cells, resulting in significant resonant enhancement of localized field and a guidance of most light through the air gap [35, 36].

As for a more practical and wide application, a flatter and broader band-stop filtering response curve with a fast roll-off would be much advantageous [37]. To gain this aim, as the general FSS-based filter design scheme does [38, 39], a third-order Butterworth band-stop filter is realized by cascading triple-layer of nanosquare unit cells with a specific separation distance D between the consecutive layers (**Figure 10a**). The corresponding equivalent circuit is modelled in **Figure 10b**. Then, the whole structure is separated into four regions along the

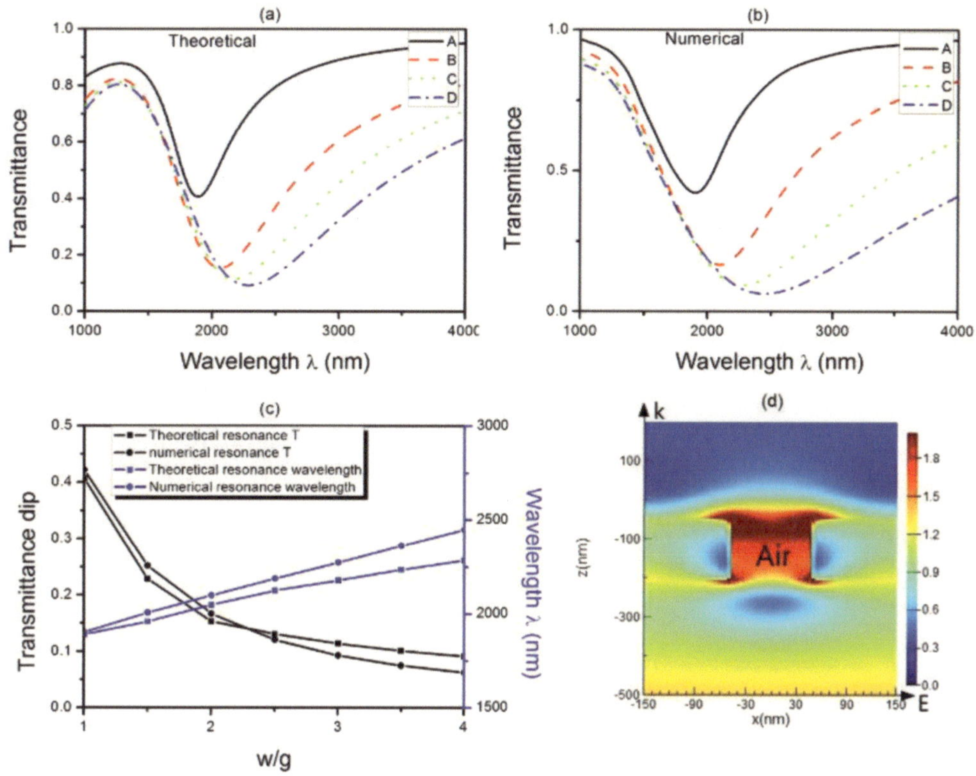

Figure 9. (a) Theoretical and (b) numerical transmittance spectra based on equivalent circuit theory and FDTD simulation for samples from A to D, respectively. (c) The resonance wavelength and transmittance dip as a function of *w/g*. (d) The representative electric field distribution for sample B at resonance wavelength (adapted from Ref. [32], **Figure 2**).

propagation direction (**Figure 10c**). Finally, along the similar procedure, the transmittance T after the third layer can be obtained step by step.

(1) As for the first layer, the theoretical S-parameters are

$$S_{11} = \frac{Z_1\|\eta - \eta}{Z_1\|\eta + \eta}, \quad S_{21} = S_{11} + 1, \quad S_{22} = S_{11}, \quad S_{12} = S_{21}. \tag{15}$$

Here, the impedance of the surrounding medium above or between the neighbouring layers is η and that of each layer is Z_i ($i = 1, 2, 3$). As for the second or third layer, the S-parameters (S_{33}, S_{43}, ...) can be get by just replacing Z_1 in Eq. (15) with Z_2 or Z_3.

(2) The reflected power P_{iR} and transmitted power P_{iF} from each layer are evaluated as follows,

$$\begin{aligned} P_{1R} &= |S_{11}|^2 P_{1F} + |S_{12}|^2 P_{2R}, & P_{2F} &= |S_{21}|^2 P_{1F} + |S_{22}|^2 P_{2R}, \\ P_{2R} &= |S_{33}|^2 P_{2F} + |S_{34}|^2 P_{3R}, & P_{3F} &= |S_{43}|^2 P_{2F} + |S_{44}|^2 P_{3R}, \\ P_{3R} &= |S_{55}|^2 P_{3F}, & P_{4F} &= |S_{65}|^2 P_{3F}. \end{aligned} \tag{16}$$

For simplicity, the absorption loss caused by the surrounding medium is assumed negligible.

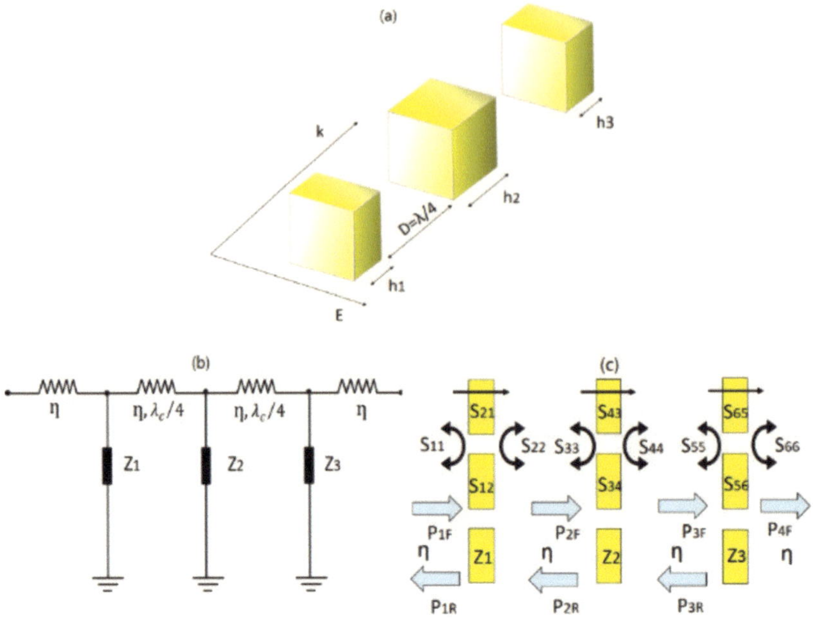

Figure 10. Schematic diagrams of (a) triple-layer FSS filter, (b) equivalent circuit and (c) *S*-parameters (adapted from Ref. [32], **Figure 3**).

(3) The total transmittance after the third layer is then written as

$$T = P_{4F}/P_{1F} = \frac{|S_{21}|^2|S_{43}|^2|S_{65}|^2}{1-|S_{11}|^2|S_{33}|^2-|S_{33}|^2|S_{55}|^2+|S_{11}|^2|S_{33}|^4|S_{55}|^2-|S_{11}|^2|S_{43}|^2|S_{55}|^2}. \tag{17}$$

As for any specific third-order Butterworth filter with desired central band-stop frequency and band-stop width, one can easily get the suitable choice of geometrical parameters for design. For example, if the band-stop edge frequencies are ω_1 and ω_2, the central frequency can be defined as $\omega_0 = \sqrt{\omega_1\omega_2}$ and band-stop width as $\omega_b = (\omega_2 - \omega_1)/\omega_0$. The reactance slope parameters of this circuit model in terms of low-pass prototype parameters G_0, G_1, ..., G_{n+1} and cut-off frequency ω_c are given as [34]

$$\frac{Z_i}{Z_0} = \frac{G_0}{\omega_c G_i \omega_b}\ if\ n = \text{even}; \quad \frac{Z_i}{Z_0} = \frac{1}{\omega_c G_0 G_i \omega_b},\ if\ n = \text{odd}. \tag{18}$$

Here, G_i is the normalized prototype element values and n is the order of the Butterworth filter, that is the number of resonators. As for a third-order (n = odd) band-stop filter ($G_1 = 1$, $G_2 = 2$, $G_3 = 1$) with the height of nanosquare cells at each layer $h_1 = 1/2h_2 = h_3$ fixed, the impedance of each layer (**Figure 10b**) can be obtained as $Z_1 = 2Z_2 = Z_3 = \eta/(\omega_c\omega_b)$.

As for this triple-layer third-order filter, if adopting parameters listed in **Table 1**, the theoretical and numerical transmittance spectra can be directly obtained (**Figure 11a**) based on Eq. (17) from an equivalent circuit theory and FDTD simulation, respectively. The comparison between

http://dx.doi.org/10.5772/67681

i	G_i	Z_i	w/g (nm)	h_i (nm)
1	1	Z	120/60	150
2	2	Z/2	120/60	300
3	1	Z	120/60	150

Table 1. Element values for low-pass prototype circuit and geometrical parameters of third-order band-stop FSS filter.

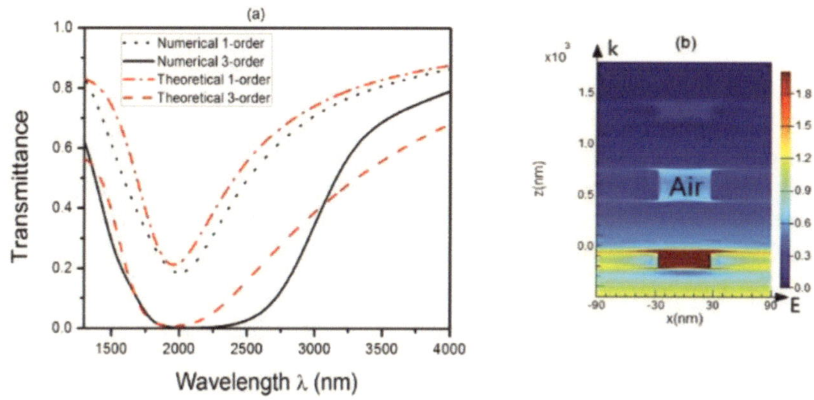

Figure 11. (a) Theoretical and numerical transmittance spectra for single-layer (first-order) or triple-layer (third-order) filter. (b) One representative electrical field distribution for a third-order filter at a wavelength of 2000 nm (adapted from Ref. [32], **Figure 4**).

single-layer and triple layer filters indicates that, for a triple-layer filter, its band-stop width and depth both become larger, and moreover, its band-stop bottom is much flatter. In addition, the transmittance is nearly zero, indicated by the electric field distribution in **Figure 11b**.

It is necessary to point out that the corresponding FDTD calculation results show a good agreement with for single-layer case, but more obvious deviation from the equivalent circuit theoretical calculation. The reason for deviation arises from the simple assumption that the triple-layer of nanosquare array is independent of each other. If the layer-layer coupling is considered the deviation maybe reduced. However, anyway, the proposed synthesis procedure is confirmed helpful to design a Butterworth band-stop filter.

3.2. Infrared broadband multi-layer MIM absorbers

Meta-material absorbers are used broadly in thermal detectors [40], imaging [41], security detection [42] and stealth devices [43]. In 2008, Landy et al. first proposed a thin perfect meta-material absorber simultaneously exciting electric and magnetic resonances (MRs) to realize the impedance match with the surrounding medium and thus eliminating any reflection and perfectly absorbing the incident waves at microwave bands [44]. Since then, applications to various wavelength regimes have been demonstrated widely by numerical simulations and experiments [45, 46]. However, the application of these perfect absorbers is limited for their narrowband and simple resonant behaviours.

The most widely used scheme [47–49] instead is to slow down the incident wave or totally absorb them by a gradually changed pyramid-shaped metal-insulator-metal (MIM) topology (**Figure 12a**). In fact, an absorber can also be thought as a filter, by operating the frequency response of the absorptance with $A = 1 - T - R$. Furthermore, each layer of the multi-layer MIM absorber projecting to the bottom plane is a nanosquare structure (**Figure 12b**), same as basic building blocks for third-order Butterworth filter (Section 3.1). Thus, naturally, the equivalent nanocircuit procedure can be directly transplanted to the design of MIM multi-layer absorber.

The first step is to evaluate the impedance Z (**Figure 12c**) of each layer, following the same procedure in Section 3.1 which can be obtained as,

$$\begin{aligned} &Z_{ITO} = Z_R \| Z_L = iw/\omega h w \varepsilon_{ITO}, &&Z_{C_1} = iw/\omega h g \varepsilon_{air}, \\ &Z_{C_2} = ig/\omega h(w+g)\varepsilon_{air}, &&Z = Z_{ITO} \| Z_{C_1} + Z_{C_2}. \end{aligned} \tag{19}$$

The second step is to cascade the impedance of each layer into a whole circuit, for example that of a triple-layer structure. Each layer is separated by one quarter of central wavelength of the incident light (**Figure 12a**). The corresponding equivalent circuit can be modelled as in **Figure 12b**, where $\eta = \sqrt{\mu/\varepsilon}$ is the intrinsic impedance of SiO_2 dielectric layers, and Z_i (i = 1, 2, 3) is the impedance of each NS layer. Within transmission line theory [34], $Z(3)$ is the effective impedance

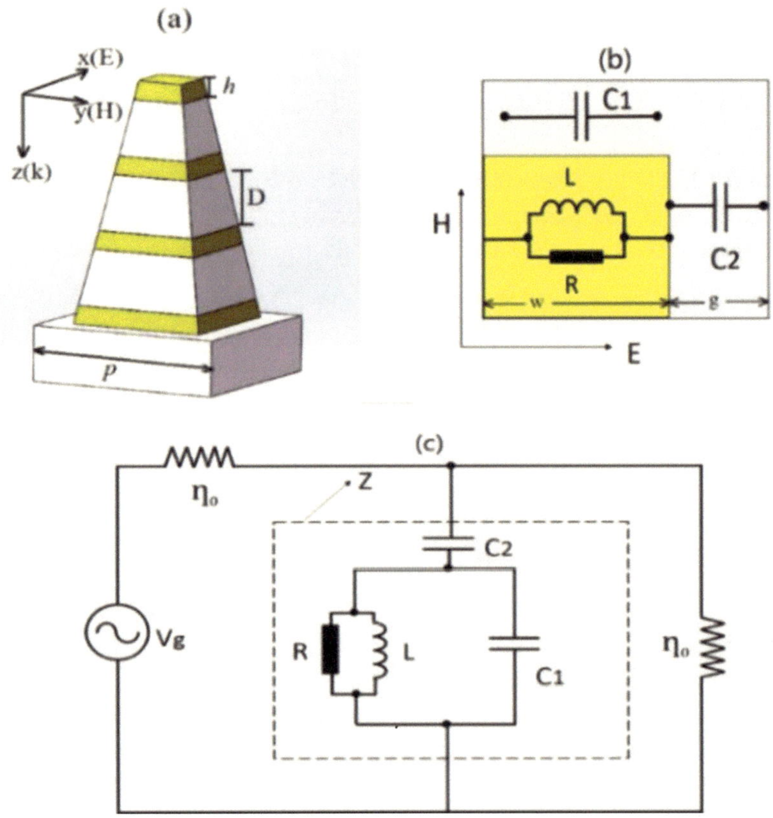

Figure 12. Schematic diagrams of (a) MIM absorber, (b) equivalent lumped circuit elements and (c) whole equivalent circuit. Here, only single nanosquare patch layer is considered (adapted from Ref. [50], **Figure 1**).

at the top interface of the third NS layer, given by $Z(3) = Z_3 \| \eta$. Similarly, for the interface at the second or first layer, the effective impedances can be written as,

$$Z(2) = Z_2 \| (\eta^2/Z(3)) , \quad Z(1) = Z_1 \| (\eta^2/Z(2)) \tag{20}$$

Then, the reflectance of this pyramid triple layer structure is given by,

$$R = \left| \frac{Z(1) - \eta_0}{Z(1) + \eta_0} \right|^2 . \tag{21}$$

Beyond the reflectance, one needs to calculate the transmittance through the bottom substrate. As shown in **Figure 13a**, based on the three different NS layers, the whole structure is separated into four independent regions from left to right. Then, the S-parameters (S_{11}, S_{21}, ...) for each layer are obtained as follows,

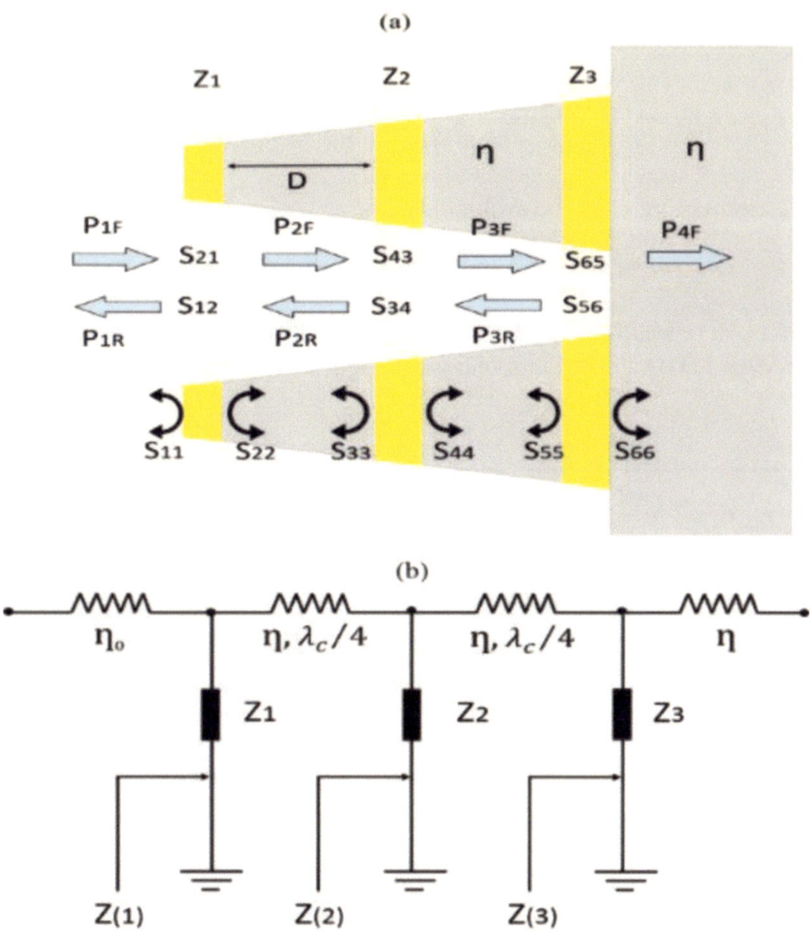

Figure 13. Schematic diagrams of (a) S-parameters used for theoretical calculation and (b) equivalent nanocircuit model for the truncated pyramid triple-layer absorber (adapted from Ref. [50], **Figure 3**).

$$S_{11} = S_{22} = \frac{Z_1 \| \eta - \eta_0}{Z_1 \| \eta + \eta_0}, \qquad S_{21} = S_{12} = S_{11} + 1,$$
$$S_{33} = S_{44} = \frac{Z_2 \| \eta - \eta}{Z_2 \| \eta + \eta}, \qquad S_{43} = S_{34} = S_{33} + 1, \tag{22}$$
$$S_{55} = S_{66} = \frac{Z_3 \| \eta - \eta}{Z_3 \| \eta + \eta}, \qquad S_{65} = S_{56} = S_{55} + 1\,.$$

For the convenience of calculation, the dielectric loss for light through SiO_2 material is neglected as long as the gap distance between two adjacent layers is large enough. Under this simplification, the reflected power $P_{i\mathrm{R}}$ and transmitted power $P_{i\mathrm{F}}$ from each NS layer are evaluated as follows,

$$\begin{aligned} P_{1R} &= |S_{11}|^2 P_{1F} + |S_{12}|^2 P_{2R}, & P_{2F} &= |S_{21}|^2 P_{1F} + |S_{22}|^2 P_{2R}, \\ P_{2R} &= |S_{33}|^2 P_{2F} + |S_{34}|^2 P_{3R}, & P_{3F} &= |S_{43}|^2 P_{2F} + |S_{44}|^2 P_{3R}, \\ P_{3R} &= |S_{55}|^2 P_{3F}, & P_{4F} &= |S_{65}|^2 P_{3F}. \end{aligned} \tag{23}$$

Solving Eqs. (22) and (23) simultaneously, the total transmittance T is finally expressed as,

$$T = P_{4F}/P_{1F} = \frac{|S_{21}|^2 |S_{43}|^2 |S_{65}|^2}{1 - |S_{11}|^2 |S_{33}|^2 - |S_{33}|^2 |S_{55}|^2 + |S_{11}|^2 |S_{33}|^4 |S_{55}|^2 - |S_{11}|^2 |S_{43}|^2 |S_{55}|^2}. \tag{24}$$

Naturally, the absorptance A can be evaluated as,

$$A = 1 - R - T. \tag{25}$$

The absorption (A), reflection (R) and transmission (T) spectra calculated by above equations and corresponding FDTD simulations, shown in **Figure 14a**, show a good agreement between

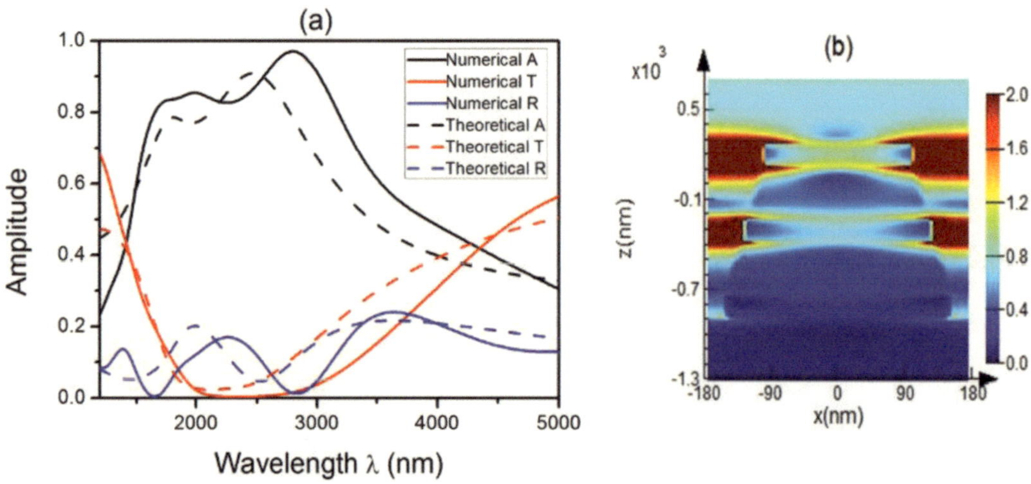

Figure 14. (a) Comparison between theoretical results from equivalent nanocircuit theory and numerical calculations from FDTD simulation. (b) Electric field distribution at a wavelength of 2500 nm. Each nanosquare S patch has a varying width w of 300, 250 and 200 nm from bottom to top (adapted from Ref. [50], **Figure 4**).

the two existing methods which further prove the feasibility of our equivalent circuit theory. The replacement of single-layer absorber by a triple-layer makes the absorptance bandwidth to become obviously wider and also the efficiency is significantly enhanced. The enhanced electric field (**Figure 14b**) concentrates at both the lateral edges of NS and air gaps between two adjacent unit cells that couples efficiently to the incident light and dissipate the energy within the metals via Ohmic loss [51]. After the third layer, nearly no light can be transmitted from the absorber.

3.3. Simplified broadband super-flat perfect absorber

In Section 3.2, a triple-layer MIM absorber design has been successfully demonstrated based on the equivalent nanocircuit theory. The proposed synthesis circuit procedure is confirmed to be feasible enough to provide us a way to predict the responses of such absorbers. It need to be emphasized that, it is in principle possible to get a perfect absorber with a 100% absorption efficiency just by adding more NS layers beyond three layers. However, with the increase of NS layer number, the equation number contained in Eqs. (22) and (23) will be added. Correspondingly, solving the multi-variable linear equations becomes more and more complicated and also time consuming which would be a big problem.

Thus, in this section, an improved equivalent nanocircuit matrix algorithm emerges as the times requires [52] which can predict the complex frequency response of multi-layer (with arbitrary layer numbers) nanostructures easily, without solving the multi-variable linear equations. One can believe that it may provide inspiring advancements in future meta-material designs.

The construction of this equivalent nanocircuit matrix algorithm derives from the design of a simplified broadband super-flat perfect absorber made of single transparent conductive oxides (TCOs) material [52]. In such an absorber design [53], in order to pursue a broadband flat response, until now, no matter how many layers, the multi-layer MIM absorber or planar multiplexed pattern absorber still cannot achieve broadband flat perfect absorption as one expects. Furthermore, there has another challenge for these kinds of broadband absorber design which is related to the fabrication process. It is hard to scale down to higher frequency mainly because of the fabrication difficulty, including lithography and alignment between neighbouring layers or resonators. Within this context, one should explore new paradigms for broadband absorber design.

Fortunately, we note that the transparent conductive oxides (TCOs), such as Al:ZnO (AZO), Ga:ZnO (GZO) and indium-tin-oxide (ITO), can play a fascinating role in the designing of broadband perfect absorber for its unique transmission or conductive property in near-infrared (NIR) region [33, 53]. As shown in **Figure 15a**, periodic arrays of truncated pyramid structure made of TCOs could work as a broadband absorber in NIR frequency, furthermore such absorbers at broadband wavelengths have continuous flat responses with near-unit light absorption. Comparing with the traditional multi-layer metal-insulator-metal (MIM) absorber, TCOs absorbers using only one single material can greatly reduce the fabrication difficulty, one do not need to consider the perfect alignment to match the relative position of each pattern in different layers.

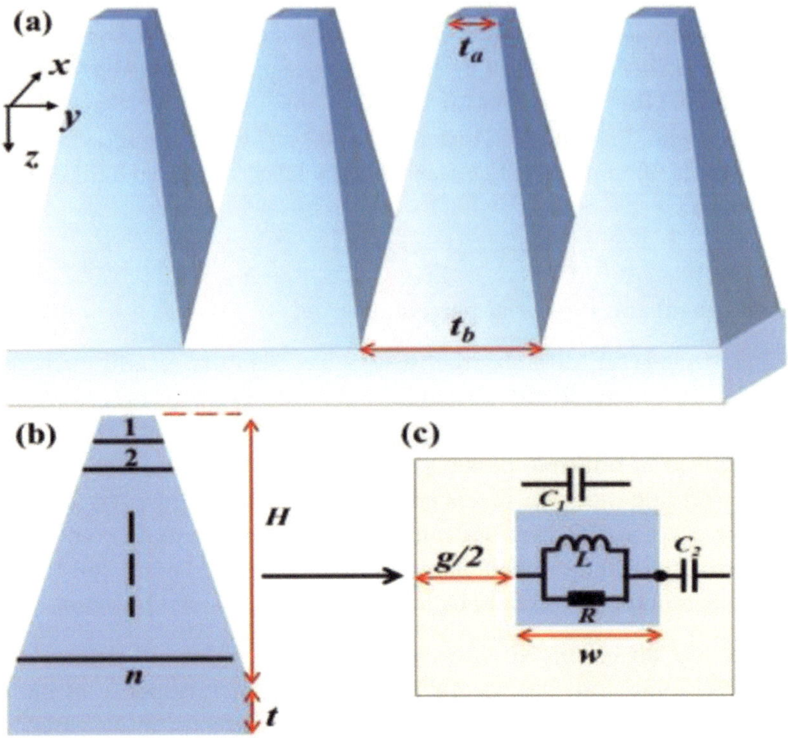

Figure 15. Schematic diagrams of (a) the super-flat absorber with atop span of $t_a = 0.1$ μm, bottom span of $t_b = 1$ μm, and ground plane thickness of $t = 0.2$ μm. (b) Side view of unit cell with $H = 1.6$ μm and (c) corresponding equivalent lumped circuit elements (adapted from Ref. [52], **Figure 1**).

The designed absorber unit consists of two TCOs elements (**Figure 15**): a truncated pyramid shaped resonator and a ground plane. The material used for the two elements can be only TCOs, i.e. materials for resonator and substrate are ITO. To establish the equivalent nanocircuit model for this absorber, one can hypothetically cut the whole pattern into n pieces along *k(z)*-direction (**Figure 15b**).

The first step is to extract the equivalent reactance of a single piece (same as those in Sections 3.1 and 3.2). The local impedances of ITO patch (Z_{ITO}), air gaps (Z_{c1}, Z_{c2}) and total effective impedance (Z_{eff}) can be calculated as,

$$\begin{aligned} Z_{\text{ITO}} &= Z_R \| Z_L = i/\omega h \varepsilon_{\text{ITO}}, \\ Z_{C_1} &= iw/\omega h g \varepsilon_{\text{air}}, \quad Z_{C_2} = ig/\omega h(w+g)\varepsilon_{\text{air}}, \\ Z_{\text{eff}} &= Z_{\text{ITO}} \| Z_{C_1} + Z_{C_2}, \quad Y = 1/Z_{\text{eff}}. \end{aligned} \tag{26}$$

When taking the single piece into the transmission line (TL) model, it can be modelled as a shunt admittance Y (=$1/Z_{eff}$), and sandwiched between two semi-infinite transmission lines with characteristic impedance Z_0 (**Figure 16a**).

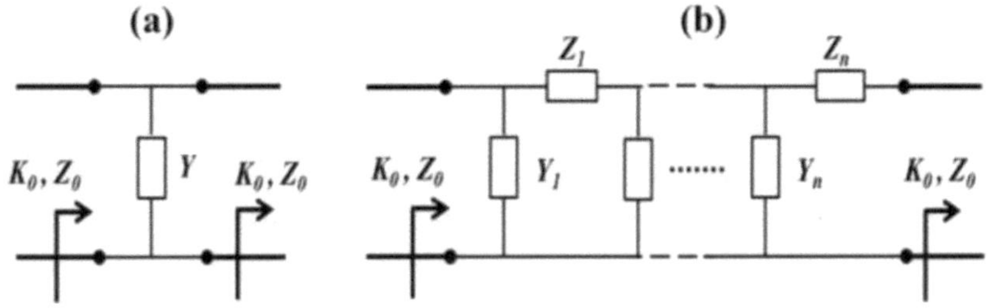

Figure 16. Transmission-line models of the truncated pyramid structure with (a) only a single piece and (b) a stack of *n* pieces (adapted from Ref. [52], **Figure 2**).

The second step is to connect each impedance into a complete TL circuit model. Series configuration of Z_i alongside with parallel configurations Y_i in layered stacked two-port networks, as shown in **Figure 16b**. The added series configuration Z_i is used to describe the conductive property between the adjacent pieces. For a thin uniform piece (P_i) of thickness *l*, the impedance Z_i can be approximately evaluated as $Z_i = 2(i\omega lp\mu_0/w_i)$ [54]. Note that in both scenarios, the realistic circuit value varies with the width of each P_i. The *ABCD*-matrix [55] for P_i of a lumped parallel electronic element combination with a series electronic element is

$$\begin{bmatrix} A_i & B_i \\ C_i & D_i \end{bmatrix} = \begin{bmatrix} 1 & 0 \\ Y_i & 1 \end{bmatrix} \begin{bmatrix} 1 & Z_i \\ 0 & 1 \end{bmatrix}. \tag{27}$$

For the whole structure with *n* pieces, it can be thought as an *n*-order filter along the transmission line, so the ABCD-matrix is expressed as,

$$\begin{bmatrix} A & B \\ C & D \end{bmatrix} = \begin{bmatrix} A_1 & B_1 \\ C_1 & D_1 \end{bmatrix} \cdots \begin{bmatrix} A_n & B_n \\ C_n & D_n \end{bmatrix}, \tag{28}$$

and the *S* matrix can be calculated as,

$$\begin{bmatrix} S_{11} & S_{12} \\ S_{21} & S_{22} \end{bmatrix} = \begin{bmatrix} \dfrac{AZ_0 + B - (CZ_0 + D)Z_0}{AZ_0 + B + (CZ_0 + D)Z_0} & \dfrac{2Z_0}{AZ_0 + B + (CZ_0 + D)Z_0} \\ \dfrac{2Z_0}{AZ_0 + B + (CZ_0 + D)Z_0} & \dfrac{-AZ_0 + B - (CZ_0 - D)Z_0}{AZ_0 + B + (CZ_0 + D)Z_0} \end{bmatrix}. \tag{29}$$

Obviously, the transmission, reflection and absorption can be obtained from the *S* matrix as,

$$\begin{aligned} T(\omega) &= |S_{21}(\omega)|^2, \quad R(\omega) = |S_{11}(\omega)|^2, \\ A(\omega) &= 1 - T(\omega) - R(\omega) \end{aligned} \tag{30}$$

In optical metatronic circuit, in order to have a parallel element between the two ports, ideally one needs to have a constant electric field across the nanoelement. Thus, each piece should be a thin slab with sub-wavelength thickness in the *z*-direction. Considering the present design, for a truncated pyramid structure with height *H* = 1.6 μm sketched in **Figure 15**, one can choose the thickness as 100 nm (cut number n =16) to get accurate results, smaller than wavelength (1–3 μm). By solving the transmission model in **Figure 16b**, the theoretical result is within a wavelength range of 1–3 μm (**Figure 17a**). It indicates that the absorption bandwidth becomes obviously wider and the efficiency is almost larger than 90%. The corresponding FDTD result (**Figure 17a**) indicates the feasibility of such a rigorous solution of TL model to describe the mechanism of this absorber. A good agreement between simulated and theoretical absorption at the wavelength larger than 1.5 μm, and only a slight deviation at the top absorption efficiency is observed. Furthermore, the simulation result shows that it has a flat response with absorption near unit between 1.4 and 2.6 μm.

Note that there is a slight deviation at the top absorption efficiency between simulation and theoretical prediction; we attribute this to the fact that there have deviations from extracted equivalent parameters. We employ the same full-wave FDTD simulations, only change is the structure from truncated pyramid to full pyramid, the truncated cone and the full cone, respectively. The four absorption spectra are compared in **Figure 17b**. We can see that the absorption bandwidth changes slightly indicating that the geometry dependence of the absorption is relatively weak. The bandwidth decreasing from pyramid to cone shape is due to the decrease of corresponding response area, when square changes to circle with the same width (diameter). It should be pointed out here that each geometric parameters used in **Figure 15** are not optimal. If we take the height, top width and bottom width altogether into account, much broader bandwidth could be expected.

Above all, under the guidance of nanocircuit theory, one can realize a broadband super-flat perfect infrared absorber in a single TCO material for its nice transmission and conductive properties. This simplified configuration without multi-layered design might releases the fabrication and design difficulties and exhibits great potentials in the applications of infrared

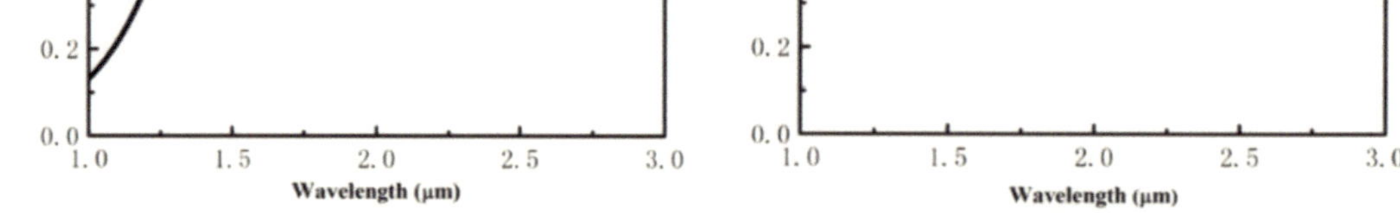

Figure 17. (a) Comparison between theoretical result from equivalent nanocircuit theory and numerical calculation from FDTD simulation. (b) Absorption curves under four different geometric structures. Truncated pyramid (t_a = 0.1 μm, t_b = 1 μm); pyramid (t_a = 0 μm, t_b = 1 μm); truncated cone (top diameter d_t = 0.1 μm, bottom diameter d_b = 1 μm); cone (top diameter d_t = 0 μm, bottom diameter d_b = 1 μm), H = 1.6 μm and t = 0.2 μm for the four structures (adapted from Ref. [52], **Figures 4** and **5**).

stealth system. Furthermore, this proposed equivalent circuit matrix algorithm is confirmed to be feasible enough to predict the complex frequency response of multi-layer nanostructures, and it can relieve the calculation from solving the multi-variable linear equations that can be easily extended to analyse other nanooptical devices.

4. Conclusions

In summary, thanks to the concept of metatronics, which gives us the possibility to transplant traditional circuit operations into a high-frequency nanodevice design. The equivalent nanocircuit (EN) theory is successfully confirmed to be feasible via the comparison with the numerical results from the rigorous FDTD calculation. With the toolbox of EN theory, an equivalent circuit matrix method can be used to conveniently predict the complicated frequency response of a complicated meta-material structure. Although here, only three application examples were demonstrated, it can be naturally and easily extended to analyse other nanooptical devices. Anyway, the interests in combining optical guiding devices with classical circuits is always high because the EN theory provides inspiring advances for designing more complex circuit systems and other related applied fields, although the deviation between results from EN theory and rigorous FDTD simulation indicates that the NE theory is still on the way of further perfection.

Acknowledgements

This work is supported by the National Natural Science Foundation of China (NNSF, Grants Nos. 11374318 and 11674312). C.L. thanks to the supports from the 100-Talents Project of Chinese Academy of Sciences.

Author details

Qing Zhang[1,2], Jun You[3] and Chengpu Liu[1]*

*Address all correspondence to: chpliu@siom.ac.cn

1 State Key Laboratory of High Field Laser Physics, Shanghai Institute of Optics and Fine Mechanics, Chinese Academy of Sciences, Shanghai, China

2 Science and Technology on Plasma Physics Laboratory, Research Center of Laser Fusion, CAEP, Sichuan, China

3 Chinese Laser Press, Shanghai, China

References

[1] A. J. Holden, J. B. Pendry, W. J. Stewart, I. Youngs, Extremely low frequency plasmons in metallic mesostructures. *Physical Review Letters* **76**(25), 4773–4776 (1996).

[2] J. B. Pendry, A. J. Holden, D. J. Robbins, W. J. Stewart, Magnetism from conductors and enhanced nonlinear phenomena. *IEEE Transactions on Microwave Theory and Techniques* **47**, 2075–2084 (1999).

[3] J. B. Pendry, Negative refraction makes a perfect lens. *Physical Review Letters* **85**, 3966–3969 (2000).

[4] A. Alù, N. Engheta, All optical metamaterial circuit board at the nanoscale. *Physical Review Letters* **103**, 143902 (2009).

[5] A. Alù, M. E. Young, N. Engheta, Design of nanofilters for optical nanocircuits. *Physical Review B* **77**, 144107 (2008).

[6] N. Engheta, Circuits with light at nanoscales: Optical nanocircuits inspired by metamaterials. *Science* **317**, 1698–1702 (2007).

[7] Y. Sun, B. Edwards, A. Alù, N. Engheta, Experimental realization of optical lumped nanocircuits at infrared wavelengths. *Nature Materials* **11**, 208–212 (2012).

[8] N. Engheta, A. Salandrino, A. Alù, Circuit elements at optical frequencies: nanoinductors, nanocapacitors and nanoresistors. *Physical Review Letters* **95**, 095504 (2005).

[9] J. Jackson. Classical Electrodynamics. (Cambridge University Press, Cambridge, 1986).

[10] H. Caglayan, S.-H. Hong, B. Edwards, C. R. Kagan, N. Engheta, Near-infrared metatronic nanocircuits by design. *Physical Review Letters* **111**, 073904 (2013).

[11] A. Alù, N. Engheta, Input impedance, nanocircuit loading, and radiation tuning of optical nanoantennas. *Physical Review Letters* **101**, 043901 (2008).

[12] M. G. Silveirinha, A. Alù, J. Li, N. Engheta, Nanoinsulators and nanoconnectors for optical nanocircuits. *Journal of Applied Physics* **103**, 064305 (2008).

[13] N. Liu, F. Wen, Y. Zhao, Y. Wang, P. Nordlander, N. J. Halas, A. Alu`, Individual nanoantennas loaded with three-dimensional optical nanocircuits. *Nano Letters* **13**, 142–147 (2012).

[14] M. L. Brongersma, Plasmonics: Engineering optical nanoantennas. *Nature Photonics* **2**, 270–272 (2008).

[15] A. Alu, N. Engheta, Tuning the scattering response of optical nanoantennas with nanocircuit loads. *Nature Photonics* **2**, 307–310 (2008).

[16] F. Monticone, N. M. Estakhri, A. Alu, Full control of nanoscale optical transmission with a composite metascreen. *Physical Review Letters* **110**, 203903 (2013).

[17] M. Agio, A. Alù, *Optical antennas*. (Cambridge University Press, New York, 2013).

[18] C. L. Holloway, E. F. Kuester, J. A. Gordon, J. O'Hara, J. Booth, D. R. Smith, An overview of the theory and applications of metasurfaces: The two-dimensional equivalents of metamaterials. *IEEE Antennas and Propagation Magazine* **54**, 10–35 (2012).

[19] H. Cheng, S. Chen, P. Yu, W. Liu, Z. Li, J. Li, B. Xie, J. Tian, Dynamically tunable broadband infrared anomalous refraction based on graphene metasurfaces. *Advanced Optical Materials* **3**, 1744–1749 (2015).

[20] A. A. High, R. C. Devlin, A. Dibos, M. Polking, D. S. Wild, J. Perczel, N. P. de Leon, M. D. Lukin, H. Park, Visible-frequency hyperbolic metasurface. *Nature* **522**, 192–196 (2015).

[21] Z. Liu, Z. Li, Z. Liu, J. Li, H. Cheng, P. Yu, W. Liu, C. Tang, C. Gu, J. Li, High-performance broadband circularly polarized beam deflector by mirror effect of multinanorod metasurfaces. *Advanced Functional Materials* **25**, 5428–5434 (2015).

[22] X. Ni, N. K. Emani, A. V. Kildishev, A. Boltasseva, V. M. Shalaev, Broadband light bending with plasmonic nanoantennas. *Science* **335**, 427–427 (2012).

[23] S. Sun, Q. He, S. Xiao, Q. Xu, X. Li, L. Zhou, Gradient-index meta-surfaces as a bridge linking propagating waves and surface waves. *Nature Materials* **11**, 426–431 (2012).

[24] G. Zheng, H. Mühlenbernd, M. Kenney, G. Li, T. Zentgraf, S. Zhang, Metasurface holograms reaching 80% efficiency. *Nature Nanotechnology* **10**, 308–312 (2015).

[25] Y. Zhao, A. Alù, Manipulating light polarization with ultrathin plasmonic metasurfaces. *Physical Review B* **84**, 205428 (2011).

[26] A. Arbabi, Y. Horie, M. Bagheri, A. Faraon, Dielectric metasurfaces for complete control of phase and polarization with subwavelength spatial resolution and high transmission. *Nature Nanotechnology* **10**, 937–43 (2015).

[27] A. Arbabi, Y. Horie, A. J. Ball, M. Bagheri, A. Faraon, Subwavelength-thick lenses with high numerical apertures and large efficiency based on high-contrast transmitarrays. *Nature Communications* **6**, 7069 (2015).

[28] E. Arbabi, A. Arbabi, S. M. Kamali, Y. Horie, A. Faraon, Multiwavelength polarization-insensitive lenses based on dielectric metasurfaces with meta-molecules. *Optica* **3**, 628–633 (2016).

[29] S. M. Kamali, E. Arbabi, A. Arbabi, Y. Horie, A. Faraon, Highly tunable elastic dielectric metasurface lenses. *Laser & Photonics Reviews* **10**, 1062–1062 (2016).

[30] M. Al-Joumayly, N. Behdad, A new technique for design of low-profile, second-order, bandpass frequency selective surfaces. *IEEE Transactions on Antennas and Propagation* **57**, 452–459 (2009).

[31] K. Sarabandi, N. Behdad, A frequency selective surface with miniaturized elements. *IEEE Transactions on Antennas and Propagation* **55**, 1239–1245 (2007).

[32] Q. Zhang, L. Bai, Z. Bai, P. Hu, C. Liu, Equivalent-nanocircuit-theory-based design to infrared broad band-stop filters. *Optics Express* **23**, 8290–8297 (2015).

[33] H. S. Jeong, H.-J. Jeon, Y. H. Kim, M. B. Oh, P. Kumar, S.-W. Kang, H.-T. Jung, Bifunctional ITO layer with a high resolution, surface nano-pattern for alignment and switching of LCs in device applications. *NPG Asia Materials* **4**, e7 (2012).

[34] G. L. Matthaei, L. Young, E. Jones, *Microwave filters, impedance-matching networks, and coupling structures*. (Artech House, Norwood, Massachusetts, 1964).

[35] Q. Zhang, P. Hu, C. Liu, Giant-enhancement of extraordinary optical transmission through nanohole arrays blocked by plasmonic gold mushroom caps. *Optics Communications* **335**, 231–236 (2015).

[36] Q. Zhang, P. Hu, C. Liu, Realization of enhanced light directional beaming via a Bull's eye structure composited with circular disk and conical tip. *Optics Communications* **339**, 216–221 (2015).

[37] J.-X. Tan, Y.-B. Xie, J.-W. Dong, H.-Z. Wang, Flat-top transmission band in periodic plasmonic ring resonators. *Plasmonics* **7**, 435–439 (2012).

[38] C. Saeidi, D. van der Weide, Nanoparticle array based optical frequency selective surfaces: Theory and design. *Optics Express* **21**, 16170–16180 (2013).

[39] C. Saeidi, D. van der Weide, Synthesizing frequency selective metasurfaces with nanodisks. *Applied Physics Letters* **103**, 183101 (2013).

[40] M. Diem, T. Koschny, C. M. Soukoulis, Wide-angle perfect absorber/thermal emitter in the terahertz regime. *Physical Review B* **79**, 033101 (2009).

[41] X. Liu, T. Starr, A. F. Starr, W. J. Padilla, Infrared spatial and frequency selective metamaterial with near-unity absorbance. *Physical Review Letters* **104**, 207403 (2010).

[42] J. F. Federici, B. Schulkin, F. Huang, D. Gary, R. Barat, F. Oliveira, D. Zimdars, THz imaging and sensing for security applications—explosives, weapons and drugs. *Semiconductor Science and Technology* **20**, S266 (2005).

[43] J. B. Pendry, D. Schurig, D. R. Smith, Controlling electromagnetic fields. *Science* **312**, 1780–1782 (2006).

[44] N. Landy, S. Sajuyigbe, J. Mock, D. Smith, W. Padilla, Perfect metamaterial absorber. *Physical Review Letters* **100**, 207402 (2008).

[45] J. Hao, J. Wang, X. Liu, W. J. Padilla, L. Zhou, M. Qiu, High performance optical absorber based on a plasmonic metamaterial. *Applied Physics Letters* **96**, 251104 (2010).

[46] X. Liu, T. Tyler, T. Starr, A. F. Starr, N. M. Jokerst, W. J. Padilla, Taming the blackbody with infrared metamaterials as selective thermal emitters. *Physical Review Letters* **107**, 045901 (2011).

[47] J. Zhou, A. F. Kaplan, L. Chen, L. J. Guo, Experiment and theory of the broadband absorption by a tapered hyperbolic metamaterial array. *ACS Photonics* **1**, 618–624 (2014).

[48] F. Ding, Y. Cui, X. Ge, Y. Jin, S. He, Ultra-broadband microwave metamaterial absorber. *Applied Physics Letters* **100**, 103506 (2012).

[49] Q. Liang, T. Wang, Z. Lu, Q. Sun, Y. Fu, W. Yu, Metamaterial-based two dimensional plasmonic subwavelength structures offer the broadest waveband light harvesting. *Advanced Optical Materials* **1**, 43–49 (2013).

[50] Q. Zhang, L. Bai, Z. Bai, P. Hu, C. Liu, Theoretical analysis and design of a near-infrared broadband absorber based on EC model. *Optics Express* **23**, 8910–8917 (2015).

[51] N. Zhang, P. Zhou, L. Zhang, X. Weng, J. Xie, L. Deng, Ultra-broadband absorption in mid-infrared spectrum with graded permittivity metamaterial waveguide structure. *Applied Physics B* **118**, 409–415 (2015).

[52] Q. Zhang, L. Bai, X. Liu, C. Liu, X. Cui, Simplified transparent conductive Oxides-based ultrabroadband absorber design. *Journal of Lightwave Technology* **34**, 1354–1359 (2016).

[53] G. V. Naik, J. Kim, A. Boltasseva, Oxides and nitrides as alternative plasmonic materials in the optical range [Invited]. *Optical Materials Express* **1**, 1090–1099 (2011).

[54] F. Abbasi, N. Engheta, Roles of epsilon-near-zero (ENZ) and mu-near-zero (MNZ) materials in optical metatronic circuit networks. *Optics Express* **22**, 25109–25119 (2014).

[55] G. L. Matthaei, L. Young, E. M. T. Jones, *Microwave filters, impedance-matching networks, and coupling structures*. (Artech House, Dedham, Massachusetts, 1980).

Chapter 11

Nanoplasmonic Waveguides Filled with Electro-Optical Materials

Montasir Qasymeh

Additional information is available at the end of the chapter

http://dx.doi.org/10.5772/67370

Abstract

In this chapter, nanoplasmonic waveguides filled with electro-optical materials are proposed and discussed. Rigorous theoretical modelling is developed to describe the propagation, generation and control of electromagnetic fields confined in these nanoscale waveguides. Two configurations are studied thoroughly. First, a metal-insulator-metal (MIM) nanoplasmonic waveguide filled with lithium niobate (i.e., $LiNbO_3$) is considered for compact terahertz (THz) generation. The waveguide is designed to generate Gaussian THz waves by the means of frequency down-conversion of two surface plasmon polariton (i.e., SPP) modes. THz generation is shown to be viable over the entire range from 1 to 10 THz by properly designing the SPP wavelengths and waveguide dimensions. Future applications of such nanoscale THz sources include nanocommunication systems and body-centric networks. Secondly, an MIM nanoplasmonic waveguide filled with doped $LiNbO_3$ is considered. The interaction between two interfering SPP modes is studied. It is shown that a strong symmetric SPP mode can be coupled to a weak antisymmetric SPP mode by the means of photorefractive effect. Future advances include implementing known photorefractive applications (such as interferometry and holography) in the nanoplasmonic field. The work of this chapter highlights the potential of functioning electro-optical materials in nanoplasmonic waveguides to achieve novel ultra-compact and efficient devices.

Keywords: plasmonic waveguides, electro-optic material, lithium niobate, terahertz, photorefractive effect

INTECH

open science | open minds

1. Introduction

Nanoplasmonic waveguides offer unprecedented tight confinement of electromagnetic waves in subwavelength structures. In principle, nanoplasmonic waveguides comprise several metal-dielectric layers. The propagating light is coupled to the free electrons in the metal, which acts like plasma at the optical frequency. It then follows that the propagating light can be strongly confined even to 100 times smaller than its wavelength. Such tight confinement has the potential to open a new world of scalability and integration. For instance, nanoplasmonic-based devices can be utilized as matching interconnectors between traditional micro-photonics and nanoelectronic devices [1–5]. This paves the way for the development of efficient, compatible and ultrafast chips [6–8]. However, the achievable tight confinement in the nanoplasmonic waveguides can be functioned to boost the propagating waves' intensities. It then follows that the nonlinear effects can be significantly enhanced, and thus, desired functionalities can be achieved. To date, several advances in plasmonic technology have been reported. These include all optical nanoplasmonic logic gates [9], electrically controlled plasmonic devices [10], plasmonic amplification and lasing [11], efficient second harmonic generation using plasmonic waveguides [12], silicon-on-insulator-compatible plasmonic devices [13], focus and enhancement of terahertz (THz) radiation using plasmonic waveguides [14], efficient and ultra-compact hybrid plasmonic devices [15], solitonic plasmonic waveguides [16] and mode conversion in plasmonic waveguides [17], to mention a few examples.

In this chapter, nanoplasmonic waveguides filled with electro-optical materials are proposed and discussed. Specifically, metal-insulator-metal (MIM) plasmonic waveguides filled with lithium niobate ($LiNbO_3$) are considered. The theoretical models that describe the nonlinear interaction of the confined electromagnetic waves in nanoplasmonic waveguides are developed. Numerical evaluations based on experimentally reported parameters are provided. Two different configurations are studied thoroughly. The first configuration is a MIM nanoplasmonic waveguide designed for compact THz generation [18]. The optical pumps are guided by means of surface plasmon polaritons (SPPs), whereas the generated THz waves are designed to diffract and propagate outside the MIM waveguide. The design of this work attains the following desired properties: first, the THz absorption is minimized. Secondly, the proposed structure is compact and relevant for nanoapplications. Thirdly, the generation efficiency is maximized by reducing losses rather than introducing resonant conditions (i.e. phase matching), and thus, wide THz generation is achieved. We note that this proposal introduces a new modality from the standpoint of applications. For instance, the proposed MIM nanostructure can be immersed inside a target (such as a cell or a biological entity), and the generated THz waves can interact with the (under-test) surrounding medium of the MIM nanostructure, while the optical modes are tightly confined inside the MIM nanostructure. Potential future applications include nanocommunication networks and body-centric systems. The second configuration considered in this work includes MIM plasmonic waveguide filled with doped $LiNbO_3$ [19]. The main interest is to study the interaction of the interfering SPP modes with the doping impurities. To model the interaction analytically, small intensity modulation interference depth is considered and a perturbation approach is employed. The evolution of the interfering modes is obtained to quantify the interaction. On considering strong pump and weak signal scenario, it is found that the two SPP modes are coupled by means of the photorefractive effect. The modal gain is calculated to characterize the photorefractive effect. First, an ideal case of lossless waveguides is considered, and the gain versus the waveguide length, the doping concentration and the input amplitudes are characterized. Secondly, the modal losses are taken into account. It is found that a weak antisymmetric mode can experience an effect gain up to certain waveguide lengths in the presence of a strong symmetric mode, considering proper doping concentration and input amplitudes. The coupling effect can be conceived either as an amplification or mode-conversion process, promising novel future application.

The remaining part of this chapter is organized as follows. The propagation modes and the dispersion relation of MIM plasmonic waveguides are introduced in Section 2. The nonlinear and the electro-optical coefficients of the $LiNbO_3$ are introduced in Section 3. THz generation in nanoplasmonic waveguides is discussed in Section 4. The photorefractive effect in nanoplasmonic waveguides is discussed in Section 5. Finally, concluding remarks are presented in Section 6.

2. MIM Plasmonic waveguides

Consider an MIM plasmonic waveguide comprising two metallic layers sandwiching a $LiNbO_3$ of thickness a, as depicted in **Figure 1**. The waveguide is two dimensional and independent of the y-axis.

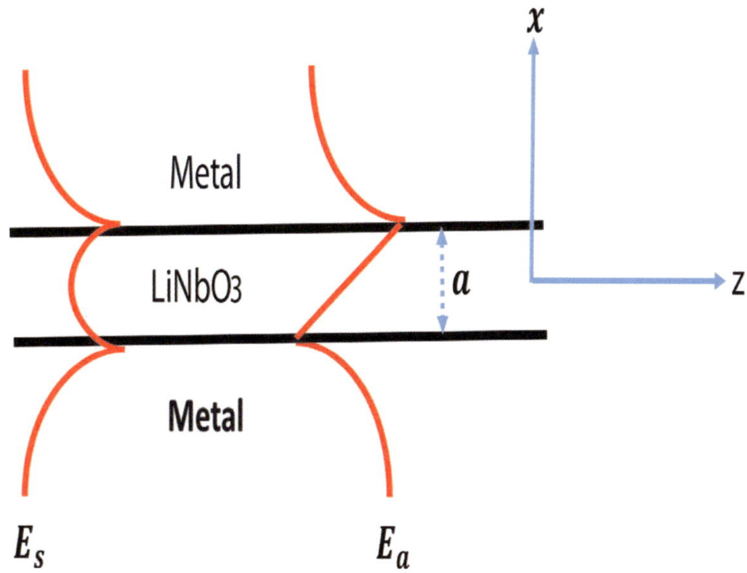

Figure 1. The MIM plasmonic waveguide filled with $LiNbO_3$. The spatial distribution of electric field x component is schematically displayed, in red, for both the symmetric and antisymmetric modes.

The waveguide has two fundamental propagating SPP modes of symmetric and antisymmetric transverse field distribution at specific frequency ω, classified according to the spatial distribution of the $x-$ electric field component. The expression of the electric field associated with the symmetric and antisymmetric modes, i.e. $\vec{E}_s$ and $\vec{E}_a$, are given by [20]:

$$\begin{aligned}\vec{E}_s(x,z,t) &= A_s(z)[v_x(x)\vec{e}_x + v_z(x)\vec{e}_z]e^{j(\beta_s z-\omega t)} + c.c\\ \vec{E}_a(x,z,t) &= A_a(z)[D_x(x)\vec{e}_x + D_z(x)\vec{e}_z]e^{j(\beta_a z-\omega t)} + c.c\end{aligned} \tag{1}$$

where A_s (and A_a) and β_s (and β_a) are the complex amplitude and propagation constant of the symmetric (and antisymmetric) propagating mode, respectively, $j = \sqrt{-1}$, $c.c.$ is the complex conjugate and $D(z)$ and $v(z)$ are the spatial distributions, given by [20].

$$\begin{aligned}v_x(x) &= \frac{\beta_s}{\omega\varepsilon_0\varepsilon_x}\cosh(k_{c_s}x),\ v_z(x) = \frac{jk_{c_s}}{\omega\varepsilon_0\varepsilon_z}\sinh(k_{c_s}x),\\ D_x(x) &= \frac{\beta_a}{\omega\varepsilon_0\varepsilon_x}\sinh(k_{c_a}x),\ D_z(x) = \frac{jk_{c_a}}{\omega\varepsilon_0\varepsilon_z}\cosh(k_{c_a}x),\end{aligned} \tag{2}$$

Here, $-\frac{a}{2} < z < \frac{a}{2}$, $\varepsilon_{x,z}$ is the $LiNbO_3$ relative permittivity for x and z polarization, respectively, and $k_{c_{s(a)}} = \sqrt{\frac{\varepsilon_z}{\varepsilon_x}\beta^2_{s(a)} - \varepsilon_z k_0^2}$ is the core transverse decay factor for the symmetric (and antisymmetric) SPP mode.

On applying the boundary conditions, the dispersion relation for the symmetric and antisymmetric modes can be, respectively, given by the following [20]:

$$tanh\left(\frac{a}{2}k_{c_s}\right) = -\frac{\varepsilon_z k_{m_s}}{\varepsilon_m k_{c_s}},\quad tanh\left(\frac{a}{2}k_{c_a}\right) = -\frac{\varepsilon_m k_{c_a}}{\varepsilon_z k_{m_a}}, \tag{3}$$

where $k_{m_{s(a)}} = \sqrt{\beta^2_{s(a)} - \varepsilon_m k_0^2}$ is the transverse decay factor in the metal for symmetric (antisymmetric) SPP mode, $k_0 = \frac{2\pi}{\lambda}$ is the propagation constant in the free space, and ε_m is the metal permittivity.

One may rewrite the dispersion relation in the following form [18]:

$$\begin{aligned}k_{c_s}\tanh\left(k_{c_s}\frac{a}{2}\right) &= -\frac{\varepsilon_z}{\varepsilon_m}\left[\frac{\varepsilon_x}{\varepsilon_z}k^2_{c_s} + (\varepsilon_x - \varepsilon_m)k_0^2\right]^{\frac{1}{2}}\\ k_{c_a}\coth\left(k_{c_a}\frac{a}{2}\right) &= -\frac{\varepsilon_z}{\varepsilon_m}\left[\frac{\varepsilon_x}{\varepsilon_z}k^2_{c_a} + (\varepsilon_x - \varepsilon_m)k_0^2\right]^{\frac{1}{2}}\end{aligned} \tag{4}$$

The dispersion form in Eq. (4) can be solved numerically for k_c, for a known frequency ω and slot thickness a. The propagation constant β can then be calculated by $\beta^2_{s,a} = \frac{\varepsilon_x}{\varepsilon_z}k^2_{c_{s,a}} + \varepsilon_x k_0^2$.

3. $LiNbO_3$ coefficients

$LiNbO_3$ is chosen as the electro-optical material for this work. Consider the indices of the $LiNbO_3$ (1, 2, 3) along the axes (x, y, z), respectively. Given that the MIM waveguide is independent of the $y-$axis, the propagating fields have x and z electric field components, denoted by E_x and E_z. The nonlinear polarization induced in the $LiNbO_3$ is given by the following [21]:

$$P_{NL_x} = 4d_{31}E_zE_x, \quad P_{NL_y} = -2d_{22}E_x^2, \quad P_{NL_z} = 2d_{31}E_x^2 + 2d_{33}E_z^2, \tag{5}$$

where P_{NL_J} is the Jth nonlinear polarization component and $d_{22} = 3$ pm/v, $d_{31} = 5$ pm/v, and $d_{33} = 33$ pm/v are the nonlinear coefficients. Several phenomena are attributed to the nonlinear polarization, including new frequency generation, frequency conversion and polarization rotation, to mention a few. The evolution of the propagating fields is governed by the nonlinear wave equation that involves the nonlinear polarization. An example of THz generation based on induced nonlinear polarization in MIM nanoplasmonic waveguide is discussed in the next section.

The electro-optic coefficients are related to the nonlinear coefficients by the following:

$$r_{It} = -\frac{4\pi}{n_x^2 n_z^2} d_{It}, \tag{6}$$

where n_x and n_z are the principal refractive indices.

The effective permittivity of the $LiNbO_3$ depends on the electric field components, by means of electro-optic effect, through the following relation [21]:

$$\begin{aligned} \frac{1}{\varepsilon_{x_{\text{eff}}}} &= \frac{1}{\varepsilon_x} + (r_{15} - r_{22})E_x + r_{13}E_z \\ \frac{1}{\varepsilon_{z_{\text{eff}}}} &= \frac{1}{\varepsilon_z} + r_{15}E_x + r_{33}E_z \end{aligned} \tag{7}$$

where ε_{eff} is the effective permittivity of the $LiNbO_3$ and ε is the intrinsic permittivity of the $LiNbO_3$.

On considering a small induced change in the permittivity, so that $(r_{15} - r_{22})E_x + r_{13}E_x \ll \frac{1}{\varepsilon_x}$ and $r_{15}E_x + r_{33}E_x \ll \frac{1}{\varepsilon_z}$, the effective permittivity can be approximated by the following [19]:

$$\begin{aligned} \varepsilon_{x_{\text{eff}}} &= \varepsilon_x - \varepsilon_x^2[(r_{15} - r_{22})E_x + r_{13}E_z], \\ \varepsilon_{z_{\text{eff}}} &= \varepsilon_z + \varepsilon_z^2[r_{15}E_x + r_{33}E_z], \end{aligned} \tag{8}$$

In Section 5, an example of photorefractive effect in MIM plasmonic waveguide based on the electro-optic effect is discussed.

4. THz generation in nanoplasmonic waveguides

In this section, an MIM plasmonic waveguide filled with $LiNbO_3$ and with two propagating SPP modes is considered. The frequency difference between the two SPP modes is properly designed for THz generation. Theoretical modelling and numerical evaluations are presented.

4.1. Field expressions

As the symmetric mode is a fundamental mode of the MIM waveguides, two symmetric modes with distinct frequencies ω_1 and ω_2 are considered for this work. It then follows that a single mode at each specific frequency can be propagated given a proper waveguide thickness a. The associated electric fields have x and z polarization components, given by the following:

$$\begin{aligned} E_{zi}(x,z,t) &= A_{si}(z)\ v_{z_i}(x)\ e^{j(\beta_{si}z-\omega_i t)}\ \vec{e}_z + c.c. \\ E_{xi}(x,z,t) &= A_{si}(z)\ v_{x_i}(x)\ e^{j(\beta_{si}z-\omega_i t)}\ \vec{e}_x + c.c. \end{aligned} \tag{9}$$

Here $i \in \{1,2\}$.

The THz wave is generated by means of difference-frequency generation of the two SPP modes enabled by the $LiNbO_3$ nonlinearity.

The MIM structure is designed to guide the SPP pumps, whereas the generated THz waves are designed to diffract and propagate outside the MIM waveguide, as shown in **Figure 2**. The thickness of the metallic layer is equal to $d = \frac{5}{k_m}$. It then follows that 99% of the optical SPP modes are confined inside the MIM waveguide, and thus one may conceive the metallic layers to be of infinite thickness. However, because THz waves are of wavelengths much greater than the MIM waveguide dimension, and given that THz waves cannot be guided as SPP modes [22], the generated THz waves will diffract into the surrounding medium and evolve as a Gaussian beam. The generated THz waves can be modelled as a transverse electromagnetic (TEM) Gaussian beam, given by [23] and [24]:

$$E_{\mathrm{THz}}(x,z,t) \ = \ A_{\mathrm{THz}}(z)G(x)e^{j(\beta_3 z-\omega_3 t)}\ \vec{e}_x \ + \ c.c, \tag{10}$$

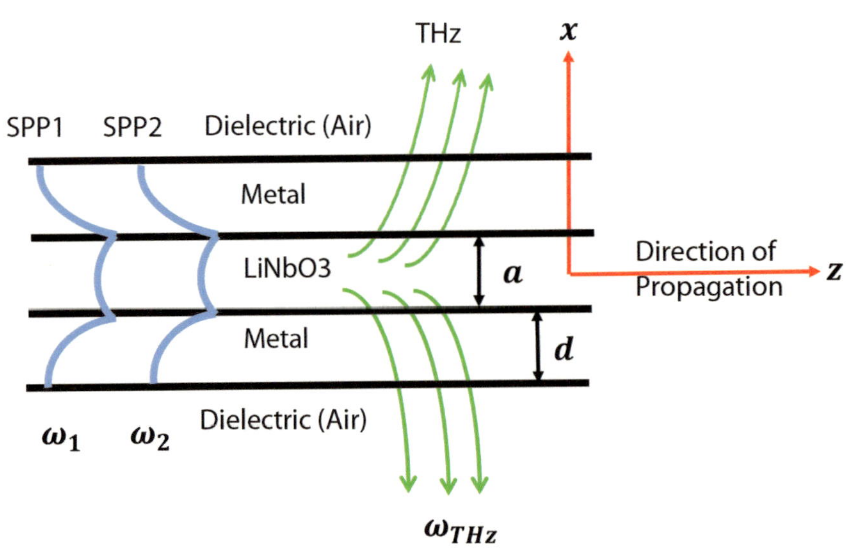

Figure 2. The MIM plasmonic waveguide for THz generation.

where $G(x) = \frac{(2/\pi)^{\frac{1}{4}}}{\sqrt{W_0\gamma(z)}}\exp\left(-\frac{x^2}{\gamma(z)W_0{}^2}\right)$, A_{THz} is the complex amplitude, W_0 is the beam spot at $z = 0$, $\gamma(z) = 1 + jz/Z_R$, and Z_R is the Ray Leigh rang given by $\frac{W_0^2\pi}{\lambda}$.

4.2. Nonlinear polarization

The induced nonlinear polarization components are at the frequencies ω_1, ω_2, $\omega_1 - \omega_2$, and $\omega_1 + \omega_2$. Here, the frequency difference between the two optical frequencies ω_1 and ω_2 is designed to lie in the THz range. It must be noted that this generated THz wave at $\omega_1 - \omega_2$ cannot be guided by the plasmonic MIM waveguide [22], as previously mentioned. The other generated harmonic at $\omega_1 + \omega_2$ lies in the optical frequency range, close to the second harmonic frequency, and is guided as SPP mode [25]. It then follows that the wave at $\omega_1 + \omega_2$ suffers from strong attenuation. Thus, the nonlinear polarization component at frequency $\omega_1 + \omega_2$ will be ignored in the following analysis. On the substitution of the fields of Eqs. (9) and (10) into Eq. (5), the nonlinear polarization components can be obtained. The nonlinear polarization components at ω_1 are as follows:

$$\begin{aligned} P_{NLz} &= 4d_{31}A_{s2}A_{THz}G(x)v_{z2}(x) \quad e^{j[(\beta_{s2}+\beta_{THz})z-\omega_1 t]}\vec{e}_z + c.c., \\ P_{NLx} &= 4d_{33}A_{s2}A_{THz}G(x)v_{x2}(x) \quad e^{j[(\beta_{s2}+\beta_{THz})z-\omega_1 t]}\vec{e}_x + c.c., \end{aligned} \tag{11}$$

the nonlinear polarization components at ω_2 are

$$\begin{aligned} P_{NLz} &= 4d_{31}A_{s1}A^*_{THz}G^*(x)v_{z1}(x) \quad e^{j[(\beta_{s1}-\beta_{THz})z-\omega_2 t]}\vec{e}_z + c.c., \\ P_{NLx} &= 4d_{33}A_{s1}A^*_{THz}G^*(x)v_{x1}(x) \quad e^{j[(\beta_{s1}-\beta_{THz})z-\omega_2 t]}\vec{e}_x + c.c., \end{aligned} \tag{12}$$

and the nonlinear polarization component at $\omega_{THz} = \omega_1 - \omega_2$ is

$$P_{NL_x} = 4A_{s1}A^*_{s2}[d_{31}v_{z_1}v^*_{z_2} + d_{33}v_{x_1}v^*_{x_2}]\times e^{j[(\beta_{s1}-\beta_{s2})z-(\omega_1-\omega_2)t]}\vec{e}_x + c.c., \tag{13}$$

The $z-$component of the nonlinear polarization at ω_{THz} is not considered because the generated THz wave is a TEM Gaussian wave that propagates in the $z-$ direction. Therefore, the THz wave has a zero $z-$component electric field. In other words, the $x-$component of the nonlinear polarization, i.e. P_{NL_x}, is the only source that contributes to the evolution of the generated THz wave.

4.3. THz generation

The evolution of the generated THz wave and the SPP modes are governed by the nonlinear wave equation, given by:

$$\nabla^2 E = \mu_0\varepsilon\frac{\partial^2 E}{\partial t^2} + \mu_0\frac{\partial^2 P_{NL}}{\partial t^2}, \tag{14}$$

On the substitution of the fields of Eqs. (9) and (10) into Eq. (14), and using the nonlinear polarization expressions of Eqs. (11–13), the corresponding slowly varying envelope approximation (SVEA) equations can be formed, given by the following [18]:

$$\frac{\partial A_{s1}}{\partial z} = j\frac{2\,\omega_1^2}{\beta_1 c^2} A_{\mathrm{THz}} A_{s1}\left(d_{31}\frac{\langle v_{z_1}/G\, v_{z_2}\rangle}{\langle v_{z_1}/v_{z_1}\rangle} + d_{33}\frac{\langle v_{x_1}/G\, v_{x_2}\rangle}{\langle v_{x_1}/v_{x_1}\rangle}\right) e^{j(\beta_{s2}+\beta_{\mathrm{THz}}-\beta_{s1})z}, \tag{15}$$

$$\frac{\partial A_{s2}}{\partial z} = j\frac{2\,\omega_2^2}{\beta_2 c^2} A_{s1} A_{\mathrm{THz}}^{*}\left(d_{31}\frac{\langle v_{z_2} G/v_{z_1}\rangle}{\langle v_{z_2}/v_{z_2}\rangle} + d_{33}\frac{\langle v_{x_2} G/v_{x_1}\rangle}{\langle v_{x_2}/v_{x_2}\rangle}\right) e^{j(\beta_{s1}-\beta_{\mathrm{THz}}-\beta_{s2})z}, \tag{16}$$

$$\frac{\partial A_{\mathrm{THz}}}{\partial z} = j\frac{2\omega_{\mathrm{THz}}^2}{c^2} A_{s1} A_2^{*}\frac{d_{31}\langle v_{z_2} G/v_{z_1}\rangle + d_{33}\langle v_{x_2} G/v_{x_1}\rangle}{\left(\beta_{\mathrm{THz}} - \frac{1}{2\gamma Y_R}\right)\langle G/G\rangle + \frac{\langle G/z^2 G\rangle}{\gamma^2 W_0^2 Y_R}} e^{j(\beta_{s1} - \beta_{s2} - \beta_{\mathrm{THz}})z}, \tag{17}$$

where $\langle \phi/\delta\rangle = \int_{\frac{a}{2}}^{+\frac{a}{2}} \phi^*(\zeta)\delta(\zeta)\partial\zeta$.

The evolution of the THz and the SPP fields can be obtained by numerically solving the SVEA Eqs. (15–17). However, various losses must first be precisely evaluated. These include SPP and THz losses. The decay factor of the SPP linear losses can be calculated from the imaginary part of the propagation constant, i.e. $\alpha_{L_i} = Im\{\beta_i\}$, by solving Eq. (4). These SPP linear losses can be taken into account by incorporating the corresponding decay factor α_{L_i} in the SVEA Eqs. (15) and (16). The SPP nonlinear losses are described by the decay factor α_{NL}, given by $\alpha_{\mathrm{NL}} = \gamma_{\mathrm{NL}}\cdot I$ [26]. Here, $\gamma_{NL} = 2.7\times 10^{-5}$ m/W is the nonlinear absorption coefficient of $LiNbO_3$ crystal, and I is the power of the SPP modes. Our investigations show that the nonlinear losses in this work are much smaller than the linear losses, and can be ignored despite the large SPP intensities because the waveguide lengths considered in this work are very small and lie in the micrometer range. However, the THz losses are also mainly linear, given the low generated intensities. The effective decay factor of the THz losses is given by the following:

$$\alpha_{\mathrm{THz}} = \left(\alpha_{\mathrm{THz\,LN}} \int_{-\frac{a}{2}}^{\frac{a}{2}} G(x)\partial x + 2\alpha_{\mathrm{THz\,Metal}} \int_{\frac{a}{2}}^{d+\frac{a}{2}} G(x)\partial x \right) \Big/ \int_{-W}^{+W} G(x)\partial x, \tag{18}$$

where $\alpha_{\mathrm{THz}_{\mathrm{Metal}}}$ is the metal THz absorption coefficient, $\alpha_{\mathrm{THz}_{\mathrm{LN}}}$ is the $LiNbO_3$ THz absorption coefficient and W is the Gaussian beam spot given by $W = W_0\sqrt{1 + \left(\frac{z}{z_R}\right)^2}$. To take the THz losses into account, one must incorporate the effective THz decay factor of Eq. (18) into the SVEA Eq. (17). Experimental values of metal and $LiNbO_3$ absorption coefficients can be found in the literature. For example, in [27] the absorption coefficient of gold is given by $\alpha_{\mathrm{THz}_{\mathrm{Metal}}} = 6\times 10^6\ \mathrm{m}^{-1}$, and in [28] the absorption coefficient of the $LiNbO_3$ is given by $\alpha_{\mathrm{THz}_{\mathrm{LN}}} = 2\times 10^4\ \mathrm{m}^{-1}$, both for the 4 THz frequency. It must be noted that the total THz losses are minimized in this work, as can be inferred from Eq. (18). This is because the generated THz waves diffract rapidly to propagate outside of the MIM plasmonic waveguide even for a few 10's of nanometers of propagation. For instance, considering a 4 THz wave, the beam spot evolves from 120 nm (which is the slot thickness, i.e. a) to approximately 10 μm by propagating only 50 nanometers.

4.4. Numerical evaluations

In the following analysis, we consider gold for the two metallic layers, a slot thickness of $a = 120$ nm, a metal thickness of $d = 110$ nm and $A_{s1}(0) = A_{s2}(0) = 1$ V/μm. In **Figure 3**, the power of a 4 THz generated wave is shown versus the waveguide length. Here, $\lambda_1 = 1550$ nm and $\lambda_2 = 1582$ nm. The power is calculated per unit width, defined by $p_i = \frac{1}{2}\varepsilon_0 c n_i |A_{si}|^2 \int_{-\infty}^{+\infty} v_i v_i^* \partial x$ and $p_{\mathrm{THz}} = \frac{1}{2}\varepsilon_0 c n_{\mathrm{THz}} |A_{\mathrm{THz}}|^2 \int_{-\infty}^{\infty} GG^* \partial x$, where the SPP power is $p_1 + p_2$.

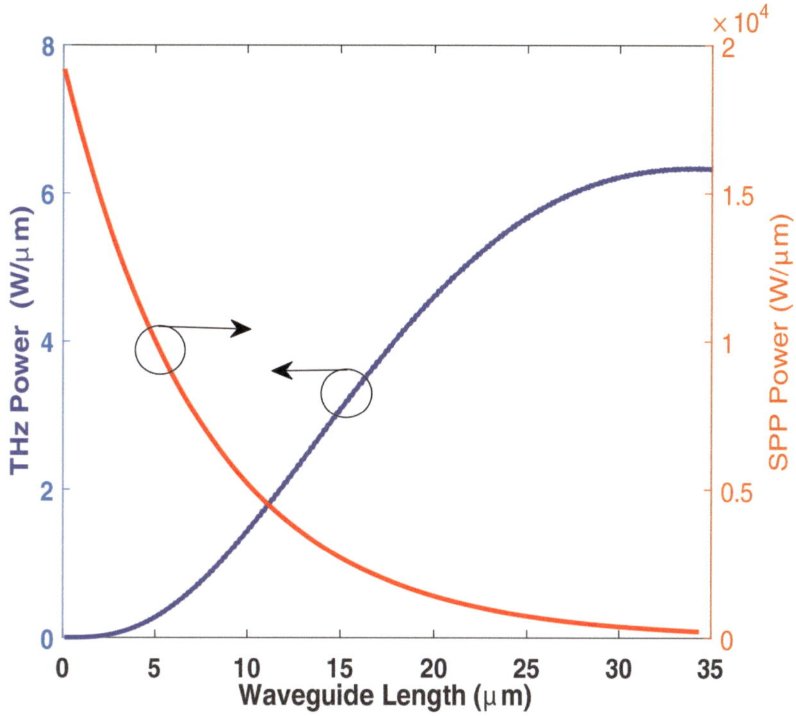

Figure 3. Power of the generated 4 THz wave, and SPPs power versus waveguide length.

As can be seen in **Figure 3**, the effective waveguide length is limited to 35 μm and most of the SPP power is dissipated by losses.

The optics-to-THz conversion efficiency can be defined as $\eta = \frac{p_{\mathrm{THz}}(L)}{p_1(0)+p_2(0)}$, where $p_{\mathrm{THz}}(L)$ is the THz power per unit width, L is the MIM waveguide length, and $p_{1,2}(0)$ is the input SPP powers per unit width.

In **Figure 4**, the conversion efficiency for a 4 THz generated wave is presented versus the input SPP power. Here, we consider a waveguide length of $L = 35$ μm, $\lambda_1 = 1550$ nm and

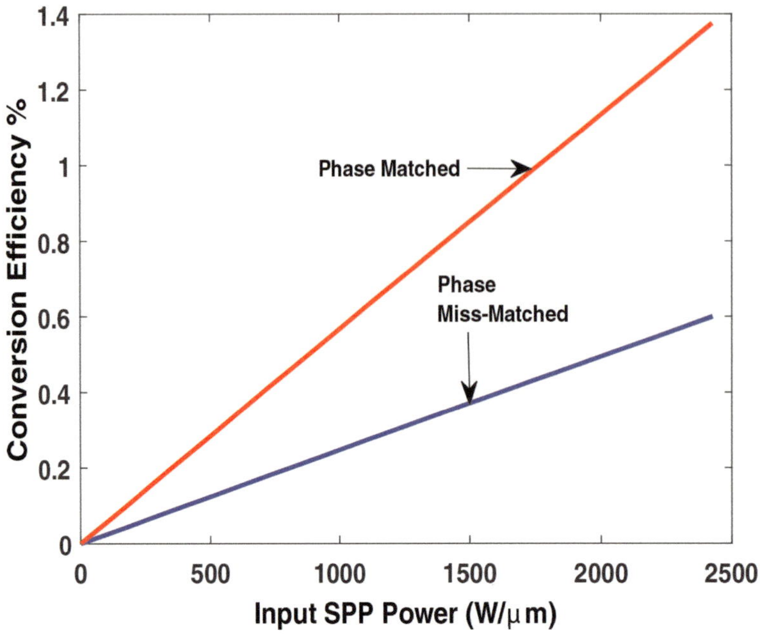

Figure 4. Conversion efficiency for a 4 THz wave versus total SPP input power. The two scenarios of phase matching and phase mismatching are shown for comparison.

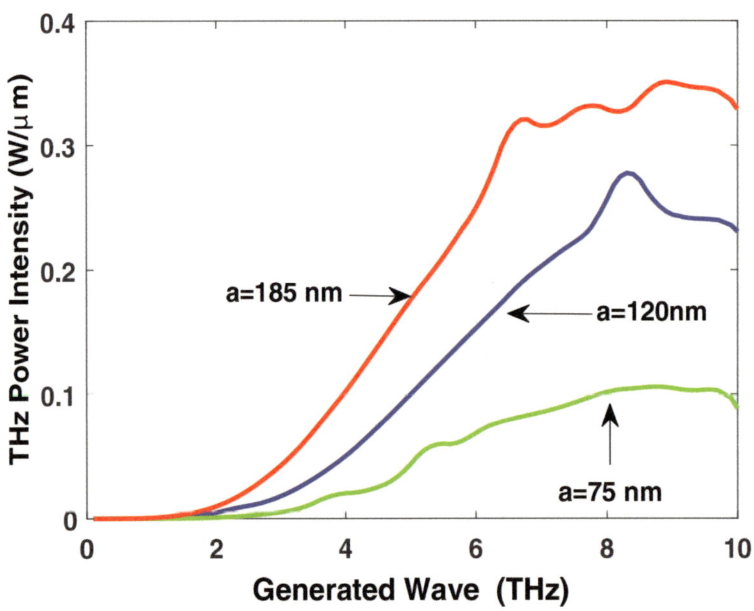

Figure 5. Power of the generated THz wave versus its frequency. Different slot thicknesses are considered. The optical wavelength λ_1 is fixed at the value $\lambda_1 = 1550$ nm. The waveguide length is $L = 35$ μm.

$\lambda_2 = 1582$ nm. Additionally, a hypothetical case of phase-matched generation is presented (by setting $\beta_1 - \beta_2 - \beta_{THz} = 0$) for comparison. As can be seen, the phase-matching enhancement is maximally approximately 4 THz because the interaction is limited by losses. For instance, for 4 THz generation and $\lambda_1 = 1550$ nm, the beat length L_B (defined by $L_B = \frac{2\pi}{\beta_{s1}-\beta_{s2}-\beta_{THZ}}$) is approximately 42 μm, whereas the SPP propagation length (defined by $L_{SPP} = (2\alpha)^{-1}$) is approximately 25 μm. Thus, it is intuitive to expect limited enhancement of phase-matched operation.

The frequency of the generated THz wave can be tuned by controlling the frequency difference between the two SPP modes, i.e. $f_1 - f_2$. In **Figure 5**, the wavelength 1 is fixed at $\lambda_1 = 1550$ nm, whereas wavelength 2 is tuned to generate different THz wave frequencies from 1 to 10 THz. As can be seen from **Figure 5**, THz waves can be generated over the entire range from 1 to 10 THz simply by tuning the frequency ω_2. Furthermore, different slot waveguide thicknesses are considered in **Figure 5**. These include $a = 75$, $a = 120$ and $a = 185$ nm. It can be seen that, for the same SPP input, smaller slot thickness result in weaker generated THz waves. This is due to higher losses for smaller slot thickness. Thus, as expected, more compact generation is achieved at the cost of higher losses.

5. Photorefractive effect in nanoplasmonic waveguides

In this section, the $LiNbO_3$ filling the MIM waveguide is doped with N_D donor and N_A acceptor atom impurities. The interaction between two (symmetric and antisymmetric) interfering SPP modes is studied.

5.1. Intensity expression

The electric fields of the two symmetric and antisymmetric propagating modes are described by Eq. (1). The optical field intensity inside of the waveguide, $I = (\vec{E}_s + \vec{E}_a)(\vec{E}_s^* + \vec{E}_a^*)$, can be formed as follows:

$$I(x,z,t) = I_0(x) + \mathrm{Re}\{I_1(x)e^{j\Delta\beta z}\} \tag{19}$$

where $\Delta\beta = \beta_a - \beta_s$, $I_o(x,z) = T_1(x)|A_a(z)|^2 + T_2(x)|A_s(z)|^2$, $I_1(x,z) = T_3(x)\, A_a(z)\, A_s^*(z)$, $T_1(x) = D_x(x)\, D_x^*(x) + D_z(x)\, D_z^*(x)$, $T_2(x) = v_x(x)\, v_x^*(x) + v_z(x)\, v_z^*(x)$, and $T_3(x) = 2D_x(x)\, v_x(x) + 2D_z(x)\, v_z(x)$.

On adapting a strong pump and weak signal scenario, i.e. $I_1 \ll I_0$, a perturbation approach can be implemented to model the evolution of the propagating fields.

5.2. Band transport model

The interaction of the interfering SPP modes with $LiNbO_3$ impurities is governed by the standard band transport model that encompasses the electron continuity equation, the current density equation and the Poisson's equation, as in the following [29]:

$$\frac{\partial N}{\partial t} - \frac{\partial N_D^i}{\partial t} = \frac{1}{e}\nabla.(J), \tag{20}$$

$$\frac{\partial N_D^i}{\partial t} = (N_D - N_D^i)SI - \gamma_R N N_D^i, \tag{21}$$

$$\nabla.\varepsilon E_{sc} = -e(N + N_A - N_D^i), \tag{22}$$

where N is the free electron density, N_D^i is the ionized donor density, e is the electron charge, $J = eN\mu E_{SC} + k_B T \mu \nabla N + \kappa(N_D - N_D^i)sI\hat{e}_z$ is the conduction current density, μ is the $LiNbO_3$ electron mobility, k_B is the Boltzmann constant, T is the temperature, κ is the photovoltaic constant, E_{SC} is the space charge electric field, and ε is the $LiNbO_3$ permittivity. The three coupled equations of the band transport model shown above can be solved analytically by obeying a perturbation approach for a small depth intensity modulation limit i.e. $I_0 \gg I_1$. Towards solving these three coupled equations (20)–(22) for a continuous wave case, and considering small intensity modulation, the electron density N and the ionized donor density N_D^i can be written in the following form:

$$\begin{aligned} N_D^i(x,z) &= N_{D0}^i(x) + \mathrm{Re}\left(N_{D1}^i(x)\, e^{j\Delta\beta z}\right) \\ N(x,z) &= N_0(x) + \mathrm{Re}\left(N_1(x)\ e^{j\Delta\beta z}\right) \end{aligned} \tag{23}$$

where N_0 and N_{D0}^i are the averages of electron and ionized donor densities, and N_1 and N_{D1}^i are the modulation amplitudes of electron and ionized donor densities, respectively.

The space charge electric field E_{SC} can be uniquely specified through Eq. (22) and the electrostatic condition, i.e. $\nabla \times E_{SC} = 0$.

Consequently, on substituting Eq. (23) into Eq. (22), and using the electrostatic condition, one can obtain the following expression [19]:

$$E_{SC}(x,z) = c_{SC}[N_1 - N_{D1}^i]e^{j\Delta\beta z}\vec{e}_z + \frac{c_{SC}}{j\Delta\beta}\frac{\partial[N_1 - N_{D1}^i]}{\partial x}e^{j\Delta\beta z}\vec{e}_x + c.c., \tag{24}$$

Where $c_{SC} = \frac{-ej\Delta\beta}{\varepsilon_0\varepsilon_x(k_{c_s}+k_{c_a})^2+\varepsilon_0\varepsilon_z(j\Delta\beta)^2}$

By substituting Eq. (23) into Eqs. (20) and (21), and considering a steady state so that the time rates of the free electrons and the ionized donor densities are zeros (i.e. $\frac{\partial N}{\partial t} = \frac{\partial N_D^i}{\partial t} = 0$) [29], the expressions for N_0, N_{D0}^i, N_1 and N_{D1}^i can be obtained. First, by substituting Eq. (23) in Eqs. (20) and (21), equating the average terms, and using $N_0 + N_A - N_{D0}^i = 0$, one obtains the following:

$$N_0 - \frac{-(N_A\gamma_R + SI_0) + \sqrt{(N_A\gamma_R + SI_0)^2 + 4\gamma_R SI_0(N_D - N_A)}}{2\gamma_R}, \tag{25}$$

$$N_{D0}^i = \frac{(N_A\gamma_R - SI_0) + \sqrt{(N_A\gamma_R + SI_0)^2 + 4\gamma_R SI_0(N_D - N_A)}}{2\gamma_R}, \tag{26}$$

Secondly, by substituting Eq. (23) into Eqs. (20) and (21) and equating the modulation terms (i.e. those with $e^{i\Delta\beta z}$), one obtains:

$$N_1(x) = \frac{a_1\, b_2 + a_2\, b_3}{a_3\, b_2 - a_2\, b_1}\, I_1(x), \tag{27}$$

$$N_{D1}^i(x) = \frac{a_1\, b_1 + a_3\, b_3}{a_2\, b_1 - a_3\, b_2}\, I_1(x), \tag{28}$$

where $a_1 = s(N_D - N_{D0}^i)$, $a_2 = N_0\gamma_R + SI_0$, $a_3 = N_{D0}^i\gamma_R$, $b_1 = \mu\left((k_{c_s} + k_{c_a})^2 + (j\Delta\beta)^2\right) \times \left(k_B T - \frac{e^2 N_0}{\varepsilon_0\varepsilon_x (k_{c_s}+k_{c_a})^2 - \varepsilon_0\varepsilon_z (j\Delta\beta)^2}\right)$, $b_2 = -\kappa s(j\Delta\beta)^2 + \frac{e^2 N_0 \mu\left((k_{c_s}+k_{c_a})^2 + (j\Delta\beta)^2\right)}{\varepsilon_0\varepsilon_x (k_{c_s}+k_{c_a})^2 - \varepsilon_0\varepsilon_z (j\Delta\beta)^2}$ and $b_3 = \kappa s(N_D - N_{D0}^i)(j\Delta\beta)$.

The spatial electric field E_{SC} can be evaluated by performing the following three steps. First, one must calculate the average electron and ionized donor densities, i.e. N_0 and N_{D0}^i, using Eqs. (25) and (26). Secondly, the modulation amplitudes of electron and ionized donor densities, i.e. N_1 and N_{D1}^i, can be calculated using Eqs. (27) and (28). Finally, the E_{SC} can be evaluated by using Eq. (24). The spatial electric field modifies the effective permittivity of the MIM waveguide, by means of electro-optic effect, causing the SPP modes to be coupled in a phase-matched fashion, a phenomenon known as the photorefractive effect.

5.3. Photorefractive effect

The evolution of the SPP modes are governed by nonlinear wave equation, given by:

$$\nabla^2 \vec{E} - \frac{1}{\varepsilon_0 c} \frac{\partial^2 \vec{D}}{\partial t^2} = 0, \tag{29}$$

where $\vec{D}$ is the displacement current given by $\vec{D} = \varepsilon_0 \varepsilon_{\text{eff}}\, \vec{E}$.

The effective permittivity can be obtained by substituting the spatial electric field, i.e. E_{SC}, of Eq. (24) into Eq. (8), yielding the following:

$$\begin{aligned} \varepsilon_{x_{\text{eff}}} &= \varepsilon_x - \varepsilon_x^1 e^{j\Delta\beta} + c.c, \\ \varepsilon_{z_{\text{eff}}} &= \varepsilon_z - \varepsilon_z^1 e^{j\Delta\beta} + c.c, \end{aligned} \tag{30}$$

where $\varepsilon_x^1 = (\varepsilon_x)^2\left(r_{13}c_{sc}(N_1 - N_{D1}^i) + (r_{15} - r_{22})\frac{c_{sc}}{j\Delta\beta}\frac{\partial(N_1 - N_{D1}^i)}{\partial x}\right)$, and $\varepsilon_z^1 = (\varepsilon_z)^2\left(r_{33}c_{sc}(N_1 - N_{D1}^i) + r_{15}\frac{c_{sc}}{j\Delta\beta}\frac{\partial(N_1 - N_{D1}^i)}{\partial x}\right)$

On the substitution of symmetric and antisymmetric mode expressions of Eq. (1) into Eq. (29) and using Eq. (30), the fields equations (for slowly varying amplitudes) can be obtained, given by the following [19]:

$$\frac{\partial A_s}{\partial z} + \frac{(j\omega)^2 C_1{}^*}{j8c^2\beta_s}\left(r_{33}\varepsilon_z^2\frac{k_{c_a}}{k_{c_s}} - r_{13}\varepsilon_x^2\frac{\beta_a}{\beta_s}\right)|A_a|^2 A_s = 0\,, \tag{31}$$

$$\frac{\partial A_a}{\partial z} + \frac{(j\omega)^2 C_1}{j8c^2\beta_a}\left(r_{13}\varepsilon_x^2\frac{\beta_s}{\beta_a} - r_{33}\varepsilon_z^2\frac{k_{c_s}}{k_{c_a}}\right)|A_s|^2 A_a = 0\,, \tag{32}$$

Where $c_1 = \frac{-2e\Delta\beta}{\varepsilon_0\varepsilon_x (k_{c_s}+k_{c_a})^2 + \varepsilon_0\varepsilon_z (j\Delta\beta)^2}\left(\frac{N_1}{I_1} - \frac{N_{D1}^i}{I_1}\right)\left(\frac{\beta_a\beta_s}{(\omega\varepsilon_0\varepsilon_x)^2} - \frac{k_{c_s}k_{c_a}}{(\omega\varepsilon_0\varepsilon_z)^2}\right)$,

As can be seen in Eqs. (31) and (32), the two SPP modes are coupled in a phase-matched fashion. The coupling is conducted by the spatial space charge electric field generated by the interfering SPP modes in the doped $LiNbO_3$. This is known as the photorefractive effect. The photorefractive effect has been known since the early 1960s [30] in bulk materials and micrometer dielectric waveguides. In this work, the photorefractive effect is investigated in nanoplasmonic waveguides. Numerical evaluations are presented in the following section.

5.4. Numerical evaluations

To illustrate the potential of the photorefractive effect in plasmonic waveguides, typical realistic values are considered. Consider $r_{13} = 11$ pm/V, $r_{33} = 34$ pm/V, $T = 300\text{K}$, $\mu = 7.4 \times 10^{-5}\text{m}^2/(V.s)$, $S = 2.34 \times 10^{-8}\text{m}^2/(s.V^2)$, $\kappa = 1.5 \times 10^{11}(\text{m/V})$ and $\gamma_R = 2.4 \times 10^{-13}\text{m}^3/\text{s}$ [29, 31, 32]. In this work, the blue wavelength $\lambda = 458$ nm is considered, given its known functionality in photorefractive devices. The $LiNbO_3$ permittivity is $\varepsilon_z = 5.177$ and $\varepsilon_x = 5.624$ at this wavelength. Also, the aluminium metal (i.e. Al) is chosen as the cladding material, that has a permittivity of $\varepsilon_m = -29 + 7j$ at $\lambda = 458$ nm.

First, the waveguide losses are ignored and the gain versus the doping concentration and the input amplitudes is characterized. Different slot thicknesses are considered. Secondly, the modal losses are taken into account, considering proper doping concentration and input amplitudes.

Our numerical investigations show that gain can be realized only for weak antisymmetric mode (the signal) co-propagating with strong symmetric mode (the pump). This is because the space charge electric field, that couples the two modes, has an antisymmetric transverse distribution, as can be inferred from Eqs. (2), (19) and (24).

In **Figure 6**, the antisymmetric gain, defined by $A_a(L)/A_a(0)$, is displayed versus the waveguide length L in (a), the doping concentration N_D in (b), the inputs ratio $A_a(0)/A_s(0)$ in (c) and the input pump amplitude $A_s(0)$ in (d). The modal losses are ignored and three different thicknesses (i.e. $a = 50$ nm, $a = 70$ nm and $a = 90$ nm) are considered. In (a), $|A_a(0)| = 0.001 \times |A_s(0)|$ and $N_D = 12 \times 10^{20}\text{m}^{-3}$. In (b), the waveguide length $L = 84$ μm, $|A_s(0)| = 1.935 \times 10^3(\text{V/m})$, and $|A_a(0)| = 0.001 \times |A_s(0)|$. In (c), the waveguide length $L = 84$ nm, $N_D = 12 \times 10^{20}\text{m}^{-3}$. In (d), the waveguide length $L = 84$ nm, $N_D = 12 \times 10^{20}\text{m}^{-3}$ and $A_a(0)/A_s(0) = 10^{-3}$.

As can be seen from the simulations in **Figure 6**, the photorefractive response, characterized by the gain, is qualitatively the same for different waveguide thicknesses, given losses are ignored. However, it can be seen from part (b) that the gain crucially depends on the doping concentration; the stronger the effect, the larger the doping concentration. Furthermore, it can be seen from part (c) that the gain does not dramatically depend on the input ratio, given that the perturbation condition is satisfied, i.e. $A(0)_a/A(0)_s \leq 1$. Finally, it can be seen from part (d) that the gain is crucially depending on the pump input amplitude (i.e. the symmetric mode input amplitude); the stronger the effect, the larger the input amplitude. These properties characterize the photorefractive effect considering ignored losses. Same properties are expected for fields propagating few micrometers of waveguide length. However, for larger waveguide lengths, such as 10's or 100's of micro-meters, altered characteristics are expected, given that losses are dominating for such waveguide length ranges [19].

http://dx.doi.org/10.5772/67370

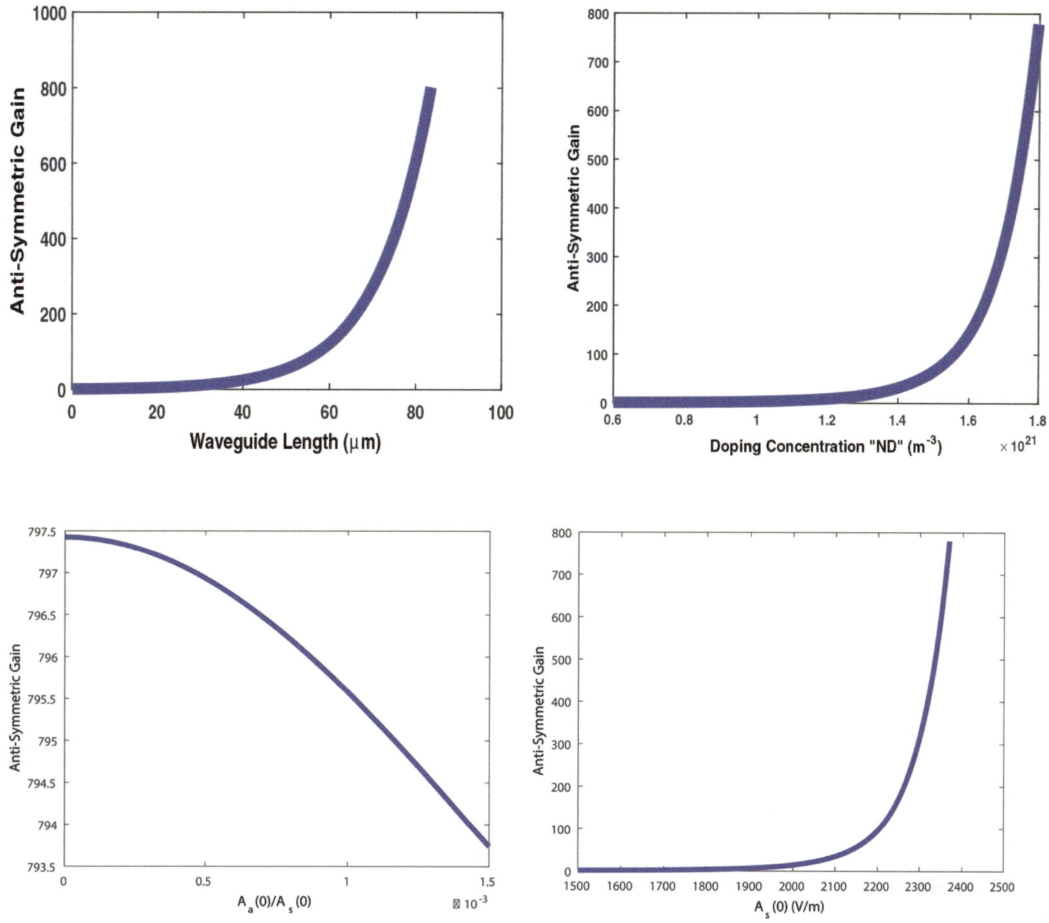

Figure 6. The antisymmetric gain (a) versus waveguide length; (b) versus the doping concentration N_D; (c) versus the input ratio $\frac{A_a(0)}{A_s(0)}$; and (d) versus the input pump field amplitude $A_s(0)$.

Losses can be taken into account by incorporating the effective decay factor, given by $\alpha_{\mathrm{eff}_{s,a}} = Im\{\beta_{s,a}\}$, in Eqs. (31) and (32).

In **Figure 7**, the antisymmetric gain is presented against the waveguide length. Here, losses are taken into account. Three different thicknesses of $a = 50$ nm, $a = 70$ nm and $a = 90$ nm are considered with input pump amplitudes of 4.7×10^3(V/m), 16.5×10^3(V/m), and 18.4×10^3(V/m) are presumed, respectively. The input pump amplitudes are chosen so that the maximum gain is identical for the three thicknesses for the purpose of comparison. As can be seen in **Figure 7**, a significant gain can be experienced despite losses with the maximum gain attainable at the same waveguide length $L = 1.1$ μm. However, a net gain is experienced up to different waveguide lengths for different thicknesses. This is because the photorefractive effect dominates the interaction for the first few micro-meters, whereas losses limit the interaction for larger waveguide lengths, with greater antisymmetric losses for larger waveguide thicknesses.

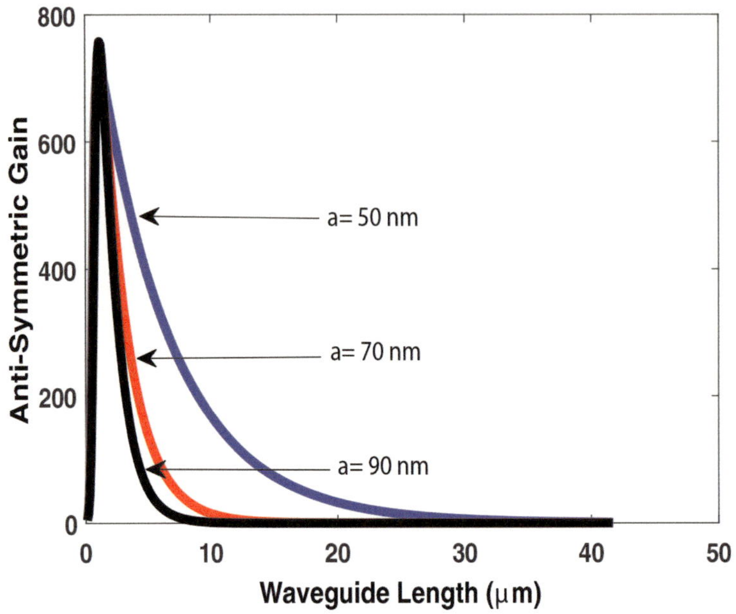

Figure 7. The antisymmetric gain versus waveguide length, for three different $LiNbO_3$ thicknesses, taking losses into account.

Figure 8 depicts the results of **Figure 7** in dB units. As can be seen, given the assumed parameters, the maximum achieved gain can be 28.8 dB for the three different thicknesses. It can also be seen that for different waveguide thicknesses, the photorefractive gain compensate losses up to different waveguide lengths. For instance, the zero dB gain can be achieved at waveguide lengths of $L = 41$ μm, $L = 16.2$ μm and $L = 10.6$ μm for waveguide thicknesses of

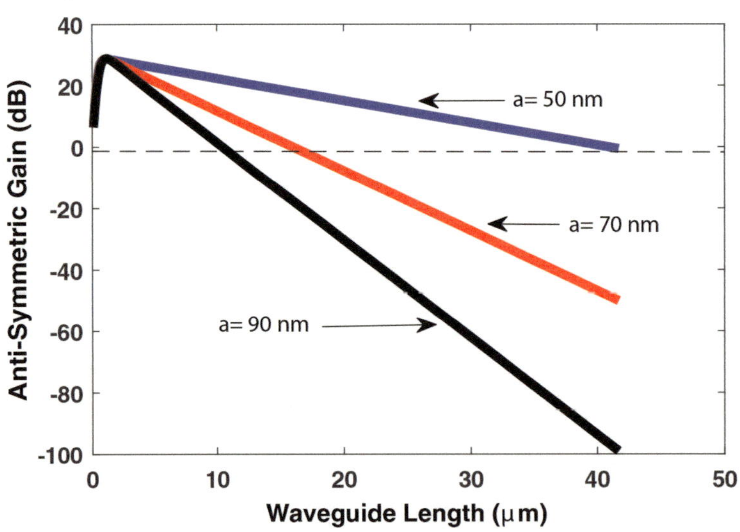

Figure 8. The antisymmetric gain in dB versus waveguide length, taking losses into account.

$a = 50$ nm, $a = 70$ nm and $a = 90$ nm, respectively, for the same reason mentioned above explained by having greater antisymmetric losses for larger waveguide thicknesses.

6. Conclusion

Nanoplasmonic waveguides filled with electro-optical material were proposed and discussed. The aim of this work is to incorporate the unique properties of plasmonic waveguides, ultra-compact structures and high internal intensities, with electro-optical material properties, thus achieving novel functionalities. Two configurations were chosen and investigated. First, an Au-$LiNbO_3$-Au nanostructure waveguide is considered for THz generation by means of difference-frequency generation of SPP modes. The SPP modes are designed to be totally confined inside the waveguide, whereas the generated THz waves are concentric with the SSP modes and contained mainly in the surrounding medium of the waveguide. Several advantages are achieved through this design. First, THz losses are minimized. Secondly, an off-resonance operation is conducted. Finally, a nanoscale and yet simple THz generation is offered. The evolution of the THz wave and the SPP modes are evaluated and found to be mainly limited by losses. Nevertheless, THz generation is shown to be viable over the entire range from 1 to 10 THz by properly designing the SPP wavelengths and waveguide dimension. Possible future applications include nanocommunication systems and body-centric networks. Secondly, an AL-$LiNbO_3$-AL nanostructure is considered with the $LiNbO_3$ being doped with donor and acceptor atom impurities. Two SPP modes with symmetric and antisymmetric spatial distribution are considered. The interaction of the interfering SPP modes with the atom impurities is modelled. It was found that the SPP modes are coupled by the means of the photorefractive effect. A net gain was shown viable for weak antisymmetric mode co-propagating with strong symmetric mode. The antisymmetric gain was studied against the doping concentration and the input amplitudes. This work opens up new opportunities to apply known photorefractive applications to nanoplasmonic devices. The two configurations discussed in this chapter demonstrate the potential of utilizing electro-optical materials in nanoplasmonic waveguides to achieve novel, efficient and ultra-compact devices.

Acknowledgement

This work was supported by the Centre on Smart Service for Smart Cities at Abu Dhabi University, UAE, under grant 19300066.

Author details

Montasir Qasymeh

Address all correspondence to: montasir.qasymeh@adu.ac.ae

Department of Electrical and Computer Engineering, Abu Dhabi University, UAE

References

[1] E. Ozbay. Plasmonic: Merging photonics and electronics at nanoscale dimensions. Science. 2006; **311**(5758):189–193.

[2] D. K. Gramotnev, and S. I. Bozhevolnyi. Plasmonics beyond the diffraction limit. Nature Photon. 2010; **4**:83–91.

[3] J. A. Schuller, E. S. Barnard, W. Cai, Y. C. Jun, J. S. White, and M. L. Brongersma. Plasmonics for extreme light concentration and manipulation. Nature Materials. 2010; **9**:193–204.

[4] N. Pleros, E. E. Kriezis, and K. Vyrsokinos. Optical interconnects using plasmonics and si-photonics. IEEE Photonics Journal. 2011; **3**(2):296–301.

[5] D.-S. Ly-Gagnon, K. C. Balram, J. S. White, P. Wahl, M. L. Brongersma, and D. A. B. Miller. Routing and photodetection in subwavelength plasmonic slot waveguides. Nanophotonic. 2012; **1**:9–16.

[6] K. F. MacDonald, Z. L. Samson, M. I. Stockman, and N. I. Zheludev. Ultrafast active plasmonics. Nature Photon. 2009; **3**:55–58.

[7] S. Yue, J. Chen, and Q. Gong. Ultrasmall and ultrafast all-optical modulation based on a plasmonic lens. Applied PhysicsLetter. 2011; **98**:#1611108.

[8] J. N. Caspers, N. Rotenberg, and H. M. Van Driel. Ultrafast silicon-based active plasmonics at telecom wavelengths. Optics Express. 2010; **18**(19):19761–19769.

[9] M. Cohen, Z. Zalevsky, and R. Shavit. Towards integrated nanoplasmonic logic circuitry. Nanoscale. 2013; **5**(12):5442–5449.

[10] W. Cai, A. P. Vasudev, and M. L. Brongersma. Electrically controlled nonlinear generation of light with plasmonics. Science. 2011; **333**:1720–1723.

[11] P. Berini, and I. D. Leon. Surface plasmon-polariton amplifiers and lasers. Nature Photon. 2011; **6**(1):16–24.

[12] J. Xhang, E. Cassan, D. Gao, and X. Zhang. Highly efficient phase-matched second harmonic generation using an asymmetric plasmonic slot waveguide configuration in hybrid polymer-silicon photonics. Optics Express. 2013; **21**(12):14876–14887.

[13] M. Z. Alam, J. Stewart Aitchison, and M. Mojahedi. Compact and silicon-on-insulator-compatible hybrid plasmonic TE-pass polarizer. Optics Letters. 2012; **37**(1):55–57.

[14] M. C. Schaafsma, H. Starmans, A. Berrier, and J. Gómez Rivas. Enhanced terahertz extinction of single plasmonic antennas with conically tapered waveguides. New Journal of Physics. 2013; **15**(1):#015006.

[15] A. Joushaghani, B. A. Kruger, S. Paradis, D. Alain, S. Aitchison, and J. K. S. Poon. Sub-volt broadband hybrid plasmonic-vanaduim dioxide switches. AppliedPhysics.Letters. 2013;**102**:#161101.

[16] J.-Y. Yan, L. L., and J. Xiao. Ring-like solitons in plasmonic fiber waveguides composed of metal-dielectric multilayers. Optics Express. 2012; **20**(3):1945–1952.

[17] X.-T. Kong, Z.-B. Li, and J.-G. Tian. Mode converter in metal-insulator-metal plasmonic waveguide designed by transformation optics. Optics Express. 2013; **21**(8):9437–9446.

[18] M. Qasymeh. Terahertz generation in nonlinear plasmonic w[aveguides. IEEE Journal of Quantum Electronics. 2016; **52**(4):#8500207.

[19] M. Qasymeh. Photorefractive Effect in Plasmonic Waveguides. IEEE Journal of Quantum Electronics. 2014; **50**(5):327–333.

[20] J. Park, K-Y. Kim, I.-M. Lee, H. Na, S-Y. Lee, and B. Lee. Trapping light in plasmonic waveguides. Optics. Express. 2010; **18**(2):598–623.

[21] K. K. Wong. Properties of Lithium Niobate. London, UK: INSPEC. 2002.

[22] D. Martin-Cano, M. L. Nesterov, A. I. Fernandez-Dominguez, F. J. Garcia-Vidal, L. Martin-Moreno, and E. Moreno. Domino plasmons for subwavelength terahertz circuitry. Optics. Express. 2010; **18**(2):754–764.

[23] J. C. Gutierrez-Vega. Fractionalization of optical beams: I. Planar analysis. Optics. Letters. 2007; **32**(11):1521–1523.

[24] M. A. Bandres and J. C. Gutierrez-Vega. Cartesian beams. Optics. Letter. 2001; **32**(23): 3459–3461.

[25] Y. Sun, Z. Zheng, J. Cheng, G. Sun, and G. Qiao. Highly efficient second harmonic generation in hyperbolic metamaterial slot waveguides with large phase matching tolerance. Optics. Express. 2015; **23**(5):6370–6378.

[26] S. M. Kostritskii, and M. Aillerie. Z-scan study of nonlinear absorption in reduced LiNbO3 crystals. Journal of. Applied. Physics. 2012; **111**:#103504.

[27] H. Yasuda, and I. Hosako. Measurement of terahertz refractive index of metal with terahertz time-domain spectroscopy. Japanese Journal of Applied. Physics. 2008; **47** (3):1632–1634.

[28] M. Unferdorben, Z. Szaller, I. Hajdara, J. Hebling, and L. Pálfalvi. Measurement of refractive index and absorption coefficient of congruent and stoichiometric lithium niobate in the terahertz range. Journal of. Infrared Milli. Terahz. Waves. 2015; **36**(12):1203–1209.

[29] P. Yeh. Introduction to Photorefractive Nonlinear Optics. New York, NY: Wiley; 1993.

[30] A. Ashkin, G. D. Boyd, J. M. Dziedzic, R. G. Smith, A. A. Ballman, J. J. Levinstein and K. Nassau. Optically induced refractive index inhomogeneities in $LiNbO_3$ and $LiTaO_3$. Applied Physics Letters. 1966; **9**(1):72–74.

[31] Y. Yang, and D. Psaltis. Photorefractive properties of lithium niobate crystals doped with manganese. Journal of Optical Society of America B. 2003; **20**(7):1491–1502.

[32] Y. Furukawa, K. Kitamura, S. Takekawa, K. Niwa, and H. Hatona. Stoichiometric Mg: $LiNbO_3$ as an effective material for nonlinear optics. Optics Letters. 1998; **23**(24):1892–1894.

Chapter 12

Impedance Matching Analysis of Cylindrical Plasmonic Nanoantennas Fed by Optical Transmission Lines

Karlo Queiroz da Costa, Janilson Leão Souza and
Victor Dmitriev

Additional information is available at the end of the chapter

http://dx.doi.org/10.5772/67414

Abstract

An impedance matching analysis of two plasmonic nanocircuits connected to cylindrical nanoantennas is presented. In the first case, a bifilar optical transmission line (OTL) with finite length is connected between two nanodipoles, where one is illuminated by an optically focused Gaussian beam (receiving dipole) and the other radiates energy received from the OTL (emitting dipole). In the second case, the OTL is fed by a voltage source on one side and connected to a dipole-loop composed antenna on the other side. These circuits are analysed electromagnetically by the linear method of moments (MoM) with equivalent surface impedance of conductors. Some results are compared using the finite element method. The results show the impedance matching characteristics of the circuits as a function of their geometries and the broadband response of the second circuit due the broadband dipole-loop antenna.

Keywords: plasmonic circuits, cylindrical nanoantennas, impedance matching, broadband nanoantennas, method of moments (MoM)

1. Introduction

Nanophotonics is the study of optical systems in the nanometre scale [1]. A sub-area of nanophotonics is the nanoplasmonics, which analyses the interaction of optical fields with metal nanostructures [2]. With the development of nanoplasmonics, the concept of nanoantennas or optical antennas has emerged naturally as metal nanostructures that receive, transmit, localize and enhances optical fields [3–6]. This definition is similar to conventional RF-microwave antennas, but the difference between these two regimes is that nanoantennas are not perfect conductors and we need to consider the finite conductivity of the metal. This characteristic leads to nanoantenna's size smaller than the wavelength of the incident wave. In other words, nanoantennas are electrically small resonant structures that can manipulate optical fields in small regions beyond the diffraction limit of light. This property of wavelength scaling for optical antennas is discussed in [7–9].

The research studies in nanoantenna field have increased mainly due to the development in modern nanofabrication techniques, such as the colloidal lithography, that is a bottom-up process, and the top-down processes like focused-ion beam (FIB) and electron-beam lithography (EBL) [10]. Some review papers on nanoantennas about theory, modelling, fabrication process and applications have been published [11–16].

The development in nanoantenna theory has also been increased due the important applications in different fields [17–28]. For example, the ability of metal nanoparticles to confine and enhance optical fields in nanometre regions is used in high-resolution microscopy, where fluorescence emission from a single molecule can be strongly enhanced [17–22]. Also, this radiation of a single emitter can be highly directed by nanoantenna arrays [23]. Other important applications are in nanobioimaging to analyse biological process [24], plasmonic photovoltaic cells [25], treatment of cancer in medicine [26], use of wireless at a nanoscale [27], plasmonic laser and optical data storage [28], and sub-wavelength integrated optical circuit [29]. In this work, we focus on the last application.

Examples of nanoantennas connected to plasmonic waveguide are presented in [29–31]. In this work, we make an alternative analysis and extend the results by using a different optical antenna with broadband characteristics. In particular, we consider two plasmonic nanocircuits connected to cylindrical nanoantennas. In the first circuit, a finite optical transmission line (OTL) is connected between two nanodipoles, where they are referred as receiving and emitting dipoles. In this circuit, the first dipole receives the energy from an optically focused Gaussian beam and delivers it to OTL, which connects to the second dipole to radiation. In the second circuit, the OTL is fed by a voltage source on one side and connected to a dipole-loop composed antenna on the other side. The analyses of these circuits are made by the linear method of moments (MoM) [32] with equivalent surface impedance of conductors [33]. We compare some results with the finite element method (FEM) [34]. The results show the impedance matching characteristics of the circuits as a function of their geometries, and the frequency response of the second circuit connected to the broadband dipole-loop antenna.

2. Theoretical model

In this section we present the geometries of the two nanocircuits, the linear method of moments (MoM) model used in the theoretical analysis, the Lorentz-Drude permittivity model and the equivalent surface impedance of the conductors, and the Gaussian beam used to feed the one circuit.

2.1. Nanocircuit geometries

The geometries of the analysed nanocircuits are presented in **Figures 1** and **2**. In the first case as given in **Figure 1**, the geometry is composed by an OTL, of length L and radius a_L, and two nanodipoles. The receiving dipole 1 is illuminated by a Gaussian beam polarized along its axis, and the emitting dipole 2 radiates the energy received from the OTL. The arm length and radius of the dipoles 1 and 2 are h_1 and a_{h1}, and h_2 and a_{h2}, respectively.

In the second case as given in **Figure 2**, the circuit is composed by an OTL and one dipole-loop combined antenna. This circuit is fed on the left side by a voltage source of width d. The dipole is connected to OTL and the rectangular loop is above the circuit, where the OTL and dipole are on the plane $z = 0$ and the loop is on the plane $z = d_e$. The geometric parameters of the dipole-loop antenna (**Figure 2**) are h_d, a_d, W_e, H_e, a_e, d_W and d_H, where the last two define the position of the loop with respect to the dipole.

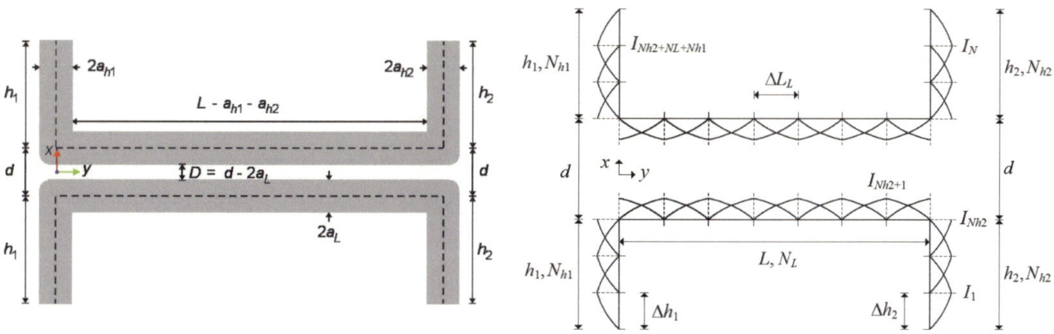

Figure 1. Left side: nanocircuit composed by OTL and two nanodipoles, where dipole 1 is illuminated by Gaussian beam and dipole 2 radiates energy of OTL. Right side: equivalent linear MoM model.

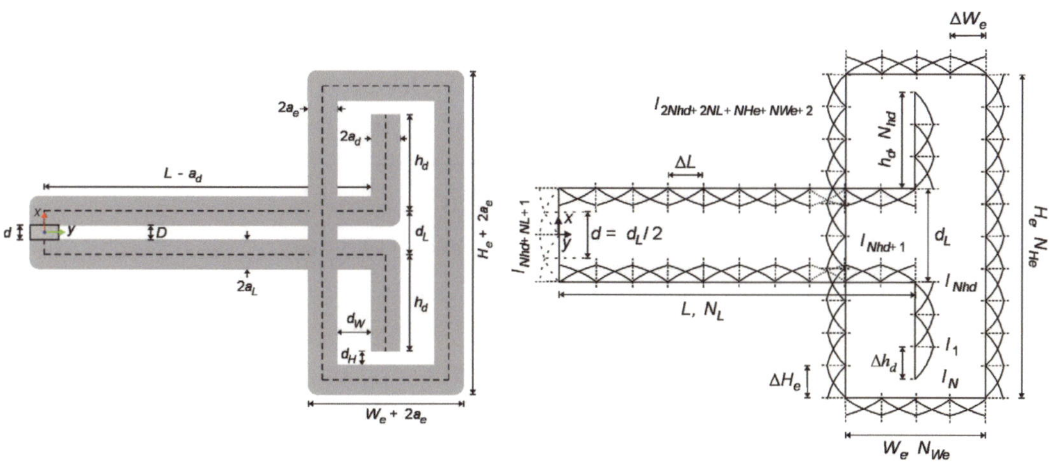

Figure 2. Left side: nanocircuit composed by OTL and one dipole-loop combined nanoantenna, where OTL is fed by a voltage source. Right side: equivalent linear MoM model.

In both circuits, the distance between the conductor's axis of OTL is d, in this case the distance between their surfaces is $D = d - 2a_L$. In these figures are also presented (right side) the equivalent linear MoM model used in the numerical analysis, which will be described in the next sections.

2.2. Method of moment model

There are different formulations of the linear approximation of MoM in the literature [32, 35–37]. Here we use the model given in Ref. [33], where the linear currents are expanded with sinusoidal basis functions. A brief description of this method is presented below, where we consider the scattering problem of **Figure 1** to explain the method.

In the scattering problem of **Figure 1**, the background medium is free-space and the conductors are in gold. The electrical permittivity of the gold conductors are represented by the Lorentz-Drude model $\varepsilon_1 = \varepsilon_0\varepsilon_{r1}$ [1], where

$$\varepsilon_{r1} = \varepsilon_\infty - \frac{\omega_{p1}^2}{\omega^2 - j\Gamma\omega} + \frac{\omega_{p2}^2}{\omega_0^2 - \omega^2 + j\gamma\omega} \tag{1}$$

$\varepsilon_\infty = 8$, $\omega_{p1} = 13.8 \times 10^{15}\ \mathrm{s}^{-1}$, $\Gamma = 1.075 \times 1014\ \mathrm{s}^{-1}$, $\omega_0 = 2\pi c/\lambda_0$, $\lambda_0 = 450$ nm, $\omega_{p2} = 45 \times 10^{14}\ \mathrm{s}^{-1}$, and $\gamma = 9 \times 10^{14}\ \mathrm{s}^{-1}$. The losses of the metal are described by surface impedance Z_s. This surface impedance can be obtained approximately by considering cylindrical waveguide with mode TM_{01} [33]:

$$Z_s = \frac{T J_0(Ta)}{2\pi a j\omega\, \varepsilon_1 J_1(Ta)} \tag{2}$$

where $T = k_0\sqrt{\varepsilon_{r1}}$ and $k_0 = \omega\sqrt{\mu_0\varepsilon_0}$, J_0 and J_1 are the Bessel functions of first kind of order zero and one, respectively, ω is the operating angular frequency, k_0 is the propagation constant in air, μ_0 is the magnetic permeability of air and ε_0 is the electrical permittivity of air.

The integral equation of the scattering problem is obtained by the boundary condition of tangential electric field at the surface's conductors $(\overline{E}_s + \overline{E}_i)\cdot\overline{a}_l = Z_s I$, where $\overline{a}_l$ is a unitary vector tangential to the surface of the metal, $\overline{E}_s$ is the scattered electric field due to the induced linear current I on the conductor, $\overline{E}_i$ the incident electric field of the Gaussian beam source, or of the voltage source in the case of **Figure 2**, and I is the induced longitudinal current in a given point of the nanocircuit. The scattered field is given by:

$$\overline{E}_s(\overline{r}) = \frac{1}{j\omega\,\varepsilon_0}\left[k_0^2 \int_l \overline{I} g(R)\, dl' + \int_l \frac{dI}{dl'}\ \nabla\ g(R)\, dl'\right] \tag{3}$$

where $g(R) = e^{-jk_0R}/4\pi R$ is the free-space Green's function, and $R = |\overline{r} - \overline{r}'|$ is the distance between source and observation points.

The numerical solution of the problem formulated by the boundary condition and Eqs. (1)–(3) is performed by linear MoM as follows. First, we discretize the linear circuit as shown in the right side of **Figure 1**, where N_L, N_{h1} and N_{h2} are the number of straight segments in L, h_1 and h_2, respectively. In the **Figure 1**, we have $N_L = 7$, $N_{h1} = N_{h2} = 3$. The discretization is uniform in L, h_1 and h_2, but the discretization length can be different, i.e. $\Delta L_L = L/N_L$, $\Delta h_1 = h_1/N_{h1}$ and

$\Delta h_2 = h_2/N_{h2}$. With this discretization, the total number of straight segments of the nanocircuit is $N_t = 2N_{h2}+2N_L+2N_{h1}$. For the method stability, we use the convergence conditions $\Delta h_2 > 2a_{h2}$, $\Delta h_1 > 2a_{h1}$, and $\Delta L_L > 2a_L$. Then the current in each segment is approximated by sinusoidal basis functions [33]. The expansion constants I_n are shown in **Figure 1** where each constant defines one triangular sinusoidal current. To determine these constants, we use $N = N_t$ -2 rectangular pulse test functions with unitary amplitude and perform the conventional testing procedure. The following linear system of equations is obtained

$$V_m = Z_s I_m \Delta_m - \sum_{n=1}^{N} Z_{mn} I_n \tag{4}$$

where $m = 1, 2, 3, \ldots, N$, Z_{mn} is the mutual impedance between sinusoidal current elements m and n, $\Delta_m = [\Delta L_m + \Delta L_{m+1}]/2$ [32], and V_m is the voltage induced in the segment m due the source field $\overline{E}_i$. The solution of Eq. (4) produces the current along the nanocircuit. With these results, it is possible to calculate near- and far-field distributions and other parameters.

2.3. Gaussian beam source

In the case of nanocircuit 1 (**Figure 1**), the incident field is a Gaussian beam. This kind of wave is obtained by solving the scalar Helmholtz wave equation with the paraxial approximation [37]. The magnetic vector potential of a Gaussian beam polarized on the x-axis and travelling in the +z direction is given by

$$\overline{A} = u_0 \sqrt{\frac{2}{\pi}} \frac{1}{w} \exp\left(\frac{-\rho^2}{w^2}\right) \exp\left[-j\left(\frac{k_0 \rho^2}{2R} - \phi\right)\right] e^{-jk_0 z} \overline{a}_x \tag{5}$$

where $u_0 = (2P\mu_0/k_0\omega)^{1/2}$, P is the power of the beam, w is the beam radius (**Figure 3**), R is the curvature radius of the phase front and ϕ is the phase of the beam. The three principal parameters that define the beam are the power P, the radius of the beam waist w_0 (**Figure 3**) and the operating wavelength λ. With the vector potential given in Eq. (5), the x-component of electric field E_x of the Gaussian beam can be obtained from the Maxwell equations [37].

The excitation beam used in **Figure 1** is focused on the receiving dipole 1 with polarization along the dipole axis (x-axis), the direction of propagation is +z, the beam axis is along the z-axis, and the minimum waist (w_0) is localized at $z = 0$, which is the plane of the nanocircuit. In all the analyses presented in this work for the nanocircuit 1 (**Figure 1**), we consider a fixed Gaussian beam with power, $P = 10^{-12}$ W, wavelength, $\lambda = 830$ nm and beam waist, $w_0 = 340$ nm. We use these values for comparisons with the results given in Ref. [29], but the analysis presented here can be applied for beams with other parameters. The field distribution of this beam is shown in **Figure 3**, where the electric field amplitude abs(E_x) is presented on the xz and xy planes and the phase distribution angle of E_x is presented on the xz plane. The phase at xy plane is constant.

3. Analysis of first nanocircuit

In this section, we analyse the first nanocircuit of **Figure 1**. In this case, we fix the Gaussian beam source given in **Figure 3** at $\lambda = 830$ nm, and analyse the impedance matching characteristic in function of the geometrical parameters of the OTL and nanodipoles.

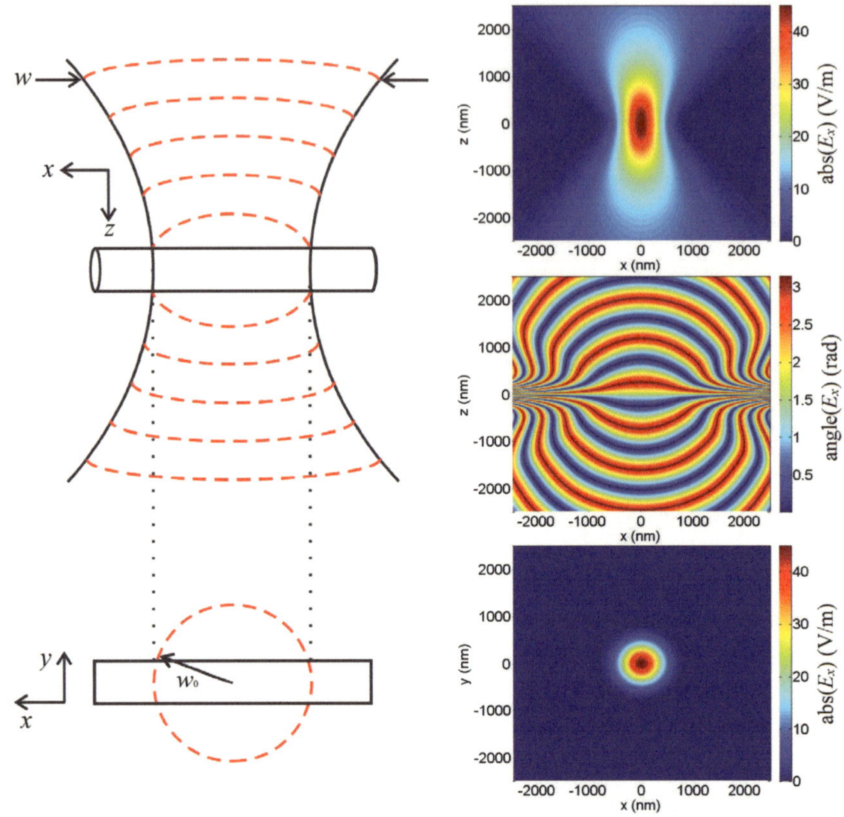

Figure 3. Field distribution of Gaussian beam with $P = 10^{-12}$ W, $\lambda = 830$ nm and beam waist $w_0 = 340$ nm at xz and xy planes.

3.1. Numerical example

Based on the theoretical model presented above, we developed a MoM code in Matlab to analyse the nanocircuit shown in **Figure 1**. In this sub-section, we present an example of simulation of the nanocircuit shown in **Figure 1** fed by the Gaussian beam depicted in **Figure 3**. **Figure 4** shows the geometry and discretization parameters used in this simulation and the result of the current distribution along the circuit. The near-field distribution for this example is given in **Figure 5**. Note that we use the total length of dipole as $H_1 = 2h_1+d$ and $H_2 = 2h_2+d$.

We observe in these results the stationary behaviour in the OTL, which is due to the mismatching in the impedances of nanodipole 2 and OTL. To make a quantitative measure of the impedance matching degree, we calculate approximately the voltage stationary wave ratio (*VSWR*) near nanodipole 2 as $VSWR = I_{max}/I_{min}$, where I_{max} and I_{min} are, respectively, the maximum and minimum current magnitude nearest to dipole 2. With this parameter, we calculate the voltage reflection coefficient as $|\Gamma_V| = (VSWR-1)/(VSWR+1)$. In this numerical example we obtained $|\Gamma_V| = 0.4$.

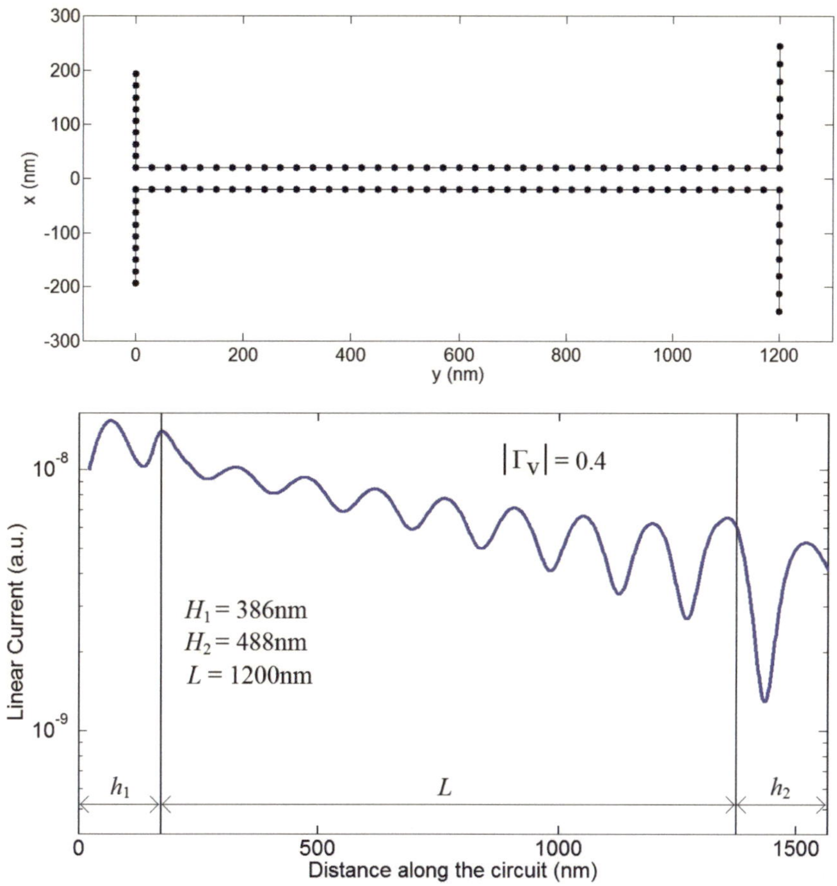

Figure 4. Up: geometry and discretization of nanocircuit with H_1 = 386 nm, H_2 = 488 nm, L = 1200 nm, a_{h1} = 10 nm, $a_L = a_{h2}$= 15 nm, D = 10 nm, N_{h1} = 8, N_{h2} = 7, N_L = 40, ΔL_L = 30 nm, Δh_1 = 21.6 nm, Δh_2 = 32 nm and N = 108. Down: amplitude of linear current distribution along nanocircuit. The voltage reflection coefficient of this circuit is $|\Gamma_V|$ = 0.4.

3.2. Impedance matching analysis

This section presents a parametric analysis of the impedance matching of the nanocircuit for different values of h_2, a_{h2}, a_L and D. In this analysis, we fixed the dimensions of the receiving dipole h_1 = 173 nm, a_{h1} = 10 nm and the length of the OTL L = 1200 nm. **Figure 6** presents the voltage reflection coefficient $|\Gamma_V|$ versus the total length of the emitting dipole $H_2 = 2h_2 + d$ for different values of D = 10, 15 and 20 nm for the four cases (a_L = 10 nm, a_{h2} = 10 nm), (a_L = 10 nm, a_{h2} = 15 nm), (a_L = 15 nm, a_{h2} = 15 nm), and (a_L = 15 nm, a_{h2} = 20nm).

Analysing these curves we come to some conclusions. We note that the nanocircuits possess in general a smaller degree of input impedance matching (higher $|\Gamma_V|$) when the gap of the OTL D is increased. The exceptions are the cases of **Figures 6a** and **b** in the range $H_2 < 400$ nm, where we have a better matching for higher D.

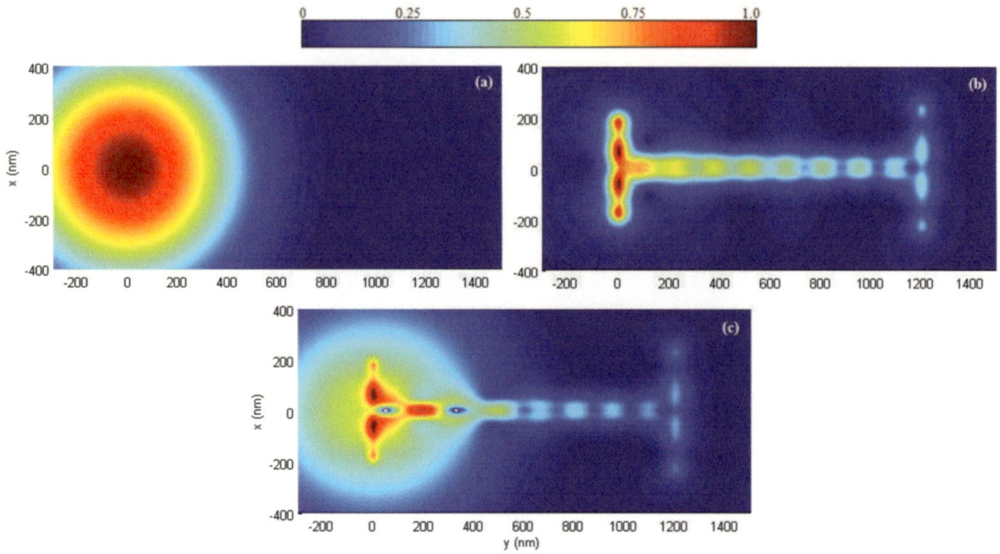

Figure 5. Normalized distribution of the electric field amplitude at plane z = 30 nm: (a) incident field of the Gaussian beam abs($E_x^{\,i}$), (b) scattered field abs($E_x^{\,s}$), and (c) total field abs(E_x).

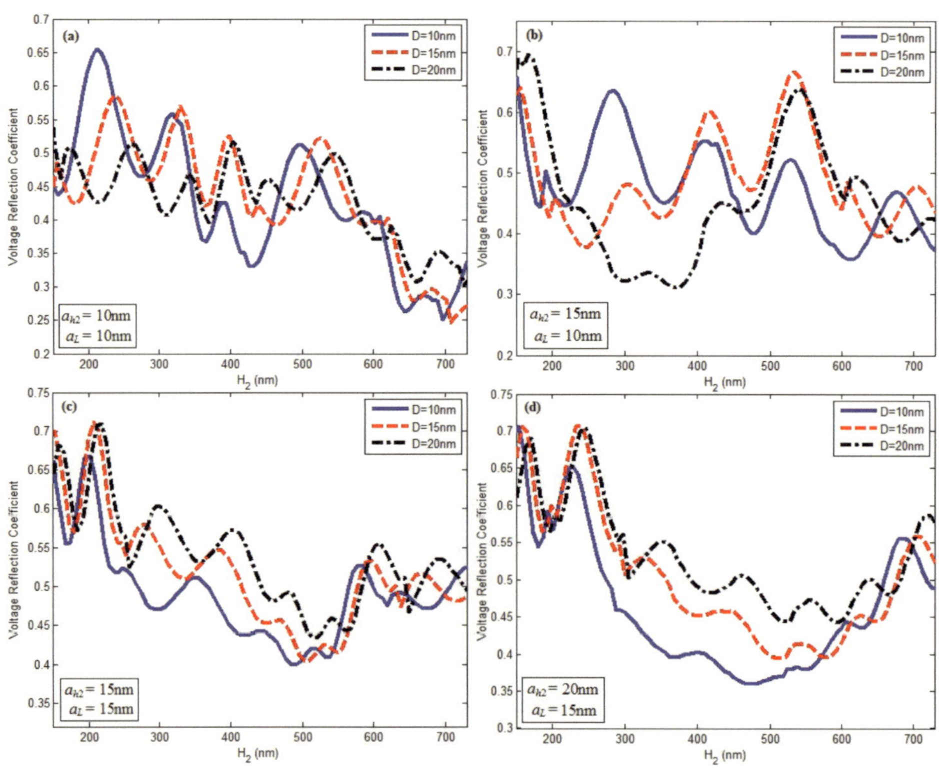

Figure 6. Voltage reflection versus $H_2 = 2h_2+d$ for different values of D = 10, 15 and 20 nm. (a) a_L = 10 nm and a_{h2} = 10 nm, (b) a_L = 10 nm and a_{h2} = 15 nm, (c) a_L = 15 nm and a_{h2} = 15 nm, and (d) a_L = 15 nm and a_{h2} = 20 nm.

We also observe that in general the impedance matching is better when $a_{h2} > a_L$. This means that the values of $|\Gamma_V|$ in **Figure 6d** are smaller than those of **Figure 6c**, and the values of $|\Gamma_V|$ in **Figure 6b** are smaller than those of **Figure 6a**. The latter comparison is only true in the range of $H_2 < 400$ nm.

All these results show that we have many situations of good matching for different values of D, a_L, a_{h2} and H_2. Min($|\Gamma_V|$) occurs for larger H_2, for examples in the cases $D = 10$ nm and $H_2 = 640$ nm in **Figure 6a**, where $|\Gamma_V| \approx 0.26$, and $D = 10$ nm and $H_2 = 610$ nm in **Figure 6b**, where $|\Gamma_V| \approx 0.35$. Other good results are for smaller H_2, for example the case of $D = 20$ nm and $H_2 = 300$ nm in **Figure 6b**, where $|\Gamma_V| \approx 0.31$.

One way to choose the best geometric parameters of the circuit is to consider the case with better impedance matching and efficiency simultaneously. The efficiency of the circuit depends mainly on the attenuation of the current along the OTL, i.e. depends on the loss constant α of the OTL. This parameter is constant for the principal mode that propagates on the OTL, and can be obtained approximately by the average inclination of the current versus distance along the OTL. In equation form we have $\alpha = \Delta I/\Delta L$, where ΔI is the variation of the average amplitude of the current in decibels (dB) along a given distance ΔL (nm) in the OTL. With this definition the unit of this parameter is dB/nm. In the numerical example presented in **Figure 4** we have $\alpha \approx 0.007393$ dB/nm. This result is very close to $\alpha = 0.007296$ dB/nm obtained in Ref. [29] where the OTL is similar to our case shown in **Figure 4** ($a_L = 15$ nm and $D = 10$ nm).

To understand better the behaviour of the impedance matching and efficiency characteristic of the results presented in **Figure 6**, we plot in **Figures 7** and **8** the current distributions for different geometric parameters. **Figure 7** shows the currents for two cases with same $a_L = 10$ nm but different voltage reflection coefficients of $|\Gamma_V| = 0.31$ and 0.67. In these results we observe a higher stationary wave for the case $|\Gamma_V| = 0.67$ than for the case $|\Gamma_V| = 0.31$. This shows that our approximate method to calculate $|\Gamma_V|$ provides a good measure of the degree of impedance matching. We also observe that the two cases present the same attenuation along the circuit (i.e. in both cases one has approximately the same loss constant $\alpha = 0.0111$ dB/nm) because the OTL are constructed with the same radius of wires, $a_L = 10$ nm.

Figure 8 presents the current distribution for two cases with good impedance matching $|\Gamma_V| = 0.26$ and 0.36, but with different loss constant of $\alpha = 0.0119$ and 0.0084 dB/nm, respectively. This difference is mainly due to the difference of the radius a_L of the OTL. For lower values of a_L the attenuation is higher in OTL and this result is similar to that observed in RF-microwave regimes. This can be explained by the surface impedance model of Eq. (2), where smaller radius produces higher Z_s and, consequently, higher loss in the conductors.

Another analysis of this circuit was done varying the dimensions of the receiving dipole 1 and fixing the dimensions of dipole 2 [32]. We observed that the dimensions of dipole 1 can modify the impedance matching and efficiency characteristics of circuit. The results show that good impedance matching does not necessarily mean a good efficiency in the receiving dipole, i.e. higher input current amplitude in dipole 1. For example, **Figure 9** presents the total electric field distribution at plane $z = 30$ nm of two opposite situations. We can see in this

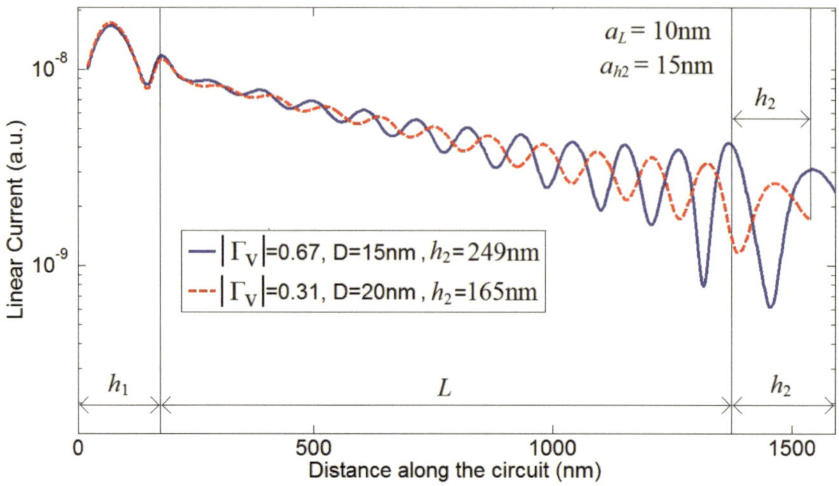

Figure 7. Current distributions of two circuits with the same values of a_L = 10 nm possessing different voltage reflection coefficient ($|\Gamma_V|$ = 0.31 and 0.67).

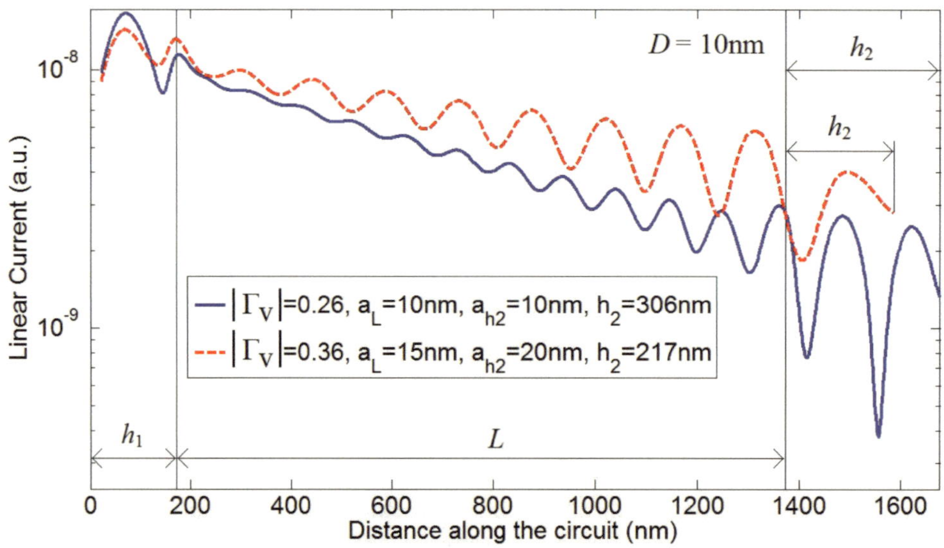

Figure 8. Current distributions of two nanocircuits with good voltage reflection coefficient ($|\Gamma_V|$ = 0.26 and 0.36) possessing different values of a_L.

figure the differences in the field intensity, showing higher values of the emitting dipole in the case when the dipole 1 operates in the first resonance (**Figure 9a**), i.e. this case presents better efficiency. However, the case of **Figure 9b** presents better impedance matching than the case of **Figure 9a**.

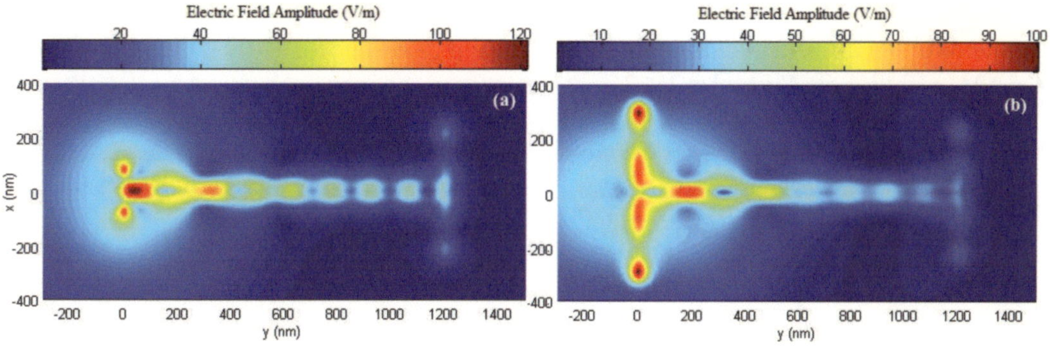

Figure 9. Total (scattered and incident) electric near-field distribution at plane z = 30 nm (a) receiving dipole at first resonance with H_1 = 176.6 nm and a_{h1} = 20 nm ($|\Gamma_V|$ = 0.42). (b) Receiving dipole at second resonance with H_1 = 637.7 nm and a_{h1} = 20 nm ($|\Gamma_V|$ = 0.21). In these simulations, we fixed the following parameters of OTL and emitting dipole: a_L =15 nm, a_{h2} = 20 nm, L = 1200 nm, and N_L = 40.

4. Analysis of second nanocircuit

This section presents the analysis of second nanocircuit of **Figure 2**, where a voltage source fed an OTL on the left side and the other side is connected to a dipole-loop combined antenna. Due the broadband characteristic of this circuit, the analysis presents the spectral response of the circuit in the range of 100–400 THz for different values of the geometrical parameters (**Figure 2**). First, we analyse the isolated broadband dipole-loop antenna, and then we consider the OTL connected to this antenna.

4.1. Analysis of isolated dipole-loop

In this section, we analyse the isolated dipole-loop antenna and its spectral response. Note that this case without the OTL is obtained when $L = 0$ in the nanocircuit given in **Figure 2**. To demonstrate the broadband characteristic of this nanoantenna, we first make a comparison with the conventional isolated nanodipole as a specific example. The geometrical parameters used in this example are: h_d = 200 nm, a_d = 20 nm, d_L = 20 nm, d_W = 50 nm, d_H = 20 nm, a_e = 20 nm, $W_e = 2a_e + 2a_d + 2d_W$, $H_e = 2h_d + d_L + 2a_e + 2d_H$ (**Figure 2**), N_h = 5, N_{He} = 13, N_{We} = 4, N_t = 46. Observe that the arm length of nanodipole is $h_d + a_d$ = 220 nm (**Figure 2**).

The first result obtained is the input impedance ($Z_{in} = R_{in} + jX_{in}$) presented in **Figure 10** for the case of isolated nanodipole (left) and dipole-loop antenna (right). This result is compared with the simulation by Comsol software, which is based on FEM. We observe that the two methods produce values with good agreement for the used frequency range. For the isolated dipole case, the first resonant frequency for the MoM is 191.9 THz and for the Comsol it is 187.7 THz, and the second resonant frequency for the MoM is 263.8 THz and for the Comsol it is 267.7 THz. In the case of dipole-loop antenna, it can be seen that the electromagnetic coupling between the dipole antenna and the loop antenna modifies the input impedance of the nanoantenna in comparison with the isolated dipole. It is further noted that the resonances are moved to the lower frequencies. The first resonant frequency ($F_{\lambda/2}$) for the MoM is 185.1 THz and for the Comsol is 174.4 THz and the second resonant frequency (F_{λ}) for the MoM is 256.4 THz and for the Comsol is 254.7 THz.

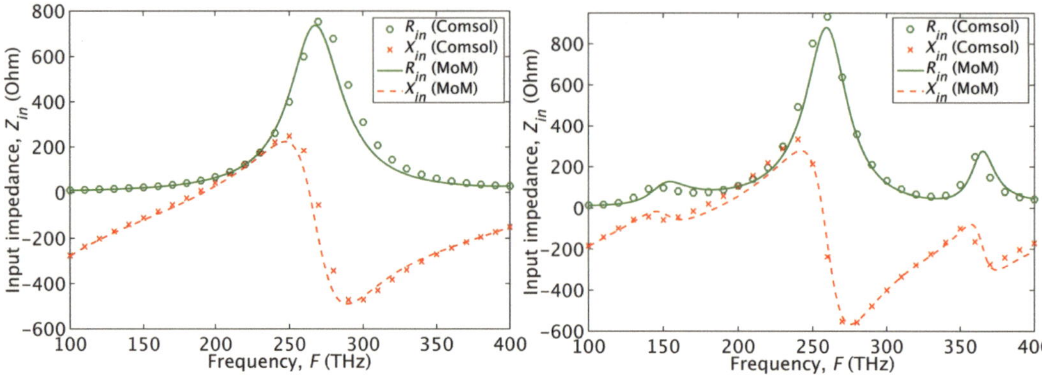

Figure 10. Input impedance of the isolated nanodipole (left) and dipole-loop antenna (right). The dimensions are: h_d = 200 nm, a_d = 20 nm, d_L = 20 nm, d_W = 50 nm, d_H = 20 nm, a_e = 20 nm.

Figure 11 shows the results of calculation of the radiation efficiency and the reflection coefficient obtained by MoM and Comsol for the isolated nanodipole (left) and dipole-loop antenna. The reflection coefficient is given by $\Gamma = |(Z_{in}-Z_0)/(Z_{in}+Z_0)|$, where Z_{in} is input impedance of the nanoantenna and Z_0 is the characteristic impedance of a given transmission line. The bandwidth is calculated by $B = 200[(F_s-F_i)/(F_s+F_i)]$, where F_s is the upper frequency and F_i is the lower frequency at the level -10 dB of the reflection coefficient.

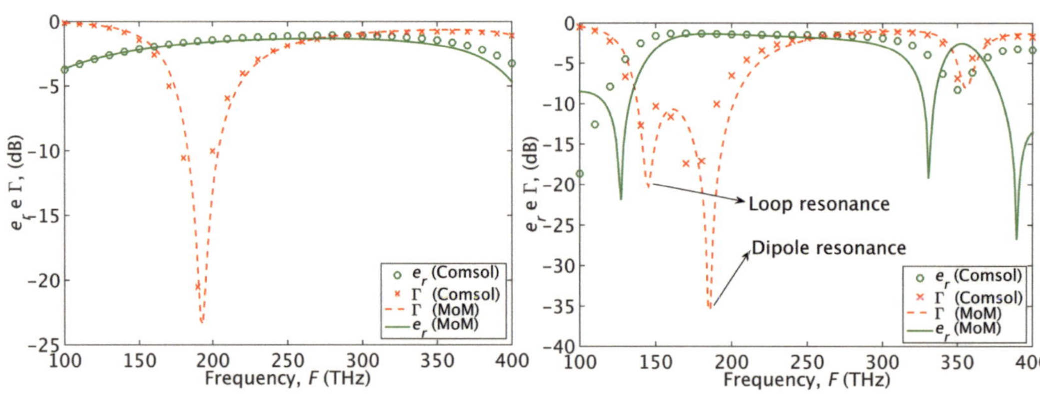

Figure 11. Radiation efficiency (e_r) and reflection coefficient (Γ) for nanodipole (left, Z_0 = 60 Ω) and dipole-loop antenna (right, Z_0 = 90 Ω) as function of frequency, calculated by MoM and Comsol.

For the isolated nanodipole (**Figure 11**, left), the maximum radiation efficiency calculated by the MoM and the Comsol are -1.06 and -1.32 dB, respectively, occurring around the second resonant frequency. However, the best input impedance matching point occurs around

http://dx.doi.org/10.5772/67414

the first resonant frequency, using $Z_0 = 60\ \Omega$. Therefore, the maximum efficiency and good impedance matching are found at different frequencies. This occurs because the characteristic impedance of an OTL is not necessarily matched to the input impedance of nanodipole with maximum efficiency [29]. The calculation of the reflection coefficient has been accomplished considering the connection of a line with $Z_0 = 60\ \Omega$, and for this impedance the final value obtained by the MoM was $B = 10.1\%$ and by Comsol $B = 11.2\%$. The results of Γ show that the isolated dipole has a narrow bandwidth.

For the dipole-loop antenna (**Figure 11**, right), we can see that insertion of the loop besides modifying the input impedance also changes the reflection coefficient causing an increase of the bandwidth of the nanoantenna to 37.1% by MoM and 35.1% by Comsol. The increased bandwidth occurs because in the compound antenna occur an overlapping of different resonances of loop and dipole, which produces a greater bandwidth. We can also observe that the resonance of the loop (near $F \cong 145$ THz) and the resonance of the dipole (near $F \cong 184$ THz). In the frequency range between 150 and 280 THz, the radiation efficiency remains almost constant reaching the maximum value of -1.276 dB by the Comsol simulation and -1.35 dB for the MoM simulation. Thus we have a broadband antenna with high efficiency and with a possibility to achieve good impedance matching with an OTL of 90 Ω.

Figure 12 shows the 3D far-field gain radiation pattern for the isolated dipole and the dipole-loop nanoantenna. The frequencies to which these diagrams were calculated correspond the bandwidth central frequencies ($F_c = (F_s + F_i)/2$) of **Figure 11**. The shape of these diagrams is approximately the same as of a small RF-microwave dipole. The maximum gain of the isolated dipole equals 1 for $F_c = 170.85$ THz and the maximum gain of the composed antenna is equal to 1.4 for $F_c = 194.97$ THz.

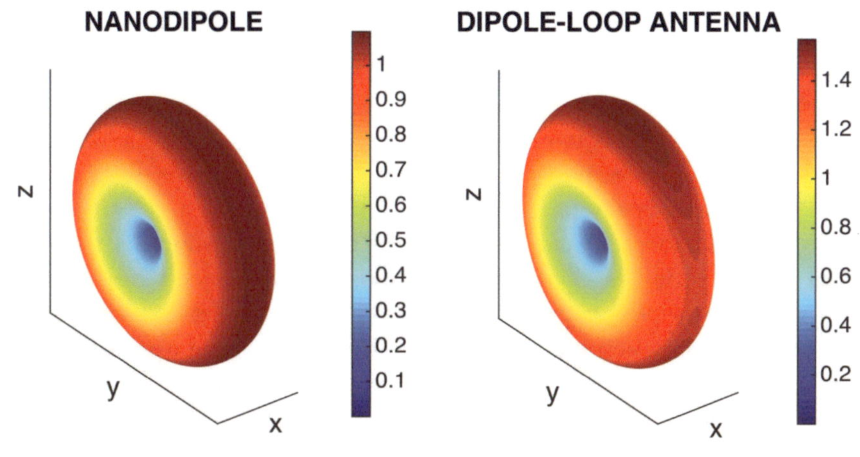

Figure 12. 3D far-field gain radiation pattern of the isolated dipole and composed antenna for the central frequencies 194.97 and 170.85 THz, respectively.

Table 1 shows a parametric analysis of bandwidth of the composed nanoantenna for $d_W = 40, 50, 60$ and 70 nm, and $d_H = 10$ and 20 nm, respectively, varying only the parameters d_W and d_H, and fixed parameters $h_d = 200$ nm, $a_d = 20$ nm, $d_L = 20$ nm and $a_e = 20$ nm. In this table are given the values of the OTL impedance which maximize the bandwidth for each simulated. We can see from this parametric analysis that it is possible to optimize the bandwidth by adjusting the dimensions of the loop element. The best case obtained was for $d_H = 10$ nm and $d_W = 70$ nm, where $B = 42\%$.

		d_W				
		30 nm	**40 nm**	**50 nm**	**60 nm**	**70 nm**
d_H	10 nm	$B = 35.4\%$	$B = 36.8\%$	$B = 38.3\%$	$B = 40\%$	$B = 42\%$
		$Z_0 = 85\ \Omega$	$Z_0 = 105\ \Omega$	$Z_0 = 100\ \Omega$	$Z_0 = 115\ \Omega$	$Z_0 = 120\ \Omega$
	20 nm	$B = 33.2\%$	$B = 34.1\%$	$B = 35.1\%$	$B = 17.7\%$	$B = 17.5\%$
		$Z_0 = 80\ \Omega$	$Z_0 = 85\ \Omega$	$Z_0 = 90\ \Omega$	$Z_0 = 80\ \Omega$	$Z_0 = 95\ \Omega$

Table 1. Results of parametric analysis of the composed nanoantenna.

4.2. Analysis of dipole-loop connected to OTL

In this section, we present the analysis of nanocircuit 2 of **Figure 2**. First, we present the variation of near- and far-field distribution in function of frequency, then the voltage reflection coefficient versus frequency is presented and finally a parametric analysis is presented.

4.2.1. Near- and far-field results

In this section, we present the variation of near and far-field distribution for a given example of nanocircuit as a function of frequency.

Figure 13 shows the current distribution along the nanocircuit for the frequencies of 100, 200 and 300 THz, for the following parameters: $h_d = 200$ nm, $L = 1200$ nm, $a_d = a_L = a_e = 20$ nm, $d_L = 60$ nm, $d = d_L - 2a_L$, $d_e = 50$ nm, $d_W = 50$ nm, $d_H = 20$ nm, $N_h = 5$, $N_{We} = 4$, $N_{He} = 13$, $N_L = 29$ and $N_t = 104$. In the figure, the circle points identify the currents of each section of the nanocircuit. The points to the left of 1 and between 4 and 5 are the currents in the nanodipole, the ones between 1 and 2 and 3 and 4 are the currents of the OTL, those between 2 and 3 represent the current of the source, and finally those between 5 and 6, 6 and 7, 7 and 8 and to the right of 8 are the currents of the loop. We can see that while the frequency increases the attenuation of the current in the OTL increases, showing that at optical frequencies the conduction losses are higher. The stationary pattern presented in the figure shows the mismatch of the impedance of the optical transmission line and the nanodipole. The same oscillatory behaviour is observed in **Figure 14**, where the distribution of the normalized electric field near a parallel plane to the nanocircuit (for the same frequencies as in **Figure 13**) is shown. These fields were calculated for the plane $z = 80$ nm. It is evident that the energy transferred from the source to the nanodipole is reduced with the increasing frequency.

Figure 15 shows the 3D far-field gain radiation diagram for this nanocircuit for $F = 100$, 200 and 300 THz. The shape of these diagrams can be explained when we pose that the nanocircuit in **Figure 2** acts as an array of two antennas spaced by the length L of the OTL, wherein the first

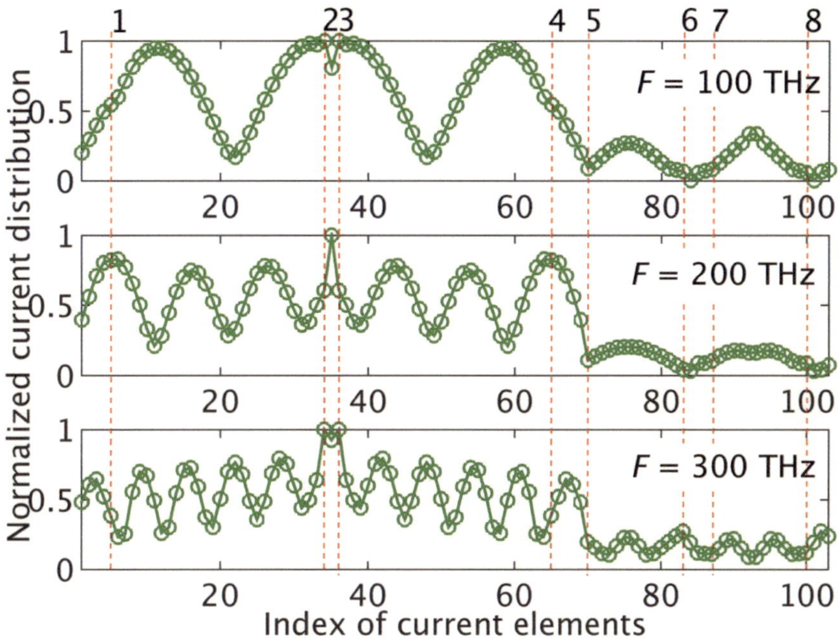

Figure 13. Normalized current distribution along the nanocircuit for F = 100, 200 and 300 THz, with the parameters: h_d = 200 nm, L = 1200 nm, $a_d = a_L = a_e$ = 20 nm, d_L = 60 nm, d_e = 50 nm, d_W = 50 nm, d_H = 20 nm, N_{hd} = 5, N_{We} = 4, N_{He} = 13, N_L = 29 and N_t =104.

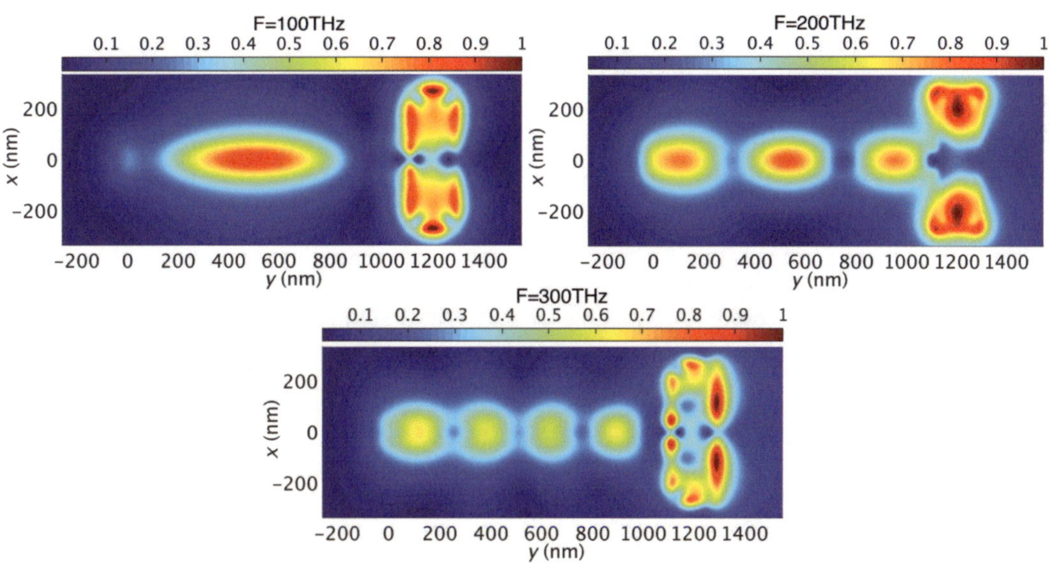

Figure 14. The electric field distribution in the plane z = 80 nm, F = 100 THz (top left), 200 THz (top right) and 300 THz (bottom), the parameters are: h_d = 200 nm, L = 1200 nm, $a_d = a_L = a_e$ = 20 nm, d_L = 60 nm, d_e = 50 nm, d_W = 50 nm, d_H = 20 nm, N_{hd} = 5, N_{We} = 4, N_{He} = 13, N_L = 29 and N_t =104.

one is the dipole-loop antenna (right side of the circuit) and the second one is an equivalent dipole formed by the voltage source (left side of the circuit), that also radiates. Thus, the resulting radiation pattern of this arrangement is the product of the radiation pattern of one element and the array factor, leading to the shape drawn in **Figure 15**. Furthermore, it can be noted that when frequency is increased the resulting pattern has more side lobes. Another observation in these diagrams is that the radiation intensity is higher in the $-z$ direction, because the loop element above the dipole acts as a reflector of waves in the downward direction.

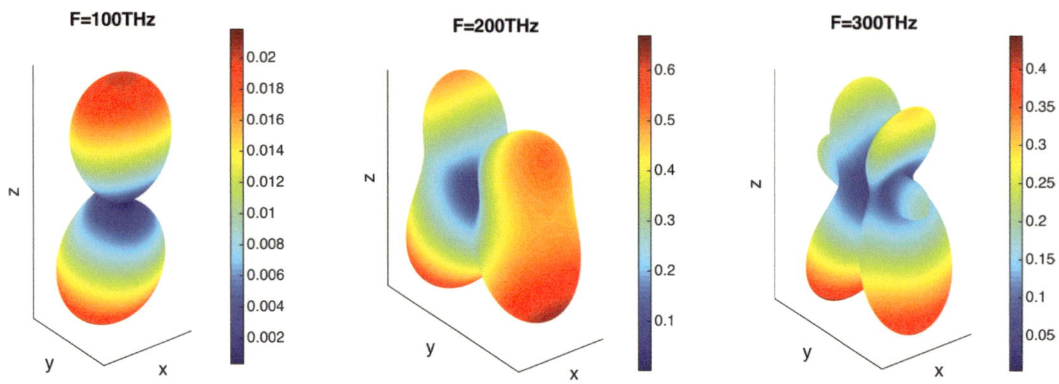

Figure 15. 3D far-field gain radiation pattern of the circuit for $F = 100$, 200 and 300 THz.

4.2.2. Voltage reflection coefficient

To analyse the impedance matching of the OTL with the antenna, it is necessary to calculate the voltage reflection coefficient as was done previously in Section 3.1. The results of calculations for the nanocircuit presented in **Figures 13** and **14** are shown in **Figure 16** for $|\Gamma_V|$ as a function of frequency for the nanocircuit with the loop and without it. In the figure, the minimum points $|\Gamma_V|$ are highlighted. They are -7.6 and -13.5 dB, at $F = 157.3$ and 383.4 THz, respectively, for the case of the nanocircuit with the loop. This figure shows that the voltage reflection coefficient decreases with the addition of the loop into the optical nanocircuit near these frequencies (157.3 and 383.4 THz). Besides, this figure explains the stationary behaviour of the current and field shown in **Figures 13** and **14**.

For the frequencies corresponding to these minimum voltage reflection coefficients for the case of the nanocircuit with loop shown in **Figure 16**, the current distribution in **Figure 17** and normalized electrical field distribution in the plane $z = 80$ nm in **Figure 18** are given.

It can be observed in **Figures 17** and **18** that with increasing of frequency the attenuation of the current and the electric field in the OTL increases due to conduction losses. It may be noted that for the frequency of 383.4 THz there is a significant drop at the standing wave rate in relation to the frequency of 157.3 THz, which presents a decrease in reflection losses on the line. On the other side, for the frequency 157.3 THz one has a higher level of current in OTL (higher transmission efficiency) than for the frequency 383.4 THz (**Figure 17**). This means, again, that a better impedance matching does not imply a higher transmission efficiency along the OTL.

http://dx.doi.org/10.5772/67414

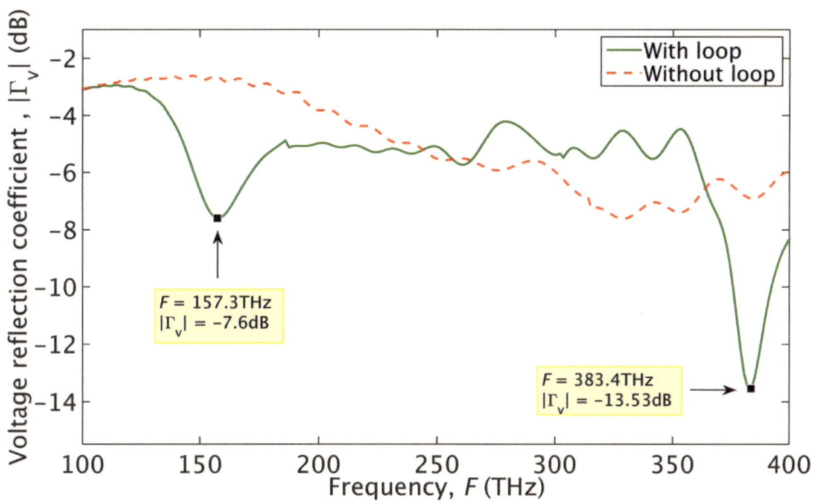

Figure 16. Voltage reflection coefficient as a function of frequency, the parameters are h_d = 200 nm, L = 1200 nm, a_d = a_L = a_e = 20 nm, d_L = 60 nm, d_e = 50 nm, d_W = 50 nm, d_H = 20 nm, N_{hd} = 5, N_{We} = 4, N_{He} = 13, N_L = 29 and N_t = 104.

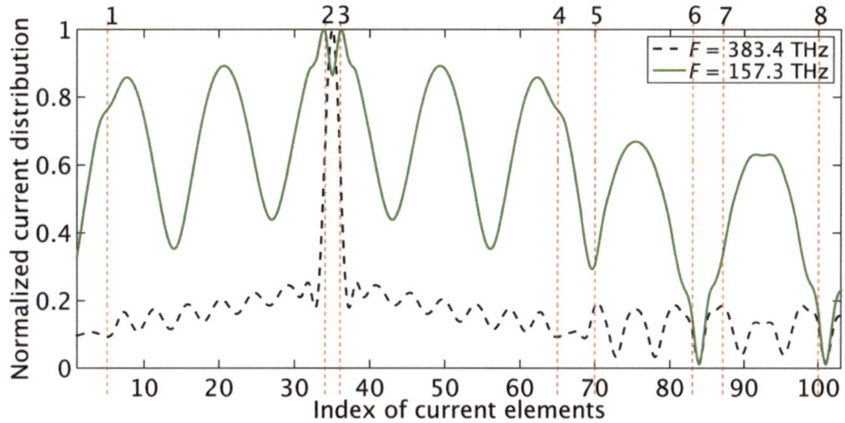

Figure 17. Current distribution for F = 157.3 and 383.4 THz, with $|\Gamma_V|$ = -7.6 and -13.5 dB, respectively.

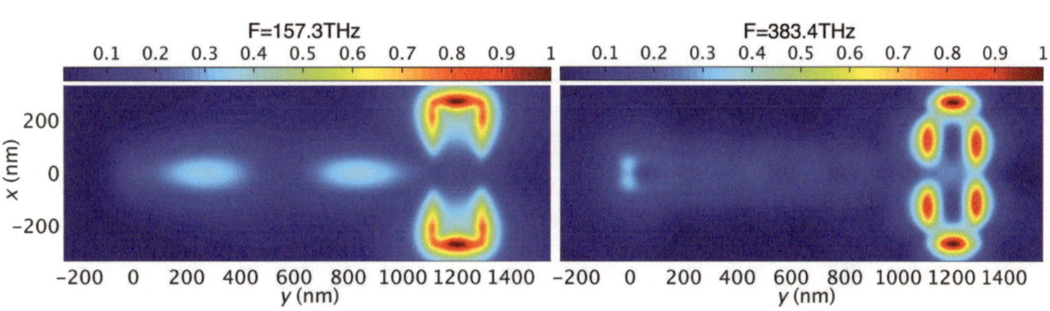

Figure 18. Electric field distribution in plane z = 80 nm for frequencies F = 157.3 (left) and 383.4 (right) THz, for cases with the reflection coefficient $|\Gamma_V|$ = -7.6 and -13.53 dB, respectively.

Figure 19 shows the 3D far-field gain radiation pattern for the nanocircuit for $F = 157.3$ and 383.4 THz. We observe again that the circuit in **Figure 2** works like an array of two dipoles spaced by the length L of the OTL. Also, the gain level of case 157.3 THz is higher than that for 383.4 THz, which is in accordance with the current levels of **Figure 17**.

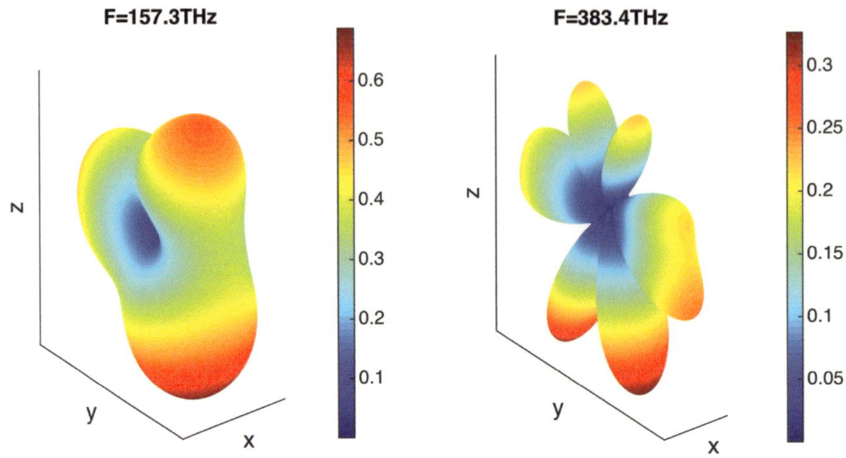

Figure 19. 3D far-field gain radiation pattern of the circuit for $F = 157.3$ and 383.4 THz.

4.2.3. Parametric analysis

Finally, a parametric analysis of the voltage reflection coefficient is shown in **Figure 20**. For the simulations we fix the following parameters: the distance between the surfaces of the OTL ($D = 20$ nm), the length of the electric dipole ($h_d = 200$ nm), the radii ($a_L = 20$ nm, $a_d = 20$ nm and $a_e = 20$ nm) and the length of the OTL ($L = 1200$ nm), and vary the parameters d_W ($d_W = 30$, 40 and 50 nm) and d_H ($d_H = 10$ and 20 nm), that consequently change the width ($W_e + 2a_e$) and the length ($H_e + 2a_e$) of the loop. In **Figure 20**, we show the results for the voltage reflection coefficient without the loop for comparison.

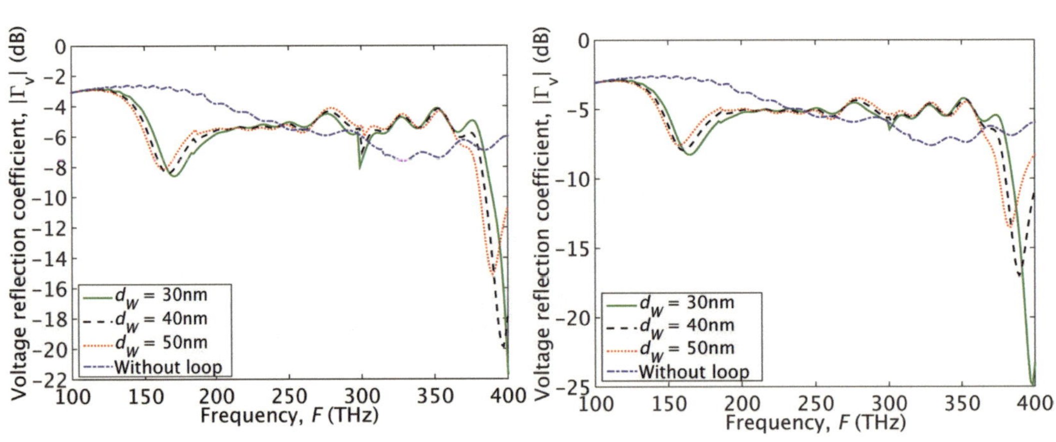

Figure 20. Voltage reflection coefficient ($|\Gamma_V|$) with the loop near the dipole for different values of d_W (30, 40 and 50 nm) with $d_H = 10$ (left) and 20 (right) nm and also $|\Gamma_V|$ without the loop.

Analysing these figures one comes to the following conclusions. In all simulated geometries of the circuit with the loop, there is an improvement in comparison with the circuit without the loop regarding impedance matching at some points as can be seen in the figures. The second conclusion is that, for smaller values of d_W, the curves of the voltage reflection coefficient ($|\Gamma_V|$) are shifted to the higher frequencies. This occurs because the impedance matching depends on the positions of the resonances of the nanoantenna, that are shifted to higher frequencies for smaller lengths of d_W. In general, the best impedance matching is obtained for smaller values of d_W. This behaviour can be explained by reduction of the reflection coefficient ($|\Gamma_V|$) of the nanoantenna when d_W decreases. But then, an increase on d_H results in increased values of $|\Gamma_V|$. This occurs because the reflection coefficient of nanoantenna increases with the increase on d_H.

5. Conclusions

An impedance matching analysis of two plasmonic nanocircuits with nanoantennas was presented. The first circuit is composed by an OTL connected between two nanodipoles, where one nanodipole is illuminated by a Gaussian beam. In the second circuit, a voltage source fed an OTL that is connected to a dipole-loop broadband nanoantenna. The linear MoM with finite surface impedance was used for numerical calculations, and some results were compared with FEM.

In the analysis of first circuit, we concluded that good impedance matching and transmission efficiency depends not only on the OTL and the emitting dipole, but also on the receiving dipole. In other words, the electromagnetic behaviour depends on the whole circuit. Also, we verified that good impedance matching does not imply on good transmission efficiency. An example of these opposite situations is presented in **Figure 9**.

In the second analysis, it was showed that a dipole-loop combined nanoantenna connected to an OTL can increase the operating bandwidth and improve the degree of impedance matching, when compared to the conventional isolated nanodipole. We obtained a best fractional bandwidth of 42%, for the dipole-loop nanoantenna, and a minimum voltage reflection coefficient of –25dB for this second nanocircuit.

Author details

Karlo Queiroz da Costa[1]*, Janilson Leão Souza[1] and Victor Dmitriev[2]

*Address all correspondence to: karlo@ufpa.br

1 Federal Institute of Education, Science and Technology of Para, Tucuruí-PA, Brazil

2 Department of Electrical Engineering, Federal University of Para, Belém-PA, Brazil

References

[1] Novotny L., Hecht B. Principles of Nano-Optics. 2nd ed. New York: Cambridge; 2012.

[2] Maier S. A. Plasmonics: Fundamentals and Applications. 1st ed. New York: Springer; 2007.

[3] Grober R. D., Schoelkopf R. J., Prober D. E. Optical antenna: towards a unity efficiency near-field optical probe. Appl. Phys. Lett. 1997;**70**(11):pp. 1354-1356.

[4] Pohl D. W. Near field optics seen as an antenna problem. In: Near Field Optics: Principles and Applications/The Second Asia-Pacific Workshop on Near Field Optics; October 20-23; Beijing, China. Singapore: World Scientific Publishing; 1999. pp. 9-21.

[5] Muhlschlegel P., Eisler H. J., Martin O. J. F., Hercht B., Pohl D. W. Resonant optical antenna. Science. 2005;**308**:pp. 1607-1608.

[6] Hecht B., Muhlschlegel P., Farahani J. N., Eisler H. J., Pohl D. W., Martin O. J. F., Biagioni P. Prospect of resonant optical antennas for nano-analysis. Chimia. 2006;**60**(11):pp. 765-769.

[7] Novotny L. Effective wavelength scaling for optical antennas. Phys. Rev. Lett. 2007;**98**:p. 266802.

[8] Bryant G. W., Abajo F. J. G., Aizpurua J. Mapping the plasmon resonances of metallic nanoantennas. Nano Lett. 2008;**8**(2):pp. 631-636.

[9] Weber D., Albella P., Gonzalez P. A., Neubrech F., Gui H., Nagao T., Hillenbrand R., Aizpurua J., Pucci A. Longitudinal and transverse coupling in infrared gold nanoantenna arrays: long range versus short range interactions regimes. Opt. Express. 2011;**19**(16):pp. 15047-15061.

[10] Giannini V., Dominguez A. I. F., Heck S. C., Maier S. A. Plasmonic nanoantennas: fundamentals and their use in controlling the radiative properties of nanoemitters. Chemical Rev. 2011;**111**(6):pp. 888-912.

[11] Bharadwaj P., Deutsch B., Novotny L. Optical antennas. Adv. Opt. Photonics. 2009;**1**:pp. 438-483.

[12] Novotny L., Hulst N. V. Antennas for light. Nat. Photonics. 2011;**5**:pp. 83-90.

[13] Lindquist N., Nagpal P., McPeak K. M., Norris D. J., Oh S. H. Engineering metallic nanostructures for plasmonics and nanophotonics. Rep. Prog. Phys. 2012;**75**:p. 036501.

[14] Biagione P., Huang J. S., Hecht B. Nanoantennas for visible and infrared radiation. Rep. Prog. Phys. 2012;**75**:p. 024402.

[15] Alu A., Engheta N. Theory, modeling and features of optical nanoantennas. IEEE Trans. Ant. Propag. 2013;**61**(4):pp. 1508-1517.

[16] Silveira G. N. M., Gabrielli L. H., Hasnain C. J. C., Figueroa H. E. H. Breakthroughs in photonics 2013: advances in nanoantennas. IEEE Phot. J. 2014;**6**(2):p. 0700706.

[17] Crozier K. B., Sundaramurthy A., Kino G, S., Quate C. F. Optical antennas: resonators for local field enhancement. J. Appl. Phys. 2003;**7**:pp. 4632-4642.

[18] Bharadwaj P., Novotny L. Spectral dependence of single molecule fluorescence enhancement. Opt. Express. 2007;**15**(21):pp. 14266-14274.

[19] Costa K. Q., Dmitriev V. Bowtie nanoantennas with polynomial sides in the excitation and emission regimes. Prog. Electro. Res. B. 2011;**32**:pp. 57-73.

[20] Taminiau T. H., Segerink F. B., Moerland R. J., Kuipers L., Hulst N. F. V. Near-field driving of a optical monopole antenna. J. Opt. A: Pure Appl. Opt. 2007;**9**:pp. S315–S321.

[21] Liaw J. W. Analysis of a bowtie nanoantenna for the enhancement of spontaneous emission. IEEE J. Sel. Top. Quantum Electron. 2008;**14**(6):pp. 1441-1447.

[22] Kinkhabwala A., Yu Z., Fan S., Avlasevich Y., Mullen K., Moerner W. E. Large single-molecule fluorescence enhancements produced by a bowtie nanoantenna. Nat. Phot. 2009;**3**:pp. 654-657.

[23] Taminiau T. H., Stefani F. D., Hulst N. F. V. Enhanced directional excitation and emission of single emitters by a nano-optical Yagi-Uda antenna. Opt. Express. 2008;**16**(14):pp. 16858-16866.

[24] Parajo M. F. G. Optical antennas focus in on biology. Nat. Phot. 2008;**2**:pp. 201-203.

[25] Atwater H. A., Polman A. Plasmonic for improved photovoltaic devices. Nat. Materials. 2010;**9**:pp. 205-213.

[26] Gupta D. D., Maltzahn G. V., Ghosh S., Bhatia S. N., Das S. K., Chakraborty S. Probing nanoantenna-directed photothermal destruction of tumors using noninvasive laser irradiation. App. Phys. Lett. 2009;**95**:p. 233701.

[27] Alu A., Engheta N. Wireless at the nanoscale: optical interconnects using matched nanoantennas. Phys. Rev. Lett. 2010;**104**(21):p. 213902.

[28] Cubukcu E., Yu N., Smythe E. J., Diehl L., Crozier K. B., Capasso F. Plasmonic laser antennas and related devices. IEEE J. Sel. Top. Quantum Electron. 2008;**14**(6):pp. 1448-1461.

[29] Huang J. S., Feichtner T., Biagione P., Hecht B. Impedance matching and emission properties of nanoantennas in an optical nanocircuit. Nano Lett. 2009;**9**(5):pp. 1897-1902.

[30] Wen J., Romanov S., Peschel U. Excitation of plasmonic gap waveguides by nanoantennas. Opt. Express. 2009;**17**(8):pp. 5925-5932.

[31] Costa K. Q., Dmitriev V., Souza J. L., Silvano G. Analysis of nanodipoles in optical nanocircuits fed by Gaussian beam. Int. J. Ant. Propag. 2014;**2014**:p. 429425.

[32] Costa K. Q., Dmitriev V. Simple and efficient computational method to analyze cylindrical plasmonic nanoantennas. Int. J. Ant. Propag. 2014;**2014**:p. 675036.

[33] Hanson G. W. On the applicability of the surface impedance integral equation for optical and near infrared copper dipole antennas. IEEE Trans. Ant. Propag. 2006;**54**(12):pp. 3677-3685.

[34] COMSOL Multiphysics 4.2a, COMSOL Inc. Available from: http://www.comsol.com./

[35] Sabaawi A. M. A., Tsimenidis C. C., Sharif B. S. Analysis and modeling of infrared solar rectennas. IEEE J. Sel. Top. Quantum Electron. 2013;**19**(3):p. 9000208.

[36] Rashidi A., Mosallaei H., Mittra R. Scattering analysis of plasmonic nanorod antennas: a novel numerically efficient computational scheme utilizing macro basis functions. J. Appl. Phys. 2011;**109**:p. 123109.

[37] Zhang K., Li D. Electromagnetic Theory for Microwaves and Optoelectronics. 2nd ed. New York: Springer; 2007.

Chapter 13

Tapered Plasmonic Nanoantennas for Energy Harvesting Applications

Youssef Mamdouh El-Toukhy,
Mohamed Farhat Othman Hameed,
Mohamed Hussein and Salah Sabry Ahmed Obayya

Additional information is available at the end of the chapter

http://dx.doi.org/10.5772/67418

Abstract

In this chapter, novel designs of tapered-dipole nanoantennas are investigated for energy harvesting applications. A full systematic analysis for the proposed structure is presented where the harvesting efficiency, return loss, radiation pattern, and near-field enhancement are calculated using a finite-element frequency domain solver. Simulation results show that the proposed nanoantennas can achieve a harvesting efficiency of 60% at a wavelength of 500 nm where the antenna input impedance is matched to that of fabricated rectifying devices. Additionally, the cross-tapered nanoantenna offers a near-field enhancement factor of 252 V/m, which is relatively high compared to previously reported nanoantennas. The spatial and spectral resonance modes are investigated, and the simulation results indicate the ability of the cross geometry to be utilized in color-sorting applications. Moreover, the particle swarm optimization technique is adapted to configure the proposed designs for maximum performance.

Keywords: nanoantenna, energy harvesting, plasmonics, finite-element frequency domain, particle swarm optimization, metal/insulator/metal diode, color-sorter

1. Introduction

The strong electromagnetic field enhancement of metallic plasmonic structures has encouraged the utilization of such devices in various applications. This enhancement is attributed to the plasmonic resonance that occurs at the near-field region of metals such as silver or gold. Plasmonic nanoantennas have gained the main interest, theoretically and experimentally, among recent researchers in the fields of energy harvesting [1–4], bio-sensing [5], plasmonic nanosensors [6], and optical communications [7]. Similar to plasmonic devices, the high-field confinement of nanoantennas is directly related to the metal refractive index and the design geometry. Other major parameters of the nanoantenna performance are harvesting efficiency, which is the efficiency by which light energy is collected, and the antenna impedance that defines the coupling efficiency between the nanoantenna and other devices.

Due to the large worldwide demand on renewable energy, an intense research work has been directed toward benefiting from the solar energy as a free and clean source. In early 1970s, Bailey has proposed the concept of rectenna in the field of solar energy harvesting [8] where rectenna stands for a rectifying diode coupled with antenna. Nanoantenna is responsible for transforming solar radiation into an alternating current (AC) electric field across the diode, which rectifies the signal to obtain direct current (DC) power. In recent years, rectennas design and fabrication techniques are getting more mature as this approach promises to offer highly efficient solar energy harvesting systems. Referring to the radiation spectrum of the sun, maximum irradiance occurs at the visible wavelength range from 400 to 700 nm. Nonetheless, extending the utilization of rectenna systems to visible frequencies (PHz) resulted in several challenges. First, the challenge of scaling down antenna dimensions to sub-micrometer is becoming more realizable with advanced technologies [1]. A second challenge is the existence of suitable rectifiers for these operation criteria. Metal\insulator\metal (MIM) diode topology is gaining interest as a candidate for high-frequency operation [9]. Additionally, the main factors contributing to the efficiency of nanorectenna systems are (1) harvesting efficiency and field confinement of the nanoantenna device; (2) the cut-off frequency of the diode and responsivity at zero voltage bias; and (3) impedance matching between the nanoantenna and the diode.

Vandenbosch and Ma [10, 11] reported an upper bound for total solar harvesting efficiency of 64%, over wavelength range 400–1200 nm, for a silver-dipole nanoantenna. This design offers a nominal input impedance of 250 Ω at resonance and harvesting efficiency of 26% at wavelength of 500 nm. Bowtie [12, 13] and Vivaldi [14] nanoantennas offered a low input impedance of about 100 Ω at resonance. However, authors study did not provide calculations for harvesting efficiency at visible light range. Other design topologies were proposed, such as spiral, log-periodic [15], flower-shaped dipole, and elliptic dipole nanoantenna [16]. While some of these designs offer higher efficiencies up to 90% at 500 nm wavelength, they suffer from very high design complexity and the lack of impedance analysis in the literature, especially over the visible light region. A tapered-dipole nanoantenna topology was first introduced in Ref. [17] where two structures were presented, two-arm and cross-arm dipole antennas with tapered end. The simulation results for these structures show high-field confinement in the antenna gap where the cross-dipole antenna exhibits the higher field enhancement. However, this study was not intended for solar harvesting application, and thus, no further analysis for the harvesting efficiency or input impedance was presented [17]. A study of the field distribution around nanoantenna and current induced on the surface was also carried out for the bowtie and spiral designs [12, 15] where simulation and experimental results showed high enhancement over the conventional dipole design [11]. However, these rectenna devices suffer from low coupling efficiency between nanoantenna and rectifier due to high impedance mismatch.

http://dx.doi.org/10.5772/67418

Moreover, cross antenna design topology has been previously investigated several times due to its compatibility with special applications such as fluorescence-sensing [17], color-sorting [18, 19], and high-harmonic generation [20]. These applications take advantage of cross antenna electromagnetic characteristics such as dual resonance, high-field confinement, and polarization control. The dual resonance characteristics of the cross antenna topology arise from the four-arm nature of the antenna and the interaction between the incident field with both x- and y-oriented arms. This phenomenon also enables the antenna to interact strongly with both polarization, horizontal and vertical. Additionally, cross nanoantennas have been previously reported to achieve higher field enhancement than the corresponding two-arm structure where Ying-Ying et al. [20] studied the field enhancement of a nanoellipsoid cross antenna which reached slightly below 200 enhancement factor. Another research work reported an enhancement factor of 320 for symmetric cross bowtie nanoantenna, on a fused silica substrate, operating as a color-sorter [19].

This chapter demonstrates two novel nanoantenna designs, which offer harvesting efficiencies up to 60% at wavelength λ of 500 nm with a field enhancement factor about 10 times greater than that of the conventional dipole design. This enhancement is attributed to high divergence of surface current over the nanoantenna surface as a result of the multiple thickness grading introduced to the design. Moreover, both nanoantennas are configured to offer a high-coupling efficiency when connected to a rectifying device by matching the antenna input impedance with that of fabricated MIM diodes operating at THz frequencies. Additionally, the radiation patterns and near-field enhancement factor are presented, which show the high performance of the cross nanoantenna structure over the two-arm dipole design. Furthermore, the near-field intensity spectrum is investigated for both designs, which indicates that the cross-tapered-dipole nanoantenna can be utilized for dual-color sorting applications where it offers a normalized near-field enhancement factor of 252 V/m. All nanoantenna designs are optimized using particle swarm optimization algorithm [21] connected to an electromagnetic solver software.

2. Design consideration

This research work focuses on overcoming the drawbacks of previously studied nanoantenna designs such as low field confinement at the gap, small nominal input impedance, and low harvesting efficiency at visible light frequencies. Moreover, this study introduces a novel design that offers relatively high manufacture feasibility as will be discussed later. For an efficient operation of rectenna system, high-intensity field should exist at the surface of the MIM diode δ_s so that sufficient electrons can tunnel through the diode. Therefore, high-field confinement at the gap of the nanoantenna represents a major requirement in the nanoantenna design. According to electromagnetic theory, electric field tends to be accumulated at the metal tips of the antenna structure [12, 22]. Based on this phenomenon, the proposed design was constructed by inserting more steps of smaller dimensions into the conventional dipole design, as shown in **Figure 1b**, which increases the number of tips and consequently the electric field confinement by the nanoantenna. This is directly related to the antenna efficiency where the harvested electric energy is transformed to AC power at the antenna port. **Figure 1** demonstrates the design parameters of the conventional dipole and the proposed tapered-dipole nanoantenna, which is considered a modified version of the conventional dipole where two more steps of smaller dimensions are inserted. Additionally, the cross structure of the same nanoantenna is investigated where the four arms of the antenna are placed in the x and y directions, horizontal and vertical orientations, as shown in **Figure 1c**. Silver material is selected for all the nanoantennas studied in this work as it is reported to offer higher efficiency when compared to other metals [11]. The proposed design parameters are chosen using particle swarm optimization (PSO) technique [21] where efficiency, impedance, and/or field enhancement are optimized. The optimization process is carried out through an external link between the algorithm and the finite-element frequency domain (FEFD) solver [23]. The PSO technique will be discussed in the next section. Additionally, a minimum feature size of 5 nm is maintained in grading adjacent dipole steps to reduce design complexity.

The suggested designs can be fabricated using top-down methods such as electron-beam lithography (EBL) and focused ion beam (FIB) techniques [14]. The EBL is employed to define desired patterns down to the nanometer scale in resists. Therefore, it is normally combined with stripping (lift-off) or etching to obtain the desired patterns in the target materials. For a specific EBL technique, reported in Ref. [24], 4–8 nm patterning and lift-off were achieved for Au nanoparticles. On contrast to EBL, the FIB technique can define patterns down to the nanoscale without using masks [25]. Most widespread instruments of FIB use gallium ion source, which is capable of fabricating sub-5 nm holes [26]. Another technique of FIB is called Helium ion beam milling, which is inherently less damaging to the sample than Ga ions but ideal for structuring thin slabs of material with high precision [27]. The helium milling was reported to achieve 5–10 nm resolution in patterning a thin gold film using milling from sides to center [27]. These fabrication methods have been commonly used in manufacturing nanoantennas with highly controllable parameters. However, the fabrication of several nanoantenna structures (such as bowtie [13], spiral [15], elliptical, and flower [16])

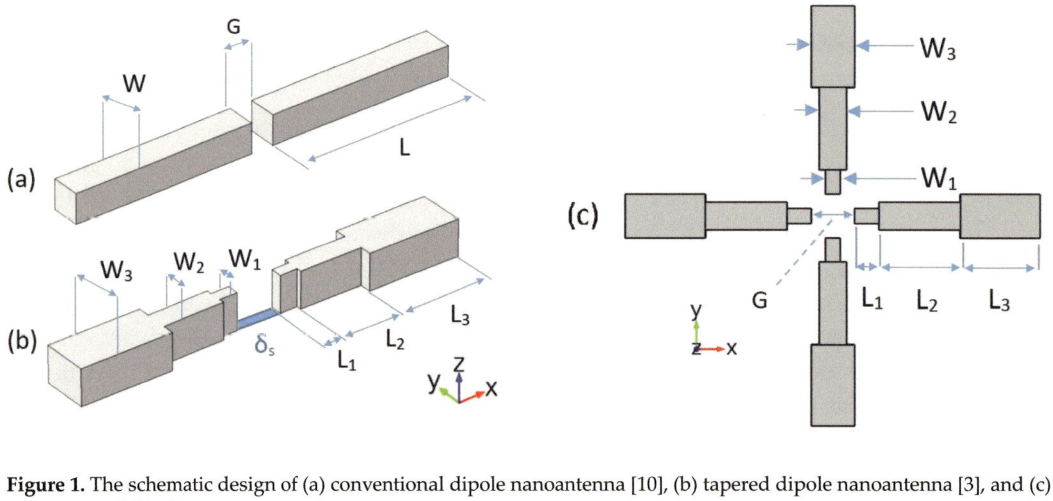

Figure 1. The schematic design of (a) conventional dipole nanoantenna [10], (b) tapered dipole nanoantenna [3], and (c) cross-tapered dipole nanoantenna.

suffers from the complexity of the lift-off process for ultra-small features and curvatures since resists are difficult to strip away. The proposed design offers high feasibility since it does not include sharp tips or curvature. Furthermore, tapered dipole and cross structures have a minimum feature size of 5 nm to be compatible with the techniques stated above and a large gap size to facilitate the fabrication process [14].

3. Particle swarm optimization

The relatively fast and accurate simulation of nanostructures operating at TeraHertz frequencies has created a space for stochastic optimizers to participate in design process refinement for various engineering electromagnetic (EM) problems where a link between the optimization algorithm and the EM solver is created. Particle swarm optimization (PSO) was previously introduced to the nanoantenna field [28] and has a great potential in the microwave antenna design recently by the authors [29, 30]. Kennedy and Eberhart were the first to introduce the PSO concept [21]. Since its creation in 1995, PSO has been reported as one of the highly efficient optimizing technique for solving complicated optimization problems even when compared to other techniques like genetic algorithms (GAs) or differential evolution [31, 32]. The PSO technique is inspired by the behavior of birds or fish, looking for food, which shows social and cooperative approach. Furthermore, for a certain EM problem, the solution is defined by N dimensions, which correspond to the design parameters to be optimized. The optimization algorithm employs the swarm intelligence to search for the optimum solution at which single or multiple performance-based parameters are maximized; in some cases, minimum value is required. These parameters are calculated by the external EM solver, which acts as fitness function evaluator; then, values are sent back to the PSO algorithm for processing.

The PSO algorithm consists of the initialization for a population (swarm) of candidate solutions called particles. The algorithm spreads these particles in the search domain, which resembles the solution space, with a prespecified velocity to find the optimal solution. Each particle possesses a memory, which is used to keep track of its previous best position. Two positions are defined after each iteration, which is the personal best position (Pbest) and the global best position (Gbest) of the whole swarm. The next step is to update the positions and velocities of the swarm particles, which represent the core of all stochastic optimization algorithms. The Gbest swarm topology [30] is utilized where Pbest is updated based on the latest fitness function value, and Gbest is obtained by sharing all known Pbest locations with the rest of the swarm and choose the location which corresponds to optimum value of the fitness function. The flow chart of the adopted PSO algorithm is shown in **Figure 2**.

At the nth iteration, the particle velocities and positions are updated by:

$$v_n = \omega * v_{n-1} + c_1 * r_1 * (Pbset_{n-1} - x_{n-1}) + c_2 * r_2 * (Gbset_{n-1} - x_{n-1}) \tag{1}$$

$$X_n = X_{n-1} + v_n \tag{2}$$

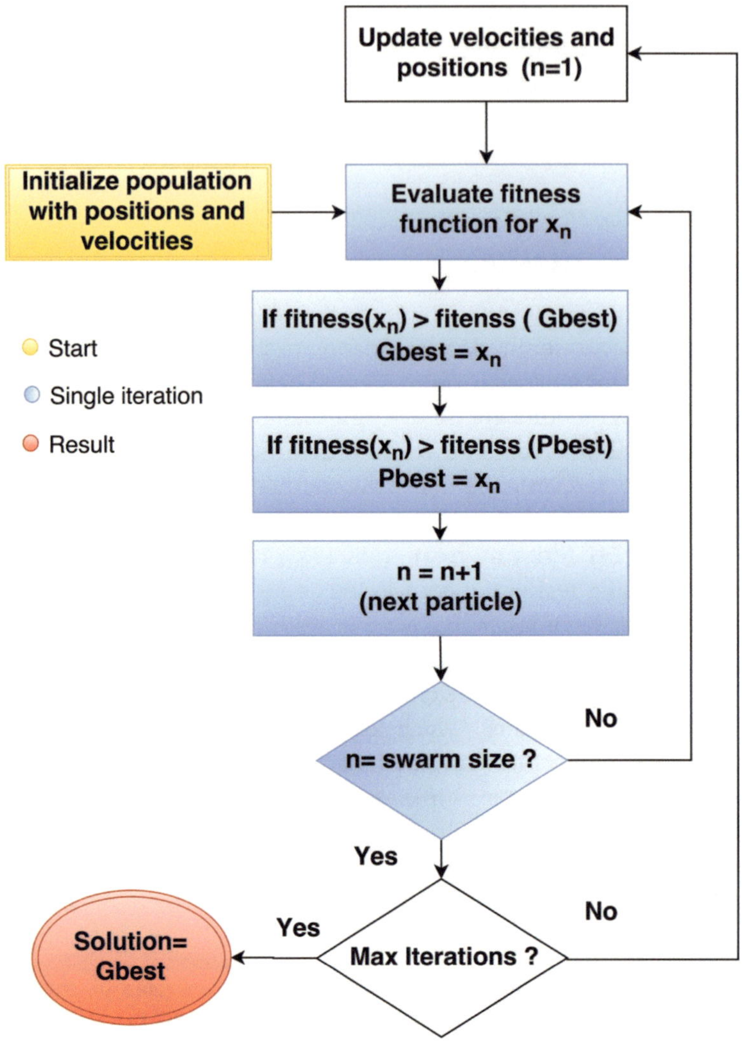

Figure 2. The flowchart of the PSO algorithm.

where r_1 and r_2 represent random numbers between 0.0 and 1.0. These random elements are introduced into the optimizer to simulate the slight unpredictable component of natural swarm behavior. Further, ω is the inertial weight, which keeps each particle in its current trajectory.

4. Simulation methodology

The proposed nanoantenna designs are analyzed using FEFD method via Comsol Multiphysics software [23] where the harvesting efficiency, return loss, and input impedance are calculated at a wavelength of 500 nm. The constructed model uses a fine mesh with a minimum and maximum element sizes of 1 and 10 nm, respectively, to resolve the skin depth of silver (3 nm

http://dx.doi.org/10.5772/67418

at 500 nm wavelength). The studied nanoantennas have a fixed thickness T of 40 nm, indicated in **Figure 1**, and are surrounded by free space everywhere as in Refs. [10, 33]. The boundary conditions are set to perfect matched layer (PML), and the permittivity of silver introduced in the model is taken from Johnson and Christy [34]. According to the reciprocity theorem, the efficiency of an antenna in the transmission mode η_{rad} is equal to the efficiency in receiving mode. The radiation efficiency of an antenna is calculated as [10]:

$$\eta_{\mathrm{rad}} = \frac{P_{\mathrm{rad}}}{P_{\mathrm{in}}} = \frac{P_{\mathrm{rad}}}{P_{\mathrm{rad}} + P_{\mathrm{loss}}} \tag{3}$$

where P_{rad} is the total radiated power, P_{in} is the power input at the antenna port, and P_{loss} represents the power dissipated in the nanoantenna material. **Table 1** shows the equations used to calculate P_{rad} and P_{loss} where S_{sc} is the scattered intensity pointing vector, J is the antenna surface current, and Q_{rl} together with Q_{ml} represent the resistive losses and magnetic losses, respectively.

$P_{rad} = \oiint S_{sc}\, ds$	$P_{\mathrm{loss}} = \iiint_{\text{antenna volume}} Q\, dV$
$S_{sc} = \frac{1}{2} Re(E \times H^{*})$	$Q = Q_{rl} + Q_{ml}$
$Q_{rl} = \frac{1}{2} Re(J \,.\, E^{*})$	$Q_{\mathrm{ml}} = \frac{1}{2} Re(j\omega B \,.\, H^{*})$

Table 1. Radiated power and power loss equations extracted from the simulated electric and magnetic fields [23, 35].

5. Results and discussion

5.1. Harvesting efficiency analysis

In this analysis, the dipole nanoantenna is excited at its gap by a voltage imposed between the two conducting arms of the dipole and thus corresponds to Thevenin equivalent circuit. This approach, also known as gap excitation, is previously reported to result in antenna impedance very similar to that calculated using conventional feeding line excitation [21]. A nominal diode impedance of 500 Ω was chosen, which corresponds to the fabricated MIM diode reported in Ref. [9]. This diode impedance acts as the Thevenin impedance of the source feeding the nanoantenna at the gap. Additionally, this analysis is carried out at λ of 500 nm where the maximum irradiance of the sun occurs. Further, to simulate the cross nanoantenna, the y-oriented and the x-oriented dipoles are shifted away from each other vertically, in z-direction, for a few nanometers so that ports could be defined in the FEFD solver. This method has been previously reported not to affect the harvesting efficiency calculations [36].

Tapered-dipole and cross nanoantennas are optimized to have resonance at λ = 500 nm with an input impedance of 500 Ω to match the diode impedance introduced to the model. The optimization process follows the PSO algorithm to maximize the specific fitness function defined by:

$$f = c_1 \times \eta_{rad} + c_2 \times |S_{11}| \quad (4)$$

where S_{11} represents a measure of the reflected power at the antenna port, which is directly related to the antenna impedance. The S_{11} is calculated in dB and has typical values of several decades in negative. Further, η_{rad} is the ratio presented in Eq. (3) and it is less than unity in magnitude. Additionally, c1 and c2 are constants that are chosen as weighting factors to compensate the magnitude variation between S_{11} and η_{rad} so that both parameters have the same influence on the fitness function. Consequently, the optimization process is not carried away by a single parameter value. These constants are chosen through an iterative process where the optimization algorithm is executed repeatedly then c1 and c2 values are adjusted after each iteration in order to optimize both S_{11} and η_{rad} fairly using the PSO algorithm. In both cases of nanoantennas, the optimizer works through a seven – dimensional solution space which resembles the structure metrics (L_1, W_1, L_2, W_2, L_3, W_3, G). Regarding the optimization of tapered-dipole nanoantenna design, the PSO algorithm converged to maximum value of fitness function in about 100 iterations. This is illustrated in **Figure 3** along with the corresponding values of S_{11} and η_{rad} at four points, which show the effectiveness of applying PSO technique in nanoantenna design problems.

For each candidate design, some constraints must be present in order to pass the values to the FEFD solver. These constraints are described as:

$$1.\ W_{i+1} > W_i \quad (5)$$

$$2.\ W_{i+1} - W_i > 5\,nm \quad (6)$$

$$3.\ \textstyle\sum_n L_i \times W_i \leq const, n = 1, 2, 3 \quad (7)$$

$$4.\ G > W_1 \quad (8)$$

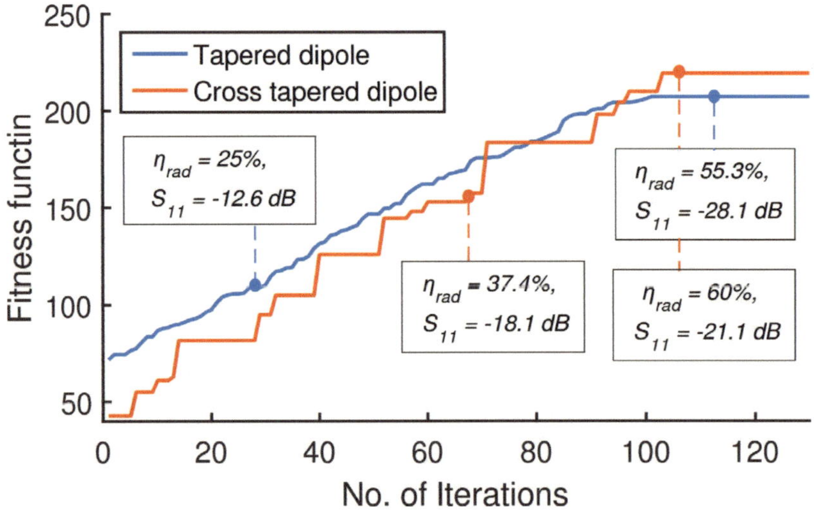

Figure 3. Fitness function (Eq. (4)) values for tapered dipole [3] and cross nanoantenna versus number of iterations of the PSO algorithm. The corresponding values of S_{11} and η_{rad} are demonstrated for four points.

where the first constraint ensures that the confinement occurs at the gap, while the second is introduced to increase design feasibility and reduces manufacture cost. Furthermore, third condition provides that optimized designs do not have exaggerated overall volume where the constant value in the equation represents a measure of the volume of the conventional dipole design in Ref. [10]. The final constraint is only applied in the cross-dipole optimization to prevent adjacent antenna arms from overlapping, which will result in a mathematical error in the FEFD solver.

The optimized nanoantenna dimensions, introduced to the FEFD analyzer, are given in **Table 2** along with the dimensions of the dipole reported in Ref. [10]. The proposed designs show high enhancement over the conventional dipole design, whereas tapered-dipole design achieves $\eta_{rad} = 55.3\ \%$ and $S_{11} = -28.1\ dB$ while the cross-dipole design offers $\eta_{rad} = 60\ \%$ and $S_{11} = -21.1\ dB$ at the same wavelength. On the other hand, the conventional dipole [4] design offers $\eta_{rad} = 26\ \%$ and a near-zero value of return loss at $\lambda = 500$ nm. This is clearly illustrated in **Figures 4** and **5** where the harvesting efficiency and return loss of the tapered-dipole designs are put in comparison with that of the conventional dipole. This enhancement is the result of the multiple tips created in the nanoantenna design, which increases the field accumulation around the antenna structure. On the other hand, the cross-tapered nanoantenna shows a slight increase in the harvesting efficiency, 5% increase. That is due to the increase in the metallic losses of the proposed cross antenna, electric and magnetic, since its volume is almost double that of the two-arm dipole antenna. Additionally, **Figure 6** shows the far-field radiation pattern for both structures, and it is shown from **Figure 6** that cross-dipole nanoantenna obtains a higher gain and a directivity in both x and y directions. This specific characteristic of the cross antenna design can be utilized in optical wireless communication [7].

Additionally, total harvesting efficiency calculations are performed for the proposed nanoantenna designs to obtain a measure of the ultimate optical efficiency over the total wavelength range of the sun irradiance. The total harvesting efficiency can be defined as [10]:

$$\eta_{\text{total}} = \frac{\int_0^\infty P(\lambda, T) \times \eta_{\text{rad}}(\lambda)\, d\lambda}{\int_0^\infty P(\lambda, T)\, d\lambda} \tag{9}$$

Antenna parameters	Conventional dipole (nm)	Tapered dipole (nm)	Cross-tapered dipole (nm)
T	40	40	40
G	10	49.8	38
L_1	250	5	7.5
W_1	40	13.4	11.5
L_2	–	70	100
W_2	–	19.5	17
L_3	–	103.5	104.4
W_3	–	51	59.6

Table 2. Deign dimensions for the proposed nanoantennas and the conventional dipole [10] for harvesting efficiency analysis.

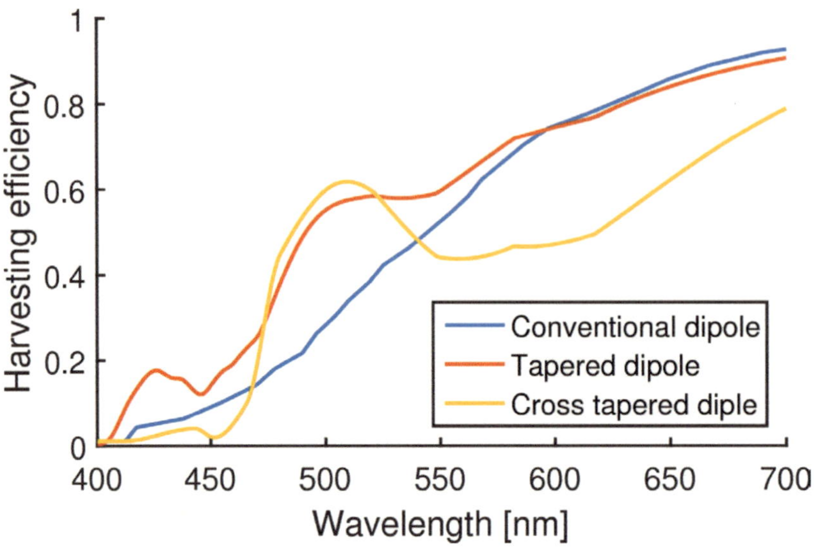

Figure 4. Harvesting efficiency versus wavelength for the conventional dipole [10], tapered dipole [3], and cross-tapered dipole nanoantennas.

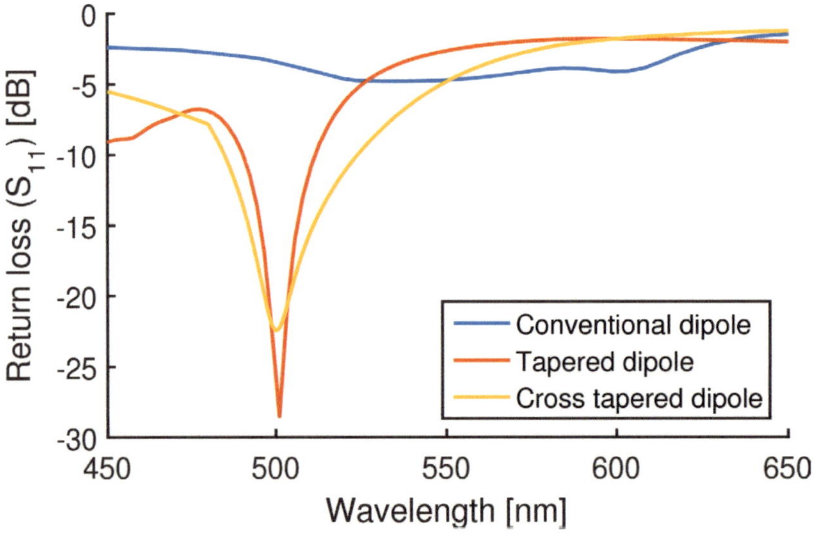

Figure 5. Wavelength dependent return loss at the nanoantenna port for conventional dipole [10], tapered dipole [3], and cross-tapered dipole nanoantennas.

where λ is the wavelength, T is the absolute temperature of the black body (in K), and P is the Planck's law for black body radiation defined in [10]. The total harvesting efficiency values for conventional dipole, tapered-dipole, and cross-tapered-dipole nanoantennas are equal to 64.1, 79.2, and 69.4%, respectively. It is worth noting that the aim of our study is to improve the nanoantenna operation at single wavelength ($\lambda = 500$ nm) where the input impedance and field confinement are optimized. However, calculating the total harvesting efficiency for the proposed design ensures a highly efficient operation at other wavelengths.

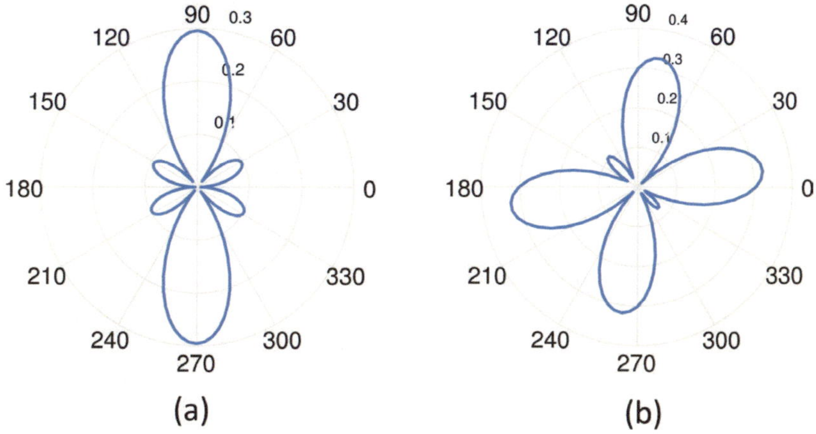

Figure 6. Far-field radiation pattern for (a) tapered dipole nanoantenna and (b) cross-tapered dipole nanoantenna.

To obtain the electric field profile of the studied nanoantenna designs, the whole structure is illuminated by a normally incident plane wave. The incident electric field has an intensity of 1 V/m and a linear polarization where, for the conventional and tapered dipole, the polarization is parallel to the antenna axis (x-axis). As for the cross-tapered-dipole nanoantenna, two polarization settings are studied, x- and y-polarization. The nanoantenna design dimensions are listed in **Table 2** as in the previous analysis. However, the gap size is held constant during this investigation at 10 nm for all design topologies due to its major effect on the field confinement through the gap. **Figure 7** illustrates the electric field profile for conventional dipole and tapered-dipole nanoantennas at $\lambda = 500$ nm. Additionally, the field profiles of the cross antenna structure are demonstrated in **Figure 8a** and **b**, which resemble x-polarized and y-polarized illuminations, respectively. As demonstrated in these figures, the electric field is accumulated with high intensity around the tips of the three nanoantennas, where a high divergence of surface current occurs [12], while the cross nanoantenna topology produces the maximum field confinement in the gap. To have a measure for this improvement, the electric field enhancement factor is calculated for the three nanoantenna structures where this factor corresponds to the maximum normalized electric field within a plane that cuts through the center of the antenna thickness [19]. As a result, the enhancement factors for the conventional dipole, tapered-dipole, and the cross-tapered-dipole nanoantennas are 12.7, 33, and 114 V/m, respectively. This high near-field intensity of the cross geometry is attributed to the field coupling between the four arms of the cross antenna structure which results in higher field confinement in the antenna gap. Another major advantage of the cross antenna structure is that it can interact efficiently with all types of polarization [19], as shown in **Figure 8**. On the contrary, two-arm dipole structures only allow for interaction with single orientation of the electric field polarization, which is parallel to the antenna axis.

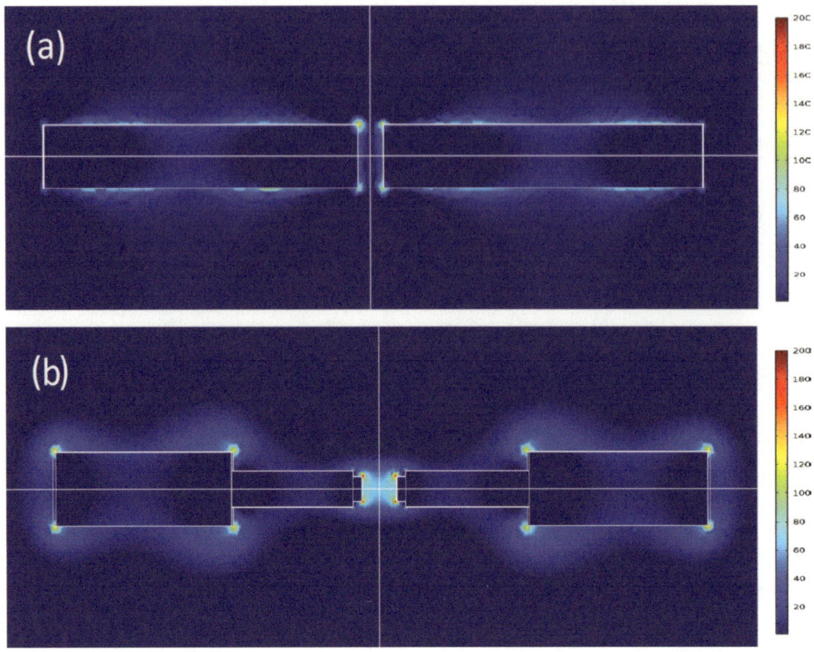

Figure 7. Electric field profile for the (a) conventional tapered dipole [10] and (b) the tapered dipole design [3].

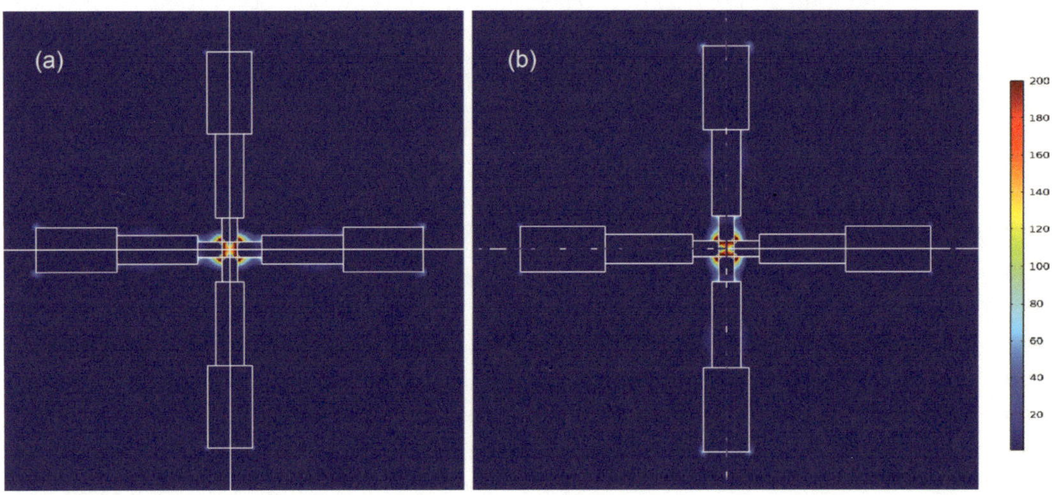

Figure 8. Electric field profile for the cross-tapered dipole nanoantenna for incident wave with polarization in (a) x-direction and (b) y-direction.

5.2. Near-field intensity spectrum analysis

In this study, the near-field intensity enhancement is investigated for the two-arm and cross-tapered-dipole nanoantennas over the range of visible wavelengths from 400 to 700 nm. The constructed model for the two antenna structures follows the same FEFD technique, meshing sizes, and boundary conditions. However, both structures are illuminated with an x-polarized electric field with amplitude of 1 V/m [19], and the permittivity of silver material is extracted from Ref. [37]. The tapered dipole and the cross-tapered dipole nanoantennas are optimized to achieve maximum near-field intensity over the visible light spectrum. Therefore, the optimization algorithm's fitness function is changed to equal to the field enhancement factor defined in the previous section. The optimized dimensions of the two-arm and cross-tapered dipole are listed in **Table 3**. It can be concluded from the optimization results, stated in **Table 3**, that maximum field enhancement occurs at the lowest values of the antenna gap. This inverse proportionality was previously reported in several studies where it was attributed to the increase of electromagnetic coupling between the metallic antenna elements as the gap size decreases [12, 19]. At this point, it is essential to notice that the optimization algorithm, in the previous section, was targeting an input impedance of 500 Ω for the two optimized nanoantenna structures in order to minimize the coupling losses for further integration with a rectifying diode. Therefore, the sizes of the antennas' gap reached values that are relatively large when compared to the gap sizes for maximum near-field intensity. This can be explained in the context of the effect of the gap size on the antenna input impedance where it was reported that the radiation resistance is directly proportional to the total electrical length of the antenna, which Is a function of total length, diameter, and gap size [37, 38]. Moreover, this study explores the utilization of the proposed nanoantenna structure to other field of applications, such as color-sorter and fluorescence-sensing [17–19].

When a dipole nanoantenna is illuminated with linearly polarized electric field, parallel to its axis, a primary plasmonic resonance occurs in the metallic nanoantenna, which is known as dipole mode [39, 40]. This dipole mode corresponds to a peak value of the near-field intensity at the resonance wavelength. **Figure 9a** shows the normalized near-field intensities for the optimized two-arm–tapered dipole, listed in **Table 2**, which indicates a maximum intensity of 51 V/m at a wavelength of 505 nm. Additionally, the field intensity distribution at the resonance wavelength, 505 nm, is presented in **Figure 9b**, which encounters a very strong electric field at the antenna gap and lower intensity at the edges of the nanoantenna outside stubs. The case is different when investigating the near-field intensity for the cross-tapered-dipole structure. Although a primary dipole plasmonic mode is excited, as in the two-arm dipole, another higher order mode exists at a second resonance wavelength [19]. This can be clearly illustrated by the

Antenna parameters	Tapered dipole (nm)	Cross-tapered dipole (nm)
T	40	40
G	5.3	18
L_1	31.3	27.6
W_1	5	17
L_2	100	94.5
W_2	15	31.8
L_3	91	92.7
W_3	60	50

Table 3. Deign dimensions for the proposed nanoantenna and the conventional dipole [10] for near-field spectrum analysis.

near-field intensity spectrum shown in **Figure 10**. It is evident from **Figure 10** that an intensity peak exists at the primary resonance wavelength of 530 nm, which resembles the dipolar mode, and a weaker blue-shifted shoulder is also present at a wavelength of 425 nm. This can be attributed to the split of the individual dipole mode into two discrete modes, which breaks the single-intensity peak into two distinct resonant peaks [40]. The values of the near-field intensity at the first and second resonance frequencies are 86 and 252 V/m, respectively, which is higher than that of the previously reported nanoantenna in the visible light region [20]. Moreover, the dual resonance characteristic of the cross nanoantenna structure allows it to be employed in fluorescence-sensing applications where the nanoantenna can be optimized to have two resonance wavelengths that correspond to the pump and emission wavelengths at shorter and longer wavelengths, respectively [17]. Consequently, a fluorescent molecule, that is coupled with the cross nanoantenna, is expected to have an enhanced emission [17].

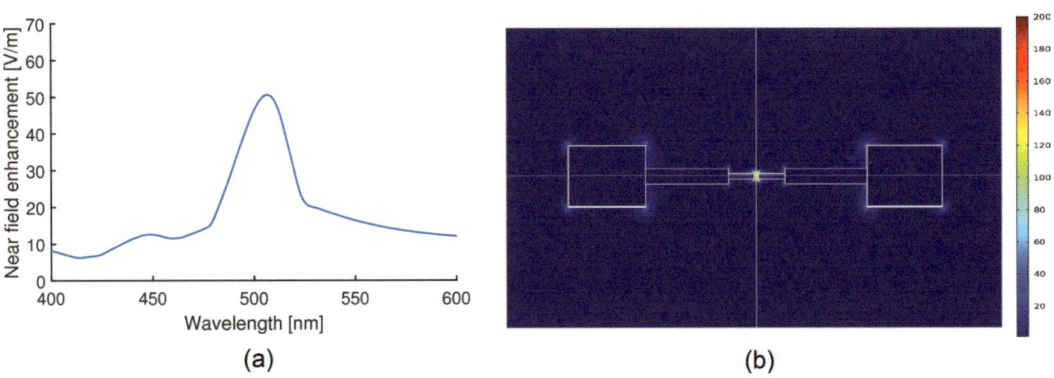

Figure 9. (a) Near-field intensity spectrum and (b) electric field profile for the optimized tapered dipole nanoantenna.

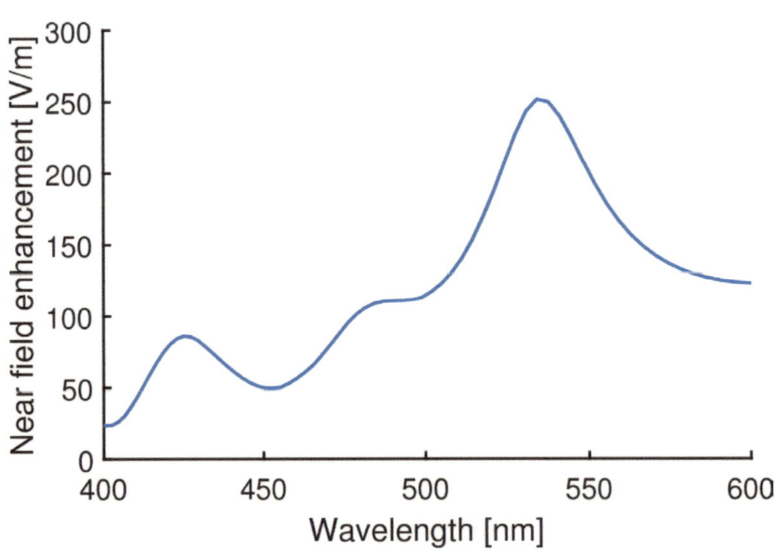

Figure 10. Near-field intensity spectrum for cross-tapered dipole nanoantenna.

Furthermore, a study of the spatial aspect of the two resonance modes of the cross nanoantenna is carried out where the field profiles associated with the two resonance wavelengths are plotted in **Figure 11**. It may be seen from **Figure 11** that the two resonant modes have totally different spatial field distributions. This can be explained through the interaction of the incident electric field with the x-oriented and y-oriented arms of the cross-tapered dipole nanoantenna. **Figure 11a** indicates that the shorter wavelength mode results from the interaction between the incident field and the y-oriented dipole of the cross antenna where the longer resonance wavelength originates from the x-oriented dipole, which is parallel to the polarization axis of the incident wave, **Figure 11b**. As a result, the proposed cross-tapered-dipole nanoantenna can be classified as a color-sorter device [19]. This class of devices is designed to control the spatial distribution of light's different wavelengths at the nanoscale. Therefore, nano-color-sorters are used to filter and steer specific spectral components of light which can be used in various applications such as ultrafast wavelength-selective photo-detection and surface-enhanced Raman spectroscopy [18]. For the specific cross-antenna dimensions, studied here, the tapered-dipole cross nanoantenna can be used to filter 425 and 530 nm wavelengths, which corresponds to violet and green colors. Moreover, the proposed nanoantenna can be optimized to filter almost any two wavelengths as it possesses many degrees of freedom, corresponding to the widths and lengths of each dipole step. Additionally, color-sorter device performance can be enhanced through the introduction of asymmetry into the nanoantenna design [19, 20, 41].

5.3. Fabrication tolerance for tapered dipole nanoantenna

While modern techniques are available for fabricating nanoantenna devices, fabrication processes at nanoscale always convey some perturbation. Therefore, a study of the fabrication tolerance is performed for tapered-dipole nanoantenna structure. This study is directed to calculate the sensitivity of the nanoantenna performance while introducing minor changes to the structure dimensions (L_1, W_1, L_2, W_2, L_3, W_3). It is worth noting that the tolerance of a specific parameter is studied while other parameters of the design are held constant at their optimum values (L_1 = 103.5 nm, W_1 = 51 nm, L_2 = 70 nm, W_2 = 19.5 nm, L_3 = 5 nm, W_3 = 13.4 nm).

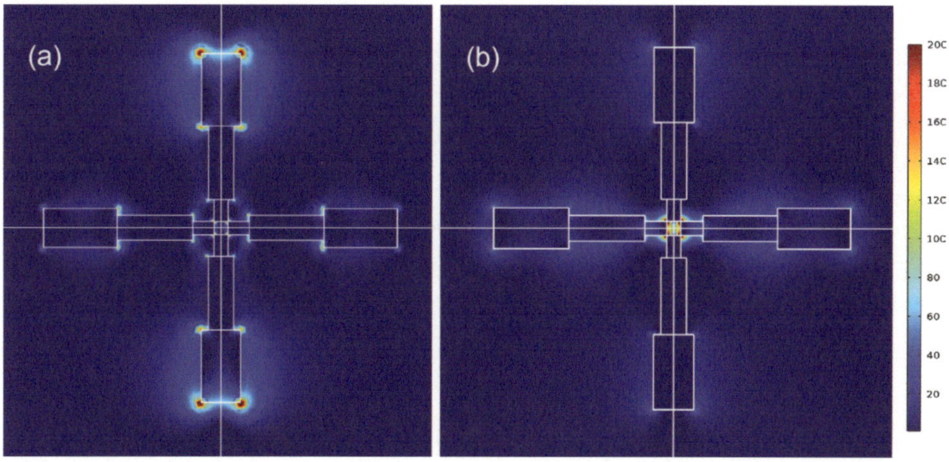

Figure 11. Electric field profile for the cross-tapered dipole nanoantenna at wavelengths of (a) 425 nm and (b) 530 nm.

Figures 12 and **13** show the variation of the harvesting efficiency and the resonance impedance with the studied parameters W_1 and L_1, respectively. Additionally, the summary of the tolerance study is listed in **Table 4**. It is evident from **Figures 12** and **13** and **Table 4** that the proposed design has a tolerance of ±5% at which the harvesting efficiency is still higher than 53.8% with a maximum deviation of ±58 Ω from the optimum input impedance of the nanoantenna. Additionally, a maximum wavelength resonance shift of 10 nm is obtained within the tolerance of ±5% of the studied parameters. Therefore, it is evident from the above results that the proposed nanoantenna design bears high robustness for fabrication imperfection.

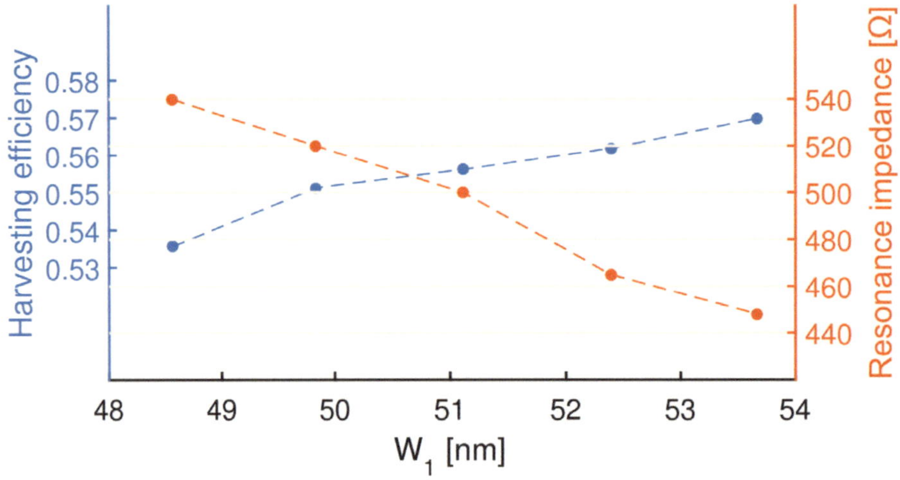

Figure 12. Values of harvesting efficiency and resonance impedance with 5% variation of W_1 parameter.

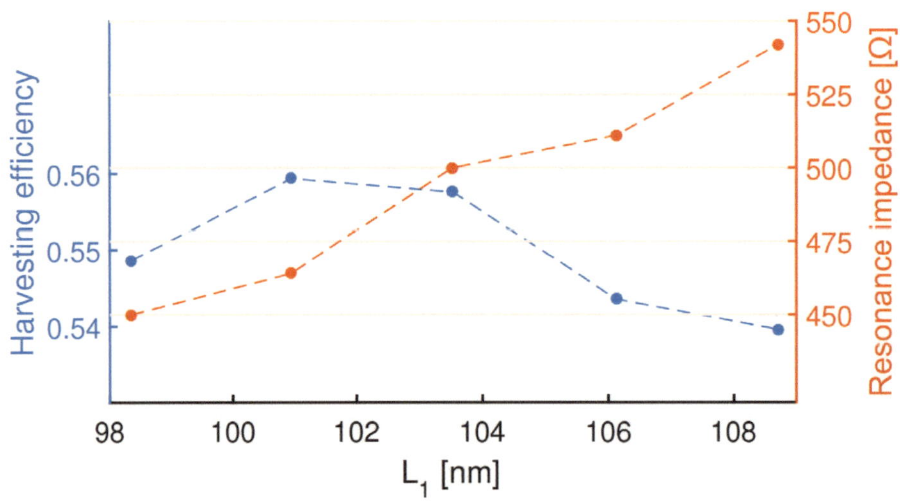

Figure 13. Values of harvesting efficiency and resonance impedance with 5% variation of L_1 parameter.

http://dx.doi.org/10.5772/67418

Parameter	Tolerance	Harvesting efficiency	Impedance
L_1	103.5 nm ± 5%	Higher than 53.9%	500 Ω ± 54 Ω
W_1	51 nm ± 5%	Higher than 53.8%	500 Ω ± 58 Ω
L_2	70 nm ± 5%	Higher than 55.2%	500 Ω ± 35 Ω
W_2	19.5 nm ± 5%	Higher than 55.25%	500 Ω ± 31 Ω
L_3	5 nm ± 5%	Higher than 55.45%	500 Ω ± 30 Ω
W_3	13.4 nm ± 5%	Higher than 55.5%	500 Ω ± 27 Ω

Table 4. Fabrication tolerance for tapered dipole design parameters [3] at $\lambda = 500$ nm.

6. Conclusion

In this chapter, a full study for novel-tapered nanoantenna designs have been implemented where antenna input impedance, return loss, harvesting efficiency, and field confinement were calculated using FEFD method. The simulation results show that the tapered design can achieve a harvesting efficiency of 60% at wavelength of 500 nm, which is higher than that of conventional dipole counterpart by 34%. Further, the antenna input impedance is tuned to match the impedance of fabricated diodes. Additionally, the cross-tapered-dipole nanoantenna achieved a near-field intensity enhancement of 252 V/m which is higher than previously reported symmetric cross nanoantenna. The spatial and spectral distinction of the resonance modes in the cross structure allows the cross-tapered-dipole to be utilized in color-sorting and fluorescence-sensing applications. Finally, the fabrication tolerance study shows that the proposed nanoantennas bear high robustness for fabrication imperfection.

Author details

Youssef Mamdouh El-Toukhy[1,2], Mohamed Farhat Othman Hameed[1,2], Mohamed Hussein[1,3] and Salah Sabry Ahmed Obayya[1,2*]

*Address all correspondence to: sobayya@zewailcity.edu.eg

1 Centre for Photonics and Smart Materials, Zewail City of Science and Technology, Giza, Egypt

2 Faculty of Engineering, Mansoura University, Mansoura, Egypt

3 Department of Physics, Faculty of Science, Ain Shams University, Cairo, Egypt

References

[1] D. K. Kotter, S. D. Novack, W. D. Slafer, and P. J. Pinhero, "Theory and manufacturing processes of solar nanoantenna electromagnetic collectors," *Journal of Solar Energy Engineering*, vol. 132, pp. 011014-011014, 2010.

[2] S. Obayya, N. Areed, M. F. O. Hameed, and M. Hussein, "Optical nano-antennas for energy harvesting," *Innovative Materials and Systems for Energy Harvesting Applications,* IGI, Italy, p. 26, 2015.

[3] Y. M. El-Toukhy, M. Hussein, M. F. O. Hameed, A. M. Heikal, M. M. Abd-Elrazzak, and S. S. A. Obayya, "Optimized tapered dipole nanoantenna as efficient energy harvester," *Optics Express,* vol. 24, pp. A1107-A1122, 2016.

[4] M. Hussein, N. F. F. Areed, M. F. O. Hameed, and S. S. A. Obayya, "Modified elliptical nanoantenna for energy harvesting applications," in *2016 IEEE/ACES International Conference on Wireless Information Technology and Systems (ICWITS) and Applied Computational Electromagnetics (ACES),* 2016, pp. 1-2.

[5] F. B. Zarrabi, M. Naser-Moghadasi, S. Heydari, M. Maleki, and A. S. Arezomand, "Cross-slot nano-antenna with graphene coat for bio-sensing application," *Optics Communications,* vol. 371, pp. 34-39, 2016.

[6] N. Verellen, P. Van Dorpe, C. Huang, K. Lodewijks, G. A. E. Vandenbosch, L. Lagae, *et al.*, "Plasmon line shaping using nanocrosses for high sensitivity localized surface plasmon resonance sensing," *Nano Letters,* vol. 11, pp. 391-397, 2011.

[7] A. M. Ghanim, M. Hussein, M. F. O. Hameed, A. Yahia, and S. S. A. Obayya, "Highly directive hybrid Yagi-Uda nanoantenna for radiation emission enhancement," *IEEE Photonics Journal,* vol. 8, pp. 1-12, 2016.

[8] R. L. Bailey, "A proposed new concept for a solar-energy converter," *Journal of Engineering for Power,* vol. 94, pp. 73-77, 1972.

[9] S. Grover and G. Moddel, "Applicability of metal/insulator/metal (MIM) diodes to solar rectennas," *IEEE Journal of Photovoltaics,* vol. 1, pp. 78-83, 2011.

[10] G. A. E. Vandenbosch and Z. Ma, "Upper bounds for the solar energy harvesting efficiency of nano-antennas," *Nano Energy,* vol. 1, pp. 494-502, 2012.

[11] Z. Ma and G. A. E. Vandenbosch, "Optimal solar energy harvesting efficiency of nano-rectenna systems," *Solar Energy,* vol. 88, pp. 163-174, 2013.

[12] M. N. Gadalla, M. Abdel-Rahman, and A. Shamim, "Design, optimization and fabrication of a 28.3 THz nano-rectenna for infrared detection and rectification," *Scientific Reports,* vol. 4, p. 4270, 03/06/online 2014.

[13] H. Fischer and O. J. F. Martin, "Engineering the optical response of plasmonic nanoantennas," *Optics Express,* vol. 16, pp. 9144-9154, 2008.

[14] Z. Iluz and A. Boag, "Dual-Vivaldi wideband nanoantenna with high radiation efficiency over the infrared frequency band," *Optics Letters,* vol. 36, pp. 2773-2775, 2011.

[15] E. Briones, J. Alda, and F. J. González, "Conversion efficiency of broad-band rectennas for solar energy harvesting applications," *Optics Express,* vol. 21, pp. A412-A418, 2013.

[16] M. Hussein, N. F. Fahmy Areed, M. F. O. Hameed, and S. S. Obayya, "Design of flower-shaped dipole nano-antenna for energy harvesting," *Optoelectronics IET,* vol. 8, pp. 167-173, 2014.

[17] J. L. Stokes, Y. Yu, Z. H. Yuan, J. R. Pugh, M. Lopez-Garcia, N. Ahmad, *et al.*, "Analysis and design of a cross dipole nanoantenna for fluorescence-sensing applications," *Journal of the Optical Society of America B*, vol. 31, pp. 302-310, 2014.

[18] E. Segal, A. Weissman, D. Gachet, and A. Salomon, "Hybridization between nanocavities for a polarimetric color sorter at the sub-micron scale," *Nanoscale*, vol. 8, pp. 15296-15302, 2016.

[19] Z. Zhang, A. Weber-Bargioni, S. W. Wu, S. Dhuey, S. Cabrini, and P. J. Schuck, "Manipulating nanoscale light fields with the asymmetric Bowtie nano-colorsorter," *Nano Letters*, vol. 9, pp. 4505-4509, 2009.

[20] Y. Ying-Ying, L. Qian-Guang, Y. Hai-Juan, and L. Xue-Chun, "The generation of MHz isolated XUV attosecond pulses by plasmonic enhancement in a tailored symmetric Ag cross nanoantenna with a few-cycle laser," *Laser Physics*, vol. 23, p. 045301, 2013.

[21] J. Kennedy and R. Eberhart, "Particle swarm optimization," in *IEEE International Conference on Neural Networks, 1995. Proceedings*, 1995, pp. 1942-1948, vol. 4.

[22] F. F. K. Hussain, A. M. Heikal, M. F. O. Hameed, J. El-Azab, W. S. Abdelaziz, and S. S. A. Obayya, "Dispersion characteristics of asymmetric channel plasmon polariton waveguides," *IEEE Journal of Quantum Electronics*, vol. 50, pp. 474-482, 2014.

[23] C. Multiphysics®, 5.1, Ed., COMSOL AB, Stockholm, Sweden, 2016.

[24] W. Hu, K. Sarveswaran, M. Lieberman, and G. H. Bernstein, "Sub-10 nm electron beam lithography using cold development of poly(methylmethacrylate)," *Journal of Vacuum Science & Technology B*, vol. 22, pp. 1711-1716, 2004.

[25] J.-T. Lv, Y. Yan, W.-K. Zhang, Y.-H. Liu, Z.-Y. Jiang, and G.-Y. Si, "Plasmonic nanoantennae fabricated by focused Ion beam milling," *International Journal of Precision Engineering and Manufacturing*, vol. 16, pp. 851-855, 2015.

[26] L. Chih Jen, A. Thomas, and B. Alexey, "Fabrication of symmetric sub-5 nm nanopores using focused ion and electron beams," *Nanotechnology*, vol. 17, p. 3264, 2006.

[27] P. F. A. Alkemade and E. van Veldhoven, "Deposition, milling, and etching with a focused helium ion beam," in *Nanofabrication: Techniques and Principles*, M. Stepanova and S. Dew, Eds., Vienna: Springer Vienna, 2012, pp. 275-300.

[28] Y. M. El-Toukhy, A. M. Heikal, M. F. O. Hameed, M. M. Abd-Elrazzak, and S. S. A. Obayya, "Optimization of nanoantenna for solar energy harvesting based on particle swarm technique," in *2016 IEEE/ACES International Conference on Wireless Information Technology and Systems (ICWITS) and Applied Computational Electromagnetics (ACES)*, 2016, pp. 1-2.

[29] N. Jin and Y. Rahmat-Samii, "Particle swarm optimization for antenna designs in engineering electromagnetics," *Journal of Artificial Evolution and Applications*, vol. 2008, 2008.

[30] S. Xu, Y. Rahmat-Samii, and D. Gies, "Shaped-reflector antenna designs using particle swarm optimization: An example of a direct-broadcast satellite antenna," *Microwave and Optical Technology Letters*, vol. 48, pp. 1341-1347, 2006.

[31] P. J. Angeline, "Evolutionary optimization versus particle swarm optimization: philosophy and performance differences," presented at the Proceedings of the 7th International Conference on Evolutionary Programming VII, 1998.

[32] J. Vesterstrom and R. Thomsen, "A comparative study of differential evolution, particle swarm optimization, and evolutionary algorithms on numerical benchmark problems," in *Proceedings of the 2004 Congress on Evolutionary Computation* (*IEEE Cat. No.04TH8753*), 2004, pp. 1980-1987, vol. 2.

[33] E. A. Soliman, M. O. Sallam, and G. A. E. Vandenbosch, "Plasmonic grid array of gold nanorods for point-to-point optical communications," *Journal of Lightwave Technology*, vol. 32, pp. 4898-4904, 2014.

[34] P. B. Johnson and R. W. Christy, "Optical constants of noble metals," *Physical Review B*, vol. 6, p. 4370, 1972.

[35] A. a. E. Alù, Nader, "Input impedance, nanocircuit loading, and radiation tuning of optical nanoantennas," *Physical Review Letters*, vol. 101, p. 4, Jul 2008.

[36] C. Y. Chin and C. F. Jou, "A slim horizontally polarized omnidirectional antenna based on turnstile slot dipole,," *Progress in Electromagnetics Research C*, vol. 50, pp. 75-85, 2014.

[37] E. D. Palik, *Handbook of Optical Constants of Solids*. Orlando: Academic Press, 1985.

[38] J.-J. Greffet and F. Marquier, "Impedance of a nanoantenna and a quantum emitter," in *Frontiers in Optics 2011/Laser Science XXVII*, San Jose, California, 2011, p. LWG4.

[39] G. W. Bryant, F. J. García de Abajo, and J. Aizpurua, "Mapping the plasmon resonances of metallic nanoantennas," *Nano Letters*, vol. 8, pp. 631-636, 2008.

[40] X. Ren, W. E. I. Sha, and W. C. H. Choy, "Tuning optical responses of metallic dipole nanoantenna using graphene," *Optics Express*, vol. 21, pp. 31824-31829, 2013.

[41] J. Chen and Z. Zhang, "Bowtie nanoantennas with symmetry breaking," *Journal of Nanophotonics*, vol. 9, pp. 093798-093798, 2014.

Section 4

Applications to Biology and Chemistry

Chapter 14

Plasmonics Devoted to Photocatalytic Applications in Liquid, Gas, and Biological Environments

Carlos J. Bueno-Alejo, Adriana Arca-Ramos and Jose L. Hueso

Additional information is available at the end of the chapter

http://dx.doi.org/10.5772/intechopen.68812

Abstract

Plasmonic nanomaterials have emerged in the last years as a very interesting option for many photocatalytic processes. Their localized surface plasmon resonance (LSPR) brings in some unique properties that overcome some of the drawbacks associated with traditional photocatalysis based on semiconductors. Even when in its infancy, many advances have been made in the field, mainly related to the synthesis of new structures with the capabilities of light absorption in the whole solar spectrum. A great number of reactions have been attempted using nanoplasmonic materials. In this chapter, we present the most recent advances made in the field of plasmonic photocatalysis comprising an introductory section to define the main types of plasmonic nanomaterials available, including the most recently labeled alternatives. Following with the major areas of catalytic application, a second section of the chapter has been devoted to liquid-phase reactions for the treatment of pollutants and a selection of organic reactions to render added-value to chemicals under mild conditions. The third part of the chapter addresses two specific applications of nanoplasmonic photocatalysts in gas-phase reactions involving the remediation of volatile organic compounds and the transformation of carbon dioxide into valuable energy-related chemicals. Finally, a fourth section of the chapter introduces the most recent applications of plasmonics in biochemical processes involving the regulation of cofactor molecules and their mimetic behavior as potential enzyme-like surrogates.

Keywords: surface plasmon, photocatalysis, NIR, pollutants, energy, VOCs, CO_2, semiconductor, mimetic enzymes

1. Introduction

Solar energy harvesting has become an important matter nowadays.The use of solar light as energy source is a cleaner and even a more economic way to provide energy to processes. In this sense, conventional photocatalysis, as it makes use of mainly semiconductors as catalyst, is not the most efficient way to harvest the light coming from the sun, since it only contains 5% of UV light, which is the kind of radiation necessary to excite those catalysts [1]. In the quest of new materials that could help to improve the efficiency of traditional photocatalyst, plasmonic nanomaterials have attracted much attention lately among the scientific community. This is so because this new family of materials offers/exhibits several advantages over their bulk counterparts that can be useful in very different fields like sensors, biomedicine, photocatalysis, etc [2]. Besides their size, what makes these materials so interesting is their characteristic localized surface plasmon resonance band (LSPR). This LSPR can be defined as the collective coherent oscillation of free charges on the surface of the nanoparticle when they are in resonance with the incident light. This resonance generates an enhancement in the electric field on the surface of the nanomaterials that may affect molecules or materials surrounding it. The LSPR can be very useful for photocatalysis in many different aspects depending mostly on the type of plasmonic material used. We have tentatively classified them into three main categories: (1) Pure plasmonic materials; (2) Hybrid plasmonic materials and (3) Alternative plasmonic materials.

1.1. Pure plasmonic materials

The photocatalysts belonging to this group are only composed by unsupported plasmonic metals and those supported on photocatalytic insulators allowing all the chemistry to happen on the metal nanomaterial. In this case, after excitation of the plasmon band, a high density of charge carriers are generated on the surface of the nanoparticle that can undergo reduction or oxidation reactions with the molecules adsorbed on their surfaces [3].

This mechanism is generally referred to as direct charge transfer since the carrier is directly transferred from higher energy levels in the metal to the lower unoccupied molecular orbital of the adsorbate initiating the chemical reaction (**Figure 1**). Another property of the plasmonic nanoparticles is the generation of heat on their vicinity. Following the excitation, the charges generated can suffer radiative or nonradiative relaxation. The former will be discussed later in this section. The nonradiative relaxation is mainly related with electron-electron and electron-phonon interactions, which give rise to the generation of heat, very localized in the surface of the nanoparticle. This heat can be transferred to the surrounding media, helping in thermally activated reactions. The temperature reached on the surface of the metal is dependent on the nature of the metal itself. For gold, it has been estimated to be as high as 500–700°C [4] and for silver, is expected to be much higher since the absorption cross section of the latter is around 10 times higher than for gold. An example of photocatalytic reaction performed by gold plasmonic nanoparticle (AuNP) is given by Hallet-Tapley et al., using them for the selective oxidation of alcohols [5]. They used green laser and LED in order to excite the plasmon band of the AuNP and oxidize selectively the alcohol molecules surrounding the AuNP. In their work, the authors proposed two possible mechanisms for the reaction, one involving a single electron transfer from gold to H_2O_2 providing the radicals that initiate a chain reaction, and the other implicating the generation of heat as the cause for the breaking of the peroxide bond. Another interesting example is the use of nanorattles composed of a gold nanosphere inside and a nanoshell of gold/silver [6]. In this case, the authors claim that the better performance of this structure comes from the formation of electromagnetic hot spots in the interface of the two different materials. These hot spots create a higher density of reactive oxygen species responsible for the oxidation reaction. In order to expand the absorption of light to the visible and even the IR part of the spectrum, many researchers have tried to synthesize anisotropic nanoparticles. One of these examples is the synthesis of hexagonal Pd nanoplates for their use in the Suzuki coupling reaction. Trinh et al. [7] were able to obtain plasmonic Pd nanoparticles with their plasmon band located along the visible and the near IR according to their aspect ratio. They used this LSPR to generate hot electrons with the energy to perform the Suzuki reaction on the surface of the catalyst.

In this group, we can also include nanomaterials consisting of metal nanoparticles supported on photochemical insulators. The latter means that since the bandgap of the material is too high, it cannot be excited in the visible or near UV part of the spectrum. Metal oxides like ZrO_2, Al_2O_3, or SiO_2 with bandgap energies above 5–6 eV are representative of this group. Liu et al. [8] performed a selective reduction of nitrocompounds to the azoxy derivatives using Ag-Cu alloy nanoparticles supported on ZrO_2. They demonstrated that by irradiating in the

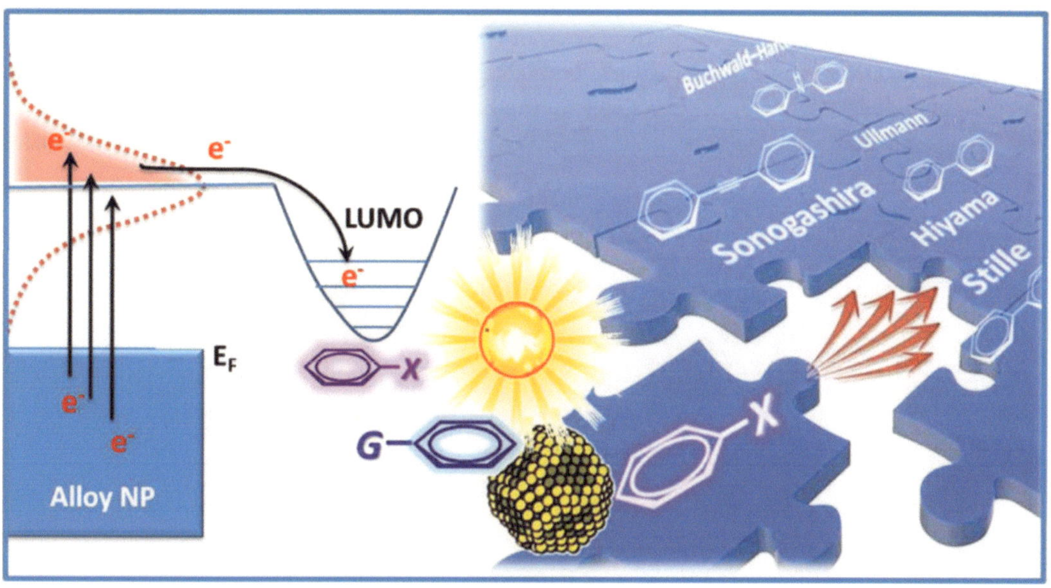

Figure 1. Typical mechanism of pure plasmonic materials. Reprinted with permission from Ref. [3]. Copyright (2014) American Chemical Society.

visible region, they could control the selectivity of the reaction only with the composition of alloy with no influence from the support.

1.2. Hybrid plasmonic materials

Semiconductors photocatalysts have dominated the field since Fujishima and Honda used TiO_2 to do water splitting with UV light in 1972 [9]. Despite the convenient properties of this semiconductor for photocatalysis, it also has some drawbacks in order to use them in real photocatalytic applications like low photon efficiency, high charge carriers recombination rate and a high bandgap (3.2 eV), that means it can only be excited with UV light. To overcome these problems, great efforts have been devoted in the past years to develop a new family of photocatalysts consisting of supporting metal plasmonic nanoparticles on the surface of semiconductors. The metal nanoparticle brings in some properties that complement and fix some of the downsides previously mentioned for the semiconductors (**Figure 2**). First, they enhance the absorption of light bythe material. They do this in different ways. The plasmon bands of the metal nanoparticles are usually in the visible region, expanding the absorption of the hybrid material to the visible or even near IR. Also, the huge absorption extinction coefficient of plasmonic materials allows them to absorb most of the light incoming to the hybrid, therefore, keeping most of the photon absorbed on the surface of the material, avoiding in a high extent the recombination of carriers that usually takes place in the semiconductor. Another feature of these materials that aids the absorption of light is the scattering. Plasmonic nanoparticles can scatter part of the light that are not able to absorb, thereby allowing its reemission and potential absorption by other nanoparticles allocated on its vicinity and increasing the pathway of light absorbed by the material. Another characteristic of these materials is the formation of what is called Schottky barrier between the metal and the semiconductor. In that interface, an electric field is built in that separates the charge carriers formed in or close to the barrier, avoiding in that way of the recombination of electron and holes, and therefore increasing the lifetime of those carriers, so they can react with other molecules adsorbed on the surface.

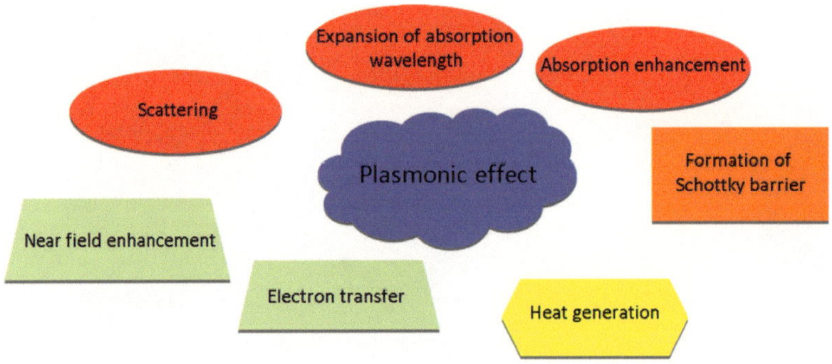

Figure 2. Different plasmonic effects in hybrid plasmonic structures. Different colors and shapes are used to identify each of them: absorption (red oval), structural (orange square), nonradiative (yellow hexagon) and radiative (green trapezium).

Besides absorption, plasmonic materials contribute to improve the photocatalytic performance of semiconductors through other mechanisms. After absorption of light, the excited electrons reach high-energy levels, so they can overcome the Schottky barrier formed between the metal and the semiconductor, and can be transferred from the nanoparticle to the semiconductor and start the photocatalytic process. This electron transfer means that a material that was only viable before for use with UV light due to its high bandgap, now can be used in almost any wavelength, because the shape and size of these metal nanoparticles can be tailored to have the plasmon band in the visible or even in the IR region of the spectrum. This is a very important feature when we think about the use of these materials under solar light illumination, since the excitation area can be expanded to the other 95% of the solar spectrum. Other mechanism by which plasmonic materials can enhance the photocatalytic performance of semiconductors is by the generation of an intensive local electric field that can promote/enhance the generation of more charge carriers on the surface of the semiconductors. Because of this electric field, plasmonic nanoparticles can also generate charge carriers without any contact with the support, which is beneficial when the contact between the metal and the semiconductor is not well enough to form a Schottky barrier, or when a layer of other material is needed to support the nanoparticles. It has also been probed that this electric field may help to polarize nonpolar molecules surrounding the metal nanoparticles and increase their adsorption to the surface of the semiconductor. Finally, another important feature of the plasmonic materials that enhances the photocatalytic response of these hybrid systems is the generation of heat on the vicinity of the nanoparticles. The high temperature generated in highly localized points can improve the reactions rate and the mass transfer in the reaction system.

Due to these properties, plasmonic hybrid materials are certainly a very interesting option to make some industrial processes environmentally and economically attractive.

1.3. Alternative plasmonic materials

Until very recently, nonmetal materials considered were not able to hold a LSPR due to the low free charge carrier concentration on their surface. However, lately many studies have probed that a whole variety of materials can be tuned to increase the free carrier concentration, and therefore be able to behave as plasmonic materials in the visible and IR regions [10]. Usually, doping the structure of metal oxides or adding oxygen vacancies increase the concentration of free electron within the metal oxide, and is a good way to get a plasmonic material. Examples of this are oxides of aluminum, zinc, cadmium, or tungsten, which after doping with different element acquire plasmonic characteristics in the visible or IR region due to the increase of the electron concentration [11]. Creating oxygen vacancies is also possible to get an analogous effect. For instance, tungsten oxide renders a plasmonic behavior in the IR, allowing their use in photocatalytic applications after the induced generation of oxygen vacancies [12]. A different circumstance happens with chalcogenides like CuS, in which the reduction of the Cu percentage creates an increase of the hole concentration, which results in a wide plasmon band in the IR region (**Figure 3**) [13]. All these materials are suitable to act as plasmonic photocatalysts with any of the mechanisms mentioned before.

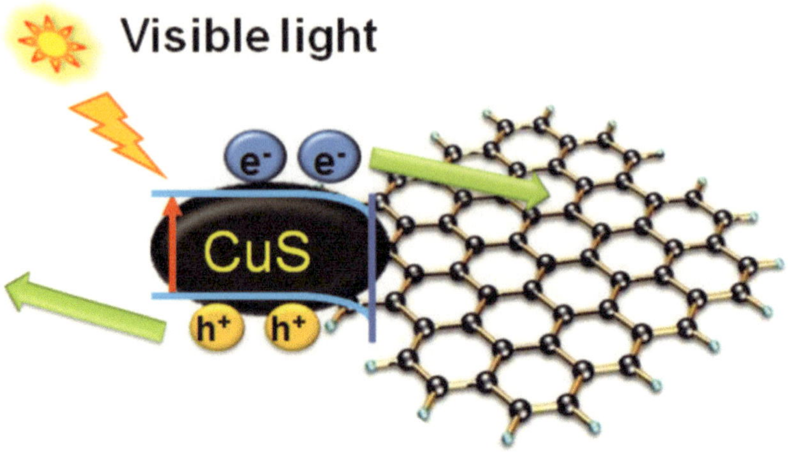

Figure 3. Schematic diagram (not to scale) to illustrate the energy level alignment and photocharge carrier dissociation at the CuS-RGO interface. Reprinted with permission from Ref. [13]. Copyright (2012) American Chemical Society.

2. Nanoplasmonic photocatalysis in liquid phase

2.1. Plasmonic-driven photocatalytic degradation of aqueous contaminants

Heterogeneous photocatalysis has been demonstrated as an economical and effective technology to remove organic pollutants from aqueous environments [14, 15]. In recent years, the LSPR effect of plasmonic materials has been applied to boost the photocatalytic performance of chemical transformations under visible light irradiation. A vast number of studies have focused on the preparation of plasmonic nanostructures and further application for the removal of pollutants from aqueous media. As there are some previous reviews reporting the use of plasmonic materials for pollutant removal [16–18], in this section we only show some recent applications on this topic, aiming at illustrating various types of plasmonic photocatalysts as well as target contaminants. These are summarized in **Table 1**.

Among the target pollutants, dyes such as rhodamine B (RhB), methylene blue (MB), or methyl orange (MO) are the most studied ones due to the ease of monitoring their removal. Nevertheless, in dye compounds the effect of photosensitization could produce extra electrons and accelerate the photocatalytic process. For this reason, the photoremediation of colorless target contaminants represents a more interesting case study [19]. For instance, the photocatalytic removal of other organic pollutants such as phenolic compounds, including bisphenol A, nitrophenol, and chlorophenols, antibiotics such as tetracycline or ciprofloxacine, organic solvent as trichloroethylene or even chlorinated paraffins have also been reported and summarized in **Table 1**. In addition to organic pollutants, this technology has been successfully applied for the treatment of hazardous inorganic compounds, such as the reduction of carcinogenic Cr (VI), from $K_2Cr_2O_7$ to Cr (III) [20].

Catalyst	Pollutant	Irradiation	Performance	Refs.
$Ag@g\text{-}C_3N_4@BiVO_4$	TC	300 *W* Xe lamp, $\lambda > 350$ nm, $\lambda > 420$ nm or $\lambda > 760$ nm	Removal of 90.8% ($\lambda > 350$ nm), 82.7% ($\lambda > 420$ nm), and 12.6 % ($\lambda > 760$ nm) in 60 min	[27]
$Ag_2SO_3/AgBr$ $GO/Ag_2SO_3/AgBr$	MO, RhB, MB	500 W Xe lamp, $\lambda > 420$ nm, 100 mW/cm^2	99.9% removal of MO in 9 min, almost complete removal of RhB and MB in 20 min	[28]
$Ag/AgCl\text{-}Bi_2WO_6@Fe_3O_4@SiO_2$	RhB, Phenol	300 W halogen tungsten projector lamp, $\lambda > 410$ nm	100% RhB removal in 90 min ~70% Phenol removal in 300 min	[19]
$Pd/PdCl_2\text{-}Bi_2MoO_6$	Phenol	300 W halogen tungsten projector lamp, $\lambda > 410$ nm	22.2% removal in 30 min and almost 100% removal in 300 min	[26]
$g\text{-}C_3N_4@Bi@Bi_2WO_6$	2,4-DCP, MO, RhB	300 W Xe lamp, $\lambda > 400$ nm	~70% removal of MO and 2,4-DCP, and almost complete removal of RhB in 120 min	[22]
Ag@AgCl/TP	RhB, X-3B, CIP, phenol	300 W Xe lamp, $\lambda > 400$ nm	Complete removal of RhB in 8 min and of X-3B in 12 min, 48% CIP removal and 51% phenol removal in 3 h	[29]
Au-Fe-doped $Bi_4Ti_3O_{12}$ (BTO) nanosheets	Phenol, BPA	400 *W* halogen lamp, VIS light	BPA removal of 72%, 87%, 99% on pure BTO, 2%Fe/BTO and Au-2%Fe/BTO, respectively, in 50 min Phenol removal of 37, 54, and 64% on BTO, 2% Fe/BTO, and Au-2% Fe/BTO, respectively, in 80 min	[30]
Bi/Bi_2WO_6	RhB, 4-CP	300 W Xe lamp, $\lambda > 400$ nm, 200 mW/cm^2	93.0% removal of RhB in 25 min, 54.4% removal of 4-CP in 120 min	[23]
$TiO_2/SiO_2/Au$ (bipyramid-like gold nanoparticles)	FA	LED lamp (400–800 nm), 156 mW/cm^2 for tests in VIS region, high-pressure Hg lamp for tests in UVA + VIS region	With UVA + VIS irradiation, 58% removal in 1 h No photocatalytic activity under VIS irradiation alone	[31]
Au/BiOCl@ mesoporous SiO_2	FAD, RhB	350 W Xe lamp, $\lambda > 420$ nm	For FAD removal, the CO_2 evolution is ca. 49-fold higher than that of N-TiO_2 NPs (used as standard VIS light photocatalyst) ~100% removal of RhB in 1 h	[32]
$RGO/CoFe_2O_4/Ag$	Short chain chlorinated paraffins	500 W Xe lamp, $\lambda > 400$ nm, 60 mW/cm^2	91.9% removal in 12 h	[33]
$Ag/Ag_2CO_3\text{-}RGO$	MO Phenol	350 W Xe lamp, $\lambda > 420$ nm 40 mW/cm^2	93% MO removal in 15 min, 93% phenol removal in 30 min	[34]
Pt–BiOBr heterostructures	PNP, TBBPA	300 W Xe lamp Simulated sunlight (320–680 nm) or VIS light (400–680 nm), 150 mW/cm^2	For PNP: Complete removal in 30 min with simulated sunlight and 99% removal in 1.5 h with VIS light For TBBPA: 100% removal in 5 min with simulated sunlight and 98% removal with VIS light	[35]

Catalyst	Pollutant	Irradiation	Performance	Refs.
$Cu_{2-x}Se$–g-C_3N_4	MB	500 W Xe lamp, Λ > 420 nm	>95% removal in 2 h	[25]
Bi/I⁻codecorated $BiOIO_3$	Phenol, 2,4-DCP, BPA, RhB, TC	500 W Xe lamp VIS light (λ ≥ 420 nm) or simulated sunlight	With λ ≥ 420 nm: 76.8% RhB removal in 4 h and >40% BPA removal in 5 h With simulated sunlight: 55% TC removal in 2h, 60% 2,4-DCP removal in 2 h, >90% phenol removal in 3 h	[24]
AgCl:Ag hollow nanocrystals	Reduction of Cr (VI) to Cr (III)	500 W Xe lamp, λ ≥ 420 nm	100% of Cr (VI) photoreduced in 10 min (only 44.7% and 16.5% of Cr (VI) was reduced over AgCl-normal and commercial P25, respectively	[20]
SiO_2-Au seeded nanoparticles	MO, TCE	Green laser λ= 532 nm	For MO, 61% removal by SiO_2-Au seeded NPs and 29% removal by bare Au seeds after 1 h laser irradiation at 2 W For TCE, 50% removal after 1 h laser exposure at 2 W	[36]

BPA: Bisphenol A; CIP: ciprofloxacin; 4-CP: 4-Chlorophenol; 2,4-DCP: 2,4-Dichlorophenol; FA: Formic acid; FAD: Formaldehyde; GO: Graphene oxide; MB: methylene blue; MO: Methyl orange, PNP: P-nitrophenol; RGO: reduced graphene oxide; RhB: rhodamine B, TBBPA: tetrabromobisphenol-A; TC: tetracycline; TCE: Tricholoroethylene, TP: titanium phosphate nanoplates; X-3B: reactive brilliant red.

Table 1. Recent studies on the utilization of plasmon-assisted photocatalysts for removal of aqueous contaminants.

Although most of plasmonic photocatalysts devoted to pollutants removal have been based on hybrid structures containing noble metals such as Au and Ag, supported on semiconductors (mostly TiO_2 or ZnO), novel plasmonic materials and structures are currently under development [21]. For instance, low-cost bismuth (Bi) has been reported to be an alternative to noble metals due to their direct plasmonic photocatalytic ability [22] which makes it a promising cocatalyst to enhance the efficiency of multiple photocatalysts, as reported in various recent studies [22–24]. Plasmonic semiconductor $Cu_{2-x}Se$ deposited on the surface of graphitic carbon nitride (g-C_3N_4) was also proposed as visible light photocatalysts for dye removal [25]. Recently, Meng and Zhang [26] developed and investigated for the first time a Pd-doped Bi_2MoO_6 photocatatalyst (Pd/$PdCl_2$-Bi_2MoO_6) which exhibited an enhanced efficiency for phenol removal due to the reduction of electron-hole recombination rate, the SPR of Pd nanoparticles, and the generation of the strong oxidizing agent Cl^0 from Cl^- on the surface of Bi_2MoO_6.

2.2. Organic synthesis

The application of heterogeneous photocatalysis in organic synthesis constitutes a more challenging issue than the degradation of organic contaminants [37] and much effort is being devoted in this field, as recently reviewed elsewhere [37–40]. The use of plasmonic nanomaterials as visible light-driven photocatalysts constitutes an interesting and appealing alternative to carry out a wide variety of chemical reactions to generate fine chemicals of industrial interest while working under mild conditions [41]. This chapter section overviews some examples of representative reactions with both industrial and fundamental interests.

2.2.1. *Oxidation reactions*

The selective oxidation of alcohols to their corresponding carbonyl counterparts is one type of mostly studied plasmon-assisted photocatalyzed reactions [16]. For instance, Au/CeO_2 was applied for the selective oxidation of benzyl alcohols to corresponding benzaldehydes in aqueous suspensions under irradiation by visible light from a green LED [42]. Hallet-Tapley et al. [43] also demonstrated the ability of Au supported on hydrotalcite, ZnO, or Al_2O_3 to selectively oxidize sec-phenethyl to acetophenone and benzyl alcohol to benzaldehyde, in presence of H_2O_2 using 530 nm LED as the photoexcitation source (**Figure 4A**). Yu et al. [44] and more recently, Chen et al. [45] supported gold catalysts for the selective aerobic oxidation of benzyl alcohol to benzaldehyde driven by visible light. Plasmonic photocatalysis has also been applied to amide production via tandem oxidation/amidation processes catalyzed by Au/SiO_2 under 532 nm laser irradiation [46].

Esterification is one of the fundamental reactions in organic synthesis, as its products are widely used as precursors and intermediates for the production of fine chemicals, fragrances, natural products, or polymers [38]. Traditionally, esters are synthesized by the reaction of activated acid derivatives with alcohols, in multi-step reaction processes that yield high amounts of undesirable byproducts. In addition, these usually require harsh conditions of temperature, pressure, and pH [38, 47]. Xiao et al. [47] reported a one-pot process for the direct oxidative esterification of aliphatic alcohols under mild conditions, using gold-palladium alloy nanoparticles (Au-Pd alloy NPs) on a phosphate-modified hydrotalcite as a recyclable photocatalyst (**Figure 4B**). It was found that the intensity and wavelength of the irradiated light could remarkably change the reactivity. The explanation was that higher irradiance provides more light excited energetic electrons, resulting in a stronger electromagnetic field in the vicinity of the NPs (LSPR field enhancement effect). On the other hand, photons with a shorter wavelength (<550 nm) are able to excite metal electrons to higher energy levels, and these electrons have more chances to transfer to the antibonding orbitals and induce reaction. Wavelength effect is more important at low temperatures, when the excited electron transfer effect dominates the photocatalytic activity, and the thermal and photothermal effect contributes much less [47].

Likewise, Zhang et al. [48] reported the visible light driven esterification from aldehydes and alcohols using supported Au nanoparticles (Au/Al_2O_3) at ambient temperatures. From the results at different wavelength ranges, it is concluded that gold plays an active role in harvesting visible light and that the LSPR effect plays a critical role in enhancing the reaction activity in the catalyzed processes.

2.2.2. *Reduction reactions*

Chemical reduction reactions constitute another interesting route for the synthesis of fine chemicals, and plasmonic materials have also been utilized in several photocatalytic reduction processes. For instance, plasmonic gold nanoparticles (Au NPs) were successfully applied as catalyst to perform the reduction of resazurin to resorufin [49] via laser drop (532 nm) or LED (530 nm) plasmon excitation. The use of Au NPs supported on different materials was also investigated for the selective reduction of organic compounds under visible light (or simulated sunlight) [50]. Au/CeO_2catalysts were found to be efficient in the reduction of nitroaromatics to azocompounds, hydrogenation of azobenzene to hydroazobenzene, reduction of ketones to alcohols, and deoxygenation of epoxides to alkenes. The SPR effect of Au NPs was reported to play a key role in assisting activation of N=O bonds, so that the main product in the reduction of nitrobenzene under visible irradiation was different from that obtained under UV irradiation [50].

It has also been reported that Cu/graphene catalyst exhibits high activity for the reductive coupling of nitroaromatics to aromatic azocompounds under solar irradiation [51]. The product selectivity changed significantly with temperature, and conversion was found light intensity dependent. The light absorbed by Cu nanoparticles was the major driving force of the reaction as the highest conversion in the visible light range was in the range 530–600 nm, where the Cu nanoparticles strongly absorb the light due to the LSPR effect [52].

Another representative example is referred to the reduction of styrene in the presence of hydrogen to yield ethylbenzene. This chemical reduction was successfully achieved in the presence of Ag–Pd nanocages under visible light irradiation. In these photocatalysts, Pd provides active sites for hydrogenation reactions, whereas Ag offers plasmonic properties to convert light into heat [53].

2.2.3. *Cross-coupling reactions*

Plasmonic materials have also been investigated as photocatalysts for cross-coupling reactions, which have been accepted as convenient one-step methods to render complex molecules of interest for the synthesis of natural and advanced materials, bioactive products, agrochemicals, or medicines [38]. Au-Pd nanostructures consisting of Au nanocrystal cores and tightly bonded Pd nanoparticles could harvest visible to near-infrared (NIR) light for Suzuki coupling of iodobenzene or bromobenzenes and aromatic boronic acids to biphenyls under solar radiation, and also under 809 nm laser irradiation [54].

In a study by Xiao et al. [3], five different cross-coupling reactions, namely the Sonogashira, Stille, Hiyama and Ullmann C-C couplings, and the Buchwald-Hartwig amination (C-N cross-coupling) were investigated to demonstrate the possibility of applying Au-Pd alloy NP photocatalyst to enhance the intrinsic catalytic activity of Pd sites under visible light irradiation and at low temperatures (**Figure 4C**). The results suggest that electrons of alloy NP are excited under light irradiation and are then transferred from the nanoparticle surface to the reactant molecules adsorbed on the nanoparticle surface, weakening the chemical bonds of the molecules and facilitating the reactions. Increasing the light intensity accelerates the reaction rate due to increased population of photoexcited electrons. The irradiation wavelength also affected the reaction rates, and ultraviolet irradiation was required to drive some reactions with certain substrates.

Hu et al. [55] developed a novel and efficient approach to synthesize interfaced dimers made of Au NPs and bimetallic nanoshells, such as Au/Ag, Pt/Ag, or Pd/Ag. The Au NP-Pd/Ag bimetallic nanoshells were then used for catalyzing Suzuki coupling reactions between phenylboronic acid and iodobenzene at 80°C, yielding biphenyl. After 45 min in the dark, the conversion of reactants was 95%. Under visible light illumination, the reaction was significantly accelerated, 95% conversion in only 15 min. The increase in the reaction rate under illumination was attributed to the strong SPRs in the Au NPs, which enabled absorption of visible light in the dimers, leading to an enhancement in catalytic performance.

Cui et al. [56] synthesized Cu_7S_4@Pd heteronanostructures and probed their photocatalytic potential for Suzuki coupling reaction, hydrogenation of nitrobenzene, and oxidation of benzyl alcohol in the NIR range, respectively. As the Cu_7S_4@Pd LSPR peak position was close to 1500 nm, a 1500 nm diode laser was used as illumination source for the photocatalytic reactions. Additionally, irradiation at 808 and 980 nm were also evaluated. The highest conversions rates were obtained for 1500 nm irradiation due to LSPR enhancement in this region. To evaluate the pragmatic feasibility of the photocatalyst, the catalytic reactions were evaluated under real sunlight irradiation (~40 mW/cm^2) as well, and high conversion rates were attained (**Figure 4D**).

In a study by Trinh et al. [7], plasmonic Pd hexagonal nanoplates, with tunable longitudinal LSPR were synthesized and applied to catalyze Suzuki coupling reaction. The catalytic activity of the Pd nanoplates was 2.5 and 2.7 times higher than that of nonplasmonic Pd nanooctahedral and Pd nanocubes, respectively, upon illumination at wavelengths in the range 300–1000 nm. These results, along with theoretical studies, revealed that Pd hexagonal nanoplates are able to harvest visible to NIR light, and that the increase in the catalytic activity on Pd nanoplates is primarily a result of the plasmonic photocatalytic effect of plasmon-induced hot electrons. Recently, it has been revealed that metal oxide nanostructures, such as WO_{3-x}, MoO_{3-x}, and TiO_{2-x}, can exhibit LSPR in the visible and NIR region due to abundant oxygen vacancies or heavy doping [57, 58]. Lou et al. [57] demonstrated that the plasmon excitation of WO_{3-x} nanowires by long wavelength irradiation can enhance main product yield and selectivity in Suzuki coupling reactions catalyzed by the attached Pd nanoparticles.

2.3. Disinfection

While TiO_2 photocatalytic inactivation of bacteria and viruses has been known for decades [59], the number of studies using plasmonic materials for water disinfection is still rather limited.

Silver halides (Ag/AgX where X = Cl, Br, I) supported on TiO_2 or carbon materials have been the most studied photocatalysts for antimicrobial applications. Tian et al. [60] prepared a AgCl@Ag@TiO_2 sandwich-structured photocatalyst and evaluated its bactericidal activities in terms of the inactivation of Gram-negative bacteria *Escherichia coli* K12 (*E. coli* K12) under visible light irradiation. In the absence of irradiation, the catalyst exhibited a negligible killing response, whereas visible light irradiation in absence of photocatalyst only achieved the inactivation of 6% *E. coli* colonies in 30 min. In contrast, when AgCl@Ag@TiO_2 was used, nearly 77% of *E. coli* K12 population was inactivated after 6 min under

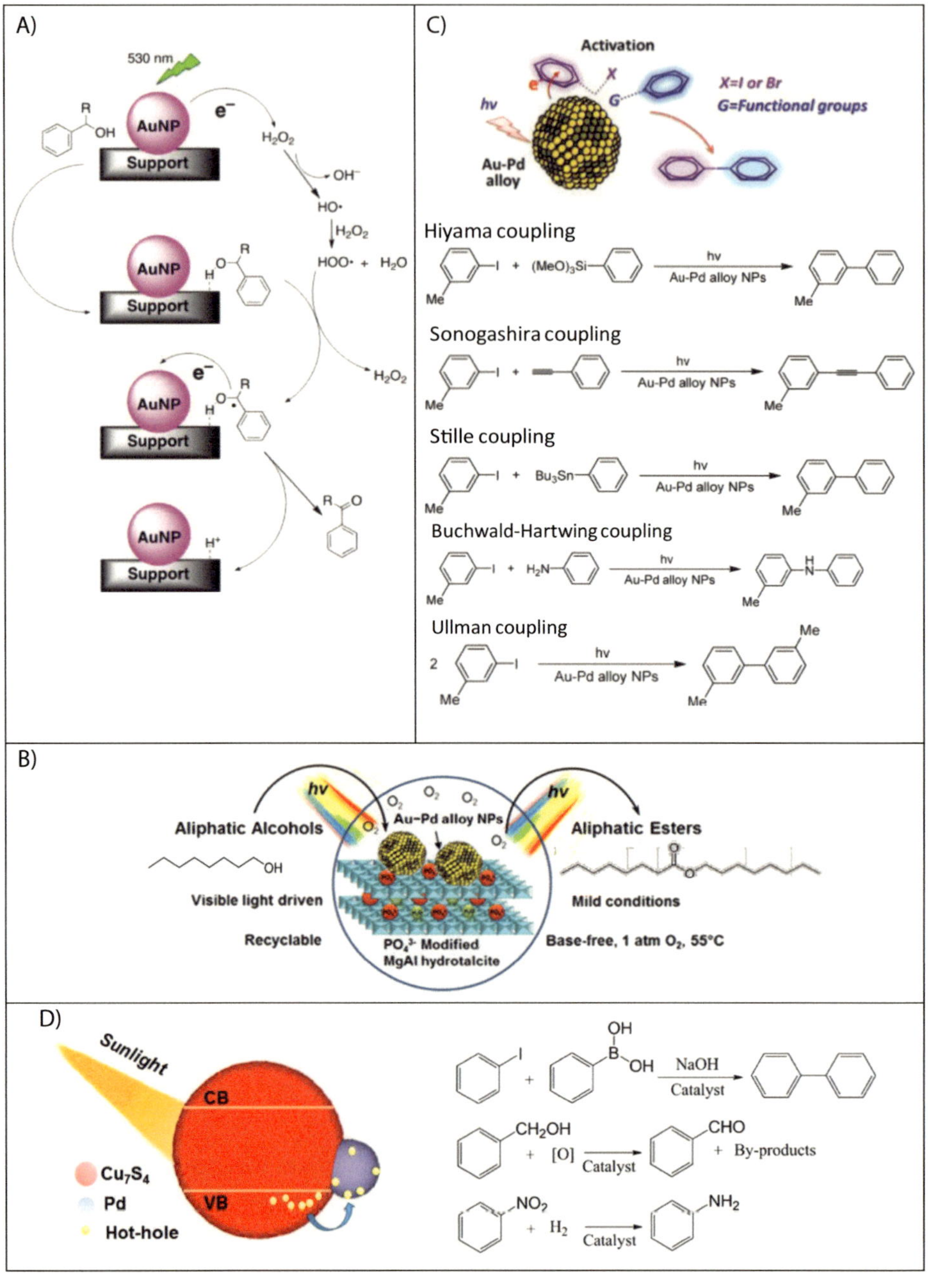

Figure 4. (A) Proposed mechanism for the plasmon-mediated oxidation of sec-phenethyl (R = CH_3) and benzyl (R = H) alcohols in the presence of supported AuNP. Reprinted with permission from Ref. [43]. Copyright (2013) American Chemical Society. (B) Direct oxidative esterification of aliphatic alcohol (1-octanol as example). Adapted with permission from Ref. [47]. Copyright (2015) American Chemical Society. (C) Scheme of cross-coupling reactions catalyzed by Au-Pd alloy NPs under visible light irradiation. Adapted with permission from Ref. [3]. Copyright (2014) American Chemical Society. (D) Schematic representation of plasmon enhanced Suzuki coupling reaction, oxidation of benzyl alcohol and hydrogenation of nitrobenzene, over Cu_7S_4@Pd heteronanostructures. Reprinted with permission from Ref. [56]. Copyright (2015) American Chemical Society.

visible light irradiation, reaching total inactivation at 15 min. Under visible light irradiation, Ag NPs produce electrons and holes, which can be separated by the SPR-induced local electromagnetic field. Due to the suitable conduction band and valence band energy levels of AgCl and TiO_2, the photo-generated electrons transfer to TiO_2 while holes transfer to AgCl. The electrons are trapped by adsorbed O_2 to form $^{\bullet}O_2^-$, while the holes combine with OH^-/or Cl^- ions to form $^{\bullet}OH$/or Cl^0 radicals. All of $^{\bullet}O_2^-$, $^{\bullet}OH$, and Cl^0 radicals are reactive species responsible for the inactivation of bacteria. Simultaneously, Shi et al. [61] studied the photocatalytic inactivation of *E. coli* by Ag/AgX-CNTs under visible light. The Ag/AgBr-CNTs exhibited an excellent photocatalytic disinfection performance, achieving the complete inactivation of 1.5×10^7 cfu/mL of *E. coli* within 40 min. This was attributed to the SPR of Ag nanoparticles and the efficient photo-generated carrier separation owing to the CNTs. Thus, a high concentration of electrons migrates to the surface of CNTs, where they can be trapped by O_2 and H_2O to produce $^{\bullet}O_2^-$ and H_2O_2.

$$O_2 + e^- \rightarrow {}^{\bullet}O_2^- \quad (1)$$

$$^{\bullet}O_2^- + H_2O \rightarrow {}^{\bullet}OOH + OH^- \quad (2)$$

$$2\,^{\bullet}OOH \rightarrow O_2 + H_2O_2 \quad (3)$$

These species, along with $^{\bullet}OH$ produced from holes in the surface of AgBr, are involved in the photocatalytic bacterial inactivation process.

More recently, Xia et al. [62] prepared a series of Ag/AgX/RGOs (RGO = reduced graphene oxide) composites and evaluated their performance for water disinfection upon visible light. The visible light irradiation without plasmonic catalyst or the photocatalyst in the absence of light irradiation showed no bactericidal effect. Upon visible light irradiation, Ag-AgBr/RGO exhibited high photocatalytic inactivation efficiency and attained complete inactivation of 7 log cfu/mL *E. coli* cells in 8 min. Remarkably, additional experiments were also carried out to evaluate the possible effect of Ag^+ ions in order to clarify the disinfection mechanism of Ag-based plasmonic photocatalyst. Authors found that light irradiation promoted the release of free Ag^+ ions from Ag-AgBr/RGO and also enhanced the antimicrobial activities of Ag^+ ions. These were able to induce the damage of metabolic process while the cell membrane damage was rather limited. The primary bactericidal effect of visible light irradiated Ag-AgBr/RGO resulted from the sustainable generation of reactive species. Plasmonic-induced H_2O_2 plays the leading role, collaborating with e-, $^{\bullet}O_2^-$, $^{\bullet}OH$ to induce damage of microbial metabolism processes, destroy the cell envelope and lead to the leakage and degradation of intracellular substances.

AuNP-based nanocomposites have also been investigated for disinfection applications. In this regard, Sarkar et al. [63] developed a plasmonic photocatalyst derived from functionalized amylopectin and *in situ* incorporated TiO_2 and AuNPs (g-AP-pAA/TiO_2-Au), and demonstrated its antimicrobial activity in the presence of both UV, and especially, visible light irradiation, as depicted in **Figure 5**.

Li et al. [64] developed a nonnoble metal plasmonic photocatalyst consisting of a TiN/TiO_2 composite and evaluated its performance for the photocatalytic disinfection of *E. coli* in a buffer solution under visible light illumination. In the absence of photocatalyst or visible light,

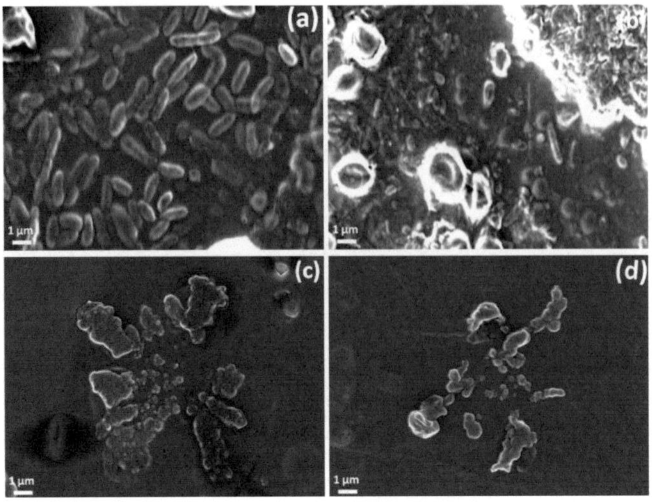

Figure 5. Field emission scanning electron microscopes images of *E. coli* (a) in absence of light source; (b) in presence of UV light; (c) in presence of g-AP-pAA/TiO_2 along with UV light; and (d) in presence of g-AP-pAA/TiO2-Au along with visible light. Reprinted with permission from Ref. [63]. Copyright (2016) American Chemical Society.

the survival ratio of *E. coli* population showed no obvious change. In the presence of TiN/TiO_2 photocatalyst under illumination ($\lambda > 400$ nm), the survival ratio of *E. coli* colonies dropped continuously upon increasing irradiation times. According to the proposed mechanism, TiN nanostructures constitute the plasmonic component in the composite to harvest visible light and generate hot electrons by the LSPR effect. These hot electrons, once they were excited above the Fermi energy level of TiN, could be injected quickly into the conduction band of TiO_2, to react with O_2 yielding $^{\bullet}O_2^-$ and subsequently $^{\bullet}OH$, which are able to disinfect microorganisms.

3. Nanoplasmonic photocatalysis in gas-phase reactions

Besides their use in aqueous media photocatalysis, plasmonic nanomaterials have been lately investigated for their use in gas-phase reactions. Due to the broad range of reactions in gas phase currently under study, in this section we will focus on two big groups of them: degradation of volatile organic compounds (VOCs) and chemical to energy conversion.

3.1. Photo-degradation of volatile organic compounds

Air pollution is of great concern nowadays due to the damage that a long exposure can do to human health. Indoor and outdoor environments are suitable to suffer from high concentration of VOCs due to human activity, so currently it is considered urgent to address this problem in a more efficient way. Currently, the methods used to eliminate VOCs vary from physical methods like filtration or adsorption, to chemical ones like UV irradiation or ozone treatment [65]. The latter, being efficient for the purpose of eliminating the hazardous materials from air, is also harmful to human health, so there is a need for new methods to eliminate VOCs from air. In this case, the use of plasmonic materials seems to be a logical step forward, not only in order to avoid exposure to dangerous atmospheres, but also because of the possibility of using solar energy instead of other expensive and unsafe source of energy. While the examples of photocatalysis for VOCs elimination are growing lately, the ones using plasmonic materials, even when growing every year, are still far from conventional semiconductors ones. One of the first examples in using plasmonic materials for VOCs remediation comes from Chen et al. [66]. They used Au NPs supported in different metal oxides to get the total oxidation of formaldehyde with visible light. In this case, the authors wanted to use a pure plasmonic material, so they attached the Au NPs to oxide supports like ZrO_2 or SiO_2, both with high band gap (5 and 9 eV, respectively) in order to avoid excitation of the support from the visible light source. The mechanism proposed involves two of the features of plasmonic materials mentioned above. In one hand, irradiating the Au NPs on their plasmon band will increase the temperature locally high enough to perform the VOCs oxidation in high extent. Besides the temperature effect, the high electric field generated on the surface of the nanoparticles would activate polar molecules as formaldehyde present on the surface of Au NPs, aiding to the total oxidation of the compound. Another example of decomposition of VOCs using plasmonic materials is the isopropanol oxidation performed by Dinh et al. [67]. In this case, the structure of the catalyst is a key for its performance. It consists of Au/TiO_2 nanostructured photocatalyst that is constructed by the three-dimensional ordered assembly of thin-shell Au/TiO_2 hollow nanospheres. Due to that composition, it can be considered a hybrid plasmonic structure with some special qualities. The authors claim that the photonic structure of the catalyst enhances the absorption of the plasmonic AuNP by the multiple scattering and the slow photon effect characteristic of these architectures. This enhancement absorption of the plasmon band provides an activity several times higher than the normal Au/TiO_2 structure.

The last example for VOCs oxidation using plasmonic materials comes from Sellappan et al. [68]. They combine gold and silver nanoparticles with TiO_2 with different configurations in order to evaluate the electron transfer after light excitation. They evaluate the photocatalysis of the degradation of methanol and ethylene under different conditions. One of those conditions is the physical contact between the metal and the semiconductor. When the nanoparticles are in contact with TiO_2, a Schottky barrier is formed, which enhances the charge separation under UV light illumination. When the metals are not in contact with the support, they also observe an enhancement in the activity of the photocatalyst that they assign to a near and far field effect of the plasmonic nanomaterial. This is so because, even when they are separated, the distance is short enough from the support to feel the high electric field generated on the metal nanoparticle.

3.2. Plasmonic-driven chemical to energy conversion processes

Global warming is nowadays one of the biggest problems to be addressed in our society. The vast amount of fossil fuel consumed during the last decades has broken the natural balance of CO_2 emission and uptake, leading to high concentrations of this gas in the atmosphere and in the oceans, and provoking a sharp climate change. Besides, due to this enormous growing in fossil fuel consumption, the society is exhausting their reserves at a very high rate creating an urgent need for alternative sources of energy. With this in mind, photocatalysis has emerged as a promising tool to kill two birds with one stone. On one hand, it can help to reduce the CO_2 emitted to the atmosphere and also in that process give products that could be able to store energy. As a first example of this, in 1979 Inoue et al. [69] attained for the first time the reduction of CO_2 with light using several semiconductors as photocatalysts. Even when the reaction was performed in liquid phase, we consider appropriate to mention it here since it was the first attempt that obtained valuable products after the reduction of CO_2 using light as energy source. Since that first work, many attempts have been made in order to synthesize photocatalysts to reduce CO_2 to different interesting products. Most of those catalysts are semiconductors like TiO_2 or ZnO which, while they possess many interesting properties to be used as photocatalysts, they also have some drawbacks that limit their use on large scale applications, as low photon absorption efficiency, high charge carrier recombination rate or being restricted to the UV region of the spectra. In order to overcome these downsides, in recent years the scientific community have paid attention to plasmonic materials as an attractive tool to be used for the reduction of CO_2. Liu et al. [70] first reported the use of plasmonic gold catalyst for the dry reforming of CO_2 with methane to obtain syngas. This reaction usually requires high temperatures (800–1000°C) and is of great interest since it would transform two greenhouse gases into molecules suitable for energy storage like CO and H_2. They prepared a catalyst consisted of Rh and Au supported on SBA-15 in order to use the good properties of Rh for this reaction and the plasmon band of Au to activate the reactants. They claim that the high electric field generated on the AuNP induces polarization on the CO_2 and the CH_4 molecules activating them, and enhances the conversion to syngas. Another interesting reaction involving the consumption of CO_2 is the reverse water gas shift reaction (RWGS). Upadhye et al. [71] probed that using AuNP supported over TiO_2 and CeO_2, they could enhance the yield of the reaction illuminating with visible light from 30 to 1300%. According to them, the LSPR of the catalyst changed the intrinsic kinetics of the reaction on the surface of the catalyst by increasing the rate constant of either the carboxyl decomposition or the hydroxyl hydrogenation, two of the key steps in this reaction. They attributed the rate enhancement either to the generation of hot electrons or the polarization effect of the high electric field generated by the plasmonic nanoparticles on the adsorbates. Not only conventional semiconductor materials have been used to support plasmonic nanoparticles, lately several authors have used graphene and derivative as active supports for photocatalysis. This is the case of Shown et al. [72] that evaluated the activity of Cu NPs supported on graphene oxide (GO) for the reduction of CO_2 (**Figure 6**). In their work, they prepared the catalyst with different Cu loading and evaluated the photoreduction using visible light, and compared the results with GO alone and with TiO_2 P25. The authors found an increase in the products yield for the CuGO hybrid material by 60 times with respect to the GO, and 240 times with respect to the P25, obtaining methanol and acetaldehyde as the main products. They attributed the enhancement to a modification in the work function of the GO by the Cu NPs that improved the charge separation.

Finally, some additional examples where CO_2 has been successfully hydrogenated to yield methane in the presence of nanoplasmonic photocatalysts are briefly described below and summarized in **Table 2**. Methane has gained an increasing interest, given its abundance and its extended use as fuel in fertilizers or as intermediate in the petrochemical industry [73]. Most of the studies report the combination of titania or P25 semiconductor supports with noble-metal-based cocatalysts, which are able to expand the band gap toward the visible range through their surface plasmon band [74], act as charge reservoirs, and also as active and selective catalytic centers [74–81]. In this regard, the presence of Pd or Pt strongly enhances the selectivity toward methane in comparison with the preferential pathway toward CO observed in TiO_2, being of paramount importance for the proper formation of organic intermediates. As a major drawback, progressive deactivation was observed upon oxidation of Pd into PdO domains [79]. Furthermore, the effective formation of Au-Cu alloys was also found extremely active, thereby outperforming the photo-

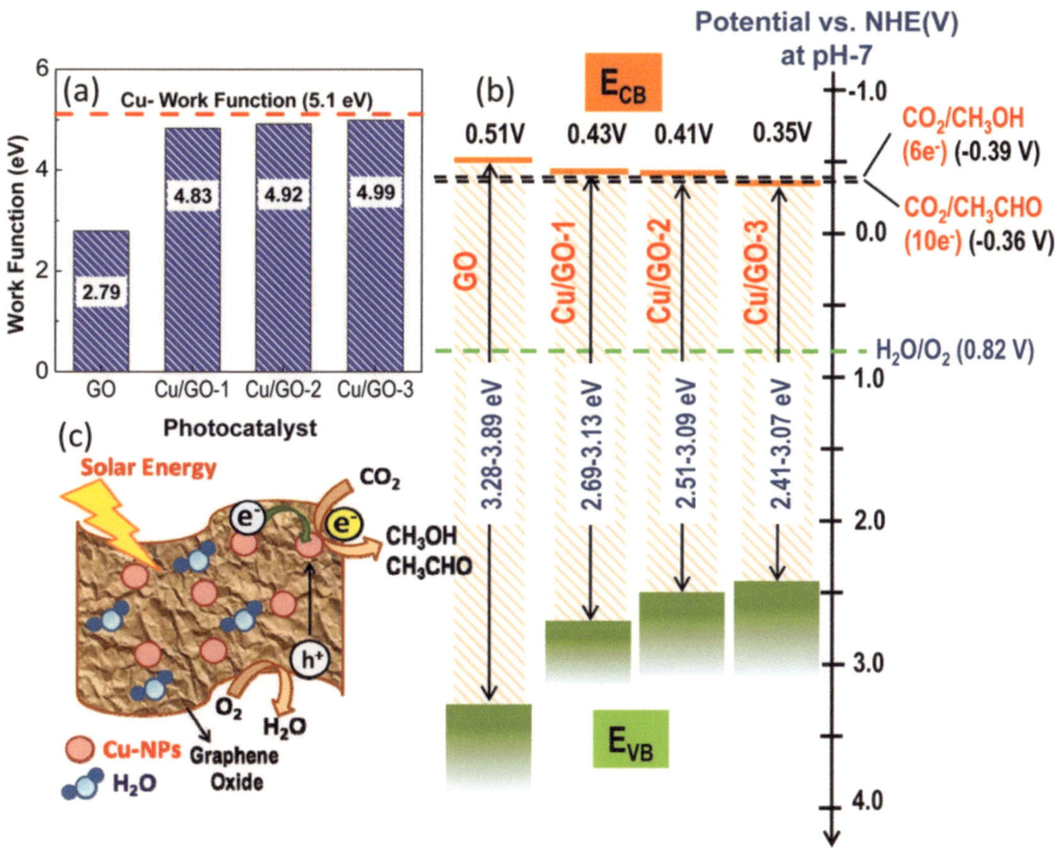

Figure 6. (a) UPS-determined work functions of GO and Cu/GO hybrids and (b) band-edge positions of pristine GO and Cu/GO hybrids as compared with CO_2/CH_3OH and CO_2/CH_3CHO formation potential. (c) Schematic photocatalytic reaction mechanism. Reprinted with permission from Ref. [72]. Copyright (2015) American Chemical Society.

catalytic conversions of their respective single metal counterparts, while expanding their response in the visible range as nicely demonstrated by transient absorption spectroscopy studies [74].

Other relevant studies additionally dealt with the use of catalytic monoliths [75], fluidized-bed reactors combined with LEDs [77] or included the use of novel semiconductor supports based on carbon nitride or reduced graphene oxide that provided additional sites for the accommodation of CO_2 and the active cocatalysts [76, 82]. Finally, Zhang et al. synthesized a hybrid material consisted of Rh nanocubes supported on Al_2O_3 and used for the selective hydrogenation of CO_2 to CH_4, avoiding almost completely the competitive production of CO (see **Figure 7**). The authors used UV and blue light coming from LED irradiators in order to excite the plasmon band of the metal nanoparticles and generate high-energy electrons that are transferred to the adsorbates [81].

Catalyst	Reductant	Cocatalyst	Remarks	Refs.
Au, In/TiO_2	H_2	Au, In	Irradiation with UV lamp (200 W; 150 mW/ cm^2) Use of monolithic reactors	[75]
Pd/TiO_2	H_2O	Pd	Irradiation with UV-LED arrays (40 pieces) (365 nm) Fluidized bed reactors + T = 140°C	[77]
Pd/TiO_2	H_2O	Pd	Irradiation at $\lambda > 310$ nm Formation of organic adsorbates is critical	[79]
Core-shell Pt/TiO_2 PtCu/TiO_2	H_2O	Cu, Pt	Light source $780 > \lambda > 320$ nm Strong influence of cocatalysts on selectivities	[80]
Au,Cu/P25	H_2O	Au, Cu, Au-Cu alloys	λ= 355 and 532 nm Performance of transient absorption experiments	[74]
Ternary composition Metal + RGO + TiO_2	H_2O	Pt, Pd, Ag, Au	Visible light irradiation (energy daylight bulb, 15 W) Pt-doped showed best activity	[76]
Ag(AgCl)$^-$ Carbon Nitride	H_2O	-	In situ generation of Ag plasmonic domains $400 < \lambda < 650$ nm 30 fold enhancement with P25	[82]
Rh cubes/Al_2O_3 Au/Al_2O_3	H_2	–	Irradiation with UV, blue and white LEDs Comparison with thermal reaction	[81]

Table 2. Summary of selected photocatalysts for conversion of CO_2 into CH_4.

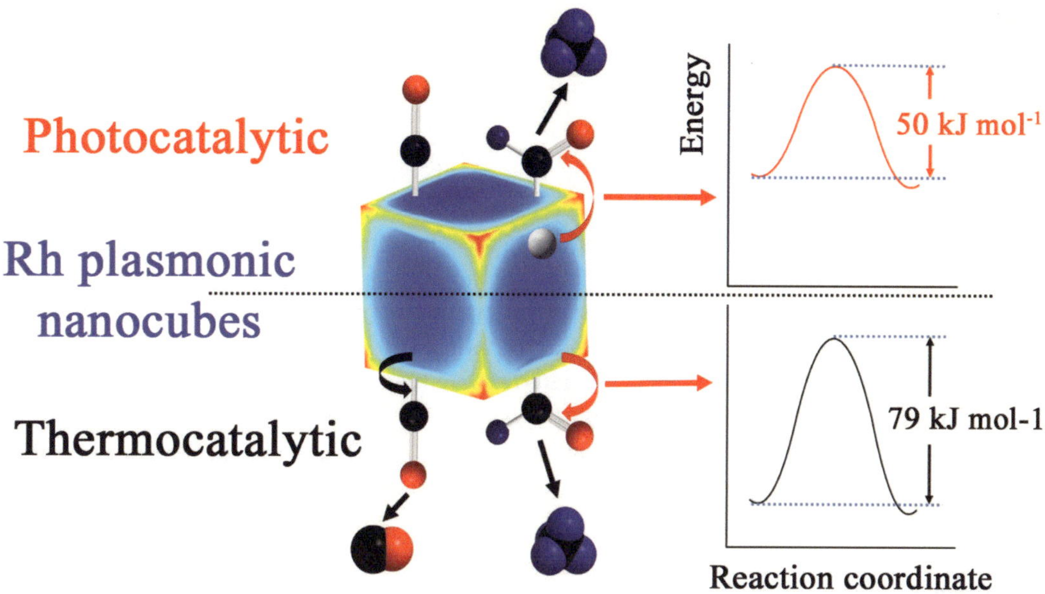

Figure 7. Representative example of recently developed plasmonic catalysts for the CO_2 hydrogenation into methane: Reaction mechanism on Rh nanocube comparing the preferential activation of CO in the thermocatalytic process in contrast to the methane pathway favored under photocatalytic conditions. Adapted with permission from Ref. [81]. Copyright (2017) Nature Publishing Group.

4. Plasmonics in biocatalytic processes

4.1. Photobiocatalysis

A promising field for the use of nanoplasmonics currently under development is the photobiocatalysis. In this relatively novel area, the main goal is obtaining photocatalysts inspired in natural photosynthetic centers. A brief overview of the current state of the art in this field devoted to regular semiconductors has been recently published elsewhere [83].More recent studies involving the use of plasmonic materials have been reported by Sanchez-Iglesias et al. Their studies focused on the evaluation of multiple gold-shaped plasmonic nanostructures (see **Figure 8**) and their effect on the effective photoregeneration of nicotinamide adenine dinucleotide (NADH) molecules. These latter molecules are extremely important in natural biochemical routes as mediating cofactors in many enzymes [84, 85]. Moreover, cofactor molecules are necessary, for instance, in the photosynthesis process as light harvesters and intervening in the reduction-oxidation balances involved in respiration [86]. The major drawbacks associated with cofactor molecules have arisen from the limited success achieved in the past year attempting their regeneration (reduction) via nonenzymatic pathways. Organic dyes, semiconductors, or polymers have been previously used as photocatalysts to regenerate NADH cofactors, but these molecules intrinsically possess a poor quenching ability to accept electrons. Therefore, the need for implementing an electron relays acting as mediator has been suggested to overcome this limitation. Up to date, the choice of suitable mediators has been rather scarce and limited to a few examples of Rh-based organometallic complexes. The use of gold plasmonic nanostructures represents a more straightforward and affordable alternative.

The first successful example of plasmonic cofactor photoregenerator was reported for gold nanorods coated with Pt domains on their tips (**Figure 8B**), thereby representing a previously defined plasmonic photocatalyst containing a heterojunction that combines a plasmonic structure (Au NRs) and a catalytic active site (Pt) to successfully carry out the regeneration (reduction) of NADH cofactor molecules [85]. A subsequent study carried out by the same authors determined an even more remarkable photo-response of gold nanostars with epitaxially grown Pt domains that was associated to the major light harvesting capacity of the star-shaped plasmonic structures (see **Figure 8A**) [84, 85]. The potential combination of these plasmonic heterojunctions with other semiconductors represents a very promising alternative to obtain suitable mediators that can help to modulate multiple biochemical processes via light-induced inputs.

4.2. Artificial enzymes

Another niche of great biotechnological interest for the potential application of nanoplasmonics is related with the search for novel artificial enzymes [86]. Natural enzymes are well-known biocatalysts that regulate every biochemical processes involving living organisms. Recently, an emerging research subfield has emerged to find novel nanomaterials that can mimic the role of natural enzymes as highly active, and specific catalysts without incurring in

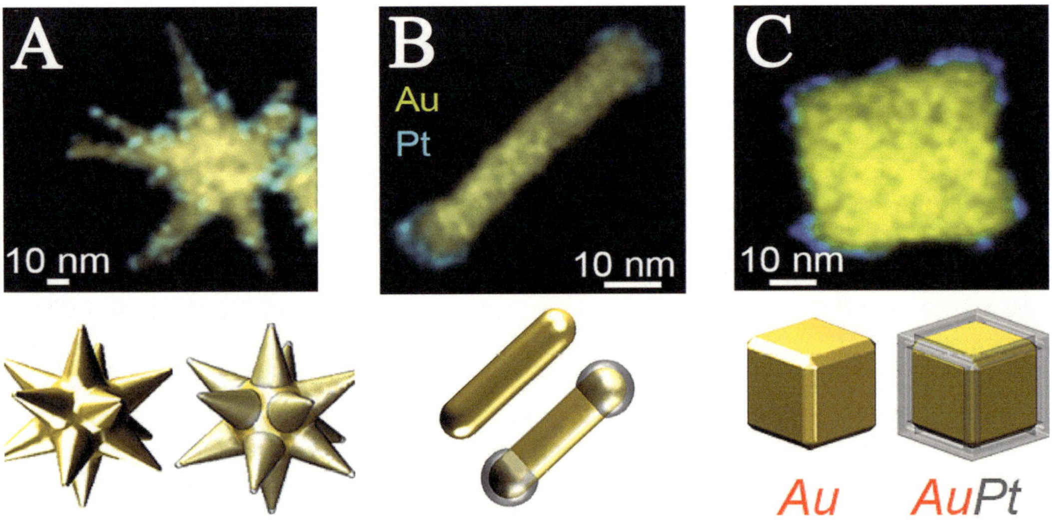

Figure 8. STEM-EDX analysis of Pt-decorated gold nanoplasmonics with different shapes and corresponding 3D models used for photoregeneration of NADH cofactor molecules: (A) stars; (B) nanorods; and (C) nanocubes. Adapted from Refs. [84, 85] with permission of The Royal Society of Chemistry.

http://dx.doi.org/10.5772/intechopen.68812

their intrinsic limitations, such as limited stability by denaturalization, strong sensibility on variations of their optimal environment, difficulties for effective retrieval and reuse, and high costs derived from the multiple synthesis and purification steps implied in their preparation. In order to circumvent these major drawbacks, the search for stable and affordable alternatives has brought the spotlight on the development of artificial systems based on inorganic/organic nanomaterials.

In this regard, it is worth mentioning that Prof. Santamaria's group recently developed and tested one of the first NIR-activated enzyme-like plasmonic photocatalysts used as a glucose-oxidase surrogate (**Figure 9**) [87]. Glucose-oxidase biomimetic systems hold a huge potential for biomass conversion, selective detection of glucose in blood at trace levels, and control/monitoring of internal metabolism in cells. Previous candidates based on photocatalytic semiconductors such as TiO_2 or ZnO rendered promising photoconversion of glucose [88, 89], but the combination of plasmonic gold nanorods cores holding excellent NIR absorption capabilites with an outer titania nanoshell proposed by Ortega-Liebana et al. provided not only an extended response toward the whole visible-NIR range, but also additional thermal stability and photoactivity toward the effective and preferential oxidation of glucose into gluconic acid and hydrogen peroxide (**Figure 9**). In addition, the activity of these core-shell nanostructures was maintained in a wide pH range and was effective at near room temperature. The authors claimed that the minimal thickness of the TiO_2 shell ensured the formation of an effective Schottky barrier at the interface between Au and TiO_2. It was also remarkable to find out that the uncoated Au NRs exhibited negligible photo-response toward the glucose oxidation, thereby corroborating the need of the semiconductor titania shell fraction to promote the generation of active radicals and the selective oxidation of the sugar molecules [87].

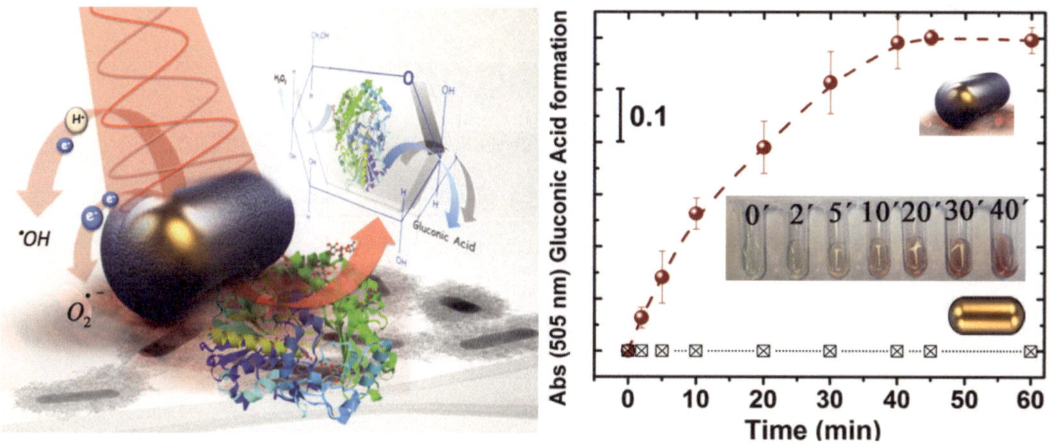

Figure 9. (Left) Schematic plot displaying the glucose-oxidase mimetic action of the titania-coated gold nanorods plasmonic photocatalyst under NIR irradiation; (Right) Colorimetric detection of the glucose when selectively converted into gluconic acid via the formation of a Fe-hydroxamate complex. Uncoated Au-NRs exhibit negligible photooxidation properties in comparison with the coated ones. Photocatalytic experiments were carried out with an 808 nm laser. Partially reprinted with permission from Ref. [87].

5. Conclusions

Plasmonic-based nanomaterials hold a very promising future as potential alternatives for the fabrication of next generation photocatalysts and processes where the wavelength irradiation range is expanded toward the visible and near-infrared windows, thereby maximizing the use of the solar irradiation. A wide variety of fields of action can be foreseen including those described in the present chapter and some additional options including the generation of biofuels from water or biomass, photoelectrocatalysis, solar cells, or photovoltaics. The use of these types of catalysts for light-triggered therapy treatments and the specific targeting for cell-mechanisms can also be envisioned as another promising area of expansion. In conclusion, the use of plasmonics is likely going to emerge with abundant and interesting breakthroughs in the next future.

Acknowledgements

The authors acknowledge the European Research Council for funding through an advanced grant research project (HECTOR-267626) and a CIG-Marie Curie Reintegration Grant (NANOLIGHT-REA 294094). The Spanish Ministry of Economy, Industry and Competitiveness is also gratefully acknowledged (project CTQ2016-77147-R). C.B.A. and A.A.R. also thank the Spanish Government and the Juan de la Cierva program for two postdoctoral fellowships. CIBER-BBN is an initiative funded by the VI National R&D&I Plan 2008-2011 financed by the Instituto de Salud Carlos III with assistance from European Regional Development Fund.

Author details

Carlos J. Bueno-Alejo[1,2], Adriana Arca-Ramos[1,2,3] and Jose L. Hueso[1,2,3]*

*Address all correspondence to: jlhueso@unizar.es

1 Institute of Nanoscience of Aragon, University of Zaragoza, Zaragoza, Spain

2 Department of Chemical Engineering and Environmental Technology, University of Zaragoza, Zaragoza, Spain

3 Networking Research Center on Bioengineering, Biomedical and Nanomedicine, CIBER-BBN, Madrid, Spain

References

[1] Zou ZG, Ye JH, Sayama K, Arakawa H. Direct splitting of water under visible light irradiation with an oxide semiconductor photocatalyst. Nature. 2001;**414**(6864):625-657. DOI: 10.1038/414625a

[2] Li XH, Zhu JM, Wei BQ. Hybrid nanostructures of metal/two-dimensional nanomaterials for plasmon-enhanced applications. Chemical Society Reviews. 2016;**45**(14):3145-3187; correction 2016;**45**(14):4032.DOI: 10.1039/c6cs90055k

[3] Xiao Q, Sarina S, Bo AX, Jia JF, Liu HW, Arnold DP, et al. Visible light-driven cross-coupling reactions at lower temperatures using a photocatalyst of palladium and gold alloy nanoparticles. Acs Catalysis. 2014;**4**(6):1725-1734. DOI: 10.1021/cs5000284

[4] Fasciani C, Alejo CJB, Grenier M, Netto-Ferreira JC, Scaiano JC. High-temperature organic reactions at room temperature using plasmon excitation: Decomposition of dicumyl peroxide. Organic Letters. 2011;**13**(2):204-207. DOI: 10.1021/ol1026427

[5] Hallett-Tapley GL, Silvero MJ, Gonzalez-Bejar M, Grenier M, Netto-Ferreira JC, Scaiano JC. Plasmon-Mediated catalytic oxidation of sec-phenethyl and benzyl alcohols. Journal of Physical Chemistry C. 2011;**115**(21):10784-10790. DOI: 10.1021/jp202769a

[6] da Silva AGM, Rodrigues TS, Correia VG, Alves TV, Alves RS, Ando RA, et al. Plasmonic nanorattles as next-generation catalysts for surface plasmon resonance-mediated oxidations promoted by activated oxygen. Angewandte Chemie-International Edition. 2016;**55**(25):7111-7115. DOI: 10.1002/anie.201601740

[7] Trinh TT, Sato R, Sakamoto M, Fujiyoshi Y, Haruta M, Kurata H, et al. Visible to near-infrared plasmon-enhanced catalytic activity of Pd hexagonal nanoplates for the Suzuki coupling reaction. Nanoscale. 2015;**7**(29):12435-12444. DOI: 10.1039/c5nr03841c

[8] Liu Z, Huang YM, Xiao Q, Zhu HY.Selective reduction of nitroaromatics to azoxy compounds on supported Ag-Cu alloy nanoparticles through visible light irradiation. Green Chemistry. 2016;**18**(3):817-825. DOI: 10.1039/c5gc01726b

[9] Fujishima A, Honda K. Electrochemical photolysis of water at a semiconductor electrode. Nature. 1972;**238**(5358):37-38. DOI: 10.1038/238037a0

[10] Manthiram K, Alivisatos AP. Tunable localized surface plasmon resonances in tungsten oxide nanocrystals. Journal of the American Chemical Society. 2012;**134**(9):3995-3998. DOI: 10.1021/ja211363w

[11] Huang QQ, Hu S, Zhuang J, Wang X. MoO_{3-x}-based hybrids with tunable localized surface plasmon resonances: Chemical oxidation driving transformation from ultrathin nanosheets to nanotubes. Chemistry-A European Journal. 2012;**18**(48):15283-15287. DOI: 10.1002/chem.201202630

[12] Yan JQ, Wang T, Wu GJ, Dai WL, Guan NJ, Li LD, et al. Tungsten oxide single crystal nanosheets for enhanced multichannel solar light harvesting. Advanced Materials. 2015;**27**(9):1580-1586. DOI: 10.1002/adma.201404792

[13] Zhang YW, Tian JQ, Li HY, Wang L, Qin XY, Asiri AM, et al. Biomolecule-assisted, environmentally friendly, one-pot synthesis of CuS/Reduced graphene oxide nanocomposites with enhanced photocatalytic performance. Langmuir. 2012;**28**(35):12893-12900. DOI: 10.1021/la303049w

[14] Ibhadon A, Fitzpatrick P. Heterogeneous photocatalysis: Recent advances and applications. Catalysts. 2013;**3**(1):189

[15] Herrmann J-M. Heterogeneous photocatalysis: Fundamentals and applications to the removal of various types of aqueous pollutants. Catalysis Today. 1999;**53**(1):115-129. DOI: http://dx.doi.org/10.1016/S0920-5861(99)00107-8

[16] Jiang R, Li B, Fang C, Wang J. Metal/semiconductor hybrid nanostructures for plasmon-enhanced applications. Advanced Materials. 2014;**26**(31):5274-5309. DOI: 10.1002/adma.201400203

[17] Fan W, Leung MKH, Yu JC, Ho WK. Recent development of plasmonic resonance-based photocatalysis and photovoltaics for solar utilization. Molecules. 2016;**21**(2):pii: E180. DOI: 10.3390/molecules21020180

[18] Xiao M, Jiang R, Wang F, Fang C, Wang J, Yu JC. Plasmon-enhanced chemical reactions. Journal of Materials Chemistry A. 2013;**1**(19):5790-5805. DOI: 10.1039/C3TA01450A

[19] Meng X, Zhang Z. Synthesis and characterization of plasmonic and magnetically separable Ag/AgCl-Bi_2WO_6@ Fe_3O_4@SiO_2 core-shell composites for visible light-induced water detoxification. Journal of Colloid and Interface Science. 2017;**485**:296-307. DOI: 10.1016/j.jcis.2016.09.045

[20] Li H, Wu T, Cai B, Ma W, Sun Y, Gan S, et al. Efficiently photocatalytic reduction of carcinogenic contaminant Cr(VI) upon robust AgCl:Ag hollow nanocrystals. Applied Catalysis B: Environmental. 2015;**164**:344-351. DOI: http://dx.doi.org/10.1016/j.apcatb.2014.09.049

[21] Guler U, Shalaev VM, Boltasseva A. Nanoparticle plasmonics: Going practical with transition metal nitrides. Materials Today. 2015;**18**(4):227-237. DOI: http://dx.doi.org/10.1016/j.mattod.2014.10.039

[22] Wang J, Tang L, Zeng G, Liu Y, Zhou Y, Deng Y, et al. Plasmonic Bi metal deposition and g-C_3N_4 coating on Bi_2WO_6 microspheres for efficient visible-light photocatalysis. ACS Sustainable Chemistry and Engineering. 2017;**5**(1):1062-1072. DOI: 10.1021/acssuschemeng.6b02351

[23] Huang Y, Kang S, Yang Y, Qin H, Ni Z, Yang S, et al. Facile synthesis of Bi/Bi_2WO_6 nanocomposite with enhanced photocatalytic activity under visible light. Applied Catalysis B: Environmental. 2016;**196**:89-99. DOI: 10.1016/j.apcatb.2016.05.022

[24] Yu S, Huang H, Dong F, Li M, Tian N, Zhang T, et al. Synchronously achieving plasmonic Bi metal deposition and I^- doping by utilizing $BiOIO_3$ as the self-sacrificing template for high-performance multifunctional applications. ACS Applied Materials and Interfaces. 2015;**7**(50):27925-27933. DOI: 10.1021/acsami.5b09994

[25] Han J, Zou HY, Liu ZX, Yang T, Gao MX, Huang CZ. Efficient visible-light photocatalytic heterojunctions formed by coupling plasmonic $Cu_{2-x}Se$ and graphitic carbon nitride. New Journal of Chemistry. 2015;**39**(8):6186-6192. DOI: 10.1039/c5nj01010a

[26] Meng X, Zhang Z. Pd-doped Bi_2MoO_6 plasmonic photocatalysts with enhanced visible light photocatalytic performance. Applied Surface Science. 2017;**392**:169-180. DOI: 10.1016/j.apsusc.2016.08.113

[27] Chen F, Yang Q, Wang Y, Zhao J, Wang D, Li X, et al. Novel ternary heterojunction photcocatalyst of Ag nanoparticles and g-C_3N_4 nanosheets co-modified $BiVO_4$ for wider spectrum visible-light photocatalytic degradation of refractory pollutant. Applied Catalysis B: Environmental. 2017;**205**:133-147. DOI: http://dx.doi.org/10.1016/j.apcatb.2016.12.017

[28] Wan Y, Liang C, Xia Y, Huang W, Li Z. Fabrication of graphene oxide enwrapped Z-scheme Ag_2SO_3/AgBr nanoparticles with enhanced visible-light photocatalysis. Applied Surface Science. 2017;**396**:48-57. DOI: 10.1016/j.apsusc.2016.10.189

[29] Ao Y, Bao J, Wang P, Wang C. A novel heterostructured plasmonic photocatalyst with high photocatalytic activity: Ag@AgCl nanoparticles modified titanium phosphate nanoplates. Journal of Alloys and Compounds. 2017;**698**:451-459. DOI: 10.1016/j.jallcom.2016.12.231

[30] Liu Y, Zhu G, Gao J, Hojamberdiev M, Zhu R, Wei X, et al. Enhanced photocatalytic activity of $Bi_4Ti_3O_{12}$ nanosheets by Fe^{3+}-doping and the addition of Au nanoparticles: Photodegradation of Phenol and bisphenol A. Applied Catalysis B: Environmental. 2017;**200**:72-82. DOI: 10.1016/j.apcatb.2016.06.069

[31] Levchuk I, Sillanpää M, Guillard C, Gregori D, Chateau D, Chaput F, et al. Enhanced photocatalytic activity through insertion of plasmonic nanostructures into porous TiO_2/SiO_2 hybrid composite films. Journal of Catalysis. 2016;**342**:117-124. DOI: 10.1016/j.jcat.2016.07.015

[32] Yan XQ, Zhu XH, Li RH, Chen WX. Au/BiOCl heterojunction within mesoporous silica shell as stable plasmonic photocatalyst for efficient organic pollutants decomposition under visible light. Journal of Hazardous Materials. 2016;**303**:1-9. DOI: 10.1016/j.jhazmat.2015.10.029

[33] Chen X, Zhao Q, Li X, Wang D. Enhanced photocatalytic activity of degrading short chain chlorinated paraffins over reduced graphene oxide/$CoFe_2O_4$/Ag nanocomposite. Journal of Colloid and Interface Science. 2016;**479**:89-97. DOI: 10.1016/j.jcis.2016.06.053

[34] Song S, Cheng B, Wu N, Meng A, Cao S, Yu J. Structure effect of graphene on the photocatalytic performance of plasmonic Ag/Ag_2CO_3-RGO for photocatalytic elimination of pollutants. Applied Catalysis B: Environmental. 2016;**181**:71-78. DOI: http://dx.doi.org/10.1016/j.apcatb.2015.07.034

[35] Guo W, Qin Q, Geng L, Wang D, Guo Y, Yang Y. Morphology-controlled preparation and plasmon-enhanced photocatalytic activity of Pt–BiOBr heterostructures. Journal of Hazardous Materials. 2016;**308**:374-385. DOI: http://dx.doi.org/10.1016/j.jhazmat.2016.01.077

[36] Gomez L, Sebastian V, Arruebo M, Santamaria J, Cronin SB. Plasmon-enhanced photocatalytic water purification. Physical Chemistry Chemical Physics. 2014;**16**(29):15111-15116. DOI: 10.1039/c4cp00229f

[37] Friedmann D, Hakki A, Kim H, Choi W, Bahnemann D. Heterogeneous photocatalytic organic synthesis: State-of-the-art and future perspectives. Green Chemistry. 2016;**18**(20):5391-5411. DOI: 10.1039/C6GC01582D

[38] Peiris S, McMurtrie J, Zhu H-Y. Metal nanoparticle photocatalysts: Emerging processes for green organic synthesis. Catalysis Science & Technology. 2016;**6**(2):320-338. DOI: 10.1039/C5CY02048D

[39] Chen J, Cen J, Xu X, Li X. The application of heterogeneous visible light photocatalysts in organic synthesis. Catalysis Science & Technology. 2016;**6**(2):349-362. DOI: 10.1039/C5CY01289A

[40] Lang X, Zhao J, Chen X. Cooperative photoredox catalysis. Chemical Society Reviews. 2016;**45**(11):3026-3038. DOI: 10.1039/C5CS00659G

[41] Cheng H, Fuku K, Kuwahara Y, Mori K, Yamashita H. Harnessing single-active plasmonic nanostructures for enhanced photocatalysis under visible light. Journal of Materials Chemistry A. 2015;**3**(10):5244-5258. DOI: 10.1039/c4ta06484d

[42] Tanaka A, Hashimoto K, Kominami H. Preparation of Au/CeO_2exhibiting strong surface plasmon resonance effective for selective or chemoselective oxidation of alcohols to aldehydes or ketones in aqueous suspensions under irradiation by green light. Journal of the American Chemical Society. 2012;**134**(35):14526-14533. DOI: 10.1021/ja305225s

[43] Hallett-Tapley GL, Silvero MJ, Bueno-Alejo CJ, Gonzalez-Bejar M, McTiernan CD, Grenier M, et al. Supported gold nanoparticles as efficient catalysts in the solvent less plasmon mediated oxidation of sec-Phenethyl and benzyl alcohol. Journal of Physical Chemistry C. 2013;**117**(23):12279-12288. DOI: 10.1021/jp311069v

[44] Yu J, Li J, Wei H, Zheng J, Su H, Wang X. Hydrotalcite-supported gold catalysts for a selective aerobic oxidation of benzyl alcohol driven by visible light. Journal of Molecular Catalysis A: Chemical. 2014;**395**:128-136. DOI: http://dx.doi.org/10.1016/j.molcata.2014.08.023

[45] Chen Y, Li W, Wang J, Yang Q, Hou Q, Ju M. Gold nanoparticle-modified TiO_2/SBA-15 nanocomposites as active plasmonic photocatalysts for the selective oxidation of aromatic alcohols. RSC Advances. 2016;**6**(74):70352-70363. DOI: 10.1039/C6RA11390G

[46] Pineda A, Gomez L, Balu AM, Sebastian V, Ojeda M, Arruebo M, et al. Laser-driven heterogeneous catalysis: Efficient amide formation catalysed by Au/SiO_2 systems. Green Chemistry. 2013;**15**(8):2043-2049. DOI: 10.1039/C3GC40166A

[47] Xiao Q, Liu Z, Bo A, Zavahir S, Sarina S, Bottle S, et al. Catalytic transformation of aliphatic alcohols to corresponding esters in O_2 under neutral conditions using visible-light irradiation. Journal of the American Chemical Society. 2015;**137**(5):1956-1966. DOI: 10.1021/ja511619c

[48] Zhang Y, Xiao Q, Bao Y, Zhang Y, Bottle S, Sarina S, et al. Direct photocatalytic conversion of aldehydes to esters using supported gold nanoparticles under visible light irradiation at room temperature. The Journal of Physical Chemistry C. 2014;**118**(33):19062-19069. DOI: 10.1021/jp505552v

[49] Alejo CJB, Fasciani C, Grenier M, Netto-Ferreira JC, Scaiano JC. Reduction of resazurin to resorufin catalyzed by gold nanoparticles: Dramatic reaction acceleration by laser or LED plasmon excitation. Catalysis Science & Technology. 2011;**1**(8):1506-1511. DOI: 10.1039/C1CY00236H

[50] Ke X, Zhang X, Zhao J, Sarina S, Barry J, Zhu H. Selective reductions using visible light photocatalysts of supported gold nanoparticles. Green Chemistry. 2013;**15**(1):236-244. DOI: 10.1039/C2GC36542A

[51] Guo X, Hao C, Jin G, Zhu H-Y, Guo X-Y. Copper nanoparticles on graphene support: An efficient photocatalyst for coupling of nitroaromatics in visible light. Angewandte Chemie International Edition. 2014;**53**(7):1973-1977. DOI: 10.1002/anie.201309482

[52] Tokarek K, Hueso JL, Kuśtrowski P, Stochel G, Kyzioł A. Green synthesis of chitosan-stabilized copper nanoparticles. European Journal of Inorganic Chemistry. 2013;**2013**(28):4940-4947. DOI: 10.1002/ejic.201300594

[53] Zhao X, Long R, Liu D, Luo B, Xiong Y. Pd-Ag alloy nanocages: Integration of Ag plasmonic properties with Pd active sites for light-driven catalytic hydrogenation. Journal of Materials Chemistry A. 2015;**3**(18):9390-9394. DOI: 10.1039/C5TA00777A

[54] Wang F, Li C, Chen H, Jiang R, Sun LD, Li Q, et al. Plasmonic harvesting of light energy for Suzuki coupling reactions. Journal of the American Chemical Society. 2013;**135**(15):5588-5601. DOI: 10.1021/ja310501y

[55] Hu YX, Liu YZ, Li Z, Sun YG. Highly asymmetric, interfaced dimers made of Au nanoparticles and bimetallic nanoshells: Synthesis and photo-enhanced catalysis. Advanced Functional Materials. 2014;**24**(19):2828-2836. DOI: 10.1002/adfm.201303557

[56] Cui J, Li Y, Liu L, Chen L, Xu J, Ma J, et al. Near-infrared plasmonic-enhanced solar energy harvest for highly efficient photocatalytic reactions. Nano Letters. 2015;**15**(10):6295-6301. DOI: 10.1021/acs.nanolett.5b00950

[57] Lou Z, Gu Q, Liao Y, Yu S, Xue C. Promoting Pd-catalyzed Suzuki coupling reactions through near-infrared plasmon excitation of WO_3-x nanowires. Applied Catalysis B: Environmental. 2016;**184**:258-263. DOI: 10.1016/j.apcatb.2015.11.037

[58] Cheng H, Kamegawa T, Mori K, Yamashita H. Surfactant-free nonaqueous synthesis of plasmonic molybdenum oxide nanosheets with enhanced catalytic activity for hydrogen generation from ammonia borane under visible light. Angewandte Chemie International Edition. 2014;**53**(11):2910-2914. DOI: 10.1002/anie.201309759

[59] Li Q, Mahendra S, Lyon DY, Brunet L, Liga MV, Li D, et al. Antimicrobial nanomaterials for water disinfection and microbial control: Potential applications and implications. Water Research. 2008;**42**(18):4591-4602. DOI: http://dx.doi.org/10.1016/j.watres.2008.08.015

[60] Tian B, Dong R, Zhang J, Bao S, Yang F, Zhang J. Sandwich-structured AgCl at Ag at TiO_2 with excellent visible-light photocatalytic activity for organic pollutant degradation and *E. coli* K12 inactivation. Applied Catalysis B: Environmental. 2014;**158-159**:76-84. DOI: 10.1016/j.apcatb.2014.04.008

[61] Shi H, Li G, Sun H, An T, Zhao H, Wong P-K. Visible-light-driven photocatalytic inactivation of *E. coli* by Ag/AgX-CNTs (X = Cl, Br, I) plasmonic photocatalysts: Bacterial performance and deactivation mechanism. Applied Catalysis B: Environmental. 2014; **158-159**:301-307. DOI: http://dx.doi.org/10.1016/j.apcatb.2014.04.033

[62] Xia D, An T, Li G, Wang W, Zhao H, Wong PK. Synergistic photocatalytic inactivation mechanisms of bacteria by graphene sheets grafted plasmonic AgAgX (X = Cl, Br, I) composite photocatalyst under visible light irradiation. Water Research. 2016;**99**:149-161. DOI: 10.1016/j.watres.2016.04.055

[63] Sarkar AK, Saha A, Tarafder A, Panda AB, Pal S. Efficient removal of toxic dyes via simultaneous adsorption and solar light driven photodegradation using recyclable functionalized Amylopectin-TiO_2-Au nanocomposite. ACS Sustainable Chemistry and Engineering. 2016;**4**(3):1679-1688. DOI: 10.1021/acssuschemeng.5b01614

[64] Li C, Yang W, Liu L, Sun W, Li Q. In situ growth of TiO_2 on TiN nanoparticles for non-noble-metal plasmonic photocatalysis. RSC Advances. 2016;**6**(76):72659-72669. DOI: 10.1039/c6ra15435b

[65] Ren H, Koshy P, Chen W-F, Qi S, Sorrell CC. Photocatalytic materials and technologies for air purification. Journal of Hazardous Materials. 2017;**325**:340-366. DOI: http://dx.doi.org/10.1016/j.jhazmat.2016.08.072

[66] Chen X, Zhu H-Y, Zhao J-C, Zheng Z-F, Gao X-P. Visible-light-driven oxidation of organic contaminants in air with gold nanoparticle catalysts on oxide supports. Angewandte Chemie. 2008;**120**(29):5433-5436. DOI: 10.1002/ange.200800602

[67] Dinh C-T, Yen H, Kleitz F, Do T-O. Three-dimensional ordered assembly of thin-shell Au/TiO_2hollow nanospheres for enhanced visible-light-driven photocatalysis. Angewandte Chemie International Edition. 2014;**53**(26):6618-6623. DOI: 10.1002/anie.201400966

[68] Sellappan R, Nielsen MG, González-Posada F, Vesborg PCK, Chorkendorff I, Chakarov D. Effects of plasmon excitation on photocatalytic activity of Ag/TiO_2 and Au/TiO_2 nanocomposites. Journal of Catalysis. 2013;**307**:214-221. DOI: http://dx.doi.org/10.1016/j.jcat.2013.07.024

[69] Inoue T, Fujishima A, Konishi S, Honda K. Photoelectrocatalytic reduction of carbon dioxide in aqueous suspensions of semiconductor powders. Nature. 1979;**277**(5698):637-638

[70] Liu HM, Meng XG, Dao TD, Zhang HB, Li P, Chang K, et al. Conversion of carbon dioxide by methane reforming under visible-light irradiation: Surface-plasmon-mediated nonpolar molecule activation. Angewandte Chemie-International Edition. 2015;**54**(39):11545-11549. DOI: 10.1002/anie.201504933

[71] Upadhye AA, Ro I, Zeng X, Kim HJ, Tejedor I, Anderson MA, et al. Plasmon-enhanced reverse water gas shift reaction over oxide supported Au catalysts. Catalysis Science & Technology. 2015;**5**(5):2590-2601. DOI: 10.1039/C4CY01183J

[72] Shown I, Hsu H-C, Chang Y-C, Lin C-H, Roy PK, Ganguly A, et al. Highly efficient visible light photocatalytic reduction of CO_2 to hydrocarbon fuels by Cu-nanoparticle decorated graphene oxide. Nano Letters. 2014;**14**(11):6097-6103. DOI: 10.1021/nl503609v

[73] Yuliati L, Itoh H, Yoshida H. Photocatalytic conversion of methane and carbon dioxide over gallium oxide. Chemical Physics Letters. 2008;**452**(1--3):178-182. DOI: http://dx.doi.org/10.1016/j.cplett.2007.12.051

[74] Baldoví HG, Neaţu Ş, Khan A, Asiri AM, Kosa SA, Garcia H. Understanding the origin of the photocatalytic CO_2 reduction by Au- and Cu-loaded TiO_2: A microsecond transient absorption spectroscopy study. The Journal of Physical Chemistry C. 2015;**119**(12):6819-6827. DOI: 10.1021/jp5106136

[75] Tahir B, Tahir M, Amin NS. Gold-indium modified TiO_2 nanocatalysts for photocatalytic CO_2 reduction with H_2 as reductant in a monolith photoreactor. Applied Surface Science. 2015;**338**:1-14. DOI: http://dx.doi.org/10.1016/j.apsusc.2015.02.126

[76] Tan L-L, Ong W-J, Chai S-P, Mohamed AR. Noble metal modified reduced graphene oxide/TiO_2 ternary nanostructures for efficient visible-light-driven photoreduction of carbon dioxide into methane. Applied Catalysis B: Environmental. 2015;**166-167**:251-259. DOI: http://dx.doi.org/10.1016/j.apcatb.2014.11.035

[77] Vaiano V, Sannino D, Ciambelli P. Steam reduction of CO_2 on Pd/TiO_2 catalysts: A comparison between thermal and photocatalytic reactions. Photochemical & Photobiological Sciences. 2015;**14**(3):550-555. DOI: 10.1039/C4PP00252K

[78] Yang K, Liu J, Si R, Chen X, Dai W, Fu X. Comparative study of Au/TiO_2 and Au/Al_2O_3 for oxidizing CO in the presence of H_2 under visible light irradiation. Journal of Catalysis. 2014;**317**:229-239. DOI: http://dx.doi.org/10.1016/j.jcat.2014.06.005

[79] Yui T, Kan A, Saitoh C, Koike K, Ibusuki T, Ishitani O. Photochemical reduction of CO_2 using TiO_2: Effects of organic adsorbates on TiO_2 and deposition of Pd onto TiO_2. ACS Applied Materials & Interfaces. 2011;**3**(7):2594-2600. DOI: 10.1021/am200425y

[80] Zhai Q, Xie S, Fan W, Zhang Q, Wang Y, Deng W, et al. Photocatalytic conversion of carbon dioxide with water into methane: Platinum and copper(I) oxide Co-catalysts with a Core–Shell structure. Angewandte Chemie International Edition. 2013;**52**(22):5776-5779. DOI: 10.1002/anie.201301473

[81] Zhang X, Li X, Zhang D, Su NQ, Yang W, Everitt HO, et al. Product selectivity in plasmonic photocatalysis for carbon dioxide hydrogenation. Nature Communications. 2017;**8**:14542. DOI: 10.1038/ncomms14542http://www.nature.com/articles/ncomms14542#supplementary-information

[82] Putri LK, Ong W-J, Chang WS, Chai S-P. Enhancement in the photocatalytic activity of carbon nitride through hybridization with light-sensitive AgCl for carbon dioxide reduction to methane. Catalysis Science & Technology. 2016;**6**(3):744-754. DOI: 10.1039/C5CY00767D

[83] Macia-Agullo JA, Corma A, Garcia H. Photobiocatalysis: The power of combining photocatalysis and enzymes. Chemistry-a European Journal. 2015;**21**(31):10940-10959. DOI: 10.1002/chem.201406437

[84] Sanchez-Iglesias A, Barroso J, Solis DM, Taboada JM, Obelleiro F, Pavlov V, et al. Plasmonic substrates comprising gold nanostars efficiently regenerate cofactor molecules. Journal of Materials Chemistry A. 2016;**4**(18):7045-7052. DOI: 10.1039/c6ta01770c

[85] Sanchez-Iglesias A, Chuvilin A, Grzelczak M. Plasmon-driven photoregeneration of cofactor molecules. Chemical Communications. 2015;**51**(25):5330-5333. DOI: 10.1039/c4cc07829b

[86] Lin YH, Ren JS, Qu XG. Catalytically active nanomaterials: A promising candidate for artificial enzymes. Accounts of Chemical Research. 2014;**47**(4):1097-1105. DOI: 10.1021/ar400250z

[87] Ortega-Liebana MC, Hueso JL, Arenal R, Santamaria J. Titania-coated gold nanorods with expanded photocatalytic response. Enzyme-like glucose oxidation under near-infrared illumination. Nanoscale. 2017;**9**(5):1787-1792. DOI: 10.1039/C6NR06300D

[88] Zhou BW, Song JL, Zhou HC, Wu TB, Han BX. Using the hydrogen and oxygen in water directly for hydrogenation reactions and glucose oxidation by photocatalysis. Chemical Science. 2016;**7**(1):463-468. DOI: 10.1039/c5sc03178h

[89] Primo A, Corma A, Garcia H. Titania supported gold nanoparticles as photocatalyst. Physical Chemistry Chemical Physics. 2011;**13**(3):886-910. DOI: 10.1039/c0cp00917b

Chapter 15

Plasmonic Nanostructures as Surface-Enhanced Raman Scattering (SERS) Substrate for Protein Biomarker Sensing

Dan Li and Yuling Wang

Additional information is available at the end of the chapter

http://dx.doi.org/10.5772/intechopen.68164

Abstract

Plasmonic nanostructures have attracted considerable interest in biomarker sensing with the goal of rapid diagnostics and personalized nanomedicine. Surface-enhanced Raman scattering (SERS) is a versatile technique for the characterization of the plasmonic effect of the metallic nanostructures as well as a sensitive read-out approach for biomarkers detection. In this contribution, we will give a review on the key optical properties of plasmonic nanostructures as SERS substrate for protein biomarkers detection. As a consequence, two approaches, label-free and SERS labels will be discussed in details for protein biomarkers sensing by using the plasmonic nanostructures as the substrate.

Keywords: plasmonic nanostructures, surface-enhanced Raman scattering (SERS), protein biomarker, label-free, SERS labels, sensing

1. Introduction

In order to understand the fundamental of plasmonic nanostructures and the related application, it is important to start with "plasmon", which was first named by Pines in the 1950s [1, 2]. As the valence electron collective oscillations resemble closely the electronic plasma oscillations observed in gaseous discharges, the term "plasmon" was thus used to describe the quantum of elementary excitation associated with this high-frequency collective motion [2]. Based on the definition, plasmon is very similar as photon, which is the quantum particle representing the elementary excitations, or modes, of the free electromagnetic field oscillations. To be simple, plasmon can be described in the classical picture as an oscillation of free electron density against the fixed positive ions in a metal [3, 4]. To visualize a plasma oscillation, imagine a cube of metal is placed in an external electric field pointing to the right. Electrons will then move to the left side (uncovering positive ions on the right side) until they cancel the field inside the metal [5]. When the electric field is switched off, and the electrons move to the right, repelled by each other and attracted to the positive ions left bare on the right side. They oscillate back and forth at the plasma frequency until the energy is lost in some kind of resistance or damping [6–10]. Plasmons are a quantization of this kind of oscillation, which plays a large role in the optical properties of metals. Light of frequency below the plasma frequency is reflected, because the electrons in the metal screen the electric field of the light. Light of frequency above the plasma frequency is transmitted, because the electrons cannot respond fast enough to screen it [5]. In most metals, the plasma frequency is in the ultraviolet, making them shiny (reflective) in the visible range [8–11]. Some metals such as copper and gold having electronic interband transitions in the visible range, whereby specific light energies (colors) are absorbed, yield distinct colors.

Surface plasmons are those plasmons that are confined to surfaces and that interact strongly with light resulting in a polariton. They occur at the interface of a vacuum or material with a positive dielectric constant and a negative dielectric constant (usually a metal or doped dielectric) [5]. Plasmonics is related to the localization, guiding and manipulation of electromagnetic waves beyond the diffraction limit and down to the nanometer-length scale [12, 13]. The key component of plasmonics is a metal, which can support surface plasmon polariton (electromagnetic waves coupled to the collective oscillations of free electrons in the metal, plasmon coupled with photons). There are two types of surface plasmon resonances (SPRs) that can be generated [12, 13]. The first one is the surface plasmon polariton (SPP), which is generated when light becomes trapped at a metal-dielectric interface. The second one is the localized surface plasmon resonance (LSPR), which happened when a surface plasmon is confined to a particle of size comparable to the wavelength of light, that is, a nanoparticle, the particle's free electrons participate in the collective oscillation. For the full detailed description of the fundamentals of plasmon and plasmonic effect, several excellent reviews are recommended [13–19].

Plasmonic nanostructures are thus the metallic nanostructures on which the electromagnetic field was generated by exciting the oscillation of surface plasmon in the metal-light dielectric interface. The study on the plasmonic nanostructures has attracted more and more interest in the research areas from the fundamentals to applications in a variety of scientific disciplines. In this chapter, we will focus on the plasmonic nanostructures as the surface-enhanced Raman scattering (SERS) substrate for protein biomarkers detection.

SERS was first discovered in the 1970s [20, 21] on electrochemical roughed silver electrode and dramatically developed with the advancement of nanotechnology. Currently, two basic principles that contribute the SERS effect are accepted, which are the chemical effect (CM) e.g., charge-transfer and the electromagnetic effect (EM) [22–28]. Basically, CM is the interaction of the adsorbate molecules and the metal surface, mostly from the first layer of the charge-transfer resonance between molecules and the metal [22–24]. Whereas EM mechanism is based on the interaction of the transition moment of an adsorbed molecule with the electric field of a surface plasmon induced by the incoming light at the metal, as discussed above [25–28].

Thus, SERS has been utilized in different areas including the catalysis, energy, and biology, and, in particular, in biomedical application. For instance, SERS has been used to monitor the catalytic reaction, image the live cell and tumor, monitor the nanoparticle distribution in live body, and so on [29–31]. Several excellent reviews have been published on SERS including the basic principle of SERS, SERS nanoparticles, SERS labels as well as SERS applications in biomedicine [32–37]. In this chapter, we will mostly focus on the optical properties (plasmonic effect) of the metallic nanostructures for SERS, with the highlight on the application of plasmonic nanostructures for protein detection by using SERS as a read-out technology.

2. Plasmonic nanostructures

In this section, optical property of plasmonic nanostructures (surface plasmon resonance) will be discussed in terms of the SERS-activity. Plenty of methods have been reported for the synthesis of plasmonic nanostructures including the wet-chemistry (seed-mediated growth), template nanoscale lithography, thin-film, and the template approach. Here in this chapter, we will mainly focus on the synthesis and characterization of plasmonic nanostructures that are widely used for SERS study. Based on the shape or geometry of the nanostructures, we will give a brief overview for the quasi-spherical metallic nanostructures, anisotropic nanoparticles/structures as well as the plasmonic nanoassemblies with the highlight on the plasmonic effect for SERS activity. As Mirkin et al. have discussed in details of the template techniques for assembly of plasmonic nanostructures [13], we will not include this technique in this chapter.

2.1. Quasi-spherical metallic nanoparticles

As gold and silver nanostructures show the most significant SERS effect due to the surface plasmon resonance generated on gold and silver surface (EM enhancement mechanism), other metals such as copper and the transit metal also show the SERS effect; however, the impact factor is very low, which depends on several factors, including the size and geometry of nanoparticles as well as the laser excitations. Numerous protocols are available for the synthesis of metallic spheres [38–41]. The simplest and most common approach is the reduction of metal salts with a variety of reducing and capping agents. To improve the stability and the enhancement of plasmonic nanoparticles, the combination of the alloy metal is an option. Thus, gold/silver nanoshells are designed and synthesized as the composite metallic nanostructures [42, 43] that show higher SERS effect compared to single metal. More importantly, by tuning the shell thickness, the LSPR of gold/silver nanoshells can be tuned (**Figure 1b**). In regard to the enhancement factor (EF) on the metallic sphere nanoparticles for SERS, it has been reported that individual spherical nanoparticles generate very low EF for SERS based on the EM calculation [34, 44]. Single molecule detection [25, 27] was reported on silver nanoparticles with the enhancement factor (EF) around 10^{12}–10^{14}, which is due to the strong surface plasmon coupling effect (called "hot-spot"). In order to get high order of EF, well-defined plasmonic nanostructures with multiple "hot-spots" are required, which will be discussed in the following session.

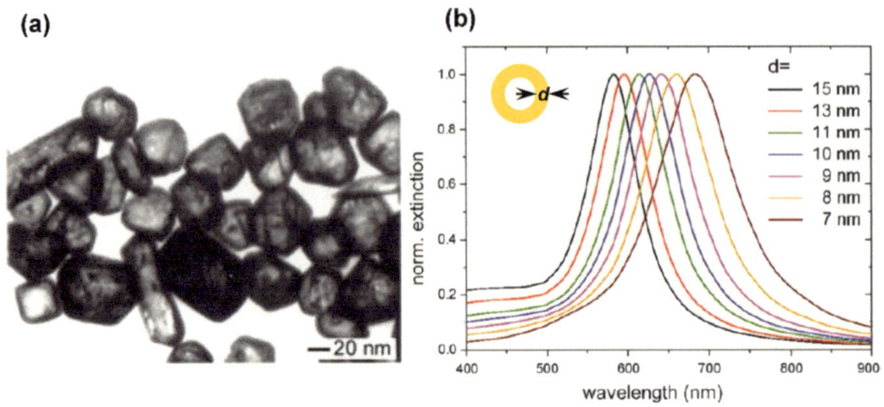

Figure 1. TEM image of gold/silver nanoshell (a) and the tunable LSPR of gold/silver nanoshells with the shell thickness (b) (cited from Ref. [33] with permission from the Royal Society of Chemistry).

2.2. Anisotropic nanoparticles

As it is hard to obtain the SERS signal in the individual metallic nanospheres, anisotropic nanoparticles such as rods, stars, cubes, prisms, and nanoplates became very important SERS substrates (**Figure 2a**), which exhibit significantly higher electromagnetic field enhancements at sharp edges ("lighting rod effect" or "plasmonic antenna effect"), making them attractive for use as plasmonic enhancers in SERS [45–49]. For the synthesis of anisotropic nanoparticles, the most widely employed approach is seed-mediated growth, which involves seed-formation and growth. Typical example is the synthesis of gold nanorods (**Figure 2a**), which starts with the quasi-spherical ~4 nm gold seeds and subsequent reduction of more metal salt with

Figure 2. Typical electronmicrographs of anisotropic nanostructures for efficient SERS. Nanorods (a), nanoflowers (b), nanoplates (c), nanoprisms (d), nanocubes (e), and nanostars (f). (cited from Ref. [33] with permission from the Royal Society of Chemistry).

a weak reducing agent such as ascorbic acid in the presence of structure-directing additives e.g., CTAB (cetyltrimethylammonium bromide), leading to the formation of gold nanorods. The aspect ratio can be controlled by the relative concentrations of the reagents. By tuning the aspect ratio of the gold nanorods, their localized surface plasmon resonance (LSPR) can be tuned, which is an important aspect for the application of gold nanorods in SERS as well as phototherapy.

2.3. Plasmonic nanoassemblies

In order to generate the hot-spots for the SERS enhancement, plasmonic clusters including the dimers, trimers, and high orders nanoaggregates have been prepared. As it is very important to understand the relationship between nanoparticle structure and the SERS activity, Van Duyne et al. have reported the SERS EF with the different nanostructures (mainly dimers and trimers) [50]. As indicated in **Figure 3**, individual trimers have been encapsulated with silica-shell to avoid the interference from the environment as well as ensure the contribution solely from the individual nanoparticles. It was demonstrated that the creation of hot spots, where two nanoparticles are in sub-nanometer proximity or have coalesced to form crevices, is the paramount to achieving maximum SERS EF. Specifically, L-shaped trimer nanoantenna comprised of three Au cores showed the EF of 10^8–10^{10} (**Figure 3a**) and the dark-field Rayleigh scattering spectrum of the L-shaped trimer contains three peaks (**Figure 3b**), corresponding to dipolar and multipolar LSPRs [50]. Single-particle SERS obtained from the trimer nanoantenna showed distinct peaks from the Raman molecule (PCEPE), which are correlated with the density functional theory (DFT) calculations (**Figure 3c–e**).

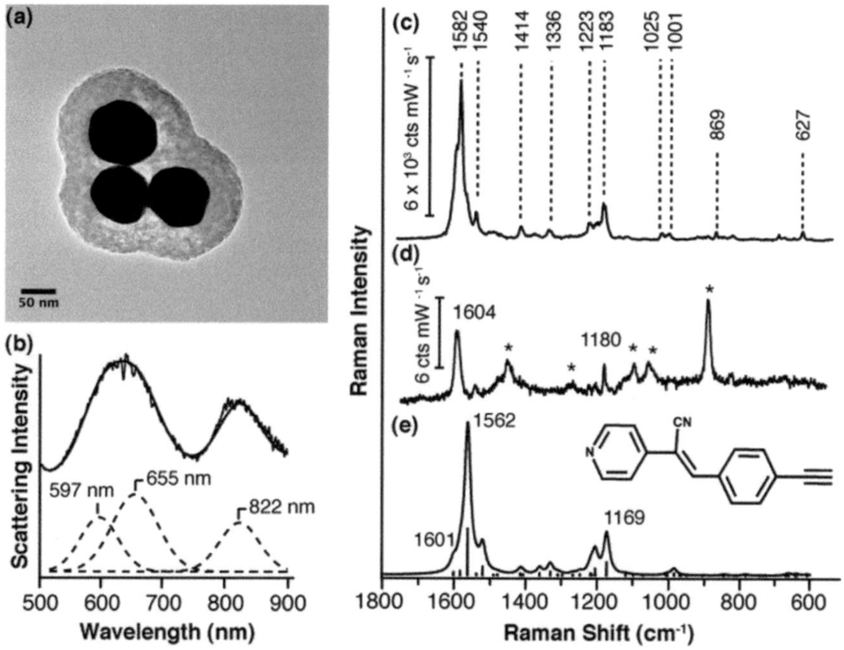

Figure 3. L-shape trimer nanoantenna (a), LSPR spectrum of the trimer obtained by dark-field Raleigh scattering microscopy (b), SERS spectrum of the Raman molecule (PCEPE) (c), normal Raman spectrum of PCEPE (d), and Raman spectrum of PECEP calculated using DFT (e). (Reproduced with permission cited from Ref. [50]).

Plasmonic nanostructures with multiple hot-spots have been reported with the satellite nanostructures by using either gold sphere or gold nanorods as the core [51–53]. To obtain the high orders of nanoassemblies, the linker is the key to connect the nanoparticles. Chemicals with dual/multifunctional groups are often used such as cystamine and dithiol-polymer. As silica surface is versatile, the functionalization of silica surface followed by the assembly of small nanoparticles is a very useful strategy. More importantly, the distance between the satellite nanoparticles with the core nanoparticles can be controlled by the silica-thickness. As reported by Gellner et al. [51], gold nanoparticles with the diameter of 80 nm were incubated with Raman reporters and encapsulated with a very thin silica shell. The ultrathin glass shell was then functionalized with a binary mixture of silanes including an aminosilane, which can adsorb the negatively charged gold nanospheres to form the 3D structures. Correlated HR-SEM/dark-field/LSPR/SERS experiments on individual 3D SERS-active superstructures together with finite element method (FEM) calculations confirmed the plasmonic coupling between the core and the satellite particles, with hot-spots occurring between core and satellites as well as between satellites [51].

3. Label-free protein biomarkers detection

SERS is a powerful vibrational spectroscopy, which can provide rich molecular information for the target, making it very useful for direct protein detection. The fingerprint information extracted from the SERS spectra of proteins can be used directly to identifying the protein confirmation, the structure as well as the component of the target protein [54–56]. Label-free SERS detection for protein originates from the chromophores such as hemoglobin, myoglobin, and cytochrome c, which showed strong SERS signals with good reproducibility due to the Raman resonance (RR) effect of the chromophore center of the proteins [57, 58]. The information related to the conformation and orientation of proteins as well as the charge-transfer processes between protein and surface can be obtained. As reported by Feng and Tachikawa [59], to determine the factors that contribute to the difference of SER(R)S signals and RR signal of metmyoglobin, they designed a Raman flow system and found that both the degree of protein-nanoparticles interaction and the laser irradiation contribute to the structural changes (**Figure 4**).

As the majority of proteins have no chromophore, the detection of proteins become much harder since the SERS signal from the native protein is very weak and most of the signals are generated from the amino acid residues and amide backbones, which are very similar for most of the proteins [60, 61]. Therefore, the key to get robust and sensitive label-free protein detection is the SERS substrate, which should have the high SERS-activity as well as proper preparation of the reproducible surface. Gold and silver nanocolloids are the widely used plasmonic nanoparticles for the label-free SERS detection [62–64]. As the gold and silver nanoparticles have relatively lower SERS enhancement effect, the procedure that could induce the aggregation, thus generate "hot-spot" nanostructures is required. Typically, the addition of aggregation agent such as salt is the simplest and easiest approach for the aggregation. As displayed in **Figure 5**, Han et al. utilized sulfate as an aggregation agent to induce strong

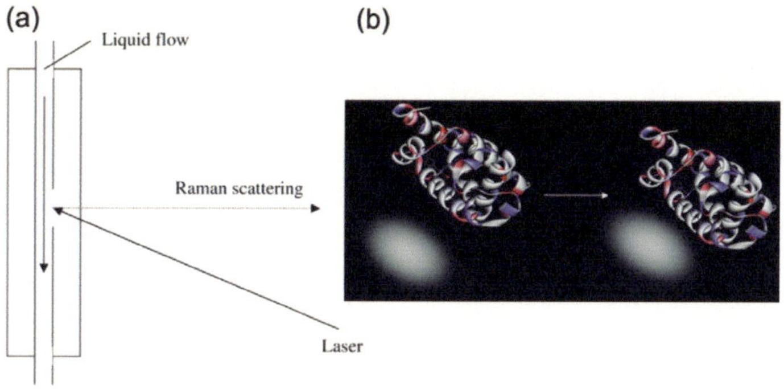

Figure 4. Raman-flow system for label-free protein detection (a) and the interaction of the protein with a silver nanoparticle monitored by SERS (b). (Reproduced with permission cited from Ref. [59]).

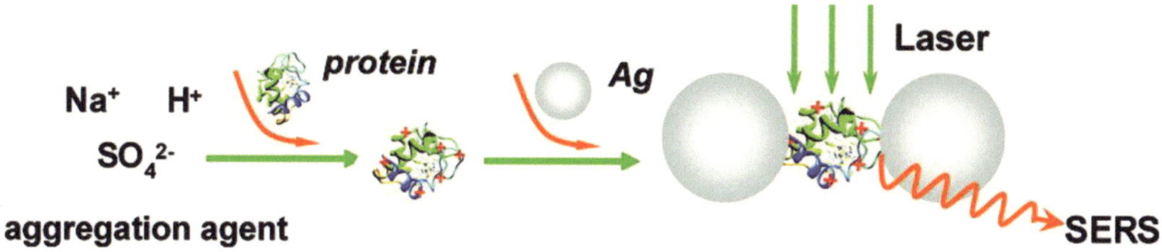

Figure 5. Scheme for aggregating nanoparticles for directly label-free SERS detection of proteins. (Reproduced with permission cited from Ref. [65]).

SERS signal due to the weak binding of SO_4^{2-} on silver surface, making it easier for protein binding [65]. With this scheme, proteins including lysozyme, ribonuclease B, avidin, catalase, and hemoglobin have been detected and analyzed [65]. As SERS signal from the protein can provide useful information in terms of the structure, constituents, conformation as well as the potential interaction of the protein with the surface, in recent years, several groups have put great effort to improve the method of SERS-based label free identification of proteins. For instance, Zhao et al. have conducted serious study on the label-free protein detection by controlling the plasmonic nanoparticles and optimizing the purification procedures of the proteins [62–65]. A typical example is the Western-blot SERS, which was based on the silver staining of the membrane after the protein separation with the gel electrophoresis, which will purify and separate the proteins on the surface, making the detection more simple and easy.

Furthermore, Ren et al. have proposed a facile method to enable reliable label-free SERS detection of the native structures of a wider range of proteins by using the iodide-modified silver colloids as illustrated in **Figure 6** [66]. The colloidal state of Ag NPs will help to keep the native structures of proteins and promote the photostability of samples. The iodide modification affords a one-atom-thick monolayer on the surface without producing interfering signal. Therefore, they demonstrated that the iodide-modified silver colloids could not only clean the surface but also avoid the strong chemical interaction between the metal surface and the proteins, and reduce the possibility of denaturation, thus make the detection reliable and reproducible.

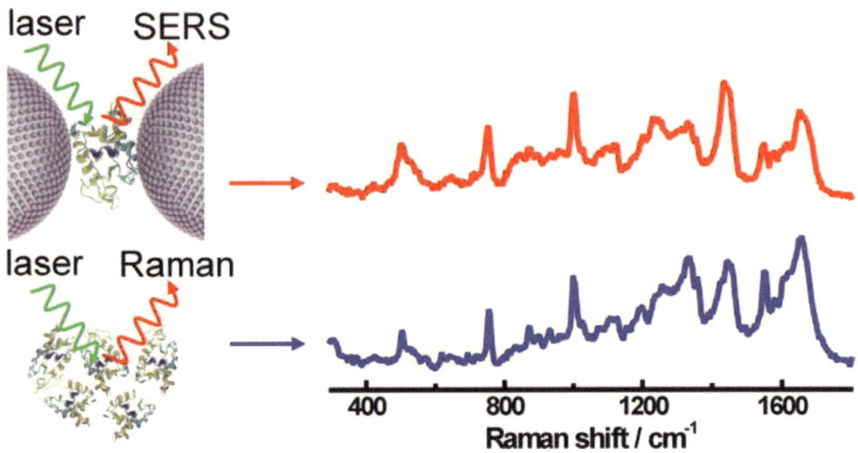

Figure 6. Label-free SERS for native protein detection on iodide-modified silver nanoparticles. (Reproduced with permission cited from Ref. [66]).

Recently, label-free SERS has been attempted for the serum protein detection, aiming for the cancer diagnosis. By inducing the aggregation of silver nanoparticles for high-quality SERS spectra, colorectal cancer has been identified based on the principal component analysis combined with linear discriminant analysis. [67–69]

4. SERS Labels for protein biomarkers sensing

Compared to label-free SERS approach, SERS labels have attracted significant attention for protein detection because of its high sensitivity comparable to fluorescence. The high sensitivity is most due to the strong plasmonic effect of the SERS substrates, which could be AuNPs, AgNPs, anisotropic nanoparticles (gold nanorods), nanoshells, plasmonic nanoassemblies, as discussed in the early session of this chapter. More importantly, SERS labels have demonstrated their unique optical properties and potential for simultaneous and multiplexed detection [33–37] owing to the advantages of SERS labels over fluorescent label including (i) the multiplexing capability for simultaneous target protein detection due to the narrow width of the vibrational Raman bands, (ii) quantification using the SERS fingerprint of the corresponding labels, (iii) the need for only a single laser excitation wavelength, and (iv) high photostability. In this section, the sensor platform using SERS labels for protein biomarker sensing will be discussed from the conventional sandwich immunoassay, dot-blot semi-sandwich immunoassay, and protein microarray to microfluidic protein assay, with the highlights on our recent works for the detection of breast cancer biomarker [70–73], pathogen antigens [74], cytokines [75, 76], and related works.

SERS sandwich immunoassay is a conventional sensor platform for the protein biomarker detection with SERS labels. Porter et al. have reported a serial of works on SERS immunoassay for various protein biomarkers detection [77–82]. The SERS labels were named as ERLs (external reporter label). Typically, ERLs were composed of metallic nanoparticles (e.g., widely used gold nanoparticles duo to the distinguished plasmonic effects) and Raman reporters (dyes or the small molecule) to indicate the presence of the target. As indicated in **Figure 7a**, detection antibodies were immobilized on the plasmonic nanoparticles by either electrostatic force or the covalent binding through the links such as DSP (dithiobis (succinimidyl propionate)) [79] and carboxyl-PEG-SH (poly (ethylene glycol) 2-mercaptoethyl ether acetic acid) [30]. To fabricate the sandwich SERS platform, the substrate could be the glass or gold film (**Figure 7b**). Additional studies have shown the higher sensitivity with gold film as the substrate because of the plasmonic coupling effect between the gold film (SPP—surface plasmon polarition) and the SERS labels (LSP—localized surface plasmon) [81, 82]. Followed by the capture of the target antigen or probes by the antibody on the surface, ERLs are bound on the surface for the signal to indicate the presence of the target proteins. With the similar platform design, plenty of proteins have been detected with few from the clinic important biomarkers [78–83].

Microassay is the platform that could detect proteins with high throughput by SERS labels using different dyes as the Raman reporters. As a typical example indicated in **Figure 8**, multiple protein targets have been immobilized on the surface, followed by the probe with the dye labeled SERS nanoparticles [83]. Upon the laser excitation, SERS signals from the dye will indicate the presence of the target proteins. Microfluidic assay is another platform with high-throughput properties for protein detection as the design for the channels can separate each of the individual proteins with different channels as well as enhance the reaction by controlling the flow conditions [74].

Although it is a high throughput, the requirement for the professional training, long-time incubation as well as the labor intensive procedures has hindered the application of the platform in the clinic setting. To improve the efficiency of the platform, a dot-bot assay was thus developed, which is targeting on the rapid and sensitive and simultaneous multiple protein

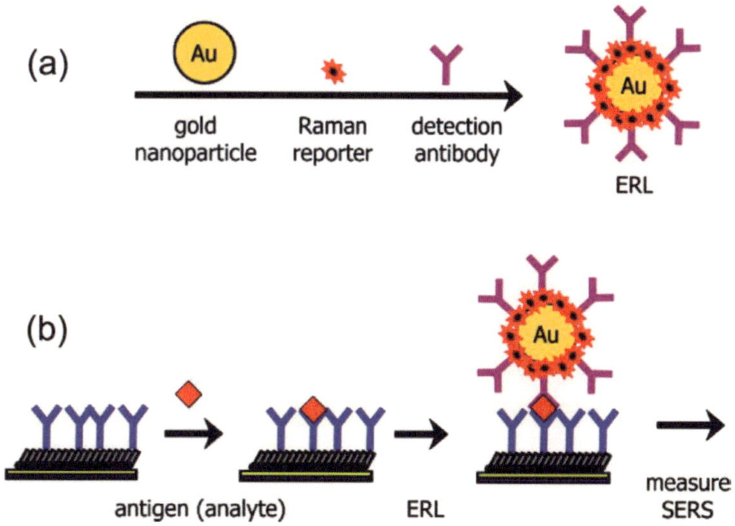

Figure 7. Schematic illustration of SERS ERLS (a) and typical SERS sandwich immunoassay for protein biomarkers detection (b). (Reproduced with permission cited from Ref. [35]).

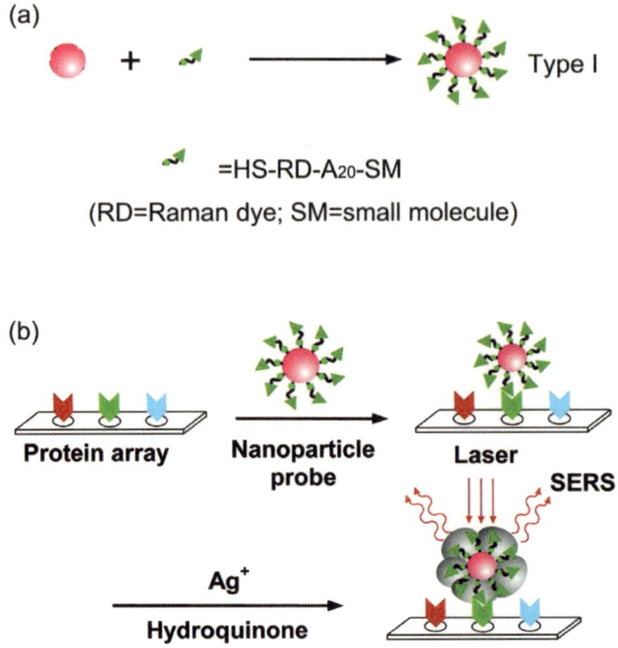

Figure 8. Dye labeled SERS nanoparticles (a) and protein array (b) for high throughput proteins detection by SERS. (Reproduced with permission cited from Ref. [83]).

detections with a simplified procedure [75]. As displayed in **Figure 9**, duplex cytokines (interleuckin-6 and interleuckin-8; IL-6 and IL-8) were detected simultaneously on the dot-blot assay with femtogram (fg) sensitivity, which was achieved by using the rational designed Au/Ag nanoshells as the plasmonic substrate (**Figure 9**). As specificity is a key issue for this study because it is hard to test the real samples without purification. Therefore, much more works were reported in this area, the limitations in the long incubation time and the labor-intensive procedures with this platform have become a big obstacle toward the application in real, in particular, for the point-of-care diagnosis.

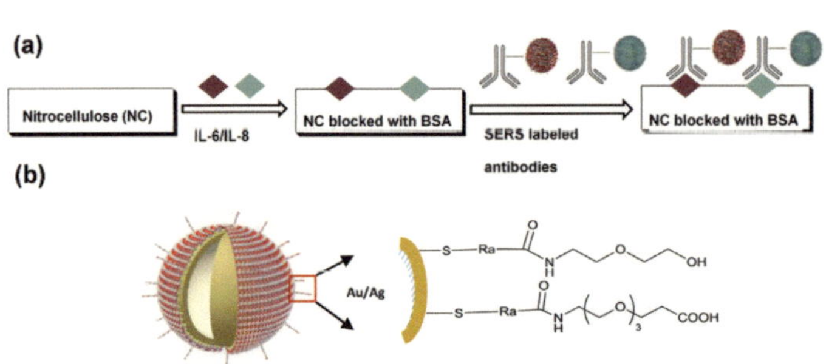

Figure 9. Scheme of the direct SERS dot-blot immunoassay platform for duplex cytokine detection (a), schematic illustration of hydrophilically stabilized Au -Ag nanoshells with Ra-MEG-OH/TEG-COOH (b). (cited from Ref. [75] with permission from the Royal Society of Chemistry).

To minimize the assay time and enable the rapid detection, several strategies have been employed [80, 84]. For instance, Driskell et al. prepared an approach to increase the flux of antigen and SERS particles to the solid-phase surface by using a rotated capture substrate. As illustrated in **Figure 10a**, by controlling the rotating rate for the capture substrate, the reaction kinetics can be improved quickly [80]. Instead of gold nanoparticles, gold nanorods were used as plasmonic nanostructures for ERLs to improve the sensitivity. The assay time was reduced from 24 hours to 25 minutes, however, in a 10-fold loss of sensitivity compared to the conventional SERS sandwich immunoassay. Further, to improve the simplicity of the assay, the syringe pump SERS immunoassay platform (**Figure 10b**) was developed to overcome diffusion-limited binding kinetics that often impedes rapid analysis in conventional SERS immunoassay. The assay time was reduced from 24 hours to 10 minutes with a 10-fold improvement in detection limit [84].

Despite these attempts being successful in reducing assay times, nonspecific adsorption of nontarget molecules still remains the biggest challenge in immunoassay for protein detection. To circumvent this problem, our group recently proposed an innovative platform that utilizes nanoscaled alternative current electrohydrodynamic (ac-EHD)-induced surface shear forces to enhance the capture efficiency as well as significantly reduce the nonspecific binding of the molecules on the surface (**Figure 11a**). Meanwhile, to improve the sensitivity, rational designed silica-coated gold/silver nanoshells have been employed as the SERS labels. It was found that the detection limit can go down to 1 fg/mL. Further, to improve the design of the channel (**Figure 11b**), simultaneous detection of four biomarkers was achieved both from the serum and patient samples.

Due to the great advantages of SERS in the sensitivity, multiplexed capability with only one single laser excitation, and photostability, it is expected that SERS labels will have more applications in the point-of-care diagnosis platforms for protein biomarkers detection by using the rationally designed plasmonic nanostructures.

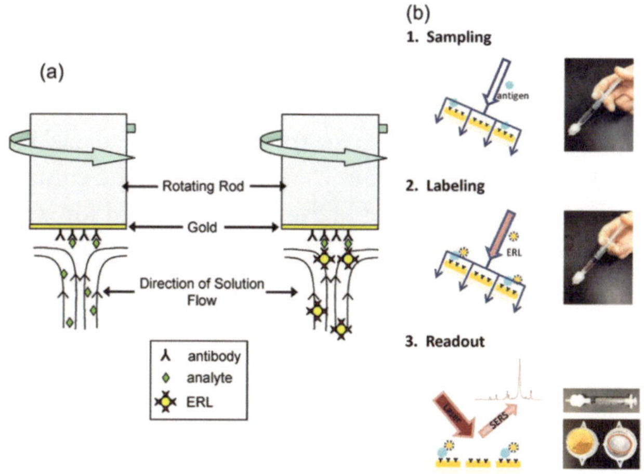

Figure 10. Two typical platforms of rotating capture substrate (a) and syringe pump (b) to enhance the assay time (Reproduced with permission cited from Refs. [80, 84]).

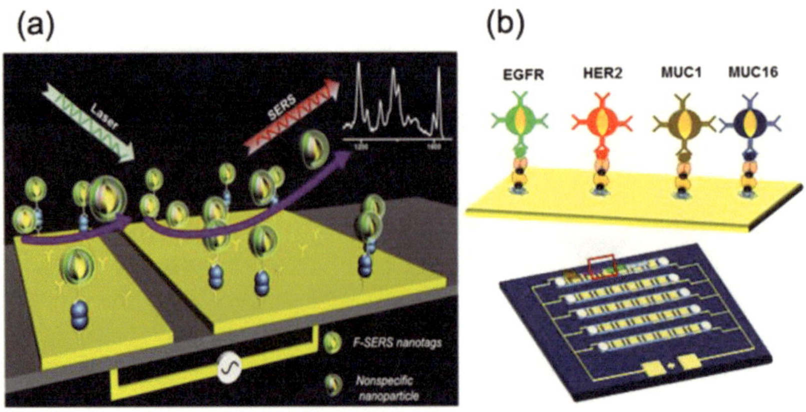

Figure 11. Scheme of the microfluidic device platform for protein biomarker sensing by using ac-EHD to enhance the assay time and minimum the nonspecific binding by applying a potential on the unsymmetry electrode pair. Dual functional Au/Ag nanoshells were used as SERS nanotags for breast cancer biomarker, HER2 detection (a) and simultaneously detecting four biomarkers in a five-channel device (b). (Reproduced with permission cited from Refs. [70, 71]).

5. Conclusions and perspectives

Plasmonic nanostructures with various size and shapes have been utilized in different areas including the catalysis, energy, and biomedicine. In this chapter, the fundamentals of plasmonic effect as well as the plasmonic nanostructures have been reviewed. Surface-enhanced Raman scattering (SERS) is an optical phenomena happened on the plasmonic nanostructures and have shown distinguished properties for protein biomarkers detection. With two typical approaches (label-free and SERS label) developed, SERS with plasmonic nanostructures has shown great potentials for proteins detection further to understand the biological system that protein involved. Meanwhile, although SERS detection for the protein biomarkers has been reported since 1990s, the application of this technology toward the real clinical sample is limited due to the rich information on the Raman peaks from the proteins, which have similar chemical bonds, thus making it very hard to be identified quickly and easily. Thanks to the development of the approach for the statistical analysis, label-free SERS detection for proteins has become feasible for clinical samples. On the contrast, SERS labels with the well-designed plasmonic nanoparticles show great potential for sensitive, reproducible, and simultaneous multiplexed detection for critical protein biomarkers. By combining the label-free SERS with SERS labels approach, we expect that SERS combined with the rational designed plasmonic nanostructures will greatly enhance the research and application in this field.

Acknowledgements

This work was supported by ARC DECRA (DE140101056) to Y.W. D. Li would like to thank the finance support from the China Scholarship Council (CSC) for her to work in Macquarie University.

Author details

Dan Li[1,2] and Yuling Wang[1*]

*Address all correspondence to: yuling.wang@mq.edu.au

1 Department of Chemistry and Biomolecular Sciences, Faculty of Science and Engineering, Macquarie University, Sydney, Australia

2 State Key Laboratory of Electroanalytical Chemistry, Changchun Institute of Applied Chemistry, Changchun, Jilin, China

References

[1] Pines D, Bohm DA. Collective description of electron interactions: II. Collective vs individual particle aspects of the interactions. Physical Review. 1952;**85**:338

[2] Pines D. Collective energy losses in solids. Review of Modern Physics. 1956;**28**:184

[3] Le Ru EC, Etchegoin PG. Principles of surface-enhanced Raman spectroscopy and related plasmonic effects. 2008, ISBN: 978-0-444-52779-0

[4] Schlücker S. Surface enhanced Raman spectroscopy: Analytical, biophysical and life science applications. 2011, ISBN: 9783527325672, DOI: 10.1002/9783527632756

[5] https://en.wikipedia.org/wiki/Plasmon

[6] Echenique JM, Pitarke, Silkin VM, Chulkov EV, Echenique PM. Surface plasmons in metallic structures. Journal of Optics A: Pure and Applied Optics 2005;**7**:S73-S84

[7] Raether H. Surface Plasmons on Smooth and Rough Surfaces and on Gratings. Springer Tracts in Modern Physics; 1988

[8] Maier SA. Plasmonics: Fundamentals and Applications. Springer: New York; 2007

[9] Kerker M. The Scattering of Light and Other Electromagnetic Radiation. Academic Press: New York; 1969

[10] Bohren CF, Huffman DR. Absorption and Scattering of Light by Small Particles. Wiley Interscience: New York; 1983

[11] Kreibig U, Vollmer M. Optical Properties of Metal Clusters. Springer: Berlin; 1995

[12] Willet KA, Van Duyne RP. Localized surface plasmon resonance spectroscopy and sensing. Annual Review Physical Chemistry. 2007;**58**:267-297

[13] Jones MR., Osberg KD, Macfarlane RJ, Langille MR, Mirkin CA. Templated techniques for the synthesis and assembly of plasmonic nanostructures. Chemical Review. 2011;**111**:3736-3827

[14] Morton SM, Silverstein DW, Jensen L. Theoretical studies of plasmonics using electronic structure methods. Chemical Review. 2011;**111**:3962-3994

[15] Giannini V, Fernandez-Dominguez AI, Heck SC, Maier SA. Plasmonic nanoantennas: Fundamentals and their use in controlling the radiative properties of nanoemitters. Chemical Review. 2011;**111**:3888-3912

[16] Mayer KM, Hafner JH. Localized surface plasmon resonance sensors. Chemical Review. 2011;**111**:3828-3857

[17] Rycenga M, Cobley CM, Zeng J, Li WG., Moran CEH, Zhang Q, Qin D, Xia Y. Controlling the synthesis and assembly of silver nanostructures for plasmonic applications. Chemical Review. 2011;**111**:3669-3712

[18] Halas NJ, Lal S, Chang W, Link S, Nordlander P. Plasmons in strongly coupled metallic nanostructures. Chemical Review. 2011;**111**:3913-3961

[19] Cortie MB, McDonagh AM. Synthesis and optical properties of hybrid and alloy plasmonic nanoparticles. Chemical Review. 2011;**111**:3713-3735

[20] Fleischmann M, Hendra PJ, McQuillan AJ. Raman spectra of pyridine adsorbed at a silver electrode original research article. Chemical Physical Letter. 1974;**26**:163-166

[21] Jeanmaire DL, Van Duyne RP. Surface Raman spectroelectrochemistry: Part I. Heterocyclic, aromatic, and aliphatic amines adsorbed on the anodized silver electrode. Journal Electroanalytical Chemistry and Interfacial Electrochemistry. 1977;**84**:1-20

[22] Ueba H. Theory of charge transfer excitation in surface enhanced Raman scattering original research article. Surface Science. 1983;**131**:347-366

[23] Otto A, Billmann J, Eickmans J, Erturk U, Pettenkofer C. The "adatom model" of SERS (Surface Enhanced Raman Scattering): The present status. Surface Science. 1984;**138**:319-338

[24] Arenas JF, Woolley MS, Otero JC, Marcos JI. Charge-transfer processes in surface-enhanced Raman scattering. Franck–Condon Active Vibrations of Pyrazine. Journal of Physical Chemistry. 1996;**100**:3199-3206

[25] Nie S, Emory SR. Probing single molecules and single nanoparticles by surface-enhanced Raman scattering. Science. 1997;**275**:1102-1106

[26] Doering WE, Nie SM. Single-molecule and single-nanoparticle SERS: Examining the roles of surface active sites and chemical enhancement. Journal of Physical Chemistry B. 2002;**106**:311-317

[27] Kneipp K, Wang Y, Kneipp H, Perelman LT, Itzkan I, Dasari RR, Feld MS. Single molecule detection using surface-enhanced Raman scattering (SERS). Review Letter 1997;**78**:1667

[28] Kneipp K, Kneipp H, Itzkan I, Dasari R, Feld MS. Ultrasensitive chemical analysis by Raman spectroscopy. Chemical Review. 1999;**99**:2957-2976

[29] Xie W, Schlucker S. Synthesis of bifunctional Au/Pt/Au Core/Shell nanoraspberries for in Situ SERS monitoring of platinum-catalyzed reactions. Journal of American Chemical Society. 2011;**133**:19302-19305

[30] Qian X, Peng XH, Ansari DO, Yin-Goen Q, Chen GZ, Shin DM, Yang L. In vivo tumor targeting and spectroscopic detection with surface-enhanced Raman nanoparticle tags. Nature Biotech. 2008;**26**:83-90

[31] Wang YL, Seebald JL, Szeto DP, Irudayaraj J. Biocompatibility and biodistribution of surface-enhanced Raman scattering nanoprobes in zebrafish embryos: In vivo and multiplex imaging. ACS Nano. 2010;**4**:4039-4053

[32] Wang YL, Wang EK. Nanoparticle SERS substrates, in Surface Enhanced Raman Spectroscopy: Analytical, Biophysical and Life Science Applications. Wiley-VCH; 2010;Chap. **2**:39-69

[33] Wang YL, Schlücker S. Rational design and synthesis of SERS labels. Analyst. 2013;**138**: 2224-2238

[34] Schlücker S. Surface-enhanced Raman spectroscopy: Concepts and chemical applications. Angewandte Chemie International Edition. 2014;**53**:4756-4795

[35] Porter MD, Lipert RJ, Siperko LM, Wang G, Narayanan R. SERS as a bioassay platform: Fundamentals, design, and applications. Chemical Society Review 2008;**37**:1001-1011

[36] Wang YQ, Yan B, Chen LX. SERS Tags: Novel Optical Nanoprobes for Bioanalysis, SERS nanotags for biomedical application. Chemical Review. 2013;**113**:1391-1428

[37] Lane LA, Qian X, Nie S. SERS nanoparticles in medicine: From label-free detection to spectroscopic tagging, Chemical Review, 2015;**115**:10489-10529

[38] Zhang Q, Xie J, Yu Y, Lee JY. Monodispersity control in the synthesis of monometallic and bimetallic quasispherical gold and silver nanoparticles. Nanoscale. 2010;**2**:1962-1975

[39] Evanoff DD Jr, Chumanov G. Synthesis and optical properties of silver nanoparticles and arrays. ChemPhysChem. 2005;**6**:1221-1231

[40] Daniel MC, Astruc D, Gold nanoparticles: assembly, supramolecular chemistry, quantum-size-related properties, and applications toward biology, catalysis and nanotechnology, Chemical Review. 2004;**104**:293-346

[41] Steinigeweg D, Schlücker S. Monodispersity and size control in the synthesis of 20-100 nm quasi-spherical silver nanoparticles by citrate and ascorbic acid reduction in glycerol-water mixtures. Chemical Communications. 2012;**48**:8682-8684

[42] Sun YG, Mayers BT, Xia YN. Template-engaged replacement reaction: A one-step approach to the large-scale synthesis of metal nanostructures with hollow interiors. Nano Letters. 2002;**25**:481-485

[43] Gellner M, Küstner B, Schlücker S. Optical properties and SERS efficiency of tuneable gold/silver nanoshells. Vibrational Spectroscopy. 2009;**50**:43-47

[44] Zhang Y, Walkenfort B, Yoon JH, Schlücker S, Xie W, Gold and silver nanoparticle monomers are non-SERS-active: a negative experimental study with silica-encapsulated Raman-reporter-coated metal colloids. Physical Chemistry Chemical Physics. 2015;**17**:21120-21126

[45] Murphy CJ, Sau TK, Gole AM, Orendorff CJ, Gao J, Gou L, Hunyadi SE, Li T. Anisotropic metal nanoparticles: Synthesis, assembly, and optical applications. Journal of Physical Chemistry B. 2005;**109**:13857-13870

[46] Xia Y, Xiong Y, Lim B, Skrabalak SE. Shape controlled synthesis of metal nanocrystals: simple chemistry meets complex physics. Angewandte Chemie International Edition. 2009;**48**:60-103

[47] Wiley BJ, Im SH, Li ZY, McLellan J, Siekkinen A, Xia Y. Maneuvering the surface plasmon resonance of silver nanostructures through shape-controlled synthesis. Journal of Physical Chemistry B. 2006;**110**:15666-15675

[48] Zou XQ, Ying EB, Dong SJ. Seed-mediated synthesis of branched gold nanoparticles with the assistance of citrate and their surface-enhanced Raman scattering properties. Nanotechnology. 2006;**17**:4758-4764

[49] Schütz M, Steinigeweg D, Salehi M, Kömpeb K, Schlücker S. Hydrophilically stabilized gold nanostars as SERS labels for tissue imaging of the tumor suppressor p63 by immuno-SERS microscopy. Chemical Commununication. 2011;**47**:4216-4218

[50] Wustholz KL, Henry AI, McMahon JM, Freeman RG, Valley N, Piotti ME, Natan MJ, Schatz GC, Van Duyne RP. Structure–activity relationships in gold nanoparticle dimers and trimers for surface-enhanced Raman spectroscopy. Journal of American Chemical Society. 2010;**132**:10903-10910

[51] Gellner M, Steinigeweg D, Ichilmann S, Salehi M, Schütz M, Kömpe K, Haase M, Schlücker S, 3D self-assembled plasmonic superstructures of gold nanospheres: Synthesis and characterization at the single-particle level. Small. 2011;**7**:3445-3451

[52] Choi I, Don H, Suseung S, YangYL, Kang T, Yi J. Core–satellites assembly of silver nanoparticles on a single gold nanoparticle via metal ion-mediated complex. Journal of American Chemical Society. 2012;**134**:12083-12090

[53] Fan Z, Tebbe M, Fery A, Agarwal S, Greiner A. Assembly of gold nanoparticles on gold nanorods using functionalized poly (N-isopropylacrylamide) as polymeric "glue". Particle & Particle Systems Characterization. 2016;**33**:698-702

[54] Deisenhofer J, Epp O, Miki K, Huber R, Michel H, X-ray structure analysis of a membrane protein complex. Electron density map at 3 A resolution and a model of the chromophores of the photosynthetic reaction center from Rhodopseudomonas viridis. Journal of Molecular Biology. 1984;**180** :385-398

[55] Picot D, Loll PJ, Garavito RM. The X-ray crystal structure of the membrane protein prostaglandin H2 synthase-1. Nature 1994;**367**:243-249

[56] Andronesi OC, Becker S, Seidel K, Heise H, Young HS, Baldus M. Determination of membrane protein structure and dynamics by magic-angle-spinning solid-state NMR spectroscopy. Journal of the American Chemical Society. 2005;**127**:12965-12974

[57] Cotton TM, Schultz SG, Van Duyne RP. Surface-enhanced resonance Raman scattering from cytochrome c and myoglobin adsorbed on a silver electrode. Journal of American Chemical Society. 1980;**102**:7960-7962

[58] Cotton TM, Van Duyne RP, Resonance Raman scattering from Rhodopseudomonas sphaeroides reaction centers absorbed on a silver electrode. FEBS Letter. 1982;**147**:81-84

[59] Feng M, Tachikawa H. Surface-enhanced resonance Raman spectroscopic characterization of the protein native structure. Journal of American Chemical Society. 2008;**130**:7443-7448

[60] Grabbe ES, Buck RP. Surface-enhanced Raman spectroscopic investigation of human immunoglobulin G adsorbed on a silver electrode. Journal of American Chemical Society.1989;**111**:8362-8366

[61] Combs ZA, Chang SH, Clark T, Singamaneni S, Anderson KD, Tsukruk VV. Label-free Raman mapping of surface distribution of protein A and IgG biomolecules. Langmuir. 2011;27:3198-3205

[62] Han XX, Jia HY, Wang YF, Lu ZC, Wang CX, Xu WQ, Zhao B, Ozaki Y. Analytical technique for label-free multi-protein detection based on western blot and surface-enhanced Raman scattering. Analytical Chemistry. 2008;**80**:2799-2804

[63] Han XX, Cai LJ, Guo J, Wang CX, Ruan WD, Han WY, Xu WQ, Zhao B, Ozaki Y. Fluorescein isothiocyanate linked immunoabsorbent assay based on surface-enhanced resonance Raman scattering. Analytical Chemistry. 2008;**80**:3020-3024.

[64] Han XX., Kitahama Y, Tanaka Y, Guo J, Xu W. Q, Zhao B, Ozaki Y. Simplified protocol for detection of protein–ligand interactions via surface-enhanced resonance Raman scattering and surface-enhanced fluorescence. Analytical Chemistry. 2008;**80**:6567-6572

[65] Han XX, Huang GG, Zhao B, Ozaki Y. Label-free highly sensitive detection of proteins in aqueous solutions using surface-enhanced Raman scattering. Analytical Chemistry. 2009;**81**:3329-3333

[66] Xu LJ, Zong C, Zheng XS, Hu P, Feng JM, Ren B. Label-free detection of native proteins by surface-enhanced Raman spectroscopy using iodide-modified nanoparticles. Analytical Chemistry. 2014;**86**:2238-2245

[67] Yu Y, Wang J, Lin J Q, Lin D, Chen WW, Feng SY, Huang ZF, Li YZ, Huang H, Shi H, Chen R. An optimized electroporation method for delivering nanoparticles into living cells for surface-enhanced Raman scattering imaging. Applied Physics Letters. 2016;**108**. DOI: 10.1063/1.4947009

[68] Wang J, Zeng YY, Lin JQ, Lin L, Wang XC, Chen GN, Huang ZF, Li BH, Zeng HS, Chen R. SERS spectroscopy and multivariate analysis of globulin in human blood. Laser Physics. 2014;**24**:6

[69] Wang J, Lin D, Lin JQ, Yu Y, Huang ZF, Chen YP, Lin JY, Feng SY, Li BH, Liu NR, Chen R. Label-free detection of serum proteins using surface-enhanced Raman spectroscopy for colorectal cancer screening. Journal of Biomedical Optics. 2014;**19**:087003

[70] Wang YL, Vaidyanathan R, Shiddiky MJA, Trau M. Enabling rapid and specific surface-enhanced Raman scattering immunoassay using nanoscaled surface shear forces. ACS Nano. 2015;**9**:6354-6362

[71] Reza KK, Wang J, Vaidyanathan R, Dey S, Wang YL, Trau M. Electrohydrodynamic-induced SERS immunoassay for extensive multiplexed biomarker sensing. Small. 2016. DOI: 10.1002/smll.201602902

[72] Wang YL, Trau M. Accurate and sensitive total genomic DNA methylation analysis from sub-nanogram input with embedded SERS nanotags. Chemical Communication. 2016;**52**:3560-3563

[73] Wang YL, Wee EJ, Trau M. Highly sensitive DNA methylation analysis at CpG resolution by surface-enhanced Raman scattering via ligase chain reaction. Chemical Communication. 2015;**51**:10953-10956

[74] Wang YL, Rauf S, Grewal YS, Spadafora LJ, Shiddiky MJA, Cangelosi GA, Schlucker S, Trau M. Duplex microfluidic SERS detection of pathogen antigens with nanoyeast single-chain variable fragments. Analytical Chemistry. 2014;**86**:9930-9938

[75] Wang YL, Salehi M, Schutz M, Schlucker S. Femtogram detection of cytokines in a direct dot-blot assay using SERS microspectroscopy and hydrophilically stabilized Au-Ag nanoshells. Chemical Communication. 2014;**50**:2711-2714

[76] Wang YL, Salehi M, Schutz M, Rudi K, Schlucker S. Microspectroscopic SERS detection of interleukin-6 with rationally designed gold/silver nanoshells. Analyst. 2013; **138**:1764-1771

[77] Ni J, Lipert RJ, Dawson GB, Porter MD. Immunoassay readout method using extrinsic Raman labels adsorbed on immunogold colloids. Analytical Chemistry. 1999;**71**:4903-4908

[78] Driskell JD, Kwarta KM, Lipert RJ, Porter MD, Neill JD, Ridpath JF. Low-level detection of viral pathogens by a surface-enhanced Raman scattering based Immunoassay. Analytical Chemistry. 2005;**77**:6147-6154

[79] Grubisha DS, Lipert RJ, Park HY, Driskell J, Porter MD. Femtomolar detection of prostate-specific antigen: An immunoassay based on surface-enhanced Raman scattering and immunogold labels. Analytical Chemistry. 2003;**75**:5936-5943

[80] Driskell JD, Uhlenkamp JM, Lipert RJ, Porter MD. Surface-enhanced Raman scattering immunoassays using a rotated capture substrate. Analytical Chemistry. 2007; **79**:4141-4148

[81] Wang G, Lipert RJ, Jain M, Kaur S, Chakraboty S, Torres MP, Batra SK, Brand RE, Porter MD. Detection of the potential pancreatic cancer marker MUC4 in serum using surface-enhanced Raman scattering. Analytical Chemistry. 2011;**83**:2554-2561

[82] Wang G, Park HY, Lipert RJ, Porter MD. Mixed monolayers on gold nanoparticle labels for multiplexed surface-enhanced Raman scattering based immunoassays. Analytical Chemistry. 2009;**81**:9643-9650

[83] Cao YC, Jin R, Nam JM, Thaxton CS, Mirkin CA. Raman dye-labeled nanoparticle probes for proteins. Journal of American Chemical Society. 2003;**125**:14676-14677

[84] Penn MA, Drake DM, Driskell JD. Accelerated surface enhanced Raman spectroscopy (SERS)-based immunoassay on a gold-plated membrane. Analytical Chemistry. 2013;**85**:8609-8617

Chapter 16

SERS Application for Analysis of Live Single Cell

Waleed Ahmed El-Said, Hyeon-Yeol Cho and
Jeong-Woo Choi

Additional information is available at the end of the chapter

http://dx.doi.org/10.5772/67593

Abstract

Monitoring changes of the protein contents and other macromolecules inside a living single cell during the key cellular processes such as cell differentiation, division, and apoptosis is a challenge for researchers. Raman spectroscopy is a powerful analytical technique for several biomedical applications that is rapid, reagent-free, and non-destructive while limited application with its weak signal. Surface-enhanced Raman scattering (SERS) technique is widely used to enhance the Raman signal (10^{9-15} fold) by using surface Plasmon resonance of noble metal nanostructures (e.g. silver, gold, copper). SERS is a non-destructive spectroscopic method applied for biomedical samples. In this chapter, we will discuss the principles and fundamentals of SERS technique, theories and different strategies to obtain SERS signals such as immobilization of metal colloids on a substrate. Also, we show the SERS applications including the identification and discrimination of different types of cells (healthy and nonhealthy cells, e.g., cancer cells), and the interaction of cells with different drugs will also be discussed on monolayer bulk cells as well as on single-cell basis and for stem cell differentiation. In addition, we show the coupling of SERS with electrochemical techniques (EC-SERS) as spectroelectrochemical technique and its applications in biology, bioanalytical, and life science.

Keywords: SERS, single cell, nanoparticles, spectroelectrochemical, microfluidics

1. Introduction

1.1. Brief overview of Raman spectroscopy

The fundamental principle of Raman spectroscopic technique is based on the inelastic scattering of photons from the targeting molecules in the sample activated by the laser source as shown in **Figure 1**. Hence, the chemical composition of the complicated subjects could be thoroughly studied by analysis of each peak from their corresponding Raman spectra, which is not available with other optical, biological, or electrical methods [1]. Therefore, numerous studies have been reported for application of Raman spectroscopy as a powerful analytical technique that enables a rapid, reagent-free, and nondestructive technique for cell analysis including the examination of cell populations in suspension [2, 3], single fixed cells [4], dried cells [5], cytospun cells [6], and living and dead cells [7–13]. Furthermore, Raman spectroscopy was reported as a candidate for monitoring the effects of different anticancer drugs on cell viability, due to the fact that different toxic agents or drugs will cause different effects on living cells and further induce changes in biochemical composition, which can readily be detected by Raman spectra without invasive procedures. However, the application of traditional Raman spectroscopy to cell-based analysis is challenged due to its weak and unstable signal.

1.2. Limitations of Raman spectroscopy

For the traditional Raman spectroscopy studies, it is well known that a typical Raman sample will produce one Raman scattered photon from 10^6 to 10^8 excitation photons [14]. Thus, strong intense light sources and efficient collection of Raman photons are required in order for an adequate number of Raman photons to be detected. Nevertheless, the number of Raman photons produced is usually small enough for measurements to be shot-noise limited, so the signal-to-noise ratio increases with the square root of the number of Raman photons detected. The weak Raman intensity causes limits to the sensitivity, and as a result, the biomedical applications of Raman spectroscopy. The Raman intensity depends directly on the intensity of the excitation source and inversely proportional to the fourth power of the excitation wavelength as shown in Eqs. (1) and (2), respectively,

$$I \alpha I_0 \tag{1}$$

$$I \alpha 1/\lambda^4 \tag{2}$$

Excitation light elastically scattered (no wavelength change) from the sample is orders of magnitude more intense than the Raman signal and must be severely attenuated. However, in case of sample fluoresces, the fluorescence intensity can easily overwhelm the Raman spectrum. On the other hand, the weak Raman signal could be easily overcome by decreasing the excitation wavelength. But unfortunately, the decreasing of the excitation wavelength increases the probability that intense fluorescence from the sample will obscure the Raman spectrum. Moreover, most biological molecules have small Raman scattering cross section, which results in very weak signals, possibility of sample photolytic damage, and the strong fluorescence background from cells and tissues. These limitations of sample damage and strong florescence background can be avoided by shifting the excitation laser wavelength from UV-visible (UV-Vis) to near-infrared (NIR) region. Therefore, Raman measurements are often made in the NIR spectral region despite the loss in sensitivity in order to avoid fluorescence because most fluorescence that does occur is outside the spectral region of the Raman spectrum. NIR laser gives very weak signal but it is fluorescence free and can penetrate much deeper into the sample. Based on all of the above, the major disadvantage of Raman is that,

http://dx.doi.org/10.5772/67593

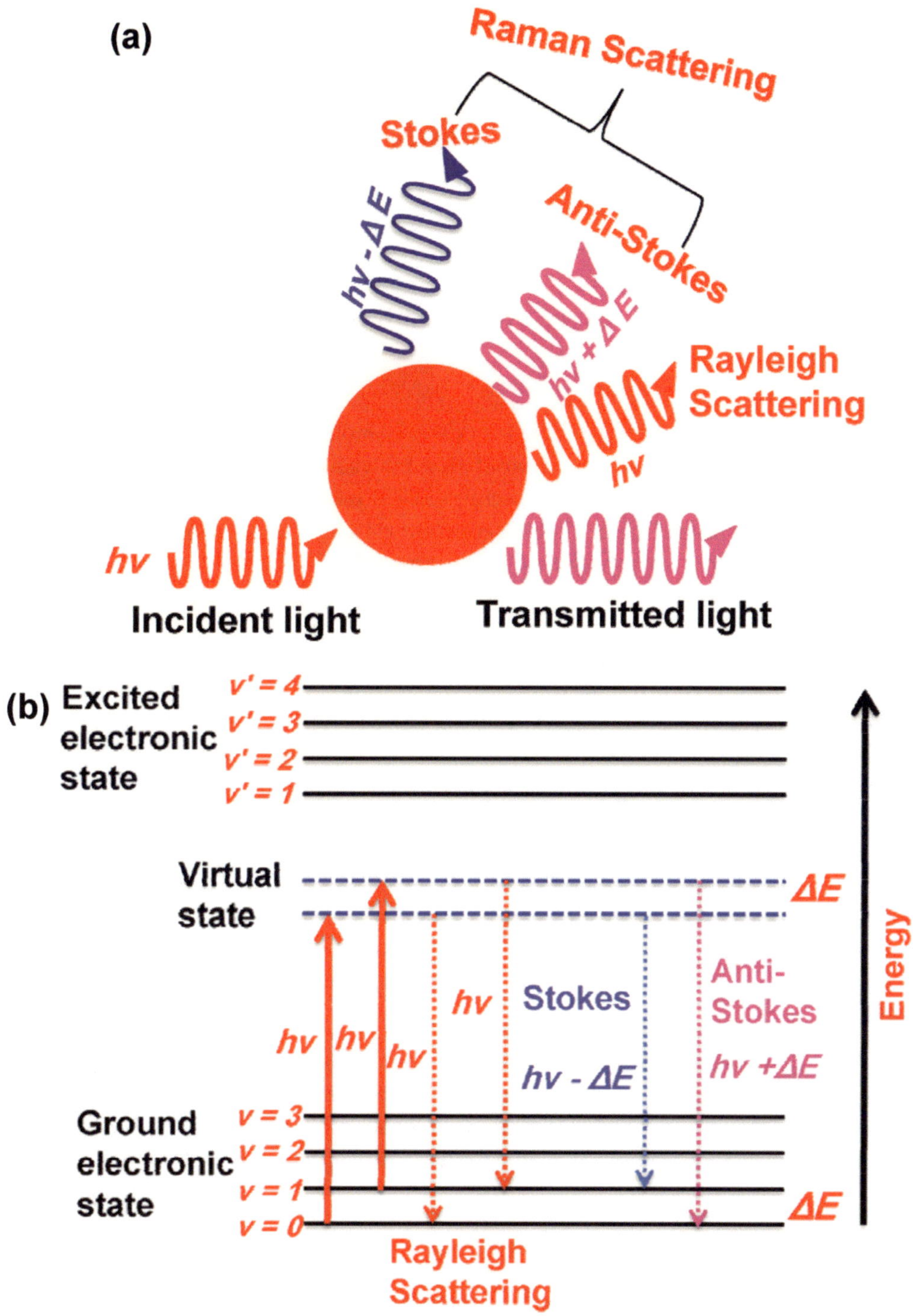

Figure 1. Interactions between photons and molecules. (a) Schematic diagram of various interactions of a molecule with monochromatic light. (b) Molecular energy diagram comparing Rayleigh scattering and Raman scattering (Stokes and anti-Stokes).

relative to UV-Vis methods, the sensitivity is poor and therefore high concentrations, long data acquisition times, or the use of specialist surface or other enhancement techniques are required. Several techniques have been reported for enhancing the Raman signals including surface-enhanced Raman scattering (SERS), surface-enhanced resonance Raman scattering (SERRS), and tip-enhanced Raman scattering (TERS). In the following sections, we focus on SERS technique including the mechanism of SERS, the applications of SERS, and its integration with other techniques.

2. Brief overview of the surface-enhanced Raman spectroscopy (SERS) phenomenon

The specificity of Raman scattering makes it a powerful molecular identification technique, but the signals are too weak for sensitive quantitative analysis especially in the biological fields. In 1974, surface-enhanced Raman spectroscopy (SERS) was observed on pyridines adsorbed on an Ag electrode roughed by oxidation-reduction cycles [15]. But they attribute the signal enhancement to the large surface area of the electrode. In 1977, it was first reported that the intensity of Raman scattering for a molecule may be dramatically increased when the molecule is placed in very close proximity to a colloidal metal NPs or roughened macroscale metal object with surface variation on the 10–100-nm scale. The enhanced re-radiated dipolar fields excite the adsorbate, and, if the resulting molecular radiation remains at or near resonance with the enhancing object, the scattered radiation will again be enhanced. Practically, SERS is a Raman spectroscopic technique that provides greatly enhanced Raman signal from analyte molecules that have been brought into close proximity to certain specially prepared metal surfaces (Ag, Cu, or Au), which is observed on micro- or nano-rough surfaces or in solution next to a nanoparticle with a diameter much smaller than the wavelength of the excitation light. When the incident light hits the surface or the particle, a surface plasmon mode is excited which locally enhances the electromagnetic energy in the vicinity of the target molecule, significantly enhancing the intensity of the inelastically scattered light and experiences a dramatic increase in the incident electromagnetic field, resulting in high Raman intensities comparable to fluorescence.

Therefore, SERS phenomenon offers an exciting opportunity to overcome the critical disadvantages of this normal Raman spectroscopy. Using the SERS technique, the Raman signal is enhanced by the structured metal surface and can be detected effectively by low laser power with short signal acquisition time available for biological applications [16].

Jeanmaire and Vanduyne [17] recognized that the large intensity is due to the electromagnetic field effect, while Albrecht and Creighton [18] proposed a charge-transfer effect (chemical enhancement (CE)). Recently, the enhancement factors in SERS can be as high as 10^9–10^{15}, which allows the technique to be sensitive enough to detect single molecules [19, 20]. Under these conditions, Raman scattering can exceed the sensitivity of fluorescence, and it has generated tremendous interest in the nanomaterials, spectroscopy, and analytical chemistry communities [21, 22]. Electromagnetic field enhancement (EFE) "field enhancement" has been reported as one of the major SERS enhancement mechanisms. Field enhancement occurred at the surface of metallic NPs as a consequence of the interaction between laser radiation and electrons on the metal surface for the activation of surface plasmons or collective oscillations of metal electrons. Aggregation of metallic NPs has been reported to generate very intense and enhanced Raman signals at the junction between two NPs, which are normally called 'hot spots' [23]. For this reason, a great deal of attention has been focused on the synthesis of shape-controlled SERS structures with different morphologies.

2.1. Mechanisms of surface enhancement Raman spectroscopy (SERS)

The mechanism of the surface enhancement effect is not totally elucidated yet; however, the SERS mechanisms were reported in the literatures to arise from two major enhancement mechanisms (electromagnetic field enhancement and chemical enhancement) [24, 25]. Before we discuss the enhancement theories, it is important to understand the nature of the roughened metal surface. Ag surfaces "or any other metals" are covered with electrons cloud that arise from the conduction electrons held in the lattice by the presence of positive charge from the Ag metal centers. At the surface, the positive charge is only on the metal side of the electrons. Therefore, the electron density extends a considerable distance from the surface and there is freedom of movement in a lateral direction along it. When a light beam interacts with these electrons, they begin to oscillate as a collective group across the surface. These oscillations are known as surface plasmons. Surface plasmons from small uniform particles or from surfaces which have a single periodic roughness feature have a resonance frequency at which they absorb and scatter light most efficiently. The frequency varies with the metal and the nature of the surface. The oscillation frequency of both Ag and Au is usually in the visible region and therefore, those metals are suitable to use with the visible and NIR laser systems for Raman scattering.

2.1.1. Electromagnetic enhancement

Electromagnetic field enhancement (EFE) "field enhancement" has been reported to play a major role of most of the observed features of SERS. EFE occurred at the surface of metallic nanostructures as a consequence of the interaction between laser radiation and electrons on the metal surface for the activation of surface plasmons or collective oscillations of metal electrons (**Figure 2**) [25, 26]. That scattered light is characterized by an electromagnetic field intensity that is extremely strong at certain portions of space near the metal nanostructures surface. A molecule present in that space is excited by an enhanced field and produces more intense Raman-scattered light than molecules outside that space. In addition, the sizes of the SERS agents are generally quite small compared to the wavelength of the excitation source. The small size of the particles allows the excitation of the metal particle's surface plasmon to be localized. The resultant electromagnetic energy density on the particle is the source of the EFE, the primary contributor to SERS. The nanostructures size and shape changes the electric field density on the nanostructures surface, which in turn changes the oscillation frequency of the electrons.

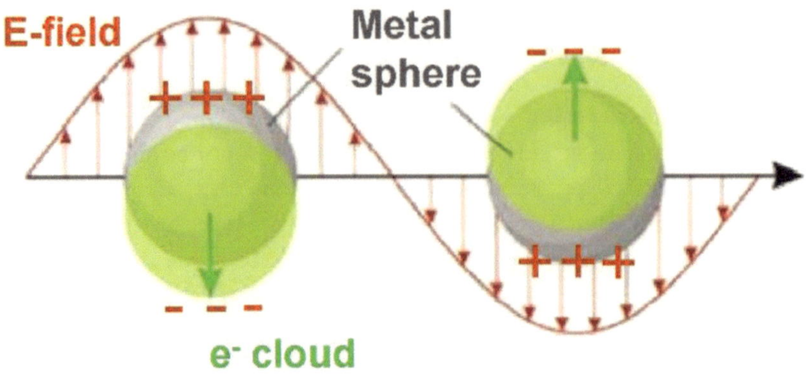

Figure 2. Electromagnetic field enhancement. Surface plasmon resonance (SPR) occurs when the oscillation of the incident light electromagnetic field induces the collective oscillation of the conduction electrons of a metal nanoparticle as the wave front of the light passes; the NP electron cloud is polarized to one side and oscillates in resonance with the light frequency.

On the other hand, on a smooth metal surface, surface plasmons exist as waves of electrons bound to the metal surface and are capable of moving only in a direction parallel to the surface, while on a roughened metal surface, the plasmons are no longer confined and the resulting electric field can radiate both in a parallel and in a perpendicular direction to the surface. Since, to get scattering, there needs to be an oscillation perpendicular to the surface plane, this is achieved by roughening the surface. This locates the plasmon in the valleys of the roughened metal surface and scattering is caused as the plasmons move up to the peaks. Hence, when an incident photon falls on the roughened surface, excitation of the plasmon resonance of the metal may occur and this allows scattering. Additionally, due to the difference in dielectric constants between the roughened surface and the surrounding media, a concentration of electric field density occurs at sharp points on the surface [24, 27].

2.1.2. *Charge-transfer mechanism*

Charge enhancement or chemical enhancement (CE) is the second mechanism of SERS, affects the latter. The chemical mechanism is much less well understood, but is often attributed to a charge-transfer intermediate state which takes place at the strong electron coupling between the metal NP and its adsorbate [28, 29]. The higher SERS enhancement of molecules is directly adsorbed to a metal relative to SERS of molecules that lie on top of a monolayer of molecules attached to a metal which cannot be explained by protest invoking the distance dependence of EFE. Also, SERS enhancement depends substantially on the chemical structure of the adsorbate, which cannot be accounted by EFE.

Basically, the charge enhancement results when molecules chemisorb directly on the roughened surface, forming an adsorbate-metal complex. As a consequence, the molecular orbitals of the adsorbate are broadened by an interaction with the conduction bands of the

http://dx.doi.org/10.5772/67593

metal surface. This results in a ready transfer of electrons and excitation from the metal to the adsorbate and vice versa. Thus, charge transfer between metal nanostructures and adsorbate can produce adsorbate electron excitation under conditions that would not occur if the adsorbates were free in solution. Relaxation of the excited electron to its ground state generates light emission (resonance Raman scattering). The excited electrons and holes in the metal nanostructures generated by surface plasmon resonance (SPR) can couple with the lowest unoccupied molecular orbital (LUMO) and highest occupied molecular orbital (HOMO) of the adsorbate as shown in **Figure 3** [30, 31]. As a consequence, the SERS spectra of chemisorbed molecules are significantly different from the Raman spectrum of the free species. The CE mechanism is restricted by the nature of molecules directly adsorbed on the metal, as opposed to the EFE, which extends a certain distance beyond the surface. Thus, it is effectively operated only on the first layer of adsorbates. There have been many experimental demonstrations confirming that both mechanisms play key roles in SERS effects [32, 33]; however, it is generally believed that electromagnetic enhancement may play a greater part than chemical enhancement [34–36]. In this context, the main analytical advantages of SERS in comparison with other optical detection methods are the inherent molecular specificity which can be obtained [25, 37], the relatively large sensitivity, and the sharpness of the spectral signals, which can be as little as 1-nm full width at half maximum [38]. This latter advantage is to be compared to conventional fluorescent labels which average about 75 nm [39] or quantum dots which average about 30 nm [40]. The relative sharpness of the spectral SERS signal can facilitate multiplexing since multi-label readouts can be carried out at single excitation wavelength [41] without being limited by spectral overlap.

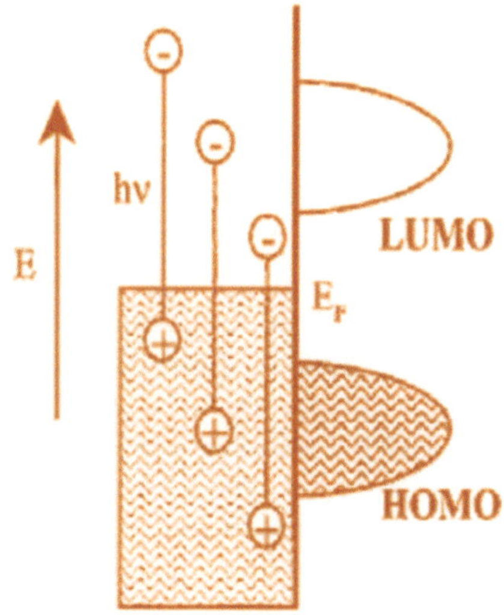

Figure 3. Chemical enhancement in SERS. Schematic diagram of relative energies of excited electron-hole pairs generated via surface plasmon resonance in the metal nanoparticle relative to the HOMO and LUMO of the chemisorbed molecule (adsorbate). E_F, Fermi level (highest energy state occupied by an electron at 0 K).

2.2. Overview of traditional and advanced SERS implementations

Broadly speaking, there are two ways to carry out the SERS detection reaction:

i. homogeneously, where the target becomes bound or absorbs onto the solution phase metallic NPs which act as Raman enhancers;

ii. heterogeneously, where the solution phase targets interact with the surface-phase SERS active sites.

The former of these has the same advantages as all homogeneous reactions (i.e., faster reaction rate and relative ease of implementation) as well as enhanced uniformity and repeatability of the SERS enhancement since the NPs can be synthesized with high uniformity. Examples of such systems include the use of metal nanoshells [42] and nanorods [43] spherical NPs, nanospheres, nanorods, or nanostars [44, 45] as SERS-active substrates. However, this homogeneous approach is disadvantaged because the Raman enhancers are dispersed in solution, thus the detection sensitivity is relatively low. Over the recent decades, many types of SERS-active surfaces have been demonstrated including electrochemically roughened electrodes [46], vapor-deposited metal island films [47], periodically aligned nanoparticles [48, 49], and lithography-produced nanostructures [50]. While these surface-phase systems can have fundamentally greater sensitivity than homogeneous ones (essentially concentrating the detection zone from 3D to 2D), the analysis time can be longer (since the molecules must diffuse to the analysis site), the chip fabrication is more complicated (since nanoscopic features must be patterned), and in some cases, it is difficult to obtain regular and repeatable SERS enhancement. If the enhancement is not consistent, then specific detection is still possible but reliable quantification is not.

2.3. Advantages of SERS method

SERS phenomenon offers an exciting opportunity to overcome the critical disadvantages of the normal Raman spectroscopy. Therefore, relatively lower laser intensity, longer wavelengths, and rapid signal acquisition times will be possible with SERS. For these reasons, NIR-SERS is becoming a useful tool for biological applications [16]. The importance of SERS is that the surface selectivity and sensitivity extend Raman utility to a wide variety of interfacial systems previously inaccessible to Raman. In addition, unlike other vibrational spectroscopies, SERS can be conducted under ambient conditions and has a broad wavenumber range. Moreover, SERS has two main advantages over fluorescence. One is its high sensitivity in comparison to fluorescence by two or three times [51, 52]. The other advantage is the multiplex detection capability due to its molecularly narrow-band spectra.

2.4. SERS-active substrates

Fabrication of SERS-active substrates was found to have a number of problems, including poor signal enhancement, uniformity, or reproducibility. Therefore, an advanced method for fabrication of the SERS-active surface is still required for more effective enhancement of Raman signals. Generally, there are two strategies to obtain the SERS signals.

First, the "average SERS" enhanced spectra [53], which were obtained from an ensemble of colloidal particles and aggregates, giving a relatively low signal, especially before aggregation. Recently, several reports described the uses of colloidal metallic (Au or Ag) NPs or nanorods to perform SERS on living cells [54–57]. The SERS signal was generated based on the direct NPs diffusion inside cells (pinocytosis process) or depend on the interaction of the antibody conjugate NPs with the protein on the cell membrane. However, it was found that the localization of colloidal particles inside a living cell is difficult to control, and also causes non-homogeneous particles aggregation that dramatically changes the efficiency of Raman signal enhancement from one point to another within the cell surface. Although the antibody-conjugated metal particles have been tried as SERS-active agents to overcome these limitations, however, the antibody was reported to cause unwanted SERS signals that hardly are distinguishable from the Raman signals originating from target molecules inside the cell [58]. Moreover, during the NPs preparing some surfactants such as cetyltrimethylammonium bromide (CTAB) or polyvinylpyrrolidone (PVP) were used [59, 60]. The existence of these species on the metal NPs surfaces will provide less active sites and the SERS signal generated by these species may severely interfere with the SERS signals of target molecules. Also, SERS-based investigation of a cell nucleus is challenged due to the existence of several cellular barriers that limit the delivery of SERS-active colloidal NPs to the cell nucleus [61]. Moreover, Au NP-targeting of cancer cell nuclei affected cellular function causing cytokinesis arrest, DNA damage, and programmed cell death, which led to failed cell division, thereby resulting in apoptosis [62].

Second, SERS intensities obtained from Ag or Au nanostructures sustain a "hot spot" (large enhancement factors), which permits the detection of a few molecules with fluctuating spectral characteristics [63, 64]. The existence of a particular hot spot (from 1 nm to several hundred nm) can lead to particularly large enhancements of Raman scattering (1014 times) because the Raman scattering rate is proportional to $|E(\omega)|2|E(\omega')|2$ at the location of the molecule, where $E(\omega)$ is the electric field component at the frequency of the incident radiation, and $E(\omega')$ is the component at the scattered frequency. A SERS-active surface that used a non-uniform distribution of Au NPs on an APTMS (3-aminopropyltrimethoxysilane)-functionalized ITO substrate was reported [65]; however, any small variation in the local arrangement of nanostructures (patterns/shapes) used as SERS-active substrates leads to critical changes of SERS signals due to the high sensitivity of the hot spots. Also, the organic linkers (e.g., APTMS) reduce the enhancing effects and interfere with the SERS signals of target molecules [53, 58]. Therefore, a geometrically well-organized and clean SERS-active substrate that allows control of both nanostructures size and shape is highly desirable [66].

2.5. Applications of SERS

SERS is one of a very few methods that can give effective, molecularly specific information about an adsorbate on a metal surface, in situ and in aqueous solutions. In early 1980s, SERS was observed on small and large biomolecules such as DNA and protein. Presently, many researchers have continued to demonstrate the great potential of SERS applications in the fields of biochemistry, biophysics, and molecular biology. In early 1990s, SERS was performed using metallic NPs on living cells [57, 67]. However, on the cell monolayer, different spectra are obtained on different spots. These distinct spectra could be reflecting the inhomogeneity of the cell composition, but also the NPs aggregated inside cells, which might be results in the signal difference due to different degrees of aggregation. Thus, if SERS is to be used as a detection technique, a careful consideration of the chemistry of the surface and the physics of surface enhancement is required, but with care and by using the technique within its limitations, good quantitative measurements can be obtained. In the following section, we represent some applications of SERS technique for single-cell analysis.

3. SERS cell-based detection

Living- or whole-cell Raman spectroscopy can serve as the basis for reliable identification of molecular events inside intact cells [68]. Breuzard et al. performed living-cell studies using SERS spectroscopy and proposed that it could be an effective method in studying the process of the anticancer drug mitoxantrone (MTX) absorption into the plasma membrane of living cells [69].

Shamsaie et al. demonstrated a cellular SERS probe that they called intracellular-grown Au NPs (IGAuNs) using NPs that grow inside MCF10 epithelial cells [70]. Since large nanoparticles are not able to pass through the nuclear membrane pores, they cannot reach the cytoplasm or the nucleus. They also show poor 'controllability' when inserted inside the cell [70, 71]. The presented IGAuNs, however, can be precisely controlled to overcome some of the inherent drawbacks, and act as a potential SERS substrate to understand intra-cellular events.

In addition, the treated cells show strong Raman signals compared with weak and noisy Raman. However, they have proposed that the cells will still live after treatment with salt solution in phosphate buffer saline (PBS) buffer without media for a long time (1 week). In our previous work, we have demonstrated the formation of different metallic NPs (Au and Ag) intracellular and extracellular condition of different human living-cell lines (cancer and normal cells) through the reduction of ions. In addition, we examined effects of these metal ions (auric chloride or silver nitrate) on cell viability as well as on cell morphology in different living-cell lines. Our results demonstrated that the treatment of different cell lines with metal ions resulted in the cell fixation and the cells could not still live after treatment. The formation of metal NPs inside the cell's nucleus as well as larger particles of different sizes and shapes in the incubation solution was demonstrated using AFM, TEM, EDX, SEM, and UV-Vis absorption techniques. **Figures 4** and **5** show the growth of Au NPs inside and outside the cells, respectively. This approach was shown to exhibit good potential for SERS detection in comparing with standard Raman spectrum (**Figure 6**) [72].

http://dx.doi.org/10.5772/67593

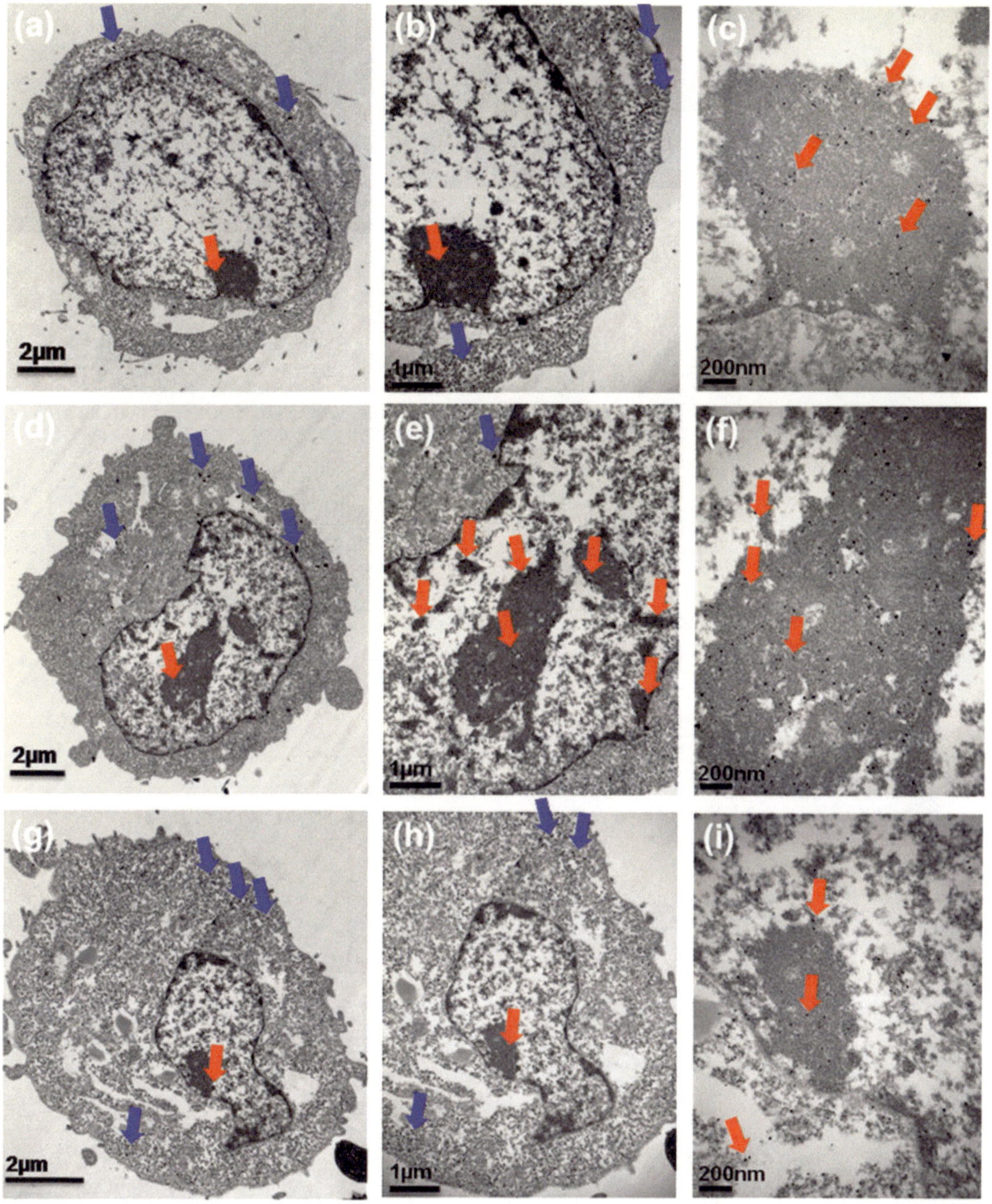

Figure 4. TEM images of negatively stained thin sections from treated HeLa cells (a–c), treated MCF-7 cells (d–f) and treated HEK293T cells (g–i) with 1 mM of auric chloride solution for 4 days. The arrows indicated the spread of black dots across the cell's nucleus and cytoplasm, respectively, that correspond to gold nanoparticles which are 10–20 nm in diameter (c, f, and i). Figure reproduced with permission from Ref. [72], © John Wiley and Sons.

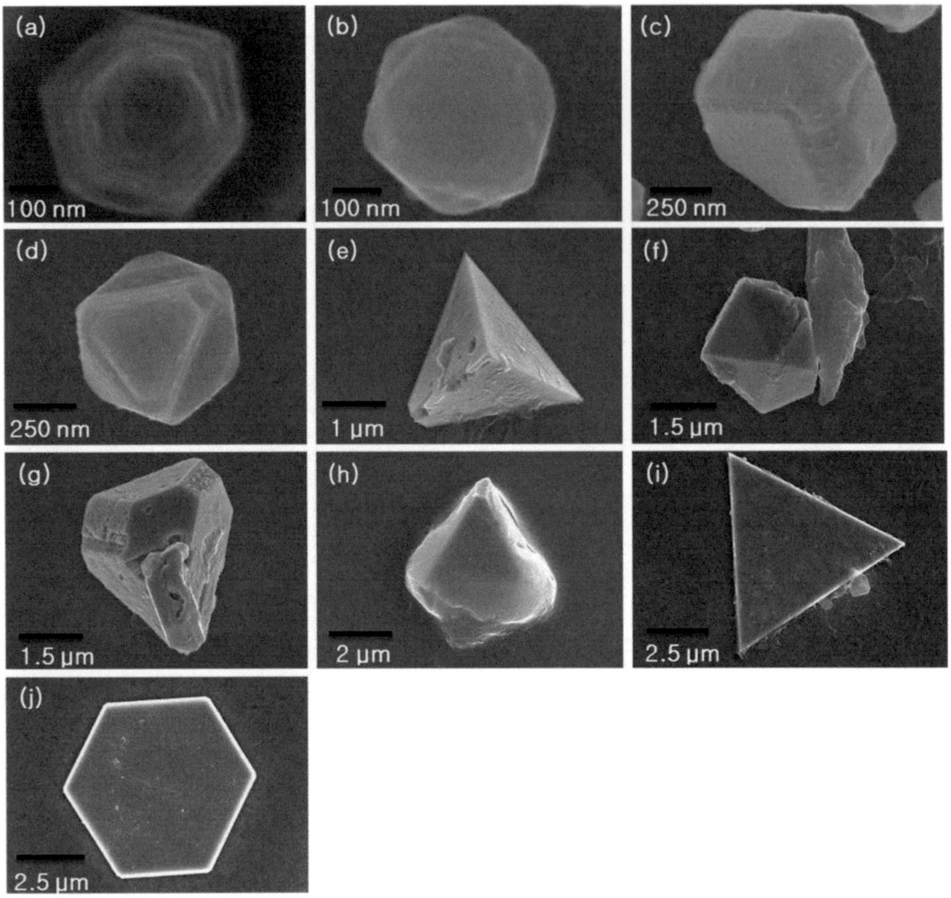

Figure 5. SEM images of selected gold particles formed in the incubation solution with HeLa cells after incubation with gold chloride solution for 4 days (a, b), 8 days (c, d), 14 days (e, f), 21 days (g, h), and 28 days (i, j). Figure reproduced with permission from Ref. [72], © John Wiley and Sons.

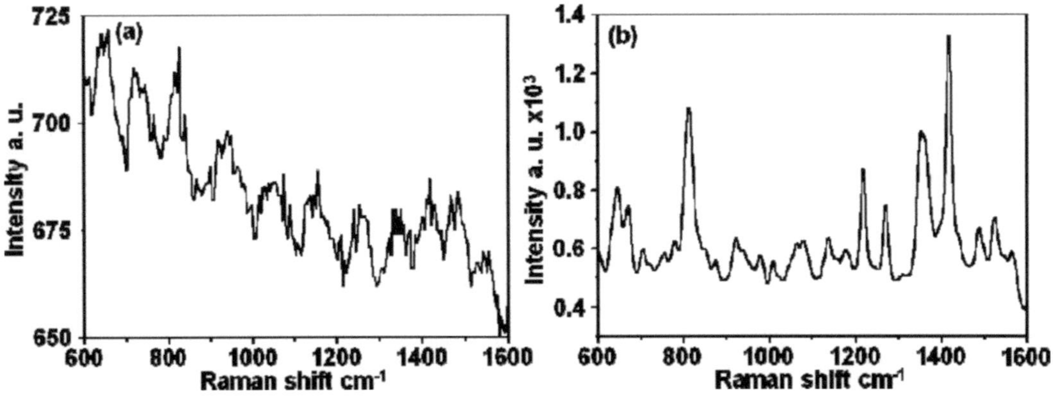

Figure 6. Raman spectrum of (a) control HEK293T cells and (b) HEK293T cells that were incubated with auric chloride for 4 days. Figure reproduced with permission from Ref. [72], © John Wiley and Sons.

4. Single molecule analysis using SERS

Single-cell technologies, such as the classical fluorescence-activated cell sorting (FACS) analysis [73] as well as Raman spectra profiling, can detect population diversity by observing distinct phenotypic parameters. Raman spectroscopy is an especially powerful analytical technique and has already been used in several studies on single cells [74]. A single-cell Raman spectrum usually contains many Raman shifts, which provide rich information of the cell components including nucleic acids, proteins, carbohydrates, and lipids, reflecting cellular genotypes, phenotypes, and physiological states [75]. Therefore, a Raman spectrum could serve as a molecular fingerprint of a single cell, enabling the distinction of various cells, including those from bacteria and animals, without prior knowledge of the cells [76].

The extremely large SERS cross sections (roughly the 10^{14}-fold enhancement discussed earlier) for near-infrared excitation of molecules attached to colloidal Ag or Au clusters [25, 77–79] make it possible to use SERS to provide molecularly specific information on a very small number of molecules. Near-infrared (NIR) excitation also offers the advantage of decreasing the fluorescence background which interferes with traditional single-molecule Raman detection [25]. This has made SERS of great practical interest to both the nanomaterials and the single-molecule spectroscopy communities [37].

4.1. SERS-based single-cell sensing

Jiang et al. have reported on developing a new SERS strategy for apoptosis detection with ultrahigh sensitivity, using silver nanoparticle (Ag NPs)-decorated silicon wafer (Ag NPs/Si), as an active and reproducible SERS in vitro platform as shown in **Figure 7** [80]. Significantly, representative Raman peaks of DNA distributed in cells could be greatly amplified and sensitively detected by the Ag NPs/Si due to SERS ultrahigh sensitivity. As a result, a whole process of apoptosis is readily analyzed at the single-cell level via recording SERS intensities of DNA distributed in the apoptotic cells. The results suggest new opportunities for novel SERS-based in vitro bio-applications, as well as provide powerful strategies for ultrasensitive apoptosis detection. Remarkably, single-cell detection of different apoptotic cells is readily realized using the Ag NPs/Si of high SERS sensitivity and excellent reproducibility (**Figure 8**). In addition, the detection process is label-free, that is, DNA-intrinsic SERS spectra are directly employed for apoptosis detection, which is free of additional chemical or biological reagents as external signal indicator. Consequently, this SERS method may serve as a potentially practical tool for ultrasensitive detection of apoptotic cells, complementing the state-of-the-art strategies for apoptosis detection (**Figure 9**). While better understanding of Ag NPs/Si in vitro behavior and the diameter effect of Ag NPs on detection sensitivity require further investigation, the present findings may open new opportunities for SERS-based in vitro-sensing applications.

Kang et al. [81] reported the application of small spherical gold nanoparticles with highly narrow intra-nanogap structures to achieve high-speed and high-resolution live-cell Raman images. They have used a specific spacer sequence of oligonucleotides (T10) on the core DNA-modified Au NP to create gold nanoparticles with very narrow intra-nanogap (Au-NNP) structures,

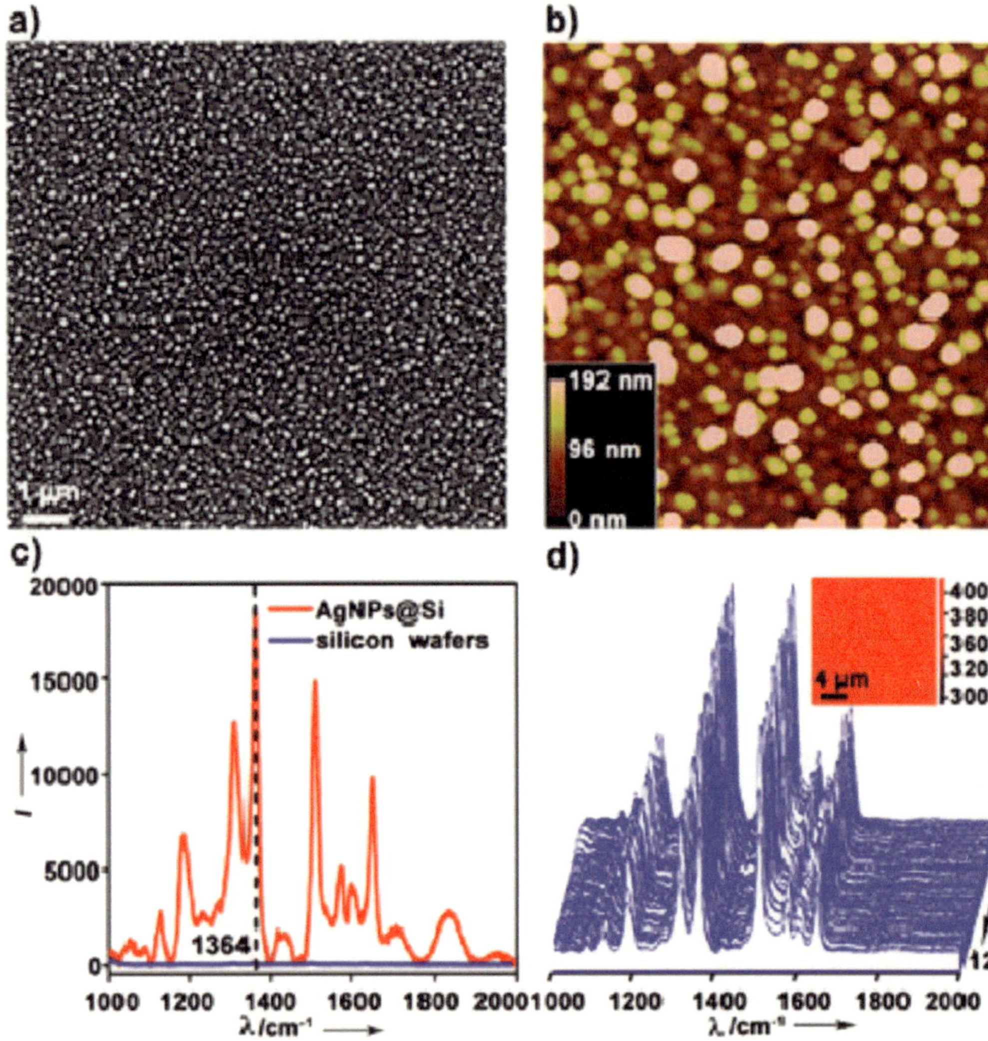

Figure 7. SEM (a) and AFM (b) images of the prepared Ag NPs/Si. The AFM image was collected from an area of 5.0 × 5.0 μm^2. (c) Raman spectra of R6G (100 μM) dispersed on the surface of a pure silicon wafer or Ag NPs/Si substrate. (d) Raman mapping spectra and corresponding mapping image (inset) of R6G dispersed on the surface of Ag NPs/Si. Insert denotes Raman intensities of the 1364 cm^{-1} peak. ($\lambda_{excitation}$ = 633 nm, acquisition time = 1 s, hole = 200 μm, slit =100 μm, grating = 600 g/mm). Reprinted with permission from Ref. [80]. © 2013 American Chemical Society.

then they have loaded three different Raman-active molecules inside the narrow intra-nanogap including 4,4'-dipyridyl (44DP), methylene blue (MB), and 4,4'-azobis (pyridine) (AB) for targeting the mitochondria, cytoplasm, and nucleus, respectively as shown in **Figures 10** and **11**, which lead to obtain high-resolution single-cell image within 30 s without inducing significant cell damage. Using oligonucleotide thiolated DNA (3'-(CH2)3-spacer sequence (A10, G10, C10, or T10)-PEG9-AAACTCTTTGCGCAC-5NA results in the formation of intra-nanogap structure of 1.2 nm in thickness. The high-resolution Raman image showed the distributions of Au NPs for their targeted sites such as cytoplasm, mitochondria, or nucleus as shown in **Figure 12**.

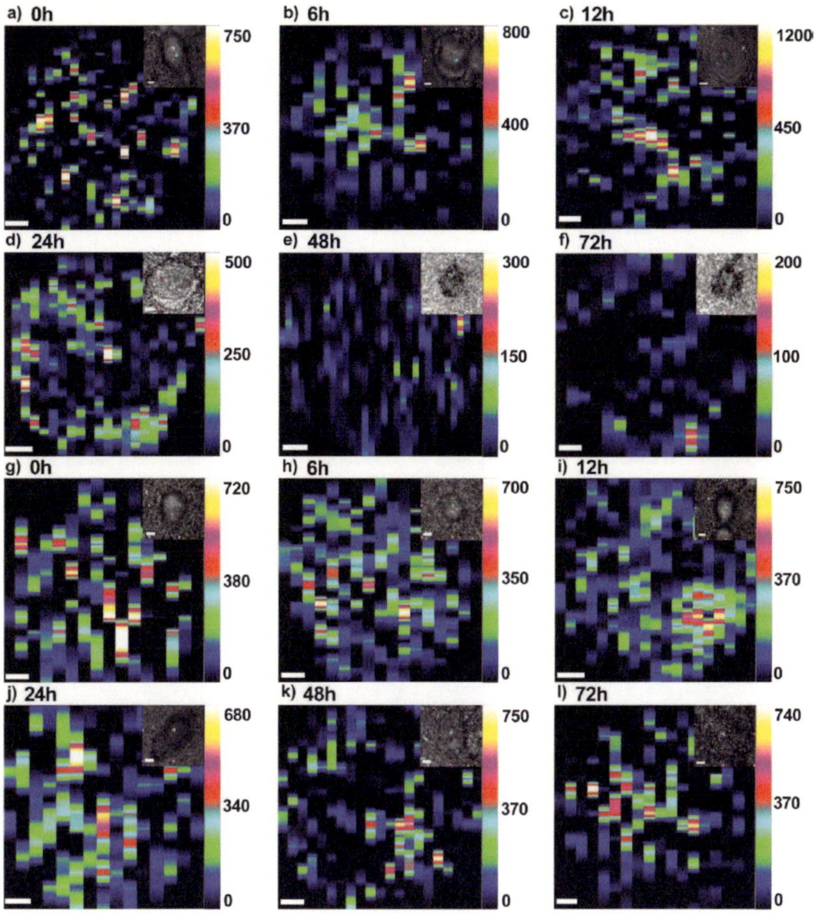

Figure 8. High-resolution Raman mapping images of the A549 cells treated with (a–f) or without (g–l) 150 μM Triton X-100 for 0, 6, 12, 24, 48, and 72 h. Insets are the corresponding Raman bright-field images. Scale bar: 2 μm. Reprinted with permission from Ref. [80]. © 2013 American Chemical Society.

The high-speed Raman-based live-cell imaging allowed monitoring rapidly changing cell morphologies during cell death induced by the addition of highly toxic KCN solution to HSC-3 cell (**Figure 13**), then the single cell was imaged repeatedly at every 2.5 min with a 200 μW of laser power and 10-ms exposure per pixel (50 × 50 pixel). The overlaid images of bright field and Raman image at $t = 0$ showed typical distributions of 44DP-coded Au-NNPs in the mitochondria as shown in **Figure 13A**. After 2.5 min of addition, it showed some changes in Raman signal distributions in the cell. After 5 min, the cell morphologies were significantly changed and the changes of Raman signal distributions were also observed (**Figure 13B**). Eventually, it was completely changed into a circular cell structure, indicative of cell death (necrosis). In the case of 10 μL of KCN solution (1.0 M) addition, it also showed the changes of cell morphologies and Raman signal distributions with time. These results demonstrated that the use of SERS-active nanoparticle can greatly improve the current temporal resolution and

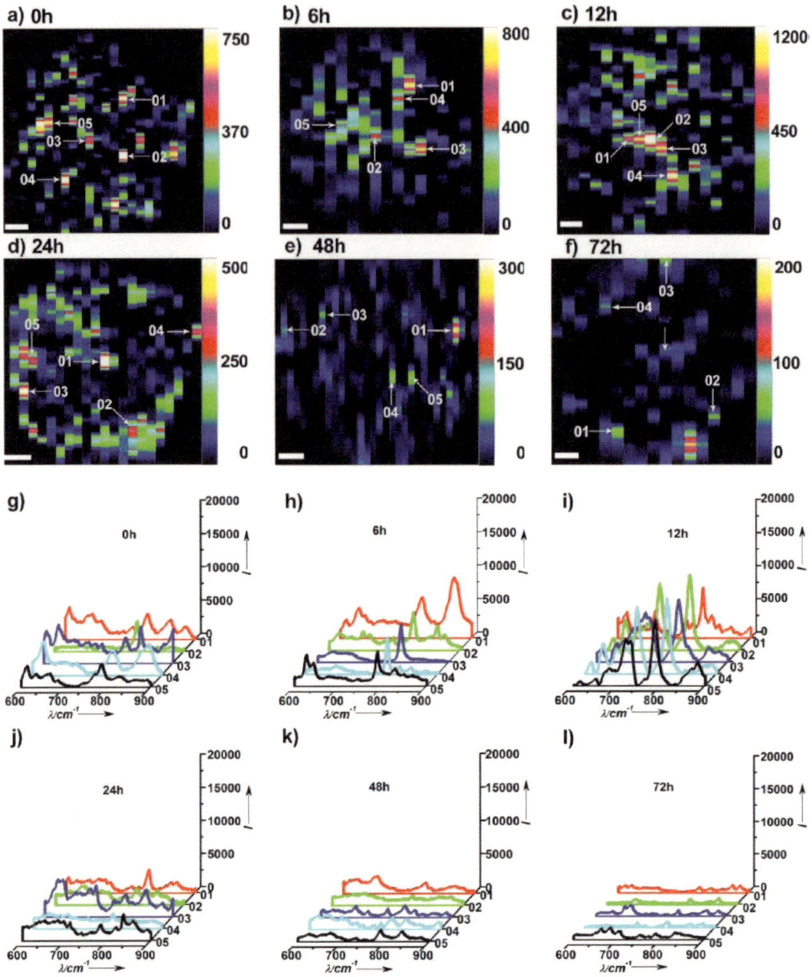

Figure 9. (a–f) Raman mapping images of the A549 cells treated with 150 μM Triton X-100 for 0, 6, 12, 24, 48, and 72 h. (g–l) Five representative Raman spectra of DNA distributed in the cell are randomly selected from the corresponding mapping images, as indicated by white arrows. Reprinted with permission from Ref. [80]. © 2013 American Chemical Society.

image quality of Raman-based cell images enough to obtain the detailed cell dynamics and/or the responses of cells to potential drug molecules. It is expected that the synthetic methods for Au-NNP would be a useful concept for the future design of SERS-active nanostructures and the high-resolution live-cell Raman-imaging method within half a minute can open new opportunities for Raman-based high-throughput and high-content drug-screening platforms.

Cooper and his coworkers [82] have reported on developing a simple microfluidic cell array to mechanically trap a living single cell (**Figure 14**) and it application as a real-time mapping of live cell, also to unequivocally distinguish between differently labeled intracellular nanoparticle SERS probe types from within a mixed population of cell types and also demonstrate the ability to distinguish different types of intracellular nanoparticle probes from within a

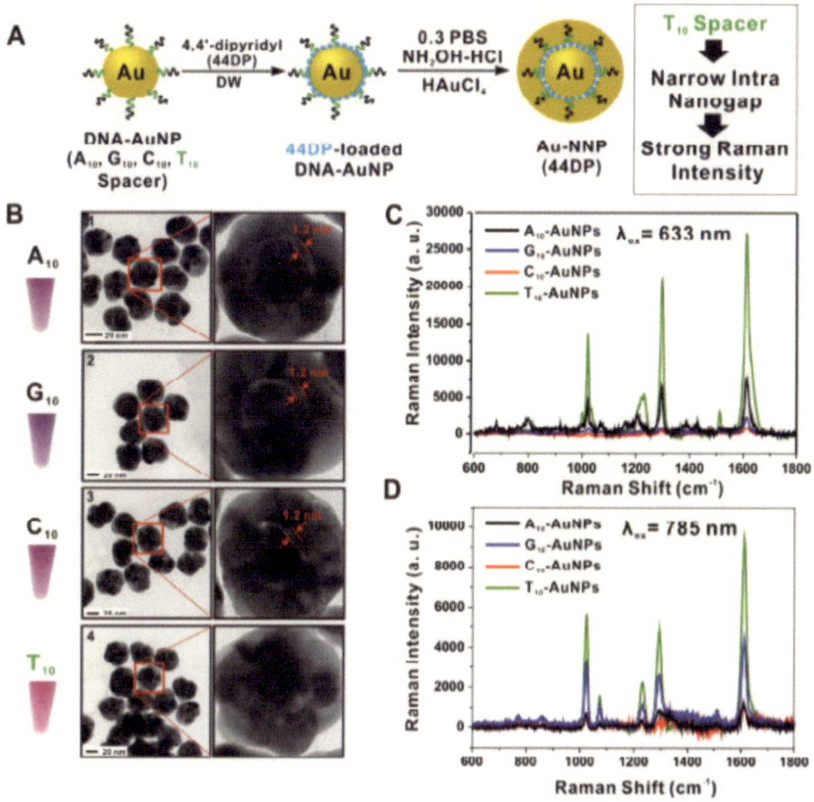

Figure 10. (A) Synthetic scheme of Raman-dye (44DP)-coded Au-NNPs using DNA-AuNPs as a core particle (four different kinds of spacer sequences such as adenine (A10), guanine (G10), cytosine (C10), and thymine (T10) in the thiolated DNA (3'-(CH2)3-spacer sequence-PEG9-AAACTCTTTGCGCAC-5') were investigated). 4,4'-dipyridyl (44DP) molecules were loaded on the surface of DNA-AuNP via electrostatic interactions, and then the Au shell was formed. (B) The solution color and HR-TEM image of 44DP-coded Au-NNPs prepared from A10-spacer DNA-AuNP (1), G10-spacer DNA-AuNP (2), C10-spacer DNA-AuNP (3), and T10-spacer DNA-AuNP (4). (C-D) Raman spectra of 44DP-coded Au-NNP solution prepared from four different spacer DNA with an excitation of 633 nm (C) and 785 nm (D). (Each spectrum was obtained using conditions (0.5 nM particle concentration, 1.0 s exposure time).) Reprinted with permission from Ref. [81]. © 2015 American Chemical Society.

single cell. These microfluidic devices were prepared from masters produced using soft lithographic techniques, and cast in polydimethylsiloxane (PDMS). Then, they have demonstrated the ability of SERS spectroscopy to detect nanoparticle probes within individual living cells, the ability to monitor an individual trapped cell containing SERS-active nanoparticles over a period of several hours, the ability to distinguish between differently labeled intracellular nanoparticle probe types from within a mixture of differently labeled cell populations, as well as the ability to distinguish between more than one nanoparticle species from within an individual cell containing a mixture of probes. To achieve the reliable SERS mapping of a single cell, they have labeled the cells after harvesting by using a colloid solution of silver NPs in 10 mM of either 4-mercaptobenzoic acid or 2-mercaptopyridine in ethanol and incubating the cells for 12 h to allow passive uptake of the colloid by the cells. Then, cell culture solutions were loaded into the PDMS chips using gas-tight Hamilton syringes controlled using

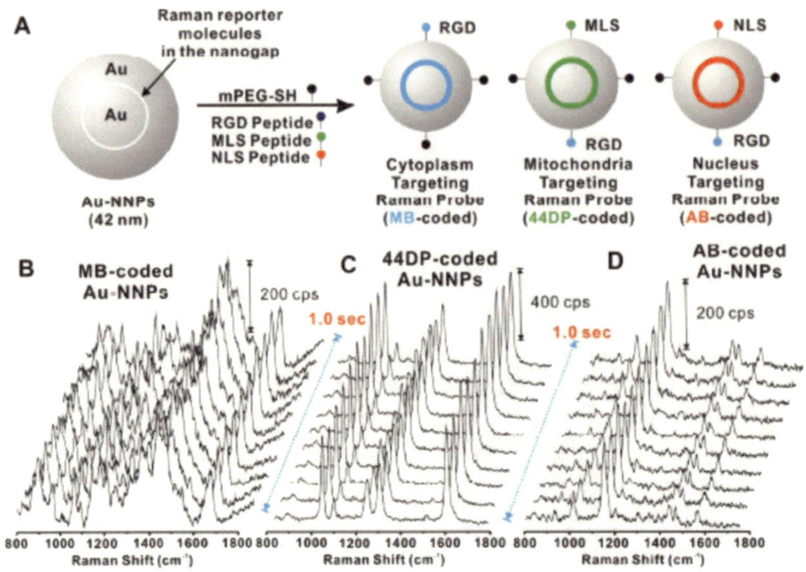

Figure 11. (A) Scheme for the PEGylation and peptide modifications on the Raman dye-coded Au-NNPs (MB-coded Au-NNP was modified with mPEG thiol and RGD peptide (RGDRGDRGDRGDPGC) so that targeting to cytoplasm, 44DP-coded Au-NNP was modified with mPEG thiol, RGD peptide, and MLS peptide (MLALLGWWWFFSRKKC) so that targeting to mitochondria, and AB-coded Au-NNP was modified with mPEG thiol, and RGD peptide and NLS peptide (CGGGPKKKRKVGG). (B–D) Representative Raman spectra of MB-coded Au-NNPs (B), 44DP-coded Au-NNPs (C), and AB-coded Au-NNPs (D) obtained from 0.5 nM particle concentration with an incident laser power of 4.0 mW and 10 ms exposure time per spectra. Reprinted with permission from Ref. [81]. © 2015 American Chemical Society.

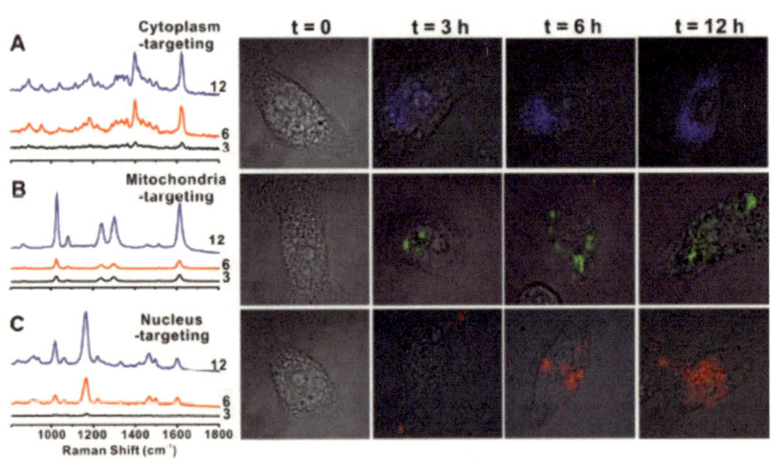

Figure 12. Time-dependent live-cell Raman images (50 × 50 pixel) after incubating with subcellular-targeting NNPs. (A) Representative Raman spectra obtained from inside cells incubated with MB-coded NNPs for cytoplasm targeting, bright-field image (t = 0), and bright-field images overlaid with Raman images (t = 3, 6, 12 h) (4.0 mW laser power, 100 ms exposure time per pixel). (B) Representative Raman spectra from cell incubated with 44DP-coded NNPs for mitochondria targeting, bright-field image (t = 0), and bright-field images overlaid with Raman images (t = 3, 6, 12 h) (200 μW laser power, 10 ms exposure time per pixel). (C) Representative Raman spectra from cell incubated with AB-coded NNPs for nucleus targeting, bright-field image (t = 0), and bright-field images overlaid with Raman images (t = 3, 6, 12 h) (200 μW laser power, 10 ms exposure time per pixel). Reprinted with permission from Ref. [81]. © 2015 American Chemical Society.

http://dx.doi.org/10.5772/67593

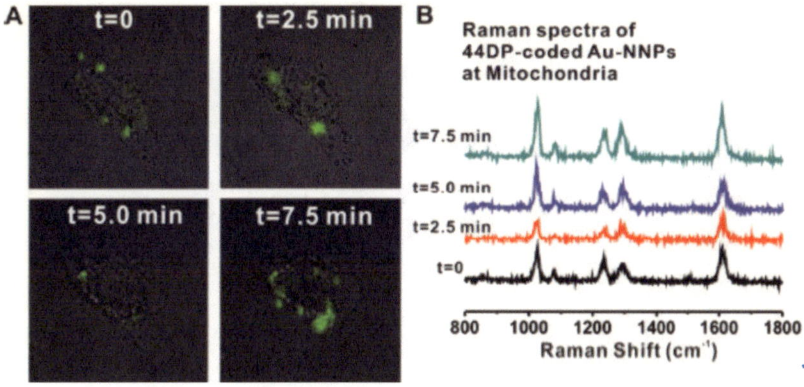

Figure 13. (A) Time-dependent Raman images overlaid with bright-field image after additions of 100 μL of potassium cyanide solution (1.0 M) into HSC-3 cell. (B) Representative Raman spectra obtained from inside cells at designated time points (0, 2.5, 5.0, and 7.5 min). Imaged at every 2.5 min after additions of KCN with an analysis conditions (50 × 50 pixel, 200 μW laser power density, 10 ms/pixel, 27.5 s (total imaging time)). Reprinted with permission from Ref. [81]. © 2015 American Chemical Society.

microsyringe pumps. SERS mapping of a single cell containing labeled nanoparticles was obtained by monitoring the SERS measurement of a single cell trapped within a microfluidic device as shown in **Figure 15**. These results demonstrate the capability of the developed chip to observe the dynamic cellular processes (such as cell division and intracellular nanoparticle redistribution and movement) in real time. This work illustrates the development of the basic technology required to facilitate the analysis of live cells using SERS.

4.2. SERS/flow cytometer for single-cell analysis

In the last years, several groups have shown different approaches to enhance Raman signals or more in general luminescent outputs like those working within the field of colloidal photonics where quantum dots are used for such a purpose [83, 84].

In 2012, Nolan and his group [85] have reported on developing NPs SERS tags as well as Raman flow cytometers for multiparameter single-cell analysis of suspension or adherent cells. SERS tags are based on plasmonically active Au nanorods whose plasmon resonance can be tuned to give optimal SERS signals at a desired excitation wavelength (**Figure 16**). Raman resonant compounds are adsorbed on the NPs to confer a unique spectral fingerprint on each SERS tag, which are then encapsulated in a polymer coating for conjugation to antibodies or other targeting molecules. Raman flow cytometry employs a high-resolution spectral flow cytometer capable of measuring the complete SERS spectra, as well as conventional flow cytometry measurements, from thousands of individual cells per minute (**Figure 17**). Automated spectral unmixing algorithms extract the contributions of each SERS tag from each cell to generate high-content, multiparameter single-cell population data. SERS-based cytometry is a powerful complement to conventional fluorescence-based cytometry. The narrow spectral features of the SERS signal enable more distinct probes to be measured in a smaller region of the optical spectrum with a single laser and detector, allowing for higher levels of multiplexing and multiparameter analysis.

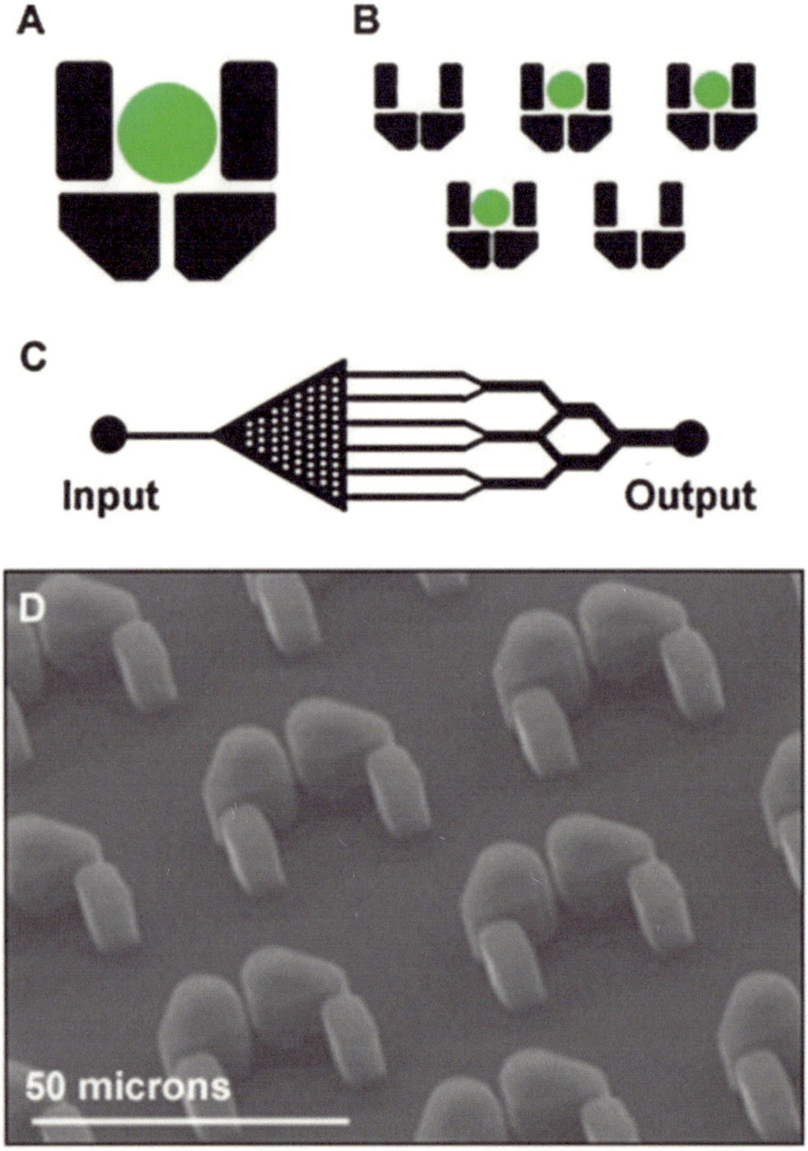

Figure 14. Schematic diagrams of a single PDMS cell trap (A), the trapping array (B), and layout of the microfluidic device (C) with the cell-trapping array represented as a black triangle with white dots. (D) Scanning electron microscope image of 18 × 18 × 10 mm PDMS cell traps on a device. Reprinted with permission from Ref. [82]. © The Royal Society of Chemistry.

They have illustrated the detection of four different SERS tags on a single particle. For either flow or image cytometry, it is feasible to consider integrating fluorescence detection with SERS detection. For example, in a multi-laser system, violet and blue excitation could be used to excite a number of popular fluorescent dyes, while red or NIR excitation could be used to excite a set of SERS tags. Such a multimodal cytometry approach would combine the versatility of fluorescence probes with the multiplexing capability of SERS probes.

http://dx.doi.org/10.5772/67593

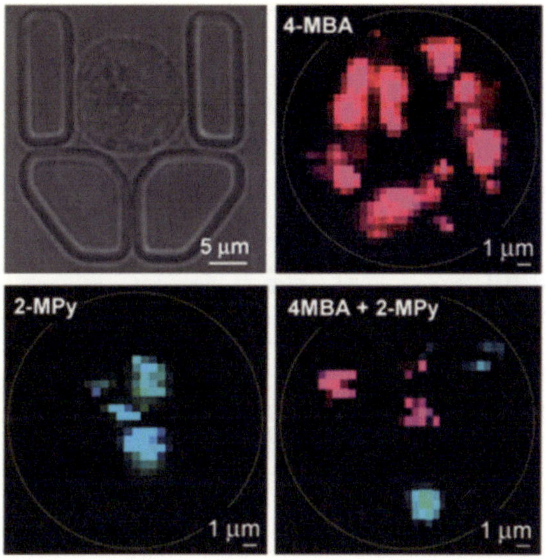

Figure 15. SERS maps of individual CHO cells containing differently labeled Ag nanoparticles. Top left: bright-field image of a CHO cell trapped in the device. The cell is roughly spherical and approximately 15 μm in diameter. Top right: a cell containing 4-MBA-coated Ag NPs. Bottom left: a cell containing 2-MPy-coated Ag nanoparticles. Bottom right: a cell containing both 4-MBA- and 2-MPy-coated Ag nanoparticles. SERS maps were constructed using reporter regions at 1050–1090 and 1560–1590 cm^{-1} to assign 4-MBA (shown in pink) and regions at 1090–1130 and 1525–1555 cm^{-1} to assign 2-MPy (shown in cyan). The broken green line shows the approximate extent of the cell in each case, which cannot itself be visualized directly during Raman imaging. Reprinted with permission from Ref. [82]. © The Royal Society of Chemistry.

SERS tags were used as tags for antibodies in cell surface receptor staining as well. They have conjugated anti-CD4 and anti-CD8 antibodies to two different carboxylated SERS tags and used these to stain SupT1 cell line that expressed both receptors. Then, the stained cells and the appropriate reference controls were analyzed using a SERS spectral flow cytometer and spectral unmixing performed. **Figure 18A** and **B** shows the light-scatter gating, individual cell and average spectra, and single parameter SERS intensity histograms for singly stained beads. In addition, **Figure 18C** showed data from cells stained with both SERS tags, illustrating the ability of SERS flow cytometry hardware and software to measure the amount of SERS-tagged antibody to cell surface receptors.

Recently, Perozziello et al. have reported a Raman/flow cytometer system that consists of a microfluidic device; this device is composed by two calcium fluoride slides among which a photosensitive resist is placed [86]. The device consists of five through-channels integrating a microfluidic trap that was fabricated by using photolithography (**Figure 19**). The trap is realized by means of a simple volumetric constriction of the channels. Flowing cells in the microchannel was blocked at each trap where Raman spectroscopy was performed on each individual cell "since the cells can flow one-by-one" that allow single-cell Raman analysis. These allow a label-free analysis, providing information about the biochemical content of membrane and cytoplasm of each cell. Furthermore, in order to measure SERS signals from single cell and to minimize the acquisition time, the device integrates a microfluidic trap with nested plasmonic nanodimers [87–91] that takes advantages of the basic principles of Raman

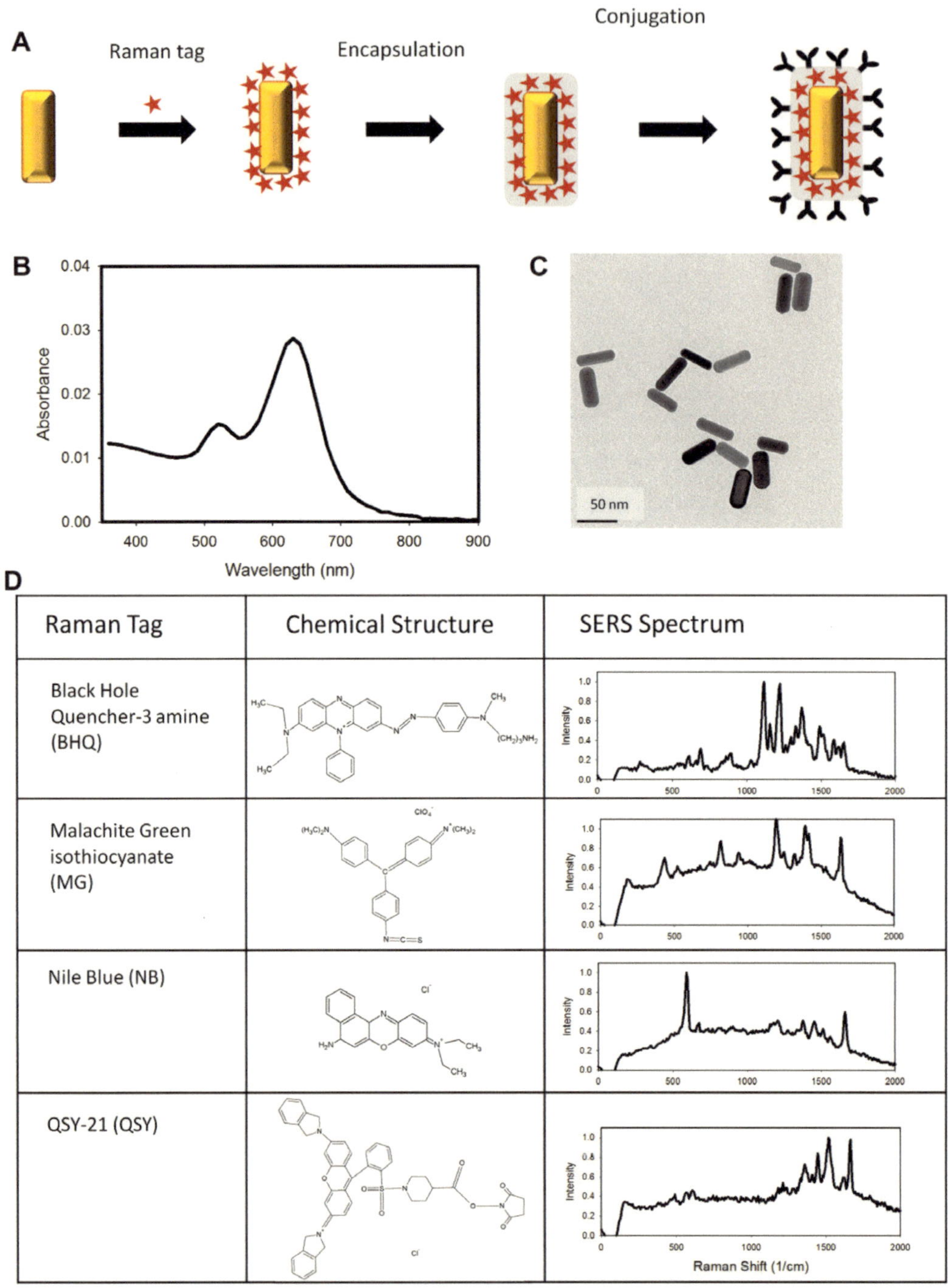

Figure 16. Schematic of nanoparticle-based SERS tags and their characteristic spectra. (A) SERS tags are typically composed of a plasmonic nanoparticle, Raman tag, surface coating, and targeting entity. (B) The SERS tag excitation is determined largely by the plasmon resonance wavelength of the plasmonic nanoparticle, which can be tuned by altering the composition, size, and shape of the nanoparticle. (C) Gold nanorods of 25–50 nm have a plasmon resonance in the red light (630 nm). (D) The Raman tag is adsorbed to the nanoparticle surface and gives each SERS tag its characteristic SERS spectrum. Reprinted with permission from Ref. [85]. © 2012 Elsevier Inc.

http://dx.doi.org/10.5772/67593

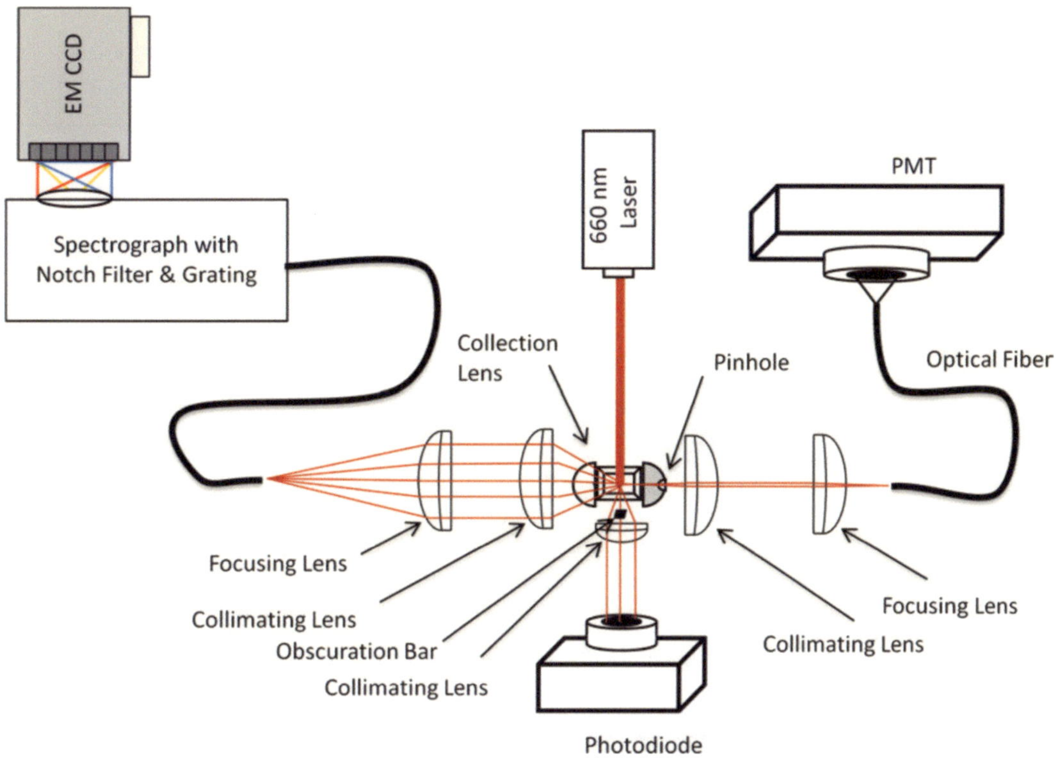

Figure 17. Schematic of a Raman flow cytometer. Excitation is provided by a solid-state laser (660 nm, 400 mW) and forward-angle light scatter is collected on a photodiode. A 90° light scatter is collected from one side of the flow cell via an optical fiber and detected with a photomultiplier tube (PMT). SERS signals are collected from the opposite side of the flow cell into and optical fiber and delivered to an imaging spectrograph coupled to a CCD detector. Particles in the probe volume are detected by forward and side scatter, which trigger the acquisition of individual particle spectra by the CCD. Reprinted with permission from Ref. [85]. © 2012 Elsevier Inc.

spectroscopy as well as the advantages of flow cytometry technique. These devices are based on the amplification of the electromagnetic field due to the polarization of the local electric field and resonance effects between the gap of the nanodimers. In this way, it is possible to obtain a field localization that depends on the size of the gap (approximately 5–10 nm). In particular, the plasmonic nanodevice allows an enhanced Raman scattering that increases the sensitivity and the spectroscopic information regarding both the cellular membrane and the cytoplasm of each cell.

Experiments were performed by using three different cell lines [red blood cells (RBCs), peripheral blood lymphocytes (PBL), and K562 tumor cells from leukemia] by using the microfluidic device integrating the nanodimers as well as at a flat surface without nanodimers.

The flow rate was set at 5 μL/min so that the cells can proceed very slowly and can be stopped in correspondence of the trap. **Figure 20** demonstrated that the cells appear under the microscope and they flow one by one due to the well-defined size of the channel that constrains them and avoid the agglomeration. This is a suitable result, ideal for the fabricated device aiming continuous analysis of a single cell. Based on the obtained results, Perozziello et al.

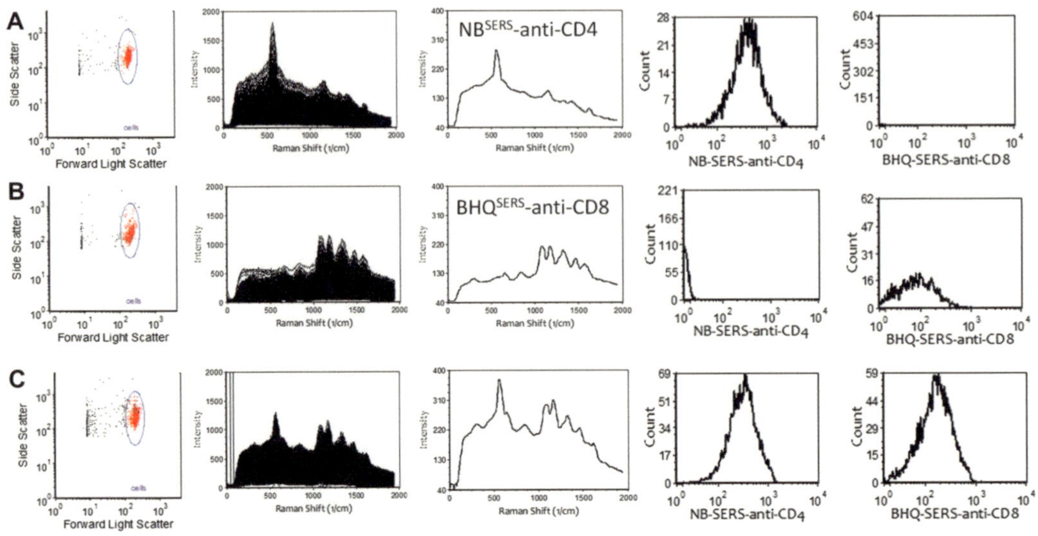

Figure 18. Cell surface immunostaining with antibody-coupled SERS tags. A cultured cell line (SupT1) was stained with (A) and anti-CD4 SERS tag, (B) and anti-CD8 SERS tag, or (C) both, and analyzed by Raman flow cytometry. Light scatter histograms with single-cell gating, individual and average spectra of gated cells, and single parameter histograms of the contributions of each SERS tags after spectral unmixing are presented. Reprinted with permission from Ref. [85]. © 2012 Elsevier Inc.

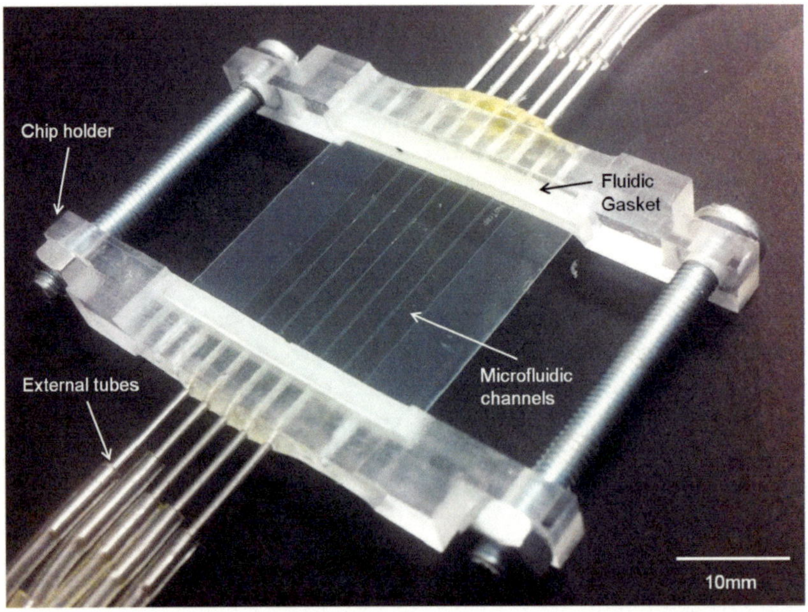

Figure 19. Assembled microfluidic device. Reprinted with permission from Ref. [86]. © 2015 Optical Society of America.

reported that the use of nanodimers results in reducing the laser power until 10% (10 mW) of the maximum power of laser to obtain an enhanced spectrum on nanodimers, while they need to use the full laser power (100%, 100 mW) to record Raman spectrum at the flat surface,

http://dx.doi.org/10.5772/67593

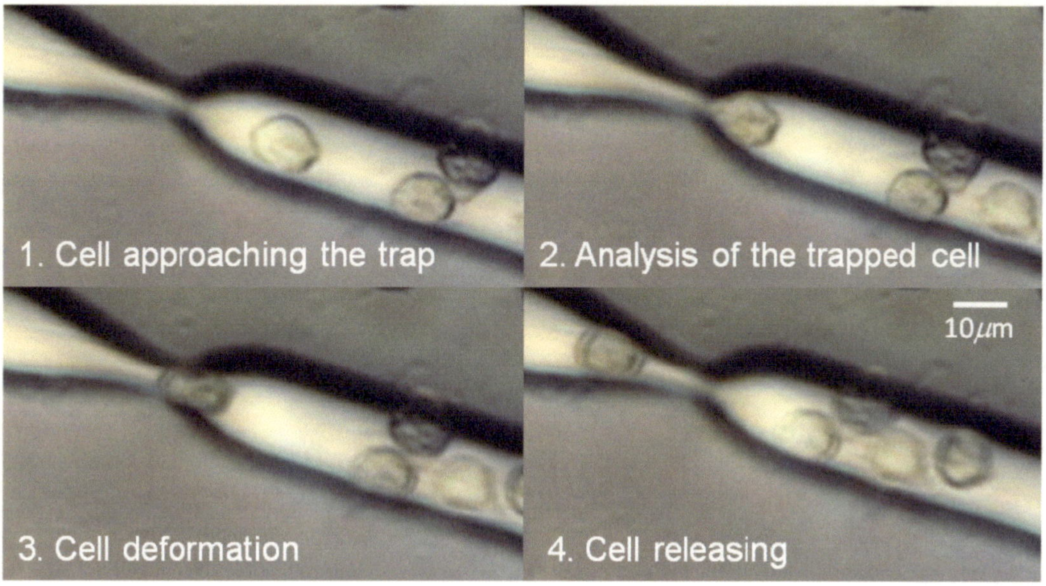

Figure 20. Sequence of the cell analysis mechanism: (1) the cell approaches the trap followed by other cells (t = 0 s); (2) the cell is trapped and the flow is stopped (t = 2 s), in this phase Raman measurements can be performed on the cell (t = 4 s); (3) the increasing pressure due to clogged channel deforms the cell forcing it through the trap (t = 6 s); (4) the cell is released continuing to move over the trap while other cells are approaching the trap aligning themselves to it (t = 8 s). Reprinted with permission from Ref. [86]. © 2015 Optical Society of America.

which is a very important point to consider when dealing with living biological samples since a high laser power can affect the viability of the cells.

Figure 21 showed the SERS measurements for RBCs in comparing with Raman measurements taken on a flat surface. The SERS spectrum shows not only a higher signal, compared to standard Raman, but above all the curve exhibits a richer peak profile due to plasmon enhancement of biomolecular vibrations occurring at the dimers locations. Furthermore, **Figure 22** shows the SERS spectra from RBC (red curve), PBL (black curve), and K562 (blue curve). The main peaks of all the three cell lines are correctly found in the Raman measurements. RBCs demonstrated the typical Raman peaks of phenylalanine at 1005 cm^{-1}, the deformation of the Cm-H bond at 1225 cm^{-1}, the half- and quarter-ring stretching of pyrimidine, respectively, at 1377 and 1397 cm^{-1}, and finally several peaks of the multi-structured band between 1540 and 1630 cm^{-1}. The relative intensities of the Raman peaks at 1565 cm^{-1} peak (CβCβ stretching), 1582 cm^{-1} (CαCm asymmetric stretching), and 1621 cm^{-1} (Ca = Cb stretching) confirmed that the RBCs are in an oxygenated state [86].

Raman signature collected on PBL cells shows characteristic peaks due to deoxyribose vibrations, at 980 and 1448 cm^{-1}, to phospholipid C-C stretching at 1067 cm^{-1}, and Amide I vibration from proteins secondary structure at 1665 cm^{-1} [92]. The 1448 cm^{-1} peak is a broad one since it is the overlapping of deoxyribose signature with C-H deformation from proteins. Finally, Raman signals coming from K562 cells show a very good agreement with the reported data from the literature [93, 94]. The SERS spectrum of the phenylalanine ring mode at 1003 cm^{-1} is observed, the PO2- backbone vibration arises at 1095 cm^{-1}, Amide III vibrations from proteins are located at 1203 and 1303 cm^{-1} where the former one (1203 cm^{-1}) overlaps with DNA bases signal,

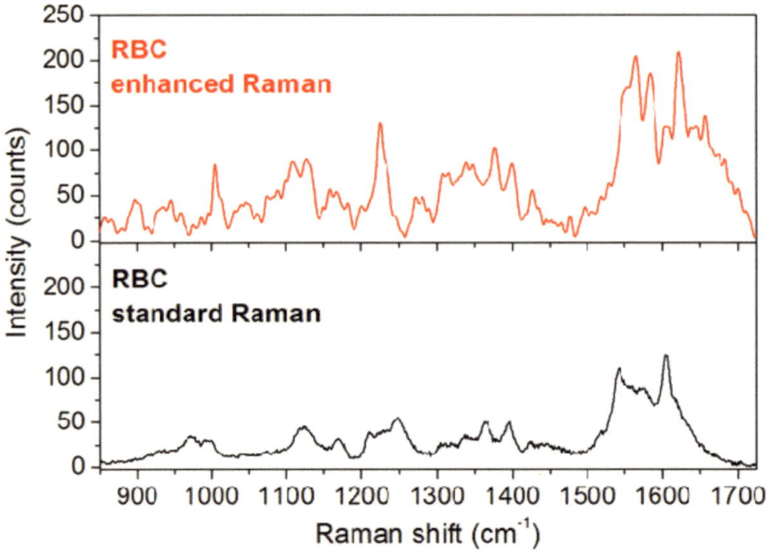

Figure 21. Raman spectra collected from red blood cells (RBC) on nanodimers (the used power for the enhanced Raman measurements is 10% of the total) and on a flat substrate (the used power for the standard Raman measurements is 100% of the total). The spectrum recorder on nanodimers exhibits a higher signal and is more resolved. Reprinted with permission from Ref. [86]. © 2015 Optical Society of America.

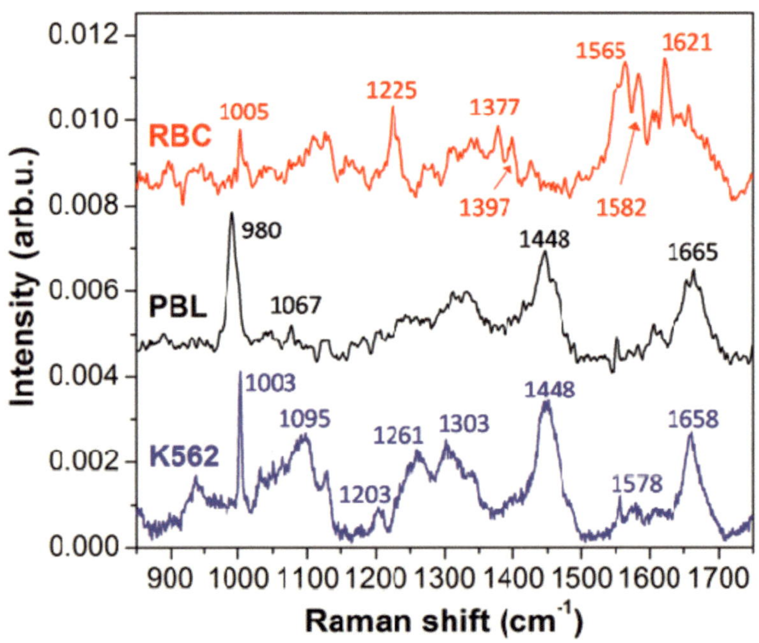

Figure 22. Raman spectra collected in three different experiments with different cell lines, in the proximity of the microfluidic traps. From top to bottom, the average spectra of red blood cells (RBC), peripheral blood lymphocytes (PBL), and leukemia cells (K562) are reported. See the text for assignment of the main peaks. Reprinted with permission from Ref. [86]. © 2015 Optical Society of America.

C-H bending and deformation are, respectively, found at 1261 and 1448 cm^{-1}, small DNA signal is also observed at 1578 cm^{-1}, and finally the Amide I strong vibration from proteins is located at 1658 cm^{-1}. Compared to PBL, the Amide I shift toward lower frequencies (from 1665 to 1658 cm^{-1}) that could be due to a larger presence of alpha helix structure of proteins. This is also supported by the small peak observed at 935 cm^{-1} (alpha helix C-C skeletal mode) for the K562 curve.

4.3. SERS-LSV spectroelectrochemical method for single-cell analysis

In our previous work [95], we have reported on the fabrication of Au nanodot array-modified ITO substrate and its application as cell culture system, SERS-active surface, and a working electrode. Moreover, we have designed a new spectroelectrochemical cell chip that combined the SERS and voltammetric methods for monitoring redox properties of single living cell. In this study, we have used PC12 cell line as an experimental modal. In order to evaluate the spectroelectrochemical assay at the single PC12 cell level (**Figure 23**), 12 Au microelectrode arrays were patterned on a 10 × 10-mm glass substrate that provided six microgaps of about 7 μm in width between each couple of microelectrode. In order to enhance the Raman signals from the single cell, we have fabricated polystyrene-assisted hexagonal array of Au nanodots inside the gap between each pair of Au microelectrodes. Then, we immobilized the single cells over the microgaps by using a polydimethylsiloxane (PDMS) microchannel (200 mm in width) that was attached. The cells were then transferred onto the chip with a new culture medium through the microchannel inlet. A sterile cell chamber with dimensions of 8 × 8 mm was developed to measure the Raman spectra of living single cells under physiological conditions (**Figure 24**).

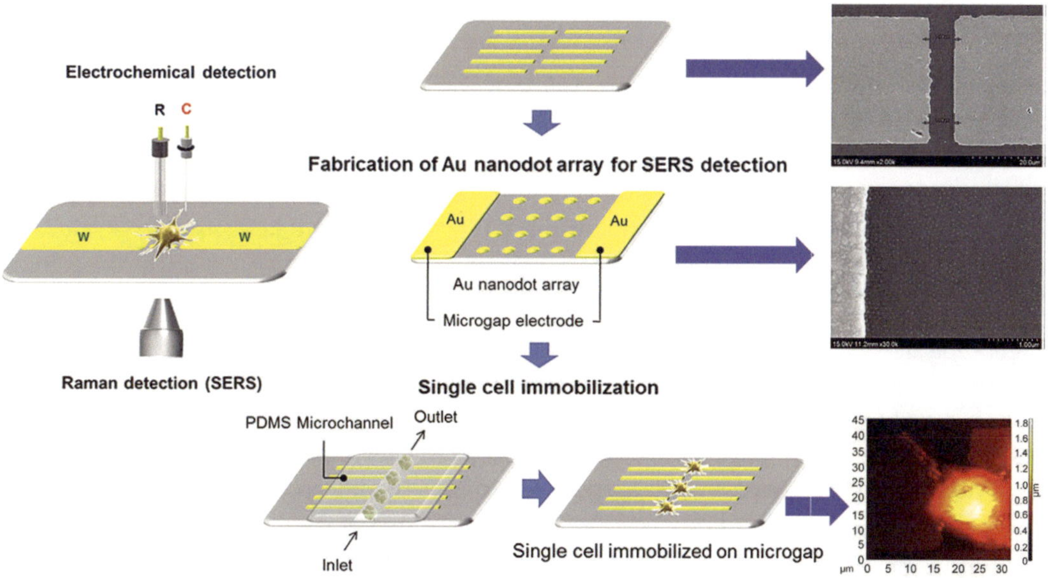

Figure 23. Schematic diagram of the immobilization of a single cell on the microgap between pairs of Au microelectrodes. Reprinted with permission from Ref. [95]. © 2015 Elsevier.

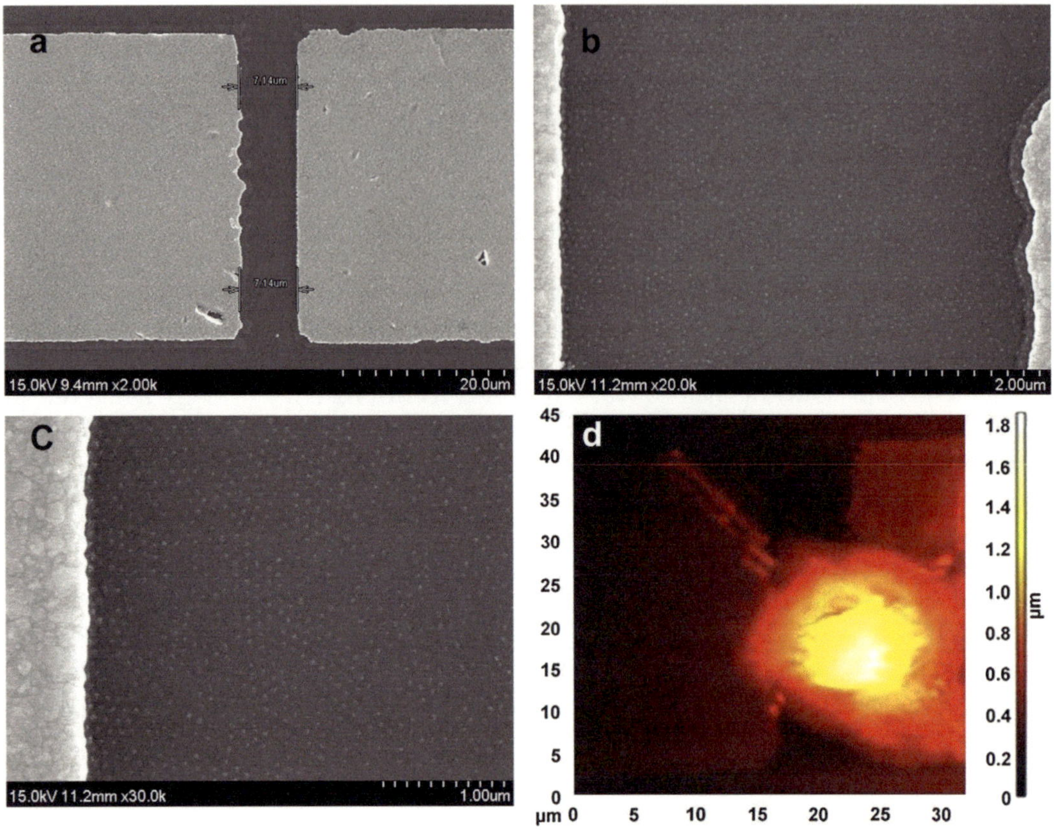

Figure 24. Fabrication of the nanodot array inside the microgap as cell-based chip for single-cell studies: (a) SEM image of the microgap between the pair of Au microelectrodes; (b and c) SEM images of Au nanodots array inside the microgap between the pair of Au microelectrodes; and (d) AFM image of PC12 cell on the microgap between the pair Au microelectrodes. Reprinted with permission from Ref. [95]. © 2015 Elsevier.

SERS-linear sweep voltammetry (LSV) technique was developed, in which LSV was used to investigate the redox behavior of neural cells (PC12), while the NIR laser source was simultaneously focused on the target cell for SERS analysis of the biochemical composition changes of living single PC12 cells during the redox processes.

Figure 25a showed the CV of the Au microelectrodes in PBS buffer solution and no redox peaks were observed. Interestingly, an irreversible oxidation process showing an anodic peak only was observed from a single PC12 cell, which was significantly distinguished with bulk PC12 cells normally showing quasi-reversible redox process. This result may be related to the low concentration of electro-active species and a lack of cell-cell interactions. The CV behavior of single PC12 cell immobilized on microgap between the pair of Au microelectrode containing hexagonal Au nanodots array showed sharp anodic peak at about 0.4 V (**Figure 25c**). This was significantly different from the cell on bare Au microelectrode that showed broad anodic peak (**Figure 25b**). These results indicated that Au nanodots array is very effective for the enhancement of the electron transfer rate. **Figure 25d** showed the LSV oxidation behavior of single PC12 cell within a potential range from −0.2 to +0.8 V, which contained anodic peak at about

+0.4 V. Moreover, many peaks in the SERS spectrum were changed relative to the control PC12 single cell (**Figure 25f**). The SERS spectra changes including a decrease in the intensity of SERS peaks at 720 cm–1 (adenine), 760 (Trp), 905 (prot. ring str.), 1001 (Phe), 1060 (str. PO2- and str. COC), 1093 (lipids: str. C═C, deoxyribose: C═O, C═C str. and str. PO2-),1125 (prot. str. CN), 1175 (prot. Tyr str. CN and CC), 1270 (T, A and amide III), 1310 (A), 1360 (prot. CH2), 1390 (T, A and G), and 1690 cm^{-1} (amide I and C═C). Additionally, the Raman peaks at 1625 cm^{-1} (C═C, Tyr and Trp), which was observed in control PC12 cells, disappeared. Also, peaks at 854 cm^{-1} (prot. Ring br. Tyr), 940 (proteins: a-helix, and deoxy and C-O-C), and 1560 cm^{-1} (proteins: amide II and Trp; C═C and/or C═O stretch of o-quinone group) shifted to 866, 952, and 1545 cm^{-1}, respectively. New Raman peaks were observed during the application of the oxidation voltage, including Raman peaks at 1330 (G), s1450 (deoxyribose), 1426, 1502, and 1595 cm^{-1} (A and G). In addition, no cathodic peak was observed when PC12 cells were subjected to reduction potential within a potential range from +0.8 to −0.2 V (**Figure 25e**). However, the Raman spectrum during the reduction process was different than the Raman spectrum of the control PC12 or PC12 cells in the oxidation state (**Figure 25f**). These changes in Raman spectra included an increase in the Raman peaks intensities at 905 (prot. ring str. CC), 1001 (Phe), 1060 (PO2- str. and C═O, C═C str.), 1093 (lipids, chain C═C str. deoxyribose: C═O, C═C str. phosphate: str. PO2_), 1125 (C-N str.), 1175 (C-H bend Tyr), 1270 (T, A and amide III), 1310 (A), 1360 (A, G, and prot. CH2 def), and 1390 cm^{-1} (T, A and G). Also, the Raman peak at 1560 cm^{-1} (C═C and/or C═O stretch of o-quinone group) shifted to 1575 cm^{-1}. Furthermore, new Raman peaks at 1230 (T, A, and amide III), 1426 (G, A and CH def), and 1502 cm^{-1} (G and A) were observed. Moreover, the Raman peaks at 1330 (G), 1595 (A and G), 1625 (C═C Tyr and Trp), and 1720 cm^{-1} (amide I and C═O ester) disappeared, which could be related to a reduction of these functional groups. This study demonstrated that the changes in the Raman spectra signals of single PC12

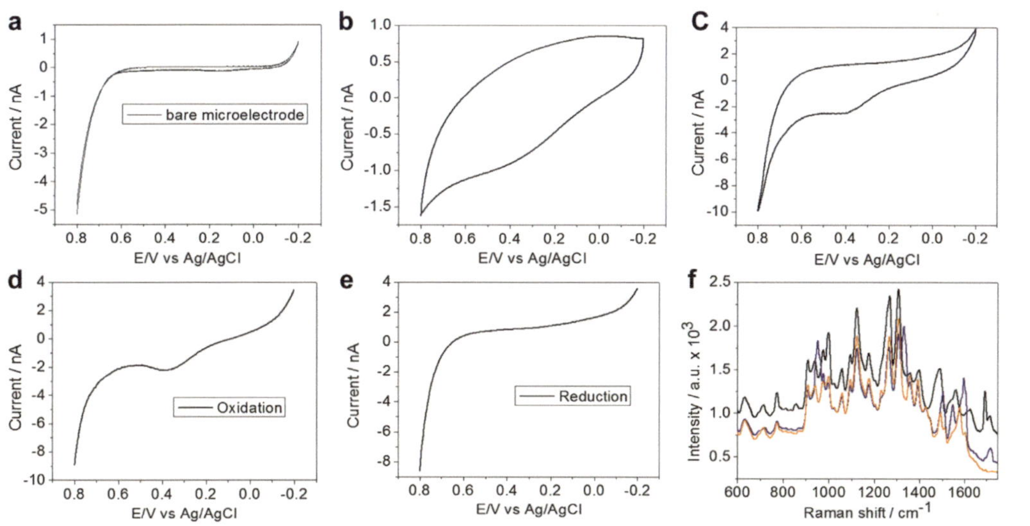

Figure 25. (a) CV background of microelectrode. (b) CV of single PC12 cell on the microgap between the pair of Au microelectrodes. (c) CV of a single PC12 cell on the microgap between the pair of Au microelectrodes containing the Au nanodots array. (d) LSV oxidation of a single PC12 cell on the microgap between the pair of Au microelectrodes. (e) LSV reduction of a single PC12 cell on the microgap between two Au microelectrodes. (f) SERS spectra of PC12 cells control (top) during reduction (middle) and oxidation (bottom). Reprinted with permission from Ref. [95]. © 2015 Elsevier.

cells during oxidation and/or reduction processes were more complicated than the changes of bulk PC12 cells. This may be due to the direct effect of voltage on the target cell [95].

5. Conclusions

Over the last couple of decades, many techniques have been used for the design and fabrication of homogeneous nanostructure-modified substrates for developing new SERS-active substrates that we reported and their application as a strong analytical tool for the analysis of both bulk and single cell. In this chapter, we have reviewed the origin of Raman spectroscopy and its limitations, advantages of SERS technique, and different mechanism theories for SERS phenomena. In addition to the uses of SERS technique, we have also studied the chemical composition of different cell components and differentiation between different cell lines, live and dead cells and to monitor the effects of drugs on different cell lines. Furthermore, we have showed the capability of SERS to combine with different techniques, for example, combining between SERS and electrochemical techniques as spectroelectrochemical tool to analyze the intracellular and extracellular state of single cell that allowed us to follow the mechanism of oxidation and reduction systematically as a function of electrode potential. Moreover, this technique may be used to monitor the biochemical changes during cell electrofusion and the electrical stimulation of differentiated neural cells. Also, we have reviewed the combination of SERS with flow cytometry measurements to analyze different single-cell lines including the uses of microfluidic system enabling easy and potential way for controlling a fast analysis cell-by-cell diagnostic tool with an amount of information that other optical techniques are lacking. Thus, this cutting-edge technology that combined the SERS and electrochemical and/or flow cytometry methods could be utilized as a noninvasive and nondestructive tool at various kinds of cellular researches.

Acknowledgements

This research was supported by the Leading Foreign Research Institute Recruitment Program, through the National Research Foundation of Korea (NRF), funded by the Ministry of Science, ICT, and Future Planning (MSIP) (2013K1A4A3055268) and by the National Research Foundation of Korea (NRF) grant funded by the Korea government (MSIP) (no. 2014R1A2A1A10051725) and Basic Science Research Program through the National Research Foundation of Korea (NRF) funded by the Ministry of Education (2016R1A6A1A03012845).

Author details

Waleed Ahmed El-Said[1], Hyeon-Yeol Cho[2] and Jeong-Woo Choi[2]*

*Address all correspondence to: jwchoi@sogang.ac.kr

1 Department of Chemistry, Faculty of Science, Assiut University, Assiut, Egypt

2 Department of Chemical and Biomolecular Engineering, Sogang University, Seoul, Republic of Korea

References

[1] Smith W.E., Dent G. Modern Raman Spectroscopy: A Practical Approach. John Wiley and Sons; 2005. 210 p. DOI: 10.1002/0470011831

[2] Short K.W., Carpenter S., Freyer J.P., Mourant J.R. Raman spectroscopy detects biochemical changes due to proliferation in mammalian cell cultures. Biophysical Journal. 2005;**88**(6):4274-4288. DOI: 10.1529/biophysj.103.038604

[3] Krishna C.M., Kegelaer G., ADT I., Rubin S., Kartha V.B., Manfait M., et al. Combined Fourier transform infrared and Raman spectroscopic approach for identification of multidrug resistance phenotype in cancer cell lines. Biopolymers. 2006;**82**:462-470. DOI: 10.1002/bip.20485

[4] Krafft C., Knetschke T., Funk R.H.W., Salzer R. Identification of organelles and vesicles in single cells by Raman microspectroscopic mapping. Vibrational Spectroscopy. 2005;**38**:85-95. DOI: dx.doi.org/10.1016/j.vibspec.2005.02.008

[5] Schuster K.C., Reese I., Urlaub E., Gapes J.R., Lendl B.B. Multidimensional information on the chemical composition of single bacterial cells by confocal Raman microspectroscopy. Analytical Chemistry. 2000;**72**(22):5529-5534. DOI: 10.1021/ac000718x

[6] Crow P., Barrass B., Kendall C., Prieto M.H., Wright M., Persad R., et al. The use of Raman spectroscopy to differentiate between different prostatic adenocarcinoma cell lines. British Journal of Cancer. 2005;**92**:2166-2170. DOI: 10.1038/sj.bjc.6602638

[7] Notingher I., Verrier S., Romanska H., Bishop A.E., Polak J.M., Hench L.L. In situ characterisation of living cells by Raman spectroscopy. Spectroscopy. 2002;**16**(2):43-51. DOI: http://dx.doi.org/10.1155/2002/408381

[8] El-Said W.A., Kim S.U., and Choi J.W. Monitoring in vitro neural stem cell differentiation based on surface-enhanced Raman spectroscopy using a gold nanostar array. Journal of Material Chemistry C. 2015;**3**:3848-3859. DOI: 10.1039/C5TC00304K

[9] El-Said W.A., Choi J.W. In situ detection of neurotransmitter release from PC12 cells using surfaceenhanced Raman spectroscopy. Biotechnology and Bioprocess Engineering. 2014;**19**(6):1069-1076. DOI: 10.1007/s12257-014-0092-7

[10] An J.H., El-Said W.A., Choi, J.W. Cell chip based monitoring of toxic effects on dopaminergic cell. Journal of Nanoscience and Nanotechnology. 2012;**12**(5):4115-4118. DOI: 10.1166/jnn.2012.5903

[11] An J.H., El-Said W.A., Choi J.W. Surface enhanced Raman scattering of neurotransmitter release in neuronal cells using antibody conjugated gold nanoparticles. Journal Nanoscience and Nanotechnology. 2011;**11**(2):1585-1588.

[12] El-Said W.A., Kim T.H., Kim H., Choi J.W. Analysis of intracellular state based on controlled 3D nanostructures mediated surface enhanced Raman scattering. PLoS One. 2011;**6**(2):e15836. DOI: 10.1371/journal.pone.0015836.g001

[13] El-Said W.A., Kim T.H., Kim H., Choi J.W. Detection of effect of chemotherapeutic agents to cancer cells on gold nanoflower patterned substrate using surface enhanced Raman

scattering and cyclic voltammetry. Biosensors and Bioelectronics. 2010;**26**(4):1486-1492. DOI: 10.1016/j.bios.2010.07.089

[14] Ellis D.I., Goodacre R. Metabolic fingerprinting in disease diagnosis: biomedical applications of infrared and Raman spectroscopy. Analyst. 2006;**131**(8):875-885. DOI: 10.1039/b602376m

[15] Fleischmann M., Hendra P.J., McQuillan A.J. Raman spectra of pyridine adsorbed at a silver electrode. Chemical Physics Letters. 1974;**26**(2):163-166. DOI: 10.1016/0009-2614(74)85388-1

[16] Volkan M., Stokes D.L., and Vo-Dinh T. Surface-enhanced Raman of dopamine and neurotransmitters using sol-gel substrates and polymer-coated fiber-optic probes. Applied Spectroscopy. 2000;**54**(12):1842-1848.

[17] Jeanmaire D.L., Vanduyne R.P. Surface Raman spectroelectrochemistry: part I. Heterocyclic, aromatic, and aliphatic amines adsorbed on the anodized silver electrode. Journal of Electroanalytical Chemistry and Interfacial Electrochemistry. 1977;**84**(1):1-20. DOI: 10.1016/S0022-0728(77)80224-6

[18] Albrecht M.G., Creighton J.A. Anomalously intense Raman spectra of pyridine at a silver electrode. Journal of the American Chemical Society. 1977;**99**(15):5215-5217. DOI: 10.1021/ja00457a071

[19] Doering W.E. and Nie S.M. Single-molecule and single-nanoparticle SERS: examining the roles of surface active sites and chemical enhancement. Journal of Physical Chemistry B. 2002;**106**(2):311-317. DOI: 10.1021/jp011730b

[20] Nie S., Emory S.R. Probing single molecules and single nanoparticles by surface-enhanced Raman scattering. Science. 1997;**275**(5303):1102-1106. DOI: 10.1126/science.275.5303.1102

[21] Moskovits M. Surface-enhanced Raman spectroscopy: a brief retrospective. Journal of Raman Spectroscopy. 2005;**36**(6-7):485-496.

[22] Kneipp K., Kneipp H., Itzkan I., Dasari R.R., Feld M.S. Surface-enhanced Raman scattering and biophysics. Journal of Physics: Condensed Matter. 2002;**14**(18):R597–R624. DOI: 10.1088/0953-8984/14/18/202

[23] Jiang J., Bosnick K., Maillard M., Brus L. Single molecule Raman spectroscopy at the junctions of large Ag nanocrystals. Journal of Physical Chemistry B. 2003;**107**(37):9964-9972. DOI: 10.1021/jp034632u

[24] Otto A., Mrozek I., Grabhorn H., Akemann W. Surface-enhanced Raman scattering. Journal of Physics: Condensed Matter. 1992;**4**(5):1143-1212. DOI: 10.1088/0953-8984/4/5/001

[25] Kneipp K., Kneipp H., Itzkan I., Dasari R.R., Feld M.S. Ultrasensitive chemical analysis by Raman spectroscopy. Chemical Reviews. 1999;**99**(10):2957-2976. DOI: 10.1021/cr980133r

[26] Tao A.R., Yang P. Polarized surface-enhanced Raman spectroscopy on coupled metallic nanowires. Journal of Physical Chemistry B. 2005;**109**(33):15687-15690. DOI: 10.1021/jp053353z

[27] Weitz D.A., Moskovits M., Creighton J.A. Surface-enhanced Raman scattering with emphasis on the liquid-solid interface. In: Hall R.B., Ellis A.B., editors. Chemistry and Structure at Interfaces, New Laser and Optical Techniques. VCH; 1986.

[28] Zou X., Dong S. Surface-enhanced Raman scattering studies on aggregated silver nanoplates in aqueous solution. Journal of Physical Chemistry B. 2006;**110**(43):21545-21550. DOI: 10.1021/jp063630h

[29] Brus L. Noble metal nanocrystals: plasmon electron transfer photochemistry and single-molecule Raman spectroscopy. Accounts of Chemical Research. 2008;**41** (12):1742-1749. DOI: 10.1021/ar800121r.

[30] Kelly K.L., Coronado E., Zhao L.L., Schatz G.C. The optical properties of metal nanoparticles: the influence of size, shape, and dielectric environment. The Journal of Physical Chemistry B. 2003;**107**(3):668-677. DOI: 10.1021/jp026731y

[31] Michaels A.M., Jiang J., Brus L. Ag nanocrystal junctions as the site for surface-enhanced Raman scattering of single rhodamine 6G molecules. Journal of Physical Chemistry B. 2000;**104**(50):11965-11971. DOI: 10.1021/jp0025476

[32] Xu Y., Wu J., Sun W., Tao D., Yang L., Song Z., et al. A new mechanism of Raman enhancement and its application. Chemistry – A European Journal. 2002;**8**(23):5323-5331. DOI: 10.1002/1521-3765(20021202)8:23<5323::AID-CHEM5323>3.0.CO;2-E

[33] Wen R., Fang Y. An investigation of the surface-enhanced Raman scattering (SERS) effect from a new substrate of silver-modified silver electrode. Journal of Colloid and Interface Science. 2005;**292**(2):469-475. DOI: 10.1016/j.jcis.2005.05.091

[34] Kambhampati P., Campion A., Song O.K. Probing photoinduced charge transfer at atomically smooth metal surfaces using surface enhanced Raman scattering. Phys Status Solidi A. 1999;**175**(1):233-239. DOI: 10.1002/(SICI)1521-396X(199909)175:1<233::AID-PSSA233>3.0.CO;2-Y

[35] Moskovits M., Tay L.L., Yang J., Haslett T. SERS and the single molecule. In: Shalaev V.M., editor. Optical Properties of Nanostructured Random Media. Springer; 2002. DOI: 215-227.

[36] Chen L., Choo J. Recent advances in surface-enhanced Raman scattering detection technology for microfluidic chips. Electrophoresis. 2008;**29**(9):1815-1828. DOI: 10.1002/elps.200700554

[37] Qian X.M., Nie S.M. Single-molecule and single-nanoparticle SERS: from fundamental mechanisms to biomedical applications. Chemical Society Reviews. 2008;**37**(5):912-920. DOI: 10.1039/b708839f

[38] Ni J., Lipert R.J., Dawson G.B., Porter M.D. Immunoassay readout method using extrinsic Raman labels adsorbed on immunogold colloids. Analytical Chemistry. 1999;**71**(21):4903-4908. DOI: 10.1021/ac990616a

[39] Isola N.R., Stokes D.L., Vo-Dinh T. Surface-enhanced Raman gene probe for HIV detection. Analytical Chemistry. 1998;**70**(7):1352-1356. DOI: 10.1021/ac970901z

[40] Zhang C.Y., Johnson L.W. Quantum-dot-based nanosensor for RRE IIB RNA-Rev peptide interaction assay. Journal of the American Chemical Society. 2006;**128**(16):5324-5325. DOI: 10.1021/ja060537y

[41] Grubisha D.S., Lipert R.J., Park H.Y., Driskell J., Porter M.D. Femtomolar detection of prostate-specific antigen: an immunoassay based on surface-enhanced Raman scattering and immunogold labels. Analytical Chemistry. 2003;**75**(21):5936-5943. DOI: 10.1021/ac034356f

[42] Jackson J.B., Westcott S.L., Hirsch L.R., West J.L., Halas N.J. Controlling the surface enhanced Raman effect via the nanoshell geometry. Applied Physics Letters. 2003;**83**(2):257-259. DOI: http://dx.doi.org/10.1063/1.1534916

[43] Nikoobakht B., El-Sayed M.A. Surface-enhanced Raman scattering studies on aggregated gold nanorods. Journal of Physical Chemistry A. 2003;**107**(18):3372-3378. DOI: 10.1021/jp026770+

[44] Orendorff, C.J., Gearheart, L., Jana N.R., Murphy C.J. Aspect ratio dependence on surface enhanced Raman scattering using silver and gold nanorod substrates. Physical Chemistry Chemical Physics. 2006;**8**(1):165-170. DOI: 10.1039/B512573A

[45] Tiwari V.S., Oleg T., Darbha G.K., Hardy W., Singh J.P., Ray P.C. Non-resonance SERS effects of silver colloids with different shapes. Chemical Physics Letters. 2007;**446**(1-3):77-82. DOI: 10.1016/j.cplett.2007.07.106

[46] Liu Y.C., Yu C.C., Sheu S.F. Improved surface-enhanced Raman scattering on optimum electrochemically roughened silver substrates. Analytica Chimica Acta. 2006;**577**(2):271-275. DOI: 10.1016/j.aca.2006.10.060

[47] Jacobson M.L., Rowlen K.L. Photo-dynamics on thin silver films. Chemical Physics Letters. 2005;**401**(1-3):52-57. DOI: 10.1016/j.cplett.2004.11.018

[48] Brolo A.G., Arctander E., Gordon R., Leathem B., Kavanagh K.L. Nanohole-enhanced Raman scattering. Nano Letters. 2004;**4**(10):2015-2018. DOI: 10.1021/nl048818w

[49] Wang H., Levin C.S., Halas N.J. Nanosphere arrays with controlled sub-10-nm gaps as surface-enhanced Raman spectroscopy substrates. Journal of the American Chemical Society. 2005;**127**(43):14992-14993. DOI: 10.1021/ja055633y

[50] Dick L.A., McFarland A.D., Haynes C.L., Van Duyne R.P. Metal film over nanosphere (MFON) electrodes for surface-enhanced Raman spectroscopy (SERS): improvements in surface nanostructure stability and suppression of irreversible loss. Journal of Physical Chemistry B. 2002;**106**(4):853-860. DOI: 10.1021/jp013638l

[51] Flauds K., Barbagallo R.P., Keer J.T., Smith W.E., Graham D. SERRS as a more sensitive technique for the detection of labelled oligonucleotides compared to fluorescence. Analyst. 2004;**129**(7):567-568. DOI: 10.1039/b406423b

[52] Cao Y.C., Jin R., Mirkin C.A. Nanoparticles with Raman spectroscopic fingerprints for DNA and RNA detection. Science. 2002;**297**(5586):1536-1540. DOI: 10.1126/science.297.5586.1536

[53] Aroca R. Surface-Enhanced Vibrational Spectroscopy. John Wiley and Sons; 2006. 260 p.

[54] Kneipp K., Haka A.S., Kneipp H., Badizadegan K., Yoshizawa N., Boone C., et al. Surface-enhanced Raman spectroscopy in single living cells using gold nanoparticles. Applied Spectroscopy. 2002;**56**(2):150-154. DOI: 10.1366/0003702021954557

[55] Tang H.W., Yang X.B., Kirkham J., Smith D.A. Chemical probing of single cancer cells with gold nanoaggregates by surface-enhanced Raman scattering. Applied Spectroscopy. 2008;**62**(10):1060-1069. DOI: 10.1366/000370208786049015

[56] Huang X., El-Sayed I.H., Qian W., El-Sayed M.A. Cancer cells assemble and align gold nanorods conjugated to antibodies to produce highly enhanced, sharp, and polarized surface Raman spectra: a potential cancer diagnostic marker. Nano Letters. 2007;**7**(6):1591-1597. DOI: 10.1021/nl070472c

[57] Morjani H., Riou J.F., Nabiev I., Lavelle F., Manfait M. Molecular and cellular interactions between intoplicine, DNA, and topoisomerase II studied by surface-enhanced Raman scattering spectroscopy. Cancer Research. 1993;**53**(20):4784-4790.

[58] Li M.D., Cui Y., Gao M.X., Luo J., Ren B., Tian Z.Q. Clean substrates prepared by chemical adsorption of iodide followed by electrochemical oxidation for surface-enhanced Raman spectroscopic study of cell membrane. Analytical Chemistry. 2008;**80**(13):5118-5125. DOI: 10.1021/ac8003083

[59] Ayyappan S., Gopalan, R.S., Subbanna, G.N., Rao C.N.R. Nanoparticles of Ag, Au, Pd, and Cu produced by alcohol reduction of the salts. Journal of Material Research. 1997;**12**(2):398-401. DOI: 10.1557/JMR.1997.0057

[60] Johnson C.J., Dujardin E., Davis S.A., Murphy C.J., Mann S. Growth and form of gold nanorods prepared by seed-mediated, surfactant-directed synthesis. Journal of Material Chemistry. 2002;**12**:1765-1770. DOI: 10.1039/b200953f

[61] Tkachenko A.G., Xie H., Coleman D., Ryan W.G.J., Anderson M.F., Franzen S., Feldheim, D.L. Multifunctional gold nanoparticle-peptide complexes for nuclear targeting. Journal of the American Chemical Society. 2003;**125**(16):4700-4701. DOI: 10.1021/ja0296935

[62] Kang B., Mackey M.A., El-Sayed M.A. Nuclear targeting of gold nanoparticles in cancer cells induces DNA damage, causing cytokinesis arrest and apoptosis. Journal of the American Chemical Society. 2010;**132**(5):1517-1519. DOI: 10.1021/ja9102698

[63] Le Ru E.C., Blackie E., Meyer M., Etchegoin P.G. Surface enhanced Raman scattering enhancement factors: a comprehensive study. Journal of Physical Chemistry C. 2007;**111**(37):13794-13803. DOI: 10.1021/jp0687908

[64] Etchegoin P.G., Meyer M., Blackie E., Le Ru E.C. Statistics of single-molecule surface enhanced Raman scattering signals: fluctuation analysis with multiple analyte techniques. Analytical Chemistry. 2007;**79**(21):8411-8415. DOI: 10.1021/ac071231s

[65] Gremlich H.U., Yan B. Infrared and Raman Spectroscopy of Biological Materials. CRC Press; 2000. 600 p.

[66] Xu H., Bjerneld E.J., Käll M., Börjesson L. Spectroscopy of single hemoglobin molecules by surface enhanced Raman scattering. Physical Review Letters. 1999;**83**:4357-4360. DOI: 10.1103/PhysRevLett.83.4357

[67] Manfait M., Morjani H., Nabiev I. Molecular events on simple living cancer cells as studied by spectrofluorometry and micro-SERS Raman spectroscopy. Journal of Cellular Pharmacology. 1992;**3**:120-125.

[68] Jarvis R.M., Goodacre R. Characterisation and identification of bacteria using SERS. Chemical Society Reviews. 2008;**37**(5):931-936. DOI: 10.1039/b705973f

[69] Breuzard G., Angiboust J.F., Jeannesson P., Manfait M., Millot J.M. Surface-enhanced Raman scattering reveals adsorption of mitoxantrone on plasma membrane of living cells. Biochemical and Biophysical Research Communications. 2004;**320**(2):615-621. DOI: 10.1016/j.bbrc.2004.05.203

[70] Shamsaie A., Jonczyk M., Sturgis J., Robinson J.P., Irudayaraj J. Intracellularly grown gold nanoparticles as potential surface-enhanced Raman scattering probes. Journal of Biomedical Optics. 2007;**12**(2):020502. DOI: 10.1117/1.2717549

[71] Chithrani B.D., Ghazani A.A., Chan W.C.W. Determining the size and shape dependence of gold nanoparticle uptake into mammalian cells. . Nano Letters. 2006;**6**(4):662-668. DOI: 10.1021/nl052396o

[72] El-Said W.A., Cho H.Y., Yea C.H., Choi J.W. Synthesis of metal nanoparticles inside living human cells based on the intracellular formation process. Advanced Materials. 2014;**26**(6):910-918. DOI: 10.1002/adma.201303699

[73] Kalyuzhnaya M.G., Zabinsky R., Bowerman S., Baker D.R., Lidstrom M.E., Chistoserdova L. Fluorescence in situ hybridization-flow cytometry-cell sorting-based method for separation and enrichment of type i and type ii methanotroph populations. Applied and Environmental Microbiology. 2006;**72**(6):4293-4301. DOI: 10.1128/AEM.00161-06

[74] Li M., Xu J., Romero-Gonzalez M., Banwart S.A., Huang W.E. Single cell Raman spectroscopy for cell sorting and imaging. Current Opinion in Biotechnology. 2012;**23**(1):56-63. DOI: 10.1016/j.copbio.2011.11.019

[75] Huang W.E., Griffiths R.I., Thompson I.P., Bailey M.J., Whiteley A.S. Raman microscopic analysis of single microbial cells. Analytical Chemistry. 2004;**76**(15):4452-4458. DOI: 10.1021/ac049753k

[76] Sun S., Wang X., Gao X., Ren L., Su X., Dongbo B., Ning K. Condensing Raman spectrum for single-cell phenotype analysis. BMC Bioinformatics. 2015;**16**:S15. DOI: 10.1186/1471-2105-16-S18-S15

[77] Kneipp K., Kneipp H., Deinum G., Itzkan I., Dasari R.R., Feld M.S. Single-molecule detection of a cyanine dye in silver colloidal solution using near-infrared surface-enhanced Raman scattering. Applied Spectroscopy. 1998;**52**(2):175-178. DOI: 10.1366/0003702981943275

[78] Michaels A.M., Nirmal M., Brus L.E. Surface enhanced Raman spectroscopy of individual rhodamine 6G molecules on large Ag nanocrystals. Journal of the American Chemical Society. 1999;**121**(43):9932-9939. DOI: 10.1021/ja992128q

[79] Haslett T.L., Tay L., Moskovits M. Can surface-enhanced Raman scattering serve as a channel for strong optical pumping?. Journal of Chemical Physics. 2000;**133**(4):1641-1646. DOI: 10.1063/1.481952

[80] Jiang X., Jiang Z., Xu T., Su S., Zhong Y., Peng F., et al. Surface-enhanced Raman scattering-based sensing in-vitro: facile and label-free detection of apoptotic cells at the single-cell Level. Analytical Chemistry. 2013;**85**(5):2809-2816. DOI: 10.1021/ac303337b

[81] Kang J.W., So P., Dasari R.R., Lim D.K. High resolution live cell Raman imaging using subcellular organelle-targeting SERS-sensitive gold nanoparticles with highly narrow intra-nanogap. Nano Letters. 2015;**15**(3):1766-1772. DOI: 10.1021/nl504444w

[82] Syme C.D., Sirimuthu N.M.S., Faley S.L., Cooper J.M. SERS mapping of nanoparticle labels in single cells using a microfluidic chip. Chemical Communication. 2010;**46**:7921-7923. DOI: 10.1039/C0CC02209H

[83] Lu L., Xu X.L., Liang W.T., Lu H.F. Raman analysis of CdSe/CdS core-shell quantum dots with different CdS shell thickness. J. Phys. Condens. Matter. 2007;**19**(40):406221. DOI: 10.1088/0953-8984/19/40/406221

[84] Zhukovsky S.V., Ozel T., Mutlugun E., Gaponik N., Eychmüller A., Lavrinenko A.V., et al. Hyperbolic metamaterials based on quantum-dot plasmon-resonator nanocomposites. Optics Express. 2014;**22**(15):18290-18298. DOI: 10.1364/OE.22.018290

[85] Nolan J.P., Duggan E., Liu E., Condello D., Dave I., Stoner S.A. Single cell analysis using surface enhanced Raman scattering (SERS) tags. Methods. 2012;**57**(3):272-279. DOI: 10.1016/j.ymeth.2012.03.024.

[86] Perozziello G., Candeloro P., De Grazia A., Esposito F., Allione M., Coluccio M.L., et al. Microfluidic device for continuous single cells analysis via Raman spectroscopy enhanced by integrated plasmonic nanodimers. Optical Express. 2016;**24**(2):A180–A190. DOI: 10.1364/OE.24.00A180

[87] Cojoc D., Garbin V., Ferrari E., Businaro L., Romanato F., Fabrizio E.D. Laser trapping and micromanipulation using optical vortices. Microelectronic Engineering. 2005;**78-79**:125-131. DOI: 10.1016/j.mee.2004.12.017

[88] Chirumamilla M., Toma A., Gopalakrishnan A., Das G., Zaccaria R.P., Krahne R., et al. 3D nanostar dimers with a sub-10-nm gap for single-/few molecule surface-enhanced Raman scattering. Advanced Materials. 2014;**26**(15):2353-2358. DOI: 10.1002/adma.201304553

[89] De Angelis F., Liberale C., Coluccio M.L., Cojoc G., Di Fabrizio E. Emerging fabrication techniques for 3D nano-structuring in plasmonics and single molecule studies. Nanoscale. 2011;**3**(7):2689-2696. DOI: 10.1039/C1NR10124B

[90] Coluccio M.L., Gentile F., Das G., Nicastri A., Perri A.M., Candeloro P., et al. Detection of single amino acid mutation in human breast cancer by disordered plasmonic self-similar chain. Science Advances. 2015;**1**(8):e1500487. DOI: 10.1126/sciadv.1500487

[91] Coluccio M.L., Gentile F., Francardi M., Perozziello G., Malara N., Candeloro P., et al. Electroless deposition and nanolithography can control the formation of materials at the nano-scale for plasmonic applications. Sensors (Basel). 2014;**14**(4):6056-6083. DOI: 10.3390/s140406056

[92] Schuster K.C., Urlau E., Gapes J.R. Single-cell analysis of bacteria by Raman microscopy: spectral information on the chemical composition of cells and on the heterogeneity in a culture. Journal of Microbiological Methods. 2000;**42**(1):29-38. DOI: 10.1016/S0167-7012(00)00169-X

[93] Tallerico R., Todaro M., Di Franco S., Maccalli C., Garofalo C., Sottile R. et al. Human NK cells selective targeting of colon cancer-initiating cells: a role for natural cytotoxicity receptors and MHC class I molecules. Journal of Immunology. 2013;**190**(5):2381-2390. DOI: 10.4049/jimmunol.1201542

[94] Ong Y.H., Lim M., Liu Q. Comparison of principal component analysis and biochemical component analysis in Raman spectroscopy for the discrimination of apoptosis and necrosis in K562 leukemia cells: errata. Optics Express. 2012;**20**(20):25041-25043. DOI: 10.1364/OE.20.025041

[95] El-Said W.A., Kim T.H., Chung Y.H., Choi J.W. Fabrication of new single cell chip to monitor intracellular and extracellular redox state based on spectroelectrochemical method. Biomaterials. 2015;**40**:80-87. DOI: 10.1016/j.biomaterials.2014.11.023

Chapter 17

Localized Surface Plasmon Resonance for Optical Fiber-Sensing Applications

Pedro J. Rivero, Javier Goicoechea and Francisco J. Arregui

Additional information is available at the end of the chapter

http://dx.doi.org/10.5772/67544

Abstract

It is well known that optical fiber sensors have attracted the attention of scientific community due to its intrinsic advantages, such as lightweight, small size, portability, remote sensing, immunity to electromagnetic interferences and the possibility of multiplexing several signals. This field has shown a dramatic growth thanks to the creation of sensitive thin films onto diverse optical fiber configurations. In this sense, a wide range of optical fiber devices have been successfully fabricated for monitoring biological, chemical, medical or physical parameters. In addition, the use of nanoparticles into the sensitive thin films has resulted in an enhancement in the response time, robustness or sensitivity in the optical devices, which is associated to the inherent properties of nanoparticles (high surface area ratio or porosity). Among all of them, the metallic nanoparticles are of great interest for sensing applications due to the presence of strong absorption bands in the visible and near-infrared regions, due to their localized surface plasmon resonances (LSPR). These optical resonances are due to the coupling of certain modes of the incident light to the collective oscillation of the conduction electrons of the metallic nanoparticles. The LSPR extinction bands are very useful for sensing applications as far as they can be affected by refractive index variations of the surrounding medium of the nanoparticles, and therefore, it is possible to create optical sensors with outstanding properties such as high sensitivity and optical self-reference. In this chapter, the attractive optical properties of metal nanostructures and their implementation into different optical fiber configuration for sensing or biosensing applications will be studied.

Keywords: LSPR, optical fiber, sensing applications

1. Introduction

In this highly technological era in which we live, we are involved in a virtuous circle in which new machines, systems and devices are continuously created. This technological evolution has produced an evident improvement in our lifestyle, making our lives easier and more comfortable, and also contributing to the wealth distribution in the world. For a high number of those new systems, devices and new applications, sensors play a key role as far as they provide the link between the real-world and the technological devices. Sensors are much more effective than our five senses, and they can provide information not only about taste, smell or touch, but also about the presence of explosives [1], certain bacteria, viruses and biological markers [2, 3], magnetic fields, radiation, traces of pollutants [4] and many other applications. Consequently, there is a continuous effort in the research of new sensing devices that allow us to get information of our environment and consequently use this information to improve the existing devices or even the discovery of new applications. There has been a dramatic evolution starting from the classic analog sensors used for decades and decades to the current electronic smart sensors and other advanced sensing devices such as optical fiber sensors (**Figure 1**).

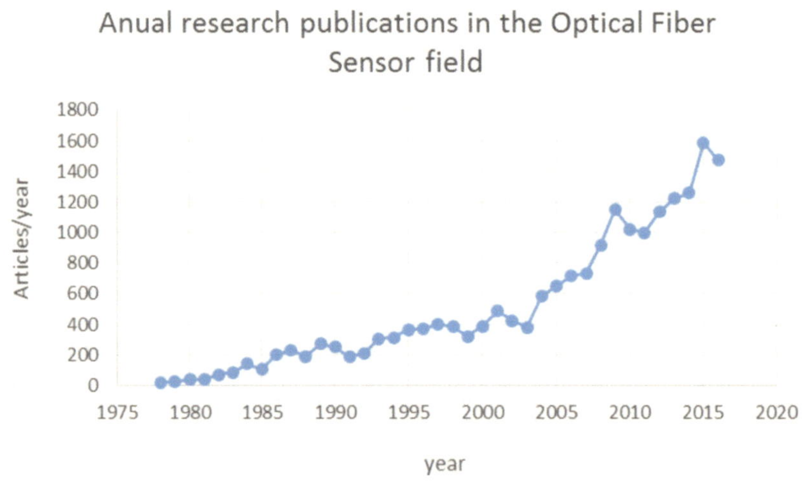

Figure 1. Number of published articles in international research journals with the words "Optical," "Fiber" and "Sensor" in their title, abstract or keywords (source: Scopus database).

The very first optical sensors were simple, mostly based on physical alterations of the natural waveguide structure of a regular optical fiber. The irruption of more advanced optical structures, such as fiber Bragg-gratings (FBGs), long-period gratings (LPGs), photonic crystal fibers (PCFs) and D-shape fibers opened the door to a new generation of optical fiber sensors based on interferometry, optical resonances, etc. Moreover, the combination of those families of optical fiber structures with coatings has made possible the apparition of more advanced optical fiber sensors.

As the sensitive coatings have been engineered, they have become more controllable and complex, and new materials have been incorporated to these devices, such as nanoparticles, fluorescent dyes, proteins, antibodies and aptamers. Additionally, the control of the structure of the coatings at the nanoscale level allows the ability to tailor at the molecular level of the composition and structure of the sensitive coatings. These nanoscaled materials have a special attraction because there are new phenomena observed when the matter is downscaled to this level. This is the case of the outstanding luminescence properties observed in semiconductor chalcogenide nanoparticles (e.g., CdTe, CdS) due to the exciton quantum confinement that has been widely used for sensor application [5]. Among these nanostructured materials, metallic nanoparticles have also been widely used because they show a very special optical resonance phenomenon known as localized surface plasmon resonance (LSPR). When light is propagated through a medium that contains certain metallic nanoparticles, the electrical field oscillation induced by light affects to the free electrons in the nanoparticles. There is a resonance condition in which the energy present in the electromagnetic wave is coupled to the electrons in the nanoparticle inducing their collective oscillation. Consequently, metallic nanoparticle dispersions show vivid and intense colors in the visible region of the electromagnetic spectrum. This resonance condition can be affected significantly by the composition of the metallic nanoparticles, by their shape and by the alteration of the effective refractive index of the medium surrounding the nanoparticle, typically in a radius of a few nanometers [6]. As a consequence, there has been a great research effort in the development of sensitive functionalization of noble metal nanoparticles in order to get improved sensing devices. The intensity and the wavelength position of LSPR attenuation bands may be affected by very small variations of the surrounding medium, and this can be used for the development of highly sensitive devices, talking even about single-molecule detection limits. LSPR has also been combined with complementary molecular identification techniques such as surface-enhanced Raman spectroscopy (SERS) [7].

Some authors have proposed different approaches using the LSPR of noble metallic nanoparticles in the research of new optical fiber sensors. Such optical fiber sensors combine the LSPR technology with the classical advantages of optical fiber sensors, such as lightweight devices, electromagnetic immunity, biocompatibility and remote sensing. Thus, taking the advantages of the LSPR resonances, it is possible to create a new generation of optical fiber sensors.

In this chapter, we will present a thorough review of the state of the art of LPSR-based optical fiber sensors. Firstly, the most used synthesis techniques of metallic nanoparticles and their possible immobilization approaches into different thin films are presented. Afterward, it will be presented the main applications for sensing physical parameters (refractive index, humidity, etc.), chemical parameters (pH, heavy metal ions, volatile and harmful compounds) and biological parameters (blood sugar, proteins, antigens and antibodies, etc.).

2. Introduction to the synthesis of metal nanoparticles

One of the most explored characteristics of the noble metal nanoparticles (essentially silver, gold or copper) is the localized surface plasmon resonance (LSPR), which is the frequency at which conduction electrons collectively oscillate in response to the alternating electric field of an incident electromagnetic radiation. As a result, an intense absorption band in the visible region with a specific coloration of the resultant nanoparticles is obtained [8, 9]. A relevant aspect in the synthesis of metal nanoparticles is that their optical properties present a great dependence with the resultant morphology. A good monitoring of the evolution of the LSPR wavelength position and the corresponding color formation makes possible to obtain nanoparticles with a desirable shape and size [10–12]. A fine control of several parameters (shape, size, surface functionalization or interparticle distance) is a challenging goal, and a large number of reports have been published with the aim to synthesize metal nanoparticles with a desired morphology, mainly for plasmonic or sensing applications [13–17].

Among all the synthetic methodologies, one of the most frequently used methods is based on the chemical reduction in metallic salts (mainly $AgNO_3$ or $HAuCl_4{\cdot}3H_2O$) by using a reducing agent, which is responsible for reducing the cationic metallic salts to produce the respective metal nanoparticles. In addition, an adequate protective agent is also used because it plays a dual role of preventing the agglomeration and maintaining the unaltered aggregation state of the metal nanoparticles. In this sense, Rivero et al. [18] presented a multicolor silver map with a long-term stability by means of a fine control of the molar ratio between both protective agent (polyacrylate, PAA) and reducing agent (dimethylaminoborane, DMAB), enabling a wide range of colors (yellow, orange, red, violet, blue, green, brown) and tunable shape and size. **Figure 2a** shows a multicolor silver map, which is formed by 56 different combinations of protective agent concentration and reducing/loading agent ratio. **Figure 2b** depicts the evolution of the LSPR absorption bands as a function of variable DMAB molar concentration.

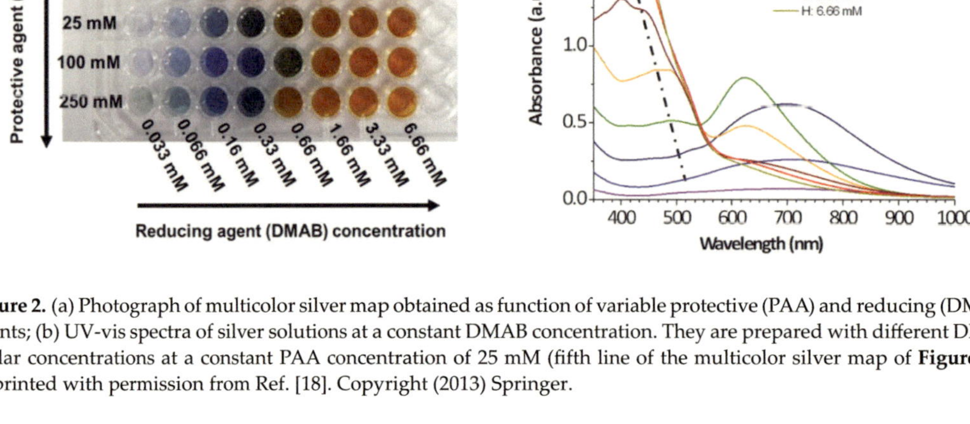

Figure 2. (a) Photograph of multicolor silver map obtained as function of variable protective (PAA) and reducing (DMAB) agents; (b) UV-vis spectra of silver solutions at a constant DMAB concentration. They are prepared with different DMAB molar concentrations at a constant PAA concentration of 25 mM (fifth line of the multicolor silver map of **Figure 1a**). Reprinted with permission from Ref. [18]. Copyright (2013) Springer.

3. Immobilization techniques of metal nanoparticles

First of all, a good immobilization of the metallic nanoparticles onto the optical fiber is necessary because this aspect is the most relevant key point for a further sensing application. The most used techniques for an adequate immobilization of the resultant metal nanoparticles are the layer-by-layer (LbL) assembly, sol-gel process or silanization process. All these processes offer a wide range of advantages, such as high versatility, easiness of implementation and simplicity. In addition, these techniques can easily incorporate metal nanoparticles in the resultant structure. However, it is necessary to remark that it can be found in the bibliography other alternative methods such as chemisorption, photodeposition or in situ synthesis of metal nanoparticles onto optical fiber, which can be also used for LSPR-sensing technology.

3.1. The sol-gel process

The sol-gel process is a synthetic route for preparing inorganic or hybrid inorganic-organic materials with a high purity through chemical reactions (known as hydrolysis, condensation and polycondensation) from a specific metal alkoxide [19–22]. All the chemical reactions follow the same evolution from a sol, a colloidal suspension of solid particles in a liquid, to produce a gel, a substance that contains a continuous solid skeleton enclosing a continuous liquid phase. A remarkable aspect is that the sol-gel process is a simple wet synthetic route, which can be easily used for loading other chemical agents. In this sense, metallic ions (precedent from metal inorganic salts) can be reduced in a later step to metal nanoparticles or well nanoparticles with a desired morphology can be perfectly introduced into the inorganic matrix during the polycondensation stage [23–25]. In addition, an important benefit is that sol-gel reactions do not employ extreme reaction conditions because the reactions take place at room or low temperature and require only moderate temperature to cure the gel.

Finally, once the sol-gel precursor is prepared and aged for a specific period of time, the substrate (i.e., optical fiber) is immersed into it using the dip-coating technique. After a fixed period of time, it is pulled out from the gel by using a determined extracting speed. Then, the substrate is dried during a fixed time and some additional processing steps can be included such as a heat treatment at controlled conditions of temperature and pressure. These steps can be repeated until the coating has acquired the desired thickness. In **Figure 3**, a summary of all the steps involved in the sol-gel technology (hydrolysis, condensation, polycondensation) (**Figure 3a**) and the critical parameters involved in the dip-coating technique (**Figure 3b**) are presented, respectively.

3.2. The layer-by-layer (LbL) assembly

Other deposition technique for immobilizing the metallic nanoparticles into thin films is the layer-by-layer (LbL) assembly, being one of the most used methods in the nanoscale level for its simplicity and versatility in comparison with other techniques [26]. In addition, other important advantage is the easiness for scaling-up with a precise control of the resultant

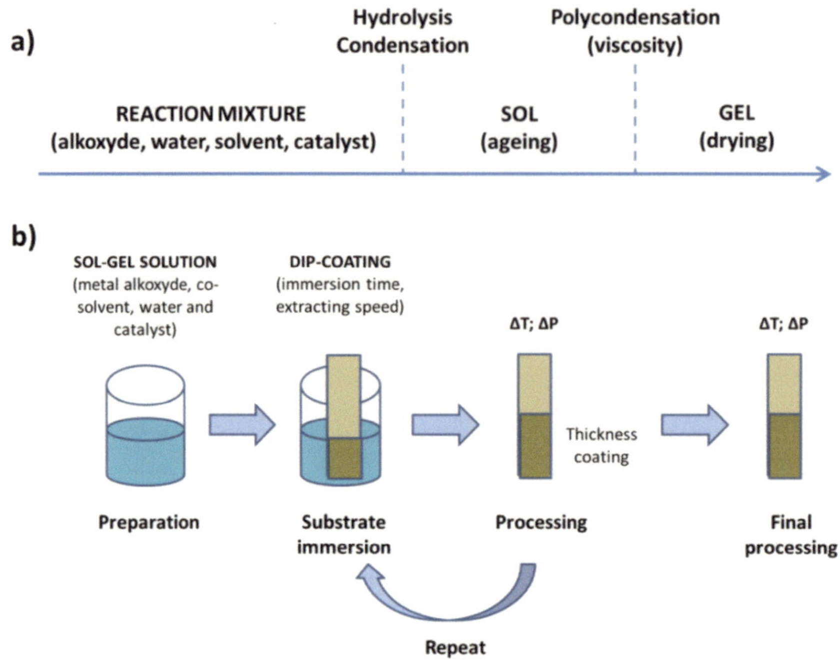

Figure 3. (a) Schematic representation of the different steps involved in the sol-gel process; (b) parameters to control in the dip-coating deposition technique.

thickness. The basis of this method is the electrostatic attraction between aqueous polyelectrolyte solutions of opposite charge, known as polycation (positive charge) and polyanion (negative charge), respectively. It is important to note that both polycations and polyanions overlap each other at the molecular level, and this produces a homogeneous optical material [27]. One of the main advantages of using polyelectrolytes is that their linear charge density can be varied over a wide range by simple adjustments of the pH or ionization degree of the dipping polyelectrolyte solutions [28, 29].

Other important advantage is that these polyelectrolytes can be used as protective agents of the metal nanoparticles. The encapsulating agent plays an important role for the synthesis of nanoparticles with a specific morphology because it limits the growth of the particles, directs their shape and provides colloidal stability [18]. Due to this, several works have been focused on the fabrication of thin films based on the incorporation of different shaped gold and silver nanoparticles using several types of polyelectrolytes [30–32]. A novel approach known as layer-by-layer embedding (LbL-E) deposition technique is based on a successive incorporation of metal nanoparticles with a well-defined morphology into LbL films, as it can be appreciated in **Figure 4**. More details about this deposition technique can be found in Refs. [31, 32].

Alternatively, other approach for incorporating metallic nanoparticles into LbL thin films is based on in situ chemical reduction from metal ions to metal nanoparticles. In order to perform this synthetic route, it is necessary to obtain a source of metal ions known as loading

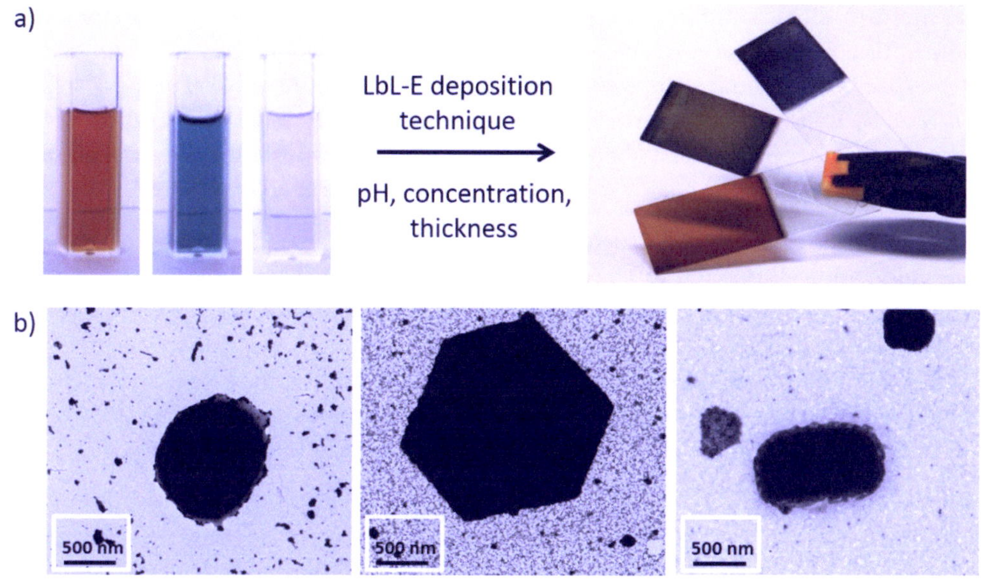

Figure 4. (a) Synthesis of silver nanoparticles with variable colors and their incorporation into polymeric thin films using LbL-E deposition technique; (b) TEM image of the AgNPs with a spherical shape (orange coloration), hexagonal shape (green coloration) and rod shape (violet coloration). Reprinted with permission from Refs. [32, 33]. Copyright (2013) Springer.

solution and a reducing solution, which is used for the chemical reduction of the metal ions. In this sense, once the metal ions have been successfully immobilized in the resultant LbL film, a further chemical reduction in the metal ions to zerovalent metal nanoparticles is performed when LbL films are immersed in a specific reducing solution. This approach based on in situ chemical reduction of the metal ions to nanoparticles into a previously host matrices used as a template can be called in situ synthesis (ISS) process. According to this, several articles can be found in the references using different loading/reducing agents as well as weak or strong polyelectrolytes in the resultant LbL films, respectively [33, 34].

3.3. Alternative methods for the immobilization of the metal nanoparticles

Other alternative methods for the immobilization of the nanoparticles consist of a previous pretreatment of the unclad surface of the optical fiber and a further silanization process to immobilize the metallic nanoparticles. For this purpose, firstly, the substrate is immersed in a piranha solution (H_2O_2:H_2SO_4; 30%:70%) for a specific period of time between 30 min and 1 h. This step is of great importance because the glass of the fiber is hydrolyzed, making possible the creation of additional SiOH sites in the outer surface of the fiber. However, this solution is extremely dangerous and must be manipulated with extreme precaution. Then, the optical fiber is rigorously rinsed in ultrapure water and placed in an oven at 100°C for 1 h in order to stabilize the new SiOH sites onto the surface of the optical fiber, being the key point for a further chemical binding. According to this, the optical fiber is immersed in a specific ORMOSIL (organically modified silicate) solution, being the most used aminosilane as well as mercaptosilane solution. These ORMOSILS are used to create a self-assembled layer onto the optical fiber surface, which is used to attach the metal nanoparticles with a desired shape by specific functional groups such mercapto (–SH) or amine (–NH_2) of the corresponding ORMOSIL. Once the nanoparticles have been successfully incubated and immobilized onto the fiber surface, the optical fibers are washed with ultrapure water to remove unbound nanoparticles.

An example of the optical fiber structure after immobilizing a specific type of nanoparticles such as gold nanorods (GNRs) can be appreciated in **Figure 5**. It can be appreciated as a thin silver layer is coated onto the end surface of fiber with the aim to make a reflective optical fiber sensor probe. Then, the silanization process is performed for a further attachment of the GNRs onto optical fiber core. Finally, the UV-vis spectrum of the GNRs shows two well-distinguished absorption bands related to the transversal plasmon resonance (530 nm) and longitudinal plasmon resonance (around 720 nm), respectively.

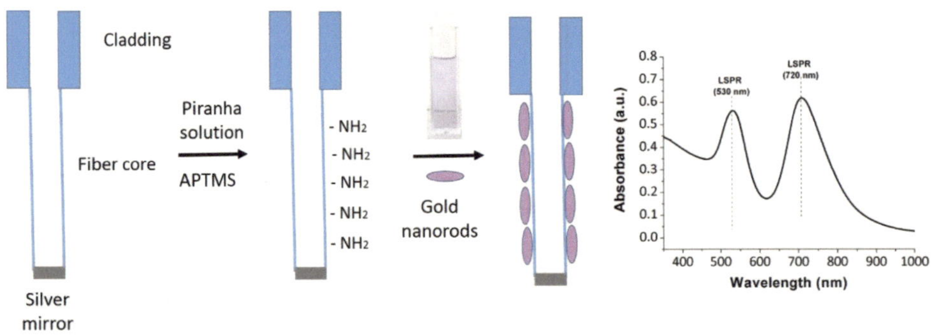

Figure 5. A schematic representation of the structure of LSPR-based optical fiber sensor after immobilization of GNRs using a silanization process.

4. Applications of the LSPR-based optical fiber sensors

In this section, once metal nanoparticles with a desired morphology have been successfully immobilized onto the optical fiber surface, potential applications of the plasmonic properties inherent to the nanoparticles are studied, specifically for sensing applications in the industry or biosensing applications in the field of clinical diagnosis or medicine. For all these purposes, the implementation of LSPR-based optical fiber sensors has attracted greater attention for all these potential applications. In the first subsection, optical fiber sensors for the detection of physical parameters such as relative humidity and refractive index are presented. In a second subsection, the design of optical fiber sensors is presented for the detection of wide variety of chemical parameters (i.e., pH, hydrogen peroxide, gases or heavy metals). Finally, optical fiber biosensors are presented for the detection of biochemical parameters (i.e., glucose, antigens, IgG) in the third subsection by using the LSPR technology.

4.1. Detection of physical parameters

4.1.1. Optical fiber refractometers

Refractometers are extensively used in fields as diverse as chemistry, biochemistry, food preservation, beverage industry or medicine to measure the surrounding medium refractive index (SMRI). In addition, they can be also combined with sensing coatings whose refractive index depends on the specific parameter under study [35, 36]. It is important to remark that different optical geometries have been studied with the aim to increase the sensitivity of the devices. However, optical fiber-based refractive index sensors are highly preferred for several applications where in situ and real-time monitoring is a key point. Due to this, recent works have demonstrated that the combination of nanodeposition techniques with optical fiber sensors is becoming a hot topic in the fabrication of thin-film-coated devices for measuring the refractive index.

As it was commented in Section 2, among all the deposition techniques at nanoscale level, one of the most used is the layer-by-layer assembly because it has the additional advantage of incorporating metal nanoparticles for plasmonic applications. In this sense, several articles related to both LbL assembly and metallic nanoparticles can be found in the references for measuring the refractive index [37–40]. According to this, Gouvêa et al. [37] have performed a dual growth kinetic study with the aim to evaluate the optimal deposition time as well as the optimal number of bilayers to fabricate a fiber optic reflection-based localized surface plasmon resonance sensor using LbL technique composed of PAMAM dendrimer/gold nanoparticles. Other interesting studies for the fabrication of optical fiber refractometers based on the successive incorporation of gold nanoparticles into LbL coatings are presented in Refs. [38, 39]. These works are based on the apparition of multiple absorption bands in the visible as a function of the thickness coating by using the LbL-E deposition technique, showing different sensitivity when the devices are exposed to variations of the refractive index from 1.33 to 1.42. In addition, as observed in Ref. [40], the utilization of both layer-by-layer polylectrolyte and metal nanoparticles within the polymeric composite films makes also possible to obtain a very high refractive index overlay with a significant sensitivity enhancement over ambient refractive index change, showing potential applications as biosensors. However, other alternative methods can be found in the references in order to immobilize the nanoparticles [41–52]. As an example, photodeposition technique has been used to immobilize AgNPs on the optical fiber end [41]. This LSPR-based optical fiber sensor has been exposed to different values of refractive index such as air (n = 1), deionized water (n = 1.33), ethanol (n = 1.36) and clove oil (n = 1.5). The experimental results indicate that there is change in the LSPR peak wavelength position when the refractive index is continuously increased (see **Figure 6**). In addition, a linear curve can be fitted according to the experimental data, showing a sensor sensitivity of 67.6 nm/RIU.

Other interesting work based on the incorporation of silver nanoparticles for the design of reflective LSPR-based optical fiber biosensors is presented in Ref. [42]. In this work, the synthesized AgNPs showed mostly a spherical shape with the peak located near 405 nm and with a particle diameter around 30 nm (see **Figure 7**). In addition, two parameters such as length of sensing area and coating time have been optimized with the aim to enhance the sensitivity of the LSPR optical sensor. The sensitivity of the sensor probe to reflective indices was 387 nm/RIU, and the total peak wavelength shift was approximately of 30 nm for a RI ranges from 1.33 to 1.40, as it can be observed in **Figure 8**.

Other characteristic example is presented in Ref. [43], where gold nanoparticles with a specific shape (nanoflower) have been successfully immobilized onto U-bent plastic optical fiber (POF) by means of a process known as chemisorption, which consists of a specific reaction

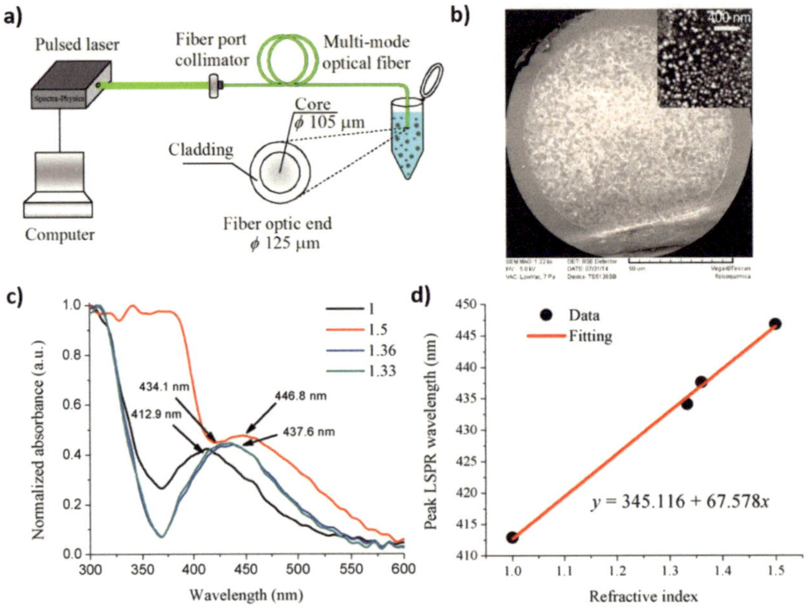

Figure 6. (a) Experimental setup for photodepositing silver nanoparticles on the optical fiber end; (b) image of the optical fiber end obtained with SEM; (c) sensor response localized in air, deionized water, ethanol and clove oil; (d) LSPR peak wavelength as a function of the refractive index. Reprinted with permission from Ref. [41].

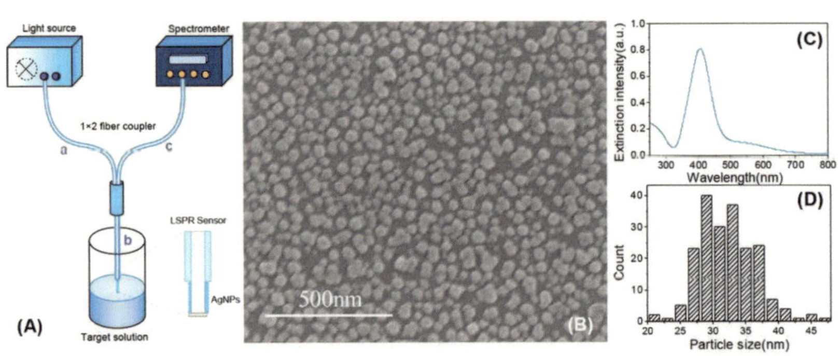

Figure 7. (A) Schematic of the experimental setup used in this study; (B) an SEM image of AgNPs after being immobilized on optical fiber; (C) absorption spectra of AgNPs solutions; (D) histogram showing the corresponding particle size distribution of AgNPs. Reprinted with permission [42].

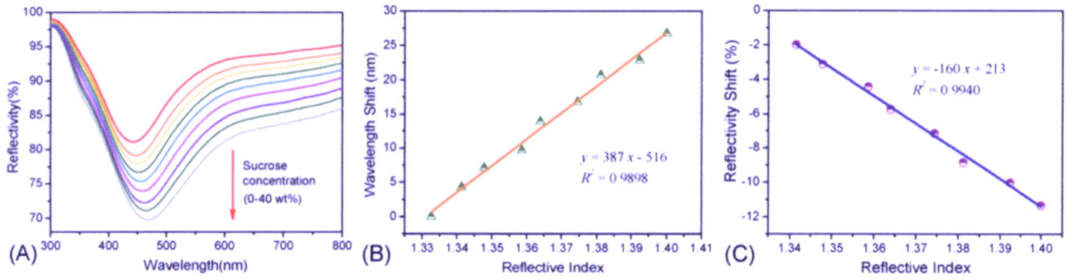

Figure 8. (A) The reflective spectra of the AgNP-based sensors of sucrose solutions with different concentrations; (B) changes in the LSPR wavelength shift to the refractive indices; (C) changes in the LSPR reflectivity shift to the refractive indices. Reprinted with permission [42].

between the amine groups on the fiber surface and negatively charged gold nanoparticles. The resulting LSPR peak wavelength position was located at 560 nm. An important consideration is that the sensitivity was found to be maximal when the bend diameter of the probe varies from 2 to 3 times the fiber diameter. The probes show sensitivity of 17.55 ($\Delta A_{560nm}/\Delta RIU$) in the visible region to refractive index changes from 1.33 to 1.47. Other approach based on a novel U-bent fiber optic probe is presented in Ref. [44]. In this work, U-bent probes with radii 0.5 and 0.75 mm have been coated with gold nanoparticles and tested with sucrose solution of variable RI with the aim to obtain information about the sensitivity and resolution. As a result, the best results were obtained for a bend radius of 0.75 mm with a sensitivity of 35 ($\Delta A_{540nm}/\Delta RIU$).

It is worth noting an interesting work of Cennamo et al. [45] using also a partially polished plastic optical fiber (POF). In this work, five-branched gold nanostars were synthesized with the peculiarity of showing three different localized surface plasmon resonances (known as LSPR1, LSPR2 and LSPR3). The strongest LSPRs (LSPR2 and LSPR3, respectively) fall in two well-separated spectral ranges. The first one (LSPR2) is located in the spectral range of 600–900 nm, whereas the second one (LSPR3) is located in the spectral range of 1100–1600 nm. However, due to the extremely strong attenuation of the employed POF in the infrared region, only the LSPR2 has been used for the RI measurements. The corresponding sensitivity is of 84 nm/RIU when the RI is ranging from 1.333 to 1.371, as it can be observed in **Figure 9**.

An interesting work is presented in Ref. [46], where a theoretical study is evaluated for the design of optical fiber sensors by using four different types of nanoparticles, such as ITO, Au, Ag and Cu. In this study, it has been assumed that all nanoparticles showed a spherical shape. The results have indicated that the sensitivity of LSPR-based fiber optic sensor increased with the increase in the thickness layer for all four materials. In addition, the simulation with ITO nanoparticles layer showed the better sensitivity in comparison with the other three materials.

A novel work for refractive index sensing is presented in Ref. [47]. In this work, a hetero-core structured fiber composed of a small section of single-mode fiber inserted in a multimode fiber is coated with gold nanoparticles using a simple, fast and low-cost method. The resultant sensor shows a sensitivity of 765 nm/RIU over a refractive index range of 1.333–1.365, and the points are fit to a linear plot with a square correlation coefficient value of 0.99, showing a

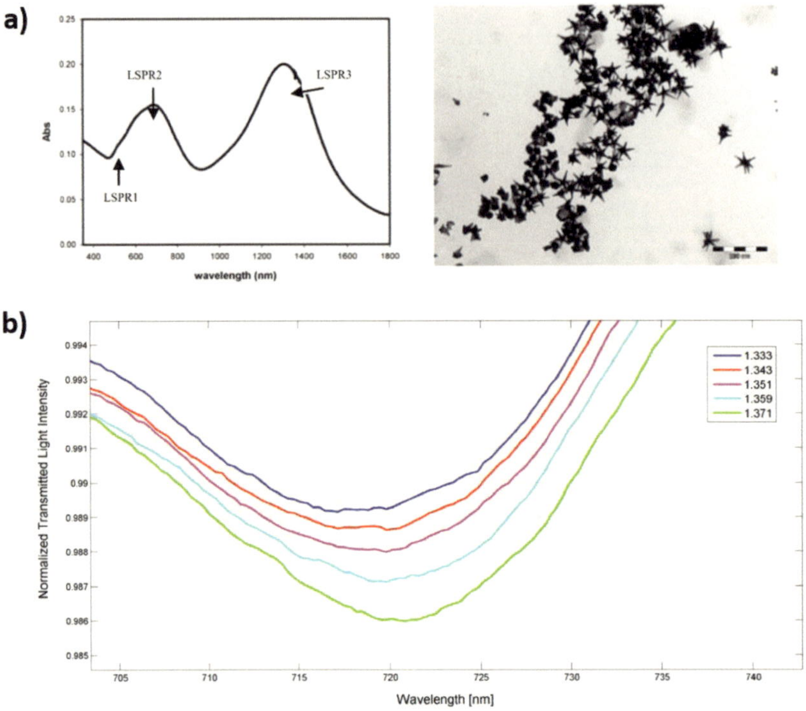

Figure 9. (a) UV-vis-NIR absorption spectrum of the five-branched gold nanostars with TEM image of the sample; (b) LSPR2 transmission spectra for different refractive indices of the aqueous solution medium. Reprinted with permission from Ref. [45].

resolution of 6.6×10^{-4}. It has to be mentioned that the sensor shows a high sensitivity, but it is necessary to obtain a better control over several parameters such as nanoparticle size, homogeneous distribution or even the cluster formation.

Interesting works with surprising results can be found in Refs. [48–52] using gold nanoparticles with different shapes. A study about the influence of the AuNPs size on the corresponding LSPR sensor sensitivity is evaluated in Ref. [48]. The main conclusion is that an increase in the sensitivity for the devices is obtained when the particle size is gradually increased. In this study, four different gold sizes (13, 20, 40 and 60 nm) have been investigated, and higher values in the resultant sensitivity have been observed (154 ± 14, 266 ± 27, 418 ± 44 and 571 ± 68 nm/RIU, respectively). A similar approach is presented in Ref. [49], although, in this work, the LSPR-based optical fiber sensitivity is evaluated using four types of gold nanorods (GNR) with different aspect ratios. The UV-vis spectra of the GNR solution reveal that by increasing the aspect ratio, a redshift of the longitudinal plasmon wavelength is appreciated. A first conclusion is that the transverse plasmon resonance (shorter wavelength location) only shows an increase in absorbance without wavelength shift when the refractive index is increased. However, the longitudinal plasmon resonance (longer wavelength location) shows both an increase in absorbance and wavelength redshift for higher values of refractive index. And a second conclusion is that by increasing the aspect ratio (2.6, 3.1, 3.7 and 4.3) of the GNR, an increase in the resulting sensitivity of the LSPR sensors has been also obtained from 269, 401, 506 and 766 nm/RIU, respectively. Similar results about the different sensitivities of both transverse and longitudinal plasmon resonances are found in Ref. [50], being the longitudinal plasmon resonance very sensitive to RI changes in comparison with transverse plasmon resonance. In this work, the aspect ratio of the GNRs synthesized is approximately 3.5 and there is a total wavelength shift of 70 nm when the RI is varied from 1.333 to 1.423, showing a sensitivity of 778 nm/RIU.

Other interesting work based on the immobilization of GNR with a specific aspect ratio of 3 shows a sensitivity of 468 nm/RIU when the sensor is exposed to RI ranges from 1.33 to 1.3749 [51]. In this work, a redshift of the longitudinal plasmon resonance from the lowest to the highest RI value is also observed, and when the RI value is decreased, it is reversed with the corresponding blueshift.

Finally, a novel LSPR-based optical fiber sensor with a considerable improvement in the sensitivity for RI measurements is presented in Ref. [52]. In this work, the use of hollow gold nanoparticles has attracted the attention due to the stronger plasmonic effect in comparison with the use of solid gold nanoparticles. A main conclusion is that by increasing the resultant degree of hollowness of the gold structures, the sensitivity has been increased from 784 to 1933 nm/RIU, respectively.

In order to have a better understanding of the different LSPR-based optical fiber sensors used as refractometer as well as the type and shape of the corresponding nanoparticles immobilized in the optical fiber with their corresponding sensitivity, a summary is shown in **Table 1**.

Type of nanoparticles	Refractive index range	Sensitivity
Silver nanoparticles	From 1 to 1.5	67.6 nm/RIU [41]
Silver nanoparticles (spherical shape)	From 1.33 to 1.40	387 nm/RIU [42]
Gold nanoparticles (nanoflower shape)	From 1.33 to 1.47	17.55 ($\Delta A_{560nm}/\Delta$RIU) [43]
Gold nanoparticles	From 1.33 to 1.40	35 ($\Delta A_{540nm}/\Delta$RIU) [44]
Gold nanoparticles (nanostar shape)	From 1.333 to 1.371	84 nm/RIU [45]
Gold nanoparticles	From 1.333 to 1.365	765 nm/RIU [47]
Gold nanoparticles	From 1.333 to 1.424	571 nm/RIU [48]
Gold nanorods	From 1.34 to 1.41	766 nm/RIU [49]
Gold nanorods	From 1.333 to 1.423	778 nm/RIU [50]
Gold nanorods	From 1.33 to 1.3749	468 nm/RIU [51]
Hollow gold nanoparticles	From 1.333 to 1.407	1933 nm/RIU [52]

Table 1. Summary of the different types of nanoparticles used for the fabrication of LSPR-based optical fiber sensor with their corresponding sensitivities to refractive index changes.

4.1.2. Optical fiber humidity sensors

Nowadays, the design of optimal relative humidity (RH)-sensing devices is a big concern among scientific sensor community. This kind of devices has a great interest in a wide range of different fields, such as food quality preservation, gas purification, medical facilities or air-conditioning regulation. Due to this high number of applications in general industry, the development of sensing devices for monitoring RH has been increasing in industrial processing and environmental control [53].

As previously commented, one of the most used techniques at nanoscale level in order to obtain self-ordered nanostructures with a precise thickness control is the layer-by-layer assembly. One advantage of using this deposition process is the ability to place different kinds of nanoparticles throughout the multilayer polymeric films. The incorporation of silver nanoparticles inside LbL polymeric coating contributes to enhance the lifetime of the devices in high RH environments. In such RH conditions, the biocide behavior of AgNPs prevents the adverse bactericidal apparition, which could damage the sensitive coating [54–56].

In this sense, several articles have been very recently published based on the combination of AgNPs-loaded polymeric thin films with optical fiber devices to monitor relative humidity changes [57–59]. In initial studies of this research group, the changes in the RH ranging from 20 to 70% have been detected by a shift of the corresponding LSPR wavelength position [57, 58]. This aspect can be clearly observed in **Figure 10** where an optical fiber sensor consisting of a thickness of 15 bilayers of PAH/PAA-AgNPs using the LbL-E deposition technique has been fabricated (**Figure 10b**), being possible to appreciate only a LSPR absorption band (around 440 nm) which it is attributed to the natural optical resonance (LSPR) of the AgNPs with mostly a spherical shape. In addition, a change in the intensity of the LSPR absorption band is observed as RH is varied with a linear response (inset **Figure 10b**). However, when the thickness coating is increased up to 25 bilayers, new absorption bands known as lossy-mode resonance (LMR) [60, 61] are appreciated in the visible region. Rivero et al. [59] were the first research group in reporting optical fiber sensors based on both localized surface plasmon resonance (LSPR) and lossy-mode resonance (LMR) for monitoring humidity changes. The experimental results indicate that the use of AgNPs makes possible the alteration of the refractive index of the polymeric overlay, allowing an improvement of the sensitivity to RH, and also improving the visibility of the LMR band with a smaller thickness in comparison with a polymeric overlay without metallic nanoparticles [62]. In addition, there is an important difference in sensitivity related to both absorption peaks for humidity changes because LSPR band only shows a variation of 3 nm, while LMR band shows a considerable shift of 50 nm when the device is tested between 20 and 70% RH (see **Figure 10c**). Due to this high difference in sensitivity of the LMR band, the response time of the device has been also evaluated by exposing to quick RH changes (**Figure 10d**).

A similar work for monitoring quick RH changes such as human breathing is also presented in Ref. [63]. However, this work shows two significant differences in comparison with Ref. [59]. Firstly, the fabrication process is different because it is based on in situ synthesis (ISS) process of AgNPs inside a polymeric coating instead of using the LbL-E deposition technique.

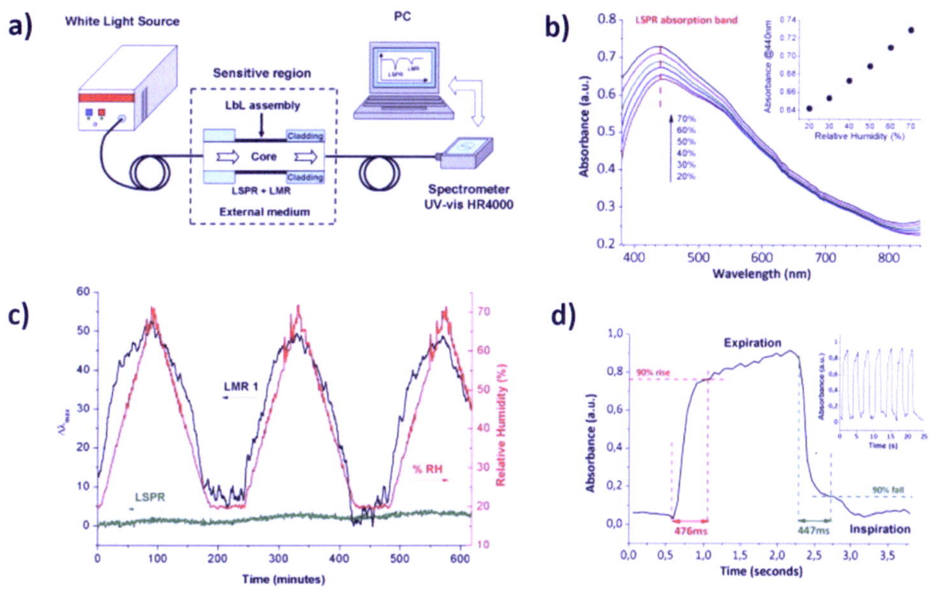

Figure 10. (a) Experimental setup used to obtain and characterize the absorption bands in the visible region; (b) UV-vis spectra when LbL-E of the silver nanoparticles is performed. The main absorption band is the LSPR corresponding to the AgNPs incorporated into the thin film. The curves plotted are 1, 3, 6, 9, 12 and 15 bilayers. (b) Spectral response of the device to RH ranges from 20 to 70% RH at a constant temperature of 25°C; (c) dynamic response of a 25 bilayers PAH/PAA-AgNPs device. The wavelength shift of both LSPR and LMR is monitored simultaneously to RH cycles from 20 to 70% RH at 25°C. (d) Response time of the device to several consecutive human breathing cycles (rise and fall). Reprinted with permission from Ref. [59] Elsevier (2012).

And secondly, the incorporation of AgNPs affects the refractive index of the overlay promoting the observation of a resonant attenuation band (LMR) in the infrared region (NIR). These results reveal that the combination of nanotechnology and biomedical science makes enable their use in practical RH monitoring applications or even it can lead to monitor high humidity changes such as human breathing due to the fast response time in the maximum sensitivity region.

4.2. Detection of chemical parameters

4.2.1. *Optical fiber pH sensors*

The polymeric structure presented in the LbL assembly (PAH/PAA) is not only sensitive to RH changes. In addition, if the specific nanocoating is dipped into liquid solutions, its thickness can be tuned by the external medium pH. This effect has been named "swelling/deswelling phenomenon" [64, 65]. As a consequence of this variation, the effective refractive index of the structure will change, producing a shift of the attenuation bands. In this sense, optical fiber resonance-based pH sensors using gold nanoparticles into polymeric layer-by-layer coatings have reported [66, 67]. These works are based on the design and fabrication of dual LSPR-LMR optical fiber pH sensors, using the LSPR band as a reference signal, and the LMR band is used as a sensing signal due to the great difference in their sensitivities to pH of the surrounding medium. In addition, although previous works have been demonstrated the possibility of combining both phenomena in the same device by using AgNPs [59], the use of metallic gold nanoparticles (AuNPs) presents several additional advantages such as a great biocompatibility, nonreactivity, chemical stability and easiness to a further functionalization. In **Figure 11**, it can be clearly observed that the AuNPs have been successfully incorporated into the coatings by using LbL-E deposition technique because when the number of bilayers is increased, the resultant color of the films is changed from transparent to purple (inset **Figure 11a**). Once the nanocoating based on AuNPs is performed onto the optical fiber core by using LbL-E deposition technique, and the two optical phenomena (LSPR and LMR, respectively) are appreciated to a desired thickness coating, the spectral response of the device is measured at different pH values ranging from 4.0 to 6.0 (see **Figure 11b**) where it can be clearly observed the great difference in sensitivity of both absorption bands to these pH changes.

Other interesting approach based on the swelling/deswelling phenomenon related to a polymeric matrix is presented by Muri and Hjelme [68] for pH-sensing applications. In this work, a pH optical fiber sensor is fabricated based on LSPR sensing in a poly(acrylamide-co-acrylic acid) hydrogel embedding gold nanoparticles (GNP). As an interesting result, the LSPR absorption band is mainly shifted by two different processes. The first one is associated with an increase in the gold nanoparticles density in hydrogel when the pH solution is decreased, and as a result, the neighboring distances between GNP are decreased that can induce electromagnetic interactions between the localized modes, giving a LSPR redshift. The second one is associated with the density of the polymer matrix of the resultant hydrogel that is also increased when the pH solution is gradually decreased. In this case, a change in the refractive index around the AuNPs makes possible a shift of the LSPR absorption band.

4.2.2. *Optical fiber sensors for detection of heavy metals*

The detection and quantification of heavy metals is a hot topic in a large number of industrial, environmental and medical applications. In addition, it has been demonstrated that elevated

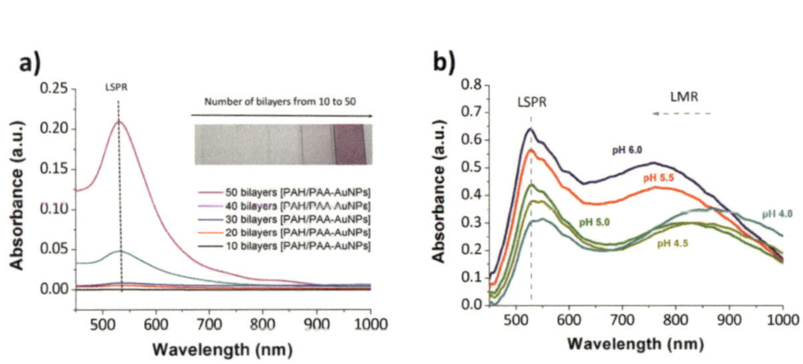

Figure 11. (a) UV-vis spectra of the nanostructured coating obtained by layer-by-layer embedding (LbL-E) deposition technique for different number of bilayers and a photograph of the resultant nanocoatings; (b) maximum wavelength variation of both LSPR and LMR absorption bands. The wavelength shift is monitored simultaneously at different pH values such as pH 4.0, pH 4.5, pH 5.0, pH 5.5 and pH 6.0, respectively. Reprinted with permission from Ref. [66]. Copyright (2016) Springer.

levels of heavy metals (lead, cadmium, mercury) in the human body can cause neurological disorders, such as birth defects, mental retardation or even death. Due to this, a great and precise control over these heavy metals is necessary for human safety o for environmental concern. For all these reasons, the development of optical fiber sensors for the detection of heavy metals can be a great alternative due to the possibility of real-time monitoring with a high degree of precision and a fast response.

An interesting study based on a fiber optic sensor for the detection of atmospheric elemental mercury can be found in Ref. [69] by using gold nanorods. In this work, the mercury readily is adsorbed on the gold nanorods and as a result, a shift related to the longitudinal plasmon resonance is appreciated. In addition, the experimental results indicate that a linear shift is obtained, which is directly proportional to the mercury concentration. The resultant optical fiber sensor can directly measure concentrations down to 1.0 $\mu g/m^3$.

Other promising works can be also found in the references for the detection of other different heavy metals by using LSPR-based fiber optical technology. As a characteristic example, Lin and Chung [70] have proposed a novel fiber optic biosensor to determine the heavy metal cadmium ion concentration. For this purpose, firstly, the preparation of gold nanoparticles and a further immobilization onto optical fiber is performed. And secondly, specific chelating agents known as phytochelatines (PCs) are immobilized onto gold nanoparticle-modified optical fiber. These agents play an important role because they are capable of chelating cadmium ions by thiolate coordination for a further sensing application. The experimental data indicate that the optical fiber sensor shows a maximal 9% change of absorbability for detecting 1–8 ppb of cadmium ions concentration. The resultant sensitivity is of 1.24 ppb with a limit of detection (LOD) of 0.16 ppb. In addition, an important consideration is that after successive cycles of deactivation and reactivation up to nine, the biosensor presents a good recovery rate with 85% original activity. In addition, one of the main characteristics of the sensor is that it can be recalibrated after storage in d-(+)-trehalose dehydrate solution. This same solution has been also used to recalibrate a biosensor for detecting lead ions [71]. However, the main difference with the previous work [70] is that a monoclonal antibody has been immobilized onto gold nanoparticle-modified optical fiber. A schematic representation of the functionalization of the optical fiber with the monoclonal antibody is presented in **Figure 12**.

As seen in this scheme, firstly, gold nanoparticles are immobilized onto optical fiber by using a silanization process with 3-(mercaptopropyl)-trimethoxysilane (MPTMS) in toluene. Then, by using a self-assembling technique, a bioactive layer composed of monoclonal antibody is immobilized by covalent coupling onto gold nanoparticle surface. For this purpose, a self-assembled layer of cystamine is initially performed to form amine functional groups onto gold nanoparticle surface. Then, the amine groups of cystamine can couple with the activated succimide esters (obtained from reaction between antibody and EDC/NHS), and as a result, the resultant antibodies are perfectly immobilized onto optical fiber surface. In addition, prior to detection process, the lead ions are treated with excess of chelators (i.e., EDTA) to form Pb-chelate complexes, Pb(II)-EDTA. Then, the monoclonal antibody can exclusively bind with this complex, causing a change in light attenuation. The experimental results indicate that the sensor shows a maximal 12% change of absorbability for detecting 10–100 ppb of

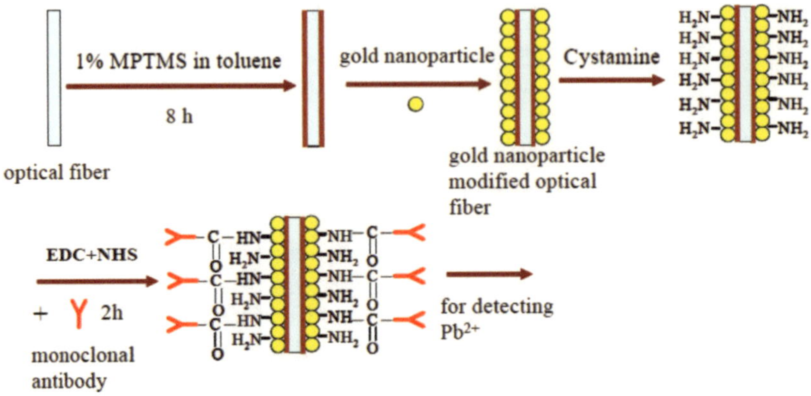

Figure 12. The chemical reactions scheme carried out for covalent binding of monoclonal antibody to the fiber-based LSPR sensor. Reprinted with permission from Ref. [71].

Pb(II)-EDTA complex concentration with a limit of detection (LOD) of 0.27 ppb. In addition, a very promising result is that the monoclonal antibody has a higher affinity to Pb(II)-EDTA in comparison with other chelate complexes such as Cu(II)-EDTA, Ni(II)-EDTA or Mg(II)-EDTA. The absorbability of these three chelate complexes is smaller (only 3% change) in comparison with Pb(II)-EDTA complex (12.2%). These experimental results can be appreciated in **Figure 13**.

4.2.3. Optical fiber sensors for detection of other chemical compounds

In this subsection, the use of LSPR-based optical fiber sensors is analyzed for the detection of different types of analytes of interest in fields as diverse as medicine, engineering, storage and food/drink industry. One of these analytes that is causing a great interest in the medical and clinical science is the hydrogen peroxide because it is a well-known reactive oxygen species (ROS) [72–74]. An excessive accumulation of this analyte in the body can trigger

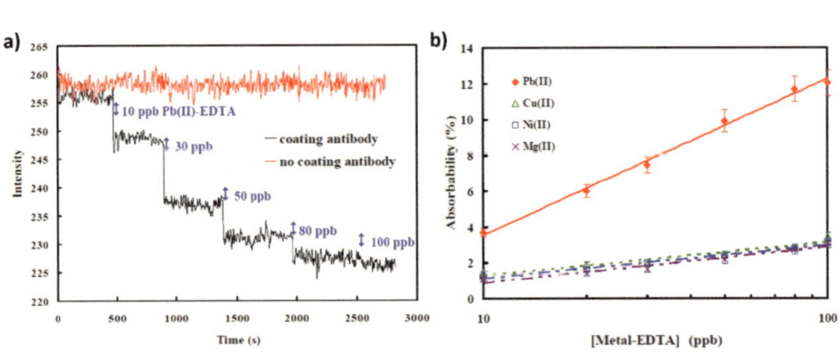

Figure 13. (a) An example of serial Pb(II)-EDTA complex response signal in the range of 10–100 ppb by two sensors with and without monoclonal antibody coating; (b) comparison of the responses between different metal-chelate complexes through the monoclonal antibody-functionalized LSPR sensor. Reprinted with permission from Ref. [71].

some diseases because the hydrogen peroxide is responsible for causing tissue damage and DNA fragmentation. There is a wide variety of techniques for its detection (spectrometry, fluorescence), although the use LSPR optical fiber sensors has the additional advantages of cost-effective, fast response and high sensitivity. An interesting approach is presented in Ref. [75] where a fiber optic sensor using the combined phenomena of both surface plasmon resonance (SPR) and localized surface plasmon resonance (LSPR) is fabricated for the detection of hydrogen peroxide. In this work, firstly, a silver layer is deposited onto an optical fiber core, and secondly, silver nanoparticles, which are embedded in a polymeric matrix of polyvinyl alcohol (PVA), are deposited over the silver-coated core of the optical fiber. The experimental results indicate that there is a redshift in the resonance wavelength when the hydrogen peroxide concentration is increased from 10^{-8} to 10^{-1} M. The resultant shift due to this combination (SPR + LSPR) is 29.401 nm, being this value higher in comparison with only SPR configuration (8.503 nm). Other approach based on this combined phenomena of SPR and LSPR can also be found in the references for the detection ascorbic acid [76].

Other interesting works based on the immobilization of metallic nanoparticles in a specific inorganic matrix are presented by using the sol-gel technology for sensing applications. As an example, sol-gel synthesis of palladium-doped silica nanocomposite (30 nm in diameter) has been successfully synthesized and incorporated onto fiber for hydrogen gas sensing [77]. The results indicate that the sensor shows response to the hydrogen gas (1% H_2 mixed with N_2 gas) and this response is reversible. Other approach based on the sol-gel technology for the detection of a harmful and corrosive analyte such as hydrofluoric acid (HF) is presented in Ref. [78]. For this purpose, a novel SiO_2 sol-gel-coating LSPR reflective sensor has been fabricated using gold nanoparticle-modified optical fibers. The results indicate that the sensors can detect HF solutions in the range from 1 to 5%. However, higher concentrations of HF are not possible to detect due to a quick erosion of the sensing area.

An interesting work for gasoline detection is presented in Ref. [79]. Gasoline is a fuel with a high volatile nature and a high degree of inflammability, which is necessary an adequate control over its storage. In this work, a comparative study about the difference in sensitivity of two different LSPR-based optical fiber sensors is presented for the detection of gasoline vapor concentration. The main difference in both sensors is that the first one is composed of only silver nanoparticles, whereas the second one is composed of only gold nanoparticles and both of them have been deposited on the decladed U-bent portion of a multimode fiber. The interaction between the sensing area and gasoline vapors at different concentrations has been monitored in a closed chamber. Finally, the experimental data indicate that the LSPR-based optical fiber based on AgNPs shows a better sensitivity than LSPR-based optical fiber based on AuNPs.

Finally, a novel work about the detection of sucrose content in fruit juices packaging industry is presented in Ref. [80] using a silicon nitride-coated LSPR-based fiber optic probe. This silicon nitride coating plays an important role because it is used to provide a protective layer on the device with the corresponding improvement in long-term stability for continuous operations. The results have demonstrated that there is 1.24 times RI sensitivity enhancement in case of metal nanoparticles (AuNPs) covered with silicon nitride film in comparison with only AuNPs-coated fiber optic probe.

4.3. Detection of biological parameters

In this section, optical fiber LSPR biosensors to perform qualitative and quantitative in real-time biomolecular sensing are presented. Several works are focused on the development of label-free biosensors because it is possible to monitor, with a high sensitivity and precision, the interaction between antibodies and their antigens in a wide number of sensing applications. As characteristic examples, biosensors for dengue diagnosis or for the detection of blood glucose are commented. In addition, the detection of the biotin-streptavidin bioconjugate pair or the IgG-anti-IgG bioreceptor-analyte pair is also analyzed for promising methods in applications as portable immunosensors.

One of the most known tropical diseases is the dengue, which can cause severe human health problems, and due to this, an early detection with an adequate therapy is one of the key factors for the human survival. For this purpose, the design and development of fiber optic sensors with AuNPs for dengue immunoassay can be found in Refs. [81, 82]. Different steps have been performed in order to immobilize a specific antibody (dengue anti-NS1) onto AuNPs deposited on the end surface of fiber [82]. Once a good immobilization of the sensing element onto AuNPs is obtained, a specular reflection LSPR optical fiber is performed for the detection of dengue NS1 antigen. The results indicate that the sensor is able to detect dengue NS1 antigen at different concentrations with a limit of detection of 1.54 nM.

Other approaches based on the same sensing mechanism by immobilization of metallic gold nanoparticles at the distal end of a multimode fiber can be found in Refs. [83–85]. As a characteristic example, the surface of gold nanorods has been functionalized with human IgG in order to create a biosensor to detect anti-human IgG [83]. The longitudinal peak has been used as sensing signal because a dependent shift was observed as a function of anti-human IgG concentration, being the limit of detection 3 nM. An improvement of the limit of detection for this same anti-human IgG is presented in Ref. [84]. In this work, a comparative study about the immobilization of two different types of gold nanoparticles such as nanospheres and nanorods onto an unclad surface of an optical fiber sensor and a further implementation as biosensors is evaluated. The results indicate that both biosensors present the same detection limit of 1.6 nM for the detection of anti-human IgG.

An interesting work for the detection of the biotin-streptavidin bioconjugate pair is presented in Ref. [85] using a reflective LSPR optical fiber. In this work, the layer-by-layer assembly (also known as ionic self-assembled multilayers, ISAM) has been used for obtaining a nanometric thickness composite, which is composed of gold nanoparticles and polyelectrolytes (see **Figure 14**). Finally, the experimental results have shown a wavelength shift as a function of the streptavidin concentration (see **Figure 15**), being the corresponding sensitivity of 800 pg/mm^2.

Other work based on the immobilization of silver nanoparticles instead of gold nanoparticles for the design of a reflective LSPR optical fiber biosensor is presented in Ref. [42]. An interesting result is that the resulting sensor shows a very high stability because a small fluctuation in resonance wavelength has been detected in a period of time of 18 days when the sensor was immersed in two different solutions such as ethanol and water (**Figure 16a**). In addition, the interaction between human IgG and rabbit anti-human IgG has been used as an analytical model in order to evaluate the antigen-antibody interaction. The results indicate that a gradual shift of LSPR peak has been observed with an increase in the functionalized time (**Figure 16c**).

Other novel works based on the immobilization of bioreceptors during analyte binding onto the surface of gold nanoparticles can be also found in Refs. [44, 86–88]. These works are based on the absorbance or wavelength changes related to the gold nanoparticles using IgG-anti-IgG

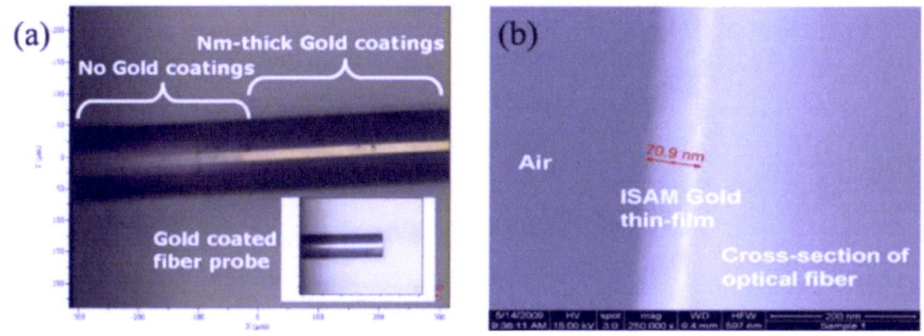

Figure 14. (a) photograph of gold metallic color in the portion of the fiber coated by ISAM; (b) cross section of the optical fiber in order to appreciate the nanometric thickness of the coating. Reprinted with permission from Ref. [85].

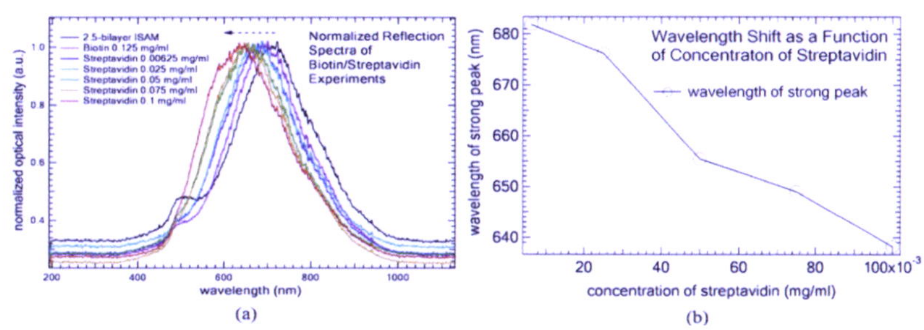

Figure 15. (a) Normalized reflection spectra measured in air as a function of the concentration of streptavidin solution using the same fiber optic probe; (b) wavelength shift as a function of concentration of streptavidin. Reprinted with permission from Ref. [85].

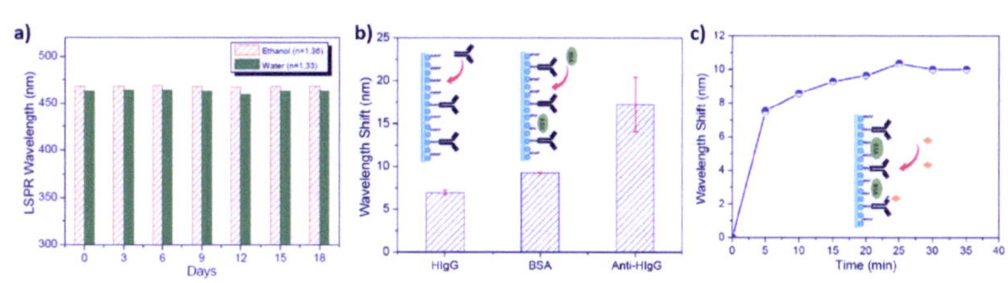

Figure 16. (a) LSPR wavelength monitoring of the AgNP-based sensor subjected in solutions with different refractive indices for a period of 18 days; (b) shifts of the LSPR spectra after different stages of surface modification relative to unmodified AgNP sensor probe; (c) LSPR peak wavelength changes during the process of anti-human IgG immobilization over a period of 35 min. Reprinted with permission from Ref. [42].

(bioreceptor-analyte pair) onto different optical fiber configurations. An approach is based on the design of novel U-bent optical fibers, which are able to detect a concentration of analyte of 0.8 nM (anti IgG) [44]. Other interesting work is presented in Ref. [86] where the optical fiber has been also bent in the form of a U-shaped probe for the detection of blood glucose. In addition, an interesting aspect is that the sensitivity of proposed biosensor has been evaluated as a function of the bending radius of the optical fiber.

Using this same configuration of U-bend fiber optic is presented in Ref. [87], although the resultant immobilization of the gold nanoparticles and IgG is performed in a multilayer structure using the layer-by-layer assembly. In this work, an exhaustive study about the plasmon penetration depth as a function of the resulting size of the AuNPs (18, 36 and 45 nm, respectively) is presented for the design of LSPR-based fiber optic biosensors. Other interesting work can be found in Ref. [88] where a tapered optical fiber sensor has been fabricated for monitoring anti-DNP antibody with a LOD of 4.8 pM.

Other optical fiber biosensor based on the layer-by-layer technique is also found in Ref. [89]. In this work, a gold nanoparticle-assembled film is used as a sensing layer, which has been built onto a trilayer polyelectrolyte structure-modified sidewall of an unclad optical fiber. The three consecutive polyelectrolytes used to form the trilayer structure are, firstly, a strong cationic polyelectrolyte (PDDA), secondly, a strong anionic polyelectrolyte (PSS) and, thirdly, a weak polyelectrolyte (PAH). After that, citrate-stabilized AuNPs have used to immobilize onto the outmost PAH layer. This multilayer structure is clearly appreciated in **Figure 17**. An important consideration of this sensing layer method is that a very time-saving compared is obtained in comparison with the assembly of AuNPs through silane-coupling process (silanization).

Once the gold nanoparticles have been successfully immobilized onto optical fiber surface, the LSPR optical fiber is modified by rabbit IgG in order to detect goat anti-rabbit IgG. In **Figure 18**, the dynamic process of goat anti-rabbit IgG adsorption is appreciated. In addition, first of all, it is necessary to remark that both rabbit IgG and goat anti-rabbit IgG cause a remarkable increase in the peak intensity. In addition, the experimental results also indicate that there is a gradual increase in the peak intensity when the goat anti-rabbit IgG concentration is increased. The lowest concentration that it can be detected for this biosensor is 11.1 ng/mL. Finally, this biosensor fabricated for this method shows potential application as a portable immunosensor.

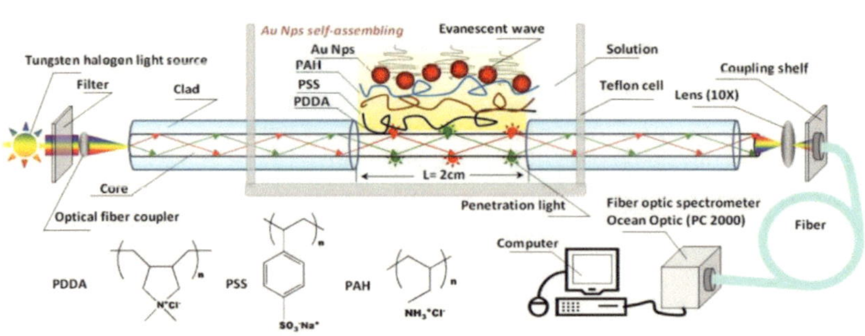

Figure 17. The experimental setup of the optical fiber LSPR sensor. Reprinted with permission from Ref. [89].

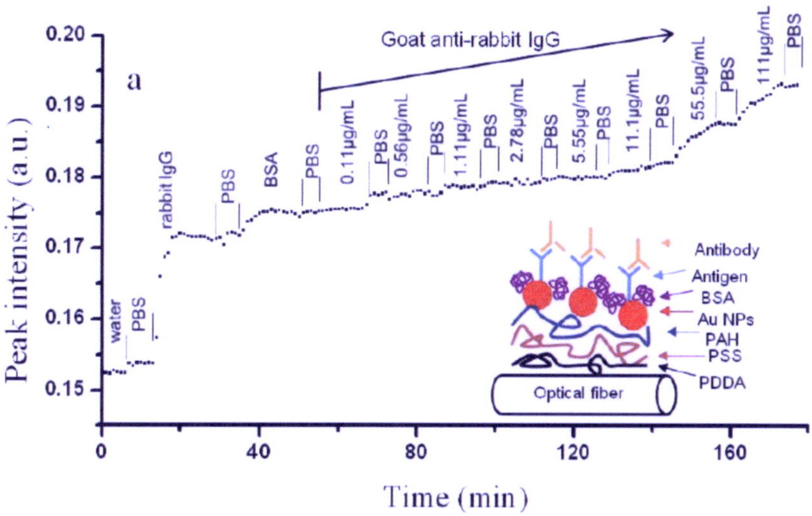

Figure 18. The LSPR biosensing of different concentrations of goat anti-rabbit IgG (0.11–100 μg/mL). Reprinted with permission from Ref. [89].

In a novel work [90], the detection limits of protein optical fiber biosensors coated with gold nanoparticles have been considerably improved. In this sense, a comparative study has been performed between a reference sensor without AuNPs and sensors composed of gold nanocages (AuNC) and nanospheres (AuNS), respectively. All the nanoparticles have been immobilized onto the fiber and the resulting sensor operated in a reflection mode. In this work, the sensor composed of AuNC showed a lower LOD and sensitivity than the sensor composed of AuNS. In addition, the LOD related to sensors composed of gold nanoparticles was three orders of magnitude greater than the reference sensor, showing the enhancement of the device due to the use of metal nanoparticles.

Finally, other promising biosensor of great and relevant interest is presented in Ref. [91]. In this work, a localized surface plasmon resonance (LSPR)-coupled fiber optic (FO) nanoprobe based on a gold nanodisk array at the fiber end facet is reported. This biosensor is capable of detecting a cancer protein biomarker, known as free prostate-specific antigen (f-PSA). The experimental data have demonstrated that the lowest limit of detection (LOD) was at 100 fg/mL (around 3 fM), showing an excellent specificity and selectivity in the resulting detection.

5. Conclusions

In this chapter, a review of the current state of the art of the LSPR-based optical fiber sensors is presented. LSPR has demonstrated to be a powerful tool for the development of high sensitivity optical sensors, and its combination with optical fibers has provided advanced sensors in many applications.

A large amount of research works reported the study about the synthesis and optical-sensing properties of immobilized metallic nanoparticles (typically gold or silver) as well as the influence of their composition, size, shape (spherical, nanorods, nanostars, nanoflowers, nanocages, etc.) or even solid or hollow core structure. Some authors have developed sensors based on thin-film-embedded metallic nanoparticles by means of layer-by-layer (LbL) and sol-gel techniques, whereas other authors have reported direct covalent binding of the nanoparticles to the optical fibers, using mostly silane-coupling processes.

Moreover, different optical configurations of the optical fiber sensors have been observed such as U-bent, transmission, reflection and optical fiber gratings. In each configuration, the metal nanoparticles have been immobilized over the optical fiber-sensitive region to detect the desired physical, chemical or biological parameters.

In summary, optical fibers provide a very powerful platform for optical sensing making possible the development of easy and robust optical measurements, remote sensing and light-weight devices. The use of LSPR in such optical fiber sensors enables to get fast and reliable measurements and very high-sensitivity devices with outstanding limit of detection (LOD) in the concentration range from nM to fM.

Acknowledgements

This work was supported by the Spanish Ministry of Economy and Competitiveness through TEC2013-43679-R Research Project.

Author details

Pedro J. Rivero[1*], Javier Goicoechea[2] and Francisco J. Arregui[2]

*Address all correspondence to: pedrojose.rivero@unavarra.es

1 Materials Engineering Laboratory, Department of Mechanical, Energetic and Materials Engineering, Public University of Navarre, Pamplona, Spain

2 Nanostructured Optical Devices Laboratory, Department of Electrical and Electronic Engineering, Public University of Navarre, Pamplona, Spain

References

[1] Federici, J. F.; Schulkin, B.; Huang, F.; Gary, D.; Barat, R.; Oliveira, F.; Zimdars, D. THz imaging and sensing for security applications—explosives, weapons and drugs. Semicond. Sci. Technol. 2005, 20, S266.

[2] Homola, J. Surface plasmon resonance sensors for detection of chemical and biological species. Chem. Rev. 2008, 108, 462-493.

[3] Fan, X.; White, I. M.; Shopova, S. I.; Zhu, H.; Suter, J. D.; Sun, Y. Sensitive optical biosensors for unlabeled targets: a review. Anal. Chim. Acta 2008, 620, 8-26.

[4] Mauter, M. S.; Elimelech, M. Environmental applications of carbon-based nanomaterials. Environ. Sci. Technol. 2008, 42, 5843-5859.

[5] Arregui, F. J.; Matias, I. R.; Goicoechea, J.; Villar, I. D. Optical fiber sensors based on nanostructured coatings. In Sensors Based on Nanostructured Materials; Ed. Springer. 2009; pp 275-301.

[6] Noguez, C. Surface plasmons on metal nanoparticles: the influence of shape and physical environment. J. Phys. Chem. C 2007, 111, 3606-3619.

[7] Anker, J. N.; Hall, W. P.; Lyandres, O.; Shah, N. C.; Zhao, J.; Van Duyne, R. P. biosensing with plasmonic nanosensors. Nat. Mater. 2008, 7, 442-453.

[8] Jeon, S.; Xu, P., Zhang, B.; MacK, N.H.; Tsai, H.; Chiang, L.Y.; Wang, H. Polymer-assisted preparation of metal nanoparticles with controlled size and morphology. J. Mater. Chem. 2011, 21, 2550-2554.

[9] Liz-Marzán L. M. Tailoring surface plasmons through the morphology and assembly of metal nanoparticles. Langmuir 2006, 22, 32-41.

[10] Zhao, J.; Zhang, X.; Yonzon, C. R.; Haes, A. J.; Van Duyne, R. P. Localized surface plasmon resonance biosensors. Nanomed 2006, 1, 219-228.

[11] Otte, M. A.; Sepúlveda, B.; Ni, W.; Juste, J. P.; Liz-Marzán, L. M.; Lechuga, L. M. Identification of the optimal spectral region for plasmonic and nanoplasmonic sensing. ACS Nano 2010, 4, 349-357.

[12] Liz-Marzán, L. M. Nanometals: formation and color. Mater. Today 2004, 7, 26-31.

[13] Angelomé, P. C.; Mezerji, H. H.; Goris, B.; Pastoriza-Santos, I.; Pérez-Juste, J.; Bals, S.; Liz-Marzán, L. M. Seedless synthesis of single crystalline Au nanoparticles with unusual shapes and tunable LSPR in the near-IR. Chem. Mater. 2012, 24, 1393-1399.

[14] Henry, A. I.; Bingham, J. M.; Ringe, E.; Marks, L. D.; Schatz, G. C.; Van Duyne, R. P. Correlated structure and optical property studies of plasmonic nanoparticles. J. Phys. Chem. C 2011, 115, 9291-9305.

[15] Mayer, K. M.; Hafner, J. H. Localized surface plasmon resonance sensors. Chem. Rev. 2011, 111, 3828-3857.

[16] Cobley, C. M.; Skrabalak, S. E.; Campbell, D. J.; Xia, Y. Shape-controlled synthesis of silver nanoparticles for plasmonic and sensing applications. Plasmonics 2009, 4, 171-179.

[17] Sepúlveda, B.; Angelomé, P. C.; Lechuga, L. M.; Liz-Marzán, L. M. LSPR-based nanobiosensors. Nano Today 2009, 4, 244-251.

[18] Rivero, P. J.; Goicoechea, J.; Urrutia, A.; Arregui, F. J. Effect of both protective and reducing agents in the synthesis of multicolor silver nanoparticles. Nanoscale Res. Lett. 2013, 8, 1-9.

[19] Hench, L. L.; West, J. K. The sol-gel process. Chem. Rev. 1990, 90, 33-72.

[20] Brinker, C. J.; Scherer, G. W. Sol → Gel → Glass: I. Gelation and gel structure. J. Non Cryst. Solids 1985, 70, 301-322.

[21] Brinker, C. J.; Scherer, G. W.; Roth, E. P. Sol → Gel → Glass: II. Physical and structural evolution during constant heating rate experiments. J. Non Cryst. Solids 1985, 72, 345-368.

[22] Scherer, G. W.; Brinker, C. J.; Roth, E. P. Sol → Gel → Glass: III. Viscous sintering. J. Non-Cryst. Solids 1985, 72, 369-389.

[23] Schmidt, H.; Scholze, H.; Kaiser, A. Principles of hydrolysis and condensation reaction of alkoxysilanes. J. Non Cryst. Solids 1984, 63, 1-11.

[24] Brinker, C. J. Hydrolysis and condensation of silicates: effects on structure. J. Non Cryst. Solids 1988, 100, 31-50.

[25] Sanchez, C.; Livage, J.; Henry, M.; Babonneau, F. Chemical modification of alkoxide precursors. J. Non Cryst. Solids 1988, 100, 65-76.

[26] Decher, G. Fuzzy nanoassemblies: toward layered polymeric multicomposites. Science 1997, 277, 1232-1237.

[27] Choi, J.; Rubner, M. F. Influence of the degree of ionization on weak polyelectrolyte multilayer assembly. Macromolecules 2005, 38, 116-124.

[28] Hammond, P. T. Form and function in multilayer assembly: new applications at the nanoscale. Adv. Mater. 2004, 16, 1271-1293.

[29] Shiratori, S. S.; Rubner, M. F. PH-dependent thickness behavior of sequentially adsorbed layers of weak polyelectrolytes. Macromolecules 2000, 33, 4213-4219.

[30] Budy, S. M.; Hamilton, D. J.; Cai, Y.; Knowles, M. K.; Reed, S. M. Polymer mediated layer-by-layer assembly of different shaped gold nanoparticles. J. Colloid. Interface Sci. 2017, 48, 336-347.

[31] Rivero, P. J.; Goicoechea, J.; Urrutia, A.; Matia, I. R.; Arregui, F. J. Multicolor Layer-by-Layer films using weak polyelectrolyte assisted synthesis of silver nanoparticles. Nanoscale Res. Lett. 2013, 8, 1-10.

[32] Rivero, P. J.; Goicoechea, J.; Matias, I. R.; Arregui, F. J. A comparative study of two different approaches for the incorporation of silver nanoparticles into layer-by-layer films. Nanoscale Res. Lett. 2014, 9, 1-11.

[33] Veletanlic, E.; Cynthia Goh, M. Polyelectrolyte multilayer films as templates for the in situ photochemical synthesis of silver nanoparticles. J. Phys. Chem. C 2009, 113, 18020-18026.

[34] Logar, M.; Jancar, B.; Šturm, S.; Suvorov, D. Weak polyion multilayer-assisted in situ synthesis as a route toward a plasmonic Ag/TiO_2 photocatalyst. Langmuir 2010, 26, 12215-12224.

[35] Arregui, F. J.; Dickerson, B.; Claus, R. O.; Matias, I. R.; Cooper, K. L. Polymeric thin films of controlled complex refractive index formed by the electrostatic self-assembled monolayer process. IEEE Photonics Technol. Lett. 2001, 13, 1319-1321.

[36] Cooper, P. R. Refractive-index measurements of liquids used in conjunction with optical fibers. Appl. Opt. 1983, 22, 3070-3072.

[37] Gouvêa, P. M. P.; Cancino-Bernardi, J.; Zucolotto, V.; Braga, A. M. B.; Carvalho, I. C. S. PAMAM dendrimer/gold nanoparticle nanocomposites for a reflection LSPR optical fiber sensor. In Proceedings of SPIE—The International Society for Optical Engineering; 2015; Vol. 9634.

[38] Rivero, P. J.; Hernaez, M.; Goicoechea, J.; Matias, I. R.; Arregui, F. J. Optical fiber refractometers based on localized surface plasmon resonance (LSPR) and lossy mode resonance (LMR). In Proceedings of SPIE—The International Society for Optical Engineering; 2014; Vol. 9157.

[39] Rivero, P. J.; Hernaez, M.; Goicoechea, J.; Matías, I. R.; Arregui, F. J. A comparative study in the sensitivity of optical fiber refractometers based on the incorporation of gold nanoparticles into layer-by-layer films. Int. J. Smart Sens. Intell. Syst. 2015, 8, 822-841.

[40] Sandhu, P.; Yang, J.; Xu, C. Q. In-fiber michelson interferometer with polymeric/nanoparticle thin-film overlay as a platform for biosensing. IEEE J. Sel. Top Quantum Electron 2010, 16, 685-690.

[41] Ortega-Mendoza, J. G.; Padilla-Vivanco, A.; Toxqui-Quitl, C.; Zaca-Morán, P.; Villegas-Hernández, D.; Chávez, F. Optical fiber sensor based on localized surface plasmon resonance using silver nanoparticles photodeposited on the optical fiber end. Sensors 2014, 14, 18701-18710.

[42] Chen, J.; Shi, S.; Su, R.; Qi, W.; Huang, R.; Wang, M.; Wang, L.; He, Z. Optimization and application of reflective LSPR optical fiber biosensors based on silver nanoparticles. Sensors 2015, 15, 12205-12217.

[43] Gowri, A.; Sai, V. V. R. Development of LSPR based U-bent plastic optical fiber sensors. Sens. Actuators B Chem. 2016, 230, 536-543.

[44] Sai, V. V. R.; Kundu, T.; Mukherji, S. Novel U-bent fiber optic probe for localized surface plasmon resonance based biosensor. Biosens. Bioelectron. 2009, 24, 2804-2809.

[45] Cennamo, N.; D'Agostino, G.; Donà, A.; Dacarro, G.; Pallavicini, P.; Pesavento, M.; Zeni, L. Localized surface plasmon resonance with five-branched gold nanostars in a plastic optical fiber for bio-chemical sensor implementation. Sensors 2013, 13, 14676-14686.

[46] Rani, M.; Sharma, N. K.; Sajal, V. Localized surface plasmon resonance based fiber optic sensor with nanoparticles. Opt. Commun. 2013, 292, 92-100.

[47] García, J. A.; Monzón-Hernández, D.; Manríquez, J.; Bustos, E. One step method to attach gold nanoparticles onto the surface of an optical fiber used for refractive index sensing. Opt. Mater. 2016, 51, 208-212.

[48] Tu, M. H.; Sun, T.; Grattan, K. T. V. Optimization of gold-nanoparticle-based optical fibre surface plasmon resonance (SPR)-based sensors. Sens. Actuators B Chem. 2012, 164, 43-53.

[49] Cao, J.; Galbraith, E. K.; Sun, T.; Grattan, K. T. V. Cross-comparison of surface plasmon resonance-based optical fiber sensors with different coating structures. IEEE Sensors J. 2012, 12, 2355-2361.

[50] Cao, J.; Galbraith, E.; Sun, T.; Grattan, K. T. V. Development and sensitivity studies of a gold nanorod platform for a localized surface plasmon resonance based optical fibre biosensor. In Proceedings of SPIE—The International Society for Optical Engineering; 2011; pp. 7753.

[51] Li, L.; Liang, Y.; Xie, L.; Lu, M.; Peng, W. Optical fiber surface plasmon resonance sensor with surface modified gold nanorods for biochemical detection. In Proceedings of SPIE—The International Society for Optical Engineering; 2014; Vol. 9277.

[52] Tu, M. H.; Sun, T.; Grattan, K. T. V. LSPR optical fibre sensors based on hollow gold nanostructures. Sens. Actuators, B Chem. 2014, 191, 37-44.

[53] Chen, Z.; Lu, C.; Grattan, K. T. V. Humidity sensors: a review of materials and mechanisms. Sensors 2005, 3, 274-295.

[54] Sharma, V. K.; Yngard, R. A.; Lin, Y. Silver nanoparticles: green synthesis and their antimicrobial activities. Adv. Colloid Interface Sci. 2009, 145, 83-96.

[55] Ilic, V.; Šaponjic, Z.; Vodnik, V.; Molina, R.; Dimitrijevic, S.; Jovancic, P.; Nedeljkovic, J.; Radetic, M. Antifungal efficiency of corona pretreated polyester and polyamide fabrics loaded with Ag nanoparticles. J. Mater. Sci. 2009, 44, 3983-3990.

[56] Rivero, P. J.; Urrutia, A.; Goicoechea, J.; Zamarreño, C. R.; Arregui, F. J.; Matías, I. R. An antibacterial coating based on a polymer/sol-gel hybrid matrix loaded with silver nanoparticles. Nanoscale Res. Lett. 2011, 6, X1–X7.

[57] Rivero, P. J.; Urrutia, A.; Goicoechea, J.; Arregui, F. J.; Matías, I. R. Humidity sensor based on silver nanoparticles embedded in a polymeric coating. In Proceedings of the 5th International Conference on Sensing Technology, ICST 2011, Palmerston North (New Zealand); 2011.

[58] Rivero, P. J.; Urrutia, A.; Goicoechea, J.; Arregui, F. J.; Matías, I. R. Humidity sensor based on silver nanoparticles embedded in a polymeric coating. Int. J. Smart Sensing Intell. Syst. 2012, 5, 71-83.

[59] Rivero, P. J.; Urrutia, A.; Goicoechea, J.; Arregui, F. J. Optical fiber humidity sensors based on localized surface plasmon resonance (LSPR) and Lossy-mode resonance (LMR) in overlays loaded with silver nanoparticles. Sens. Actuators B Chem. 2012, 173, 244-249.

[60] Del Villar, I.; Hernaez, M.; Zamarreno, C. R.; Sánchez, P.; Fernández-Valdivielso, C.; Arregui, F. J.; Matias, I. R. Design rules for Lossy mode resonance based sensors. Appl. Opt. 2012, 51, 4298-4307.

[61] Del Villar, I.; Zamarreño, C. R.; Sanchez, P.; Hernaez, M.; Valdivielso, C. F.; Arregui, F. J.; Matias, I. R. Generation of Lossy mode resonances by deposition of high-refractive-index coatings on uncladded multimode optical fibers. J. Optics 2010, 12, 095503.

[62] Zamarreño, C. R.; Hernáez, M.; Del Villar, I.; Matías, I. R.; Arregui, F. J. Optical fiber pH sensor based on Lossy-mode resonances by means of thin polymeric coatings. Sens. Actuators B Chem. 2011, 155, 290-297.

[63] Rivero, P. J.; Urrutia, A.; Goicoechea, J.; Matias, I. R.; Arregui, F. J. A Lossy mode resonance optical sensor using silver nanoparticles-loaded films for monitoring human breathing. Sens. Actuators B Chem. 2012, 187, 40-44.

[64] Goicoechea, J.; Zamarreño, C. R.; Matías I. R.; Arregui, F. J. Optical fiber pH sensors based on layer-by-layer electrostatic self-assembled Neutral Red. Sens. Actuators B Chem. 2008, 132, 305-311.

[65] Itano, K.; Choi, J.; Rubner, M. F. Mechanism of the pH-induced discontinuous swelling/deswelling transitions of poly(allylamine hydrochloride)-containing polyelectrolyte multilayer films. Macromolecules 2005, 38, 3450-3460.

[66] Rivero, P. J.; Goicoechea, J.; Hernaez, M.; Socorro; A. B.; Matias, I. R.; Arregui, F. J. Optical fiber resonance based-pH sensors using gold nanoparticles into polymeric layer-by-layer coatings. Microsyst. Technol. 2016, 22, 1821-1829.

[67] Socorro, A.B.; Rivero, P. J.; Hernaez, M.; Goicoechea, J.; Matias, I. R.; Arregui, F. J. Optical fiber pH sensor based on gold nanoparticles into polymeric coatings. In Proceedings of SPIE—The International Society for Optical Engineering; 2015; Vol. 9517.

[68] Muri H. I. D. I.; Hjelme D. R. Novel localized surface plasmon resonance based optical fiber sensor. In Progr Biomed Opt Imaging Proc SPIE; 2016; pp 9702.

[69] Crosby, J. S.; Lucas, D.; Koshland, C. P. Fiber optic based evanescent wave sensor for the detection of elemental mercury utilizing gold nanorods. Sens. Actuators B Chem. 2013, 181, 938-942.

[70] Lin, T. J.; Chung, M. F. Detection of cadmium by a fiber-optic biosensor based on localized surface plasmon resonance. Biosens. Bioelectron. 2009, 24, 1213-1218.

[71] Lin, T. J.; Chung, M. F. Using monoclonal antibody to determine lead ions with a localized surface plasmon resonance fiber-optic biosensor. Sensors 2008, 8, 582-593.

[72] Teerasong, S.; Sani, M.; Numsawat, P.; Martchoo, R.; Chompoosor, A.; Nacapricha, D. A silver nanoparticle thin film modified glass substrate as a colourimetric sensor for hydrogen peroxide. J. Exp. Nanosci. 2015, 10, 1327-1335.

[73] Gavrilenko, N. A.; Saranchina, N. V.; Gavrilenko, M. A. Colorimetric sensor based on silver nanoparticle—embedded polymethacrylate matrix. Adv. Mater. Res. 2014, 1040, 923-927.

[74] Hu, L.; Yuan, Y.; Zhang, L.; Zhao, J.; Majeed, S.; Xu, G. Copper nanoclusters as peroxidase mimetics and their applications to H_2O_2 and glucose detection. Anal. Chim. Acta 2013, 762, 83-86.

[75] Bhatia, P.; Yadav, P.; Gupta, B. D. Surface plasmon resonance based fiber optic hydrogen peroxide sensor using polymer embedded nanoparticles. Sens. Actuators B Chem. 2013, 182, 330-335.

[76] Shrivastav, A. M.; Usha, S. P.; Gupta, B. D. A Localized and propagating SPR, and molecular imprinting based fiber-optic ascorbic acid sensor using an in situ polymerized polyaniline-Ag nanocomposite. Nanotechnology 2016, 27, 345501.

[77] Guo, H.; Tao, S. Sol-gel synthesis of palladium-doped silica nanocomposite fiber using triton X-100 micelle template and the application for hydrogen gas sensing. IEEE Sens. J. 2007, 7, 323-328.

[78] Chen, I.; Lin, S. S.; Lin, T. J.; Du, J. K.. Detection of hydrofluoric acid by a SiO_2 sol-gel coating fiber-optic probe based on reflection-based localized surface plasmon resonance. Sensors 2011, 11, 1907-1923.

[79] Paul, D.; Dutta, S.; Biswas, R. LSPR enhanced gasoline sensing with a U-bent optical fiber. J. Phys. D 2016, 49, 305104.

[80] Chauhan, S. K.; Punjabi, N.; Sharma, D. K.; Mukherji, S. A silicon nitride coated LSPR based fiber-optic probe for possible continuous monitoring of sucrose content in fruit juices. Sens. Actuators B Chem. 2014, 222, 1240-1250.

[81] Camara, A. R.; Dias, A. C. M. S.; Gouvêa, P. M. P.; Braga, A. M. B.; Dutra, R. F.; De Araujo, R. E.; Carvalho, I. C. S. Fiber optic sensor with au nanoparticles for dengue immunoassay. In Optics InfoBase Conference Papers; 2013.

[82] Camara, A. R.; Gouvêa, P. M. P.; Dias, A. C. M. S.; Braga, A. M. B.; Dutra, R. F.; De Araujo, R. E.; Carvalho, I. C. S. Dengue immunoassay with an LSPR fiber optic sensor. Opt. Express 2013, 21, 27023-27031.

[83] Cao, J.; Sun, T.; Grattan, K. T. V. Development of gold nanorod-based localized surface plasmon resonance optical fiber biosensor. In Proceedings of SPIE—The International Society for Optical Engineering; 2012; Vol. 8421.

[84] Cao, J.; Tu, M. H.; Sun, T.; Grattan, K. T. V. Wavelength-based localized surface plasmon resonance optical fiber biosensor. Sens. Actuators B Chem. 2013, 181, 611-619.

[85] Wan, M.; Luo, P.; Jin, J.; Xing, J.; Wang, Z.; Wong, S. T. C. Fabrication of localized surface plasmon resonance fiber probes using ionic self-assembled gold nanoparticles. Sensors 2010, 10, 6477-6487.

[86] Srivastava, S. K.; Arora, V.; Sapra, S.; Gupta, B. D. Localized surface plasmon resonance-based fiber optic U-shaped biosensor for the detection of blood glucose. Plasmonics 2012, 7, 261-268.

[87] Bharadwaj, R.; Mukherji, S.; Mukherji, S. Probing the localized surface plasmon field of a gold nanoparticle-based fibre optic biosensor. Plasmonics 2016, 11, 753-761.

[88] Lin, H. Y.; Huang, C. H.; Cheng, G. L.; Chen, N. K.; Chui, H. C. Tapered optical fiber sensor based on localized surface plasmon resonance. Opt. Express 2012, 20, 21693-21701.

[89] Shao, Y.; Xu, S.; Zheng, X.; Wang, Y.; Xu, W. Optical fiber LSPR biosensor prepared by gold nanoparticle assembly on polyelectrolyte multilayer. Sensors 2010, 10, 3585-3596.

[90] Lepinay, S.; Staff, A.; Ianoul, A.; Albert, J. Improved detection limits of protein optical fiber biosensors coated with gold nanoparticles. Biosens. Bioelectron. 2014, 52, 337-344.

[91] Sanders, M.; Lin, Y.; Wei, J.; Bono, T.; Lindquist, R. G. An enhanced LSPR fiber-optic nanoprobe for ultrasensitive detection of protein biomarkers. Biosens. Bioelectron. 2014, 61, 95-101.

Section 5

Application - Fabrication for Plasmonics

Chapter 18

Laser Annealing as a Platform for Plasmonic Nanostructuring

Nikolaos Kalfagiannis,
Demosthenes C. Koutsogeorgis,
Elefterios Lidorikis and Panos Patsalas

Additional information is available at the end of the chapter

http://dx.doi.org/ 10.5772/67670

Abstract

Nanoconstruction of metals is a significant challenge for the future manufacturing of plasmonic devices. Such a technology requires the development of ultra-fast, high-throughput and low cost fabrication schemes. Laser processing can be considered as such and can potentially represent an unrivalled tool towards the anticipated arrival of modules based in metallic nanostructures, with an extra advantage: the ease of scalability. Specifically, laser nanostructuring of either thin metal films or ceramic/metal multilayers and composites can result on surface or subsurface plasmonic patterns, respectively, with many potential applications. In this chapter, the photo-thermal processes involved in surface and subsurface nanostructuring are discussed and processes to develop functional plasmonic nanostructures with pre-determined morphology are demonstrated. For the subsurface plasmonic conformations, the temperature gradients that are developed spatially across the metal/dielectric structure during the laser processing can be utilized. For the surface plasmonic nanoassembling, the ability to tune the laser's wavelength to either match the absorption spectral profile of the metal or to be resonant with the plasma oscillation frequency can be utilised, i.e. different optical absorption mechanisms that are size-selective can be probed. Both processes can serve as a platform for stimulating further progress towards the engineering of large-scale plasmonic devices.

Keywords: laser annealing, laser induce self-assembly, plasmonics, surface plasmon resonance, plasmonic nanoparticles

1. Introduction

Plasmonic nanoparticles have become an increasingly common research area as well as becoming a key component in many important applications, such as solar energy harvesting [1–6], chemical sensing via surface-enhanced spectroscopies [7, 8–14], cancer treatment [15, 16] and optical encoding of information [17–20] to name a few. The main reason behind their adaptability, to these and other prominent applications, is their unique optical properties that allow for the manipulation of light below the diffraction limit. Among metals, Au and Ag are considered more suitable for plasmonic nanoparticles (PNPs) due to their stability when formed into nanoparticles and their strong LSPR absorption bands in the visible region of the spectrum [21, 22]. Under irradiation, the free electrons in a PNP are driven by the alternating electric field and collectively oscillate at a resonant frequency in a phenomenon known as localized surface plasmon resonance (LSPR) [23–27]. The LSPR phenomenon induced in PNPs depends strongly on their composition, size, shape, geometrical distribution and on the refractive index of the employed dielectric environment [28–30]. To this end, efforts are undertaken in order to control these structural parameters depending on the growth conditions [22, 31–34].

A critical parameter in delivering practical plasmonic devices is the material preparation methods, which should allow the production of nanostructures with tunable plasmonic properties. So far, nanomaterials and nanodevice manufacturing have traditionally followed two distinct routes: (a) the top-down approach, where a process starts from a uniform material and subsequently finer and finer tools are employed to create smaller structures, like lithographic processes [35–37] and/or ion beam nanofabrication [38] and (b) the bottom-up approach, where smaller components of atomic or molecular dimensions self-assemble together, according to a natural physical principle or an externally applied driving force, to give rise to larger and more organized systems, like atomic layer deposition [39], cold welding [40], flash thermal annealing [41, 42], pattern transfer [43] and template stripping [44–46]. Practically, the top-down approach offers unequalled control and reproducibility down to a few nanometres in feature size but at high cost for large area manufacturing, while the bottom-up route applies for macroscopic scale nanopatterning albeit without the fine feature and reproducibility control.

The use of laser radiation can be viewed as an intermediate route for nanotechnology manufacturing. Light-matter interactions offer precise energy delivery and control over the physicochemical processes in the nanoworld. Amongst the methodologies followed for creating nanoparticle arrangements and/or nanocomposite thin films, laser annealing (LA) has been proven to be simple and versatile, providing freedom of design, fast processing, compatibility with large-scale manufacturing and allowing for the use of inexpensive flexible substrates [17, 19, 47], due to the ultra-fast and macroscopically cold process. For example, LA of a thin metal film results into nanoparticle arrangements on top of a substrate [11, 13, 14, 47–54], while embedded nanoparticles in a dielectric matrix can be fabricated in large scale through the LA of a stack of alternating ultrathin layers of a metal and a dielectric [17, 18].

One of the key design issues in LA is the spatial distribution of the energy delivery (from the laser pulse) and the precise temporal temperature profile emerged in the structure (caused by this energy). Understanding the dependence of the temperature profile on the nanoparticle/substrate geometries can provide valuable insights in order to be able to design the appropriate LA process and get custom-made plasmonic devices. In the systems of interest, several effects such as material absorption, plasmon resonances and interference contribute to how the absorption is distributed, while the material thermal properties and the laser pulse profile determine the temperature transients developed. Here, the investigations required to extract the underline physical understanding of all these photothermal processes, involved in surface and sub-surface laser nanostructuring, are summarised.

2. UV laser annealing experimental set-up

A typical laser processing apparatus, such as the one used for the experimental part of the work that will be presented later, is shown in **Figure 1**. It comprises three main components: (i) the two laser sources, (ii) the laser beam delivery system and (iii) the sample manipulation stage.

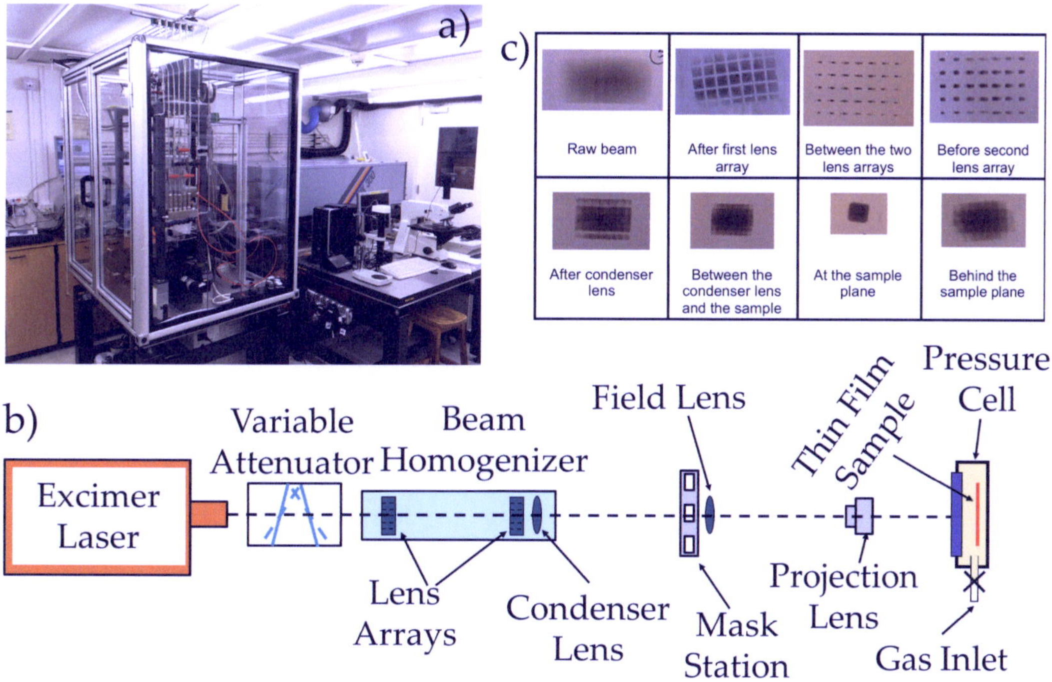

Figure 1. (a) Digital photo of the laser annealing system at Nottingham Trent University, (b) schematic representation of the beam delivery path and (c) snapshots of the laser spot at different positions of the beam path.

(i) The two laser sources are an excimer laser (LAMBDA PHYSIK LPX 305i) which is capable of providing unpolarized light pulses of 25 ns in duration and up to 1200 mJ/pulse at 248nm (KrF) and an excimer laser (LAMBDA PHYSIK LPX 205i) capable of providing unpolarized light pulses of 20 ns up to 700 mJ/pulse at 193 nm (ArF). The raw beam, in both cases, is approximately 30 × 15 mm² with a Gaussian-like profile on the long axis and an almost top-hat profile on the short axis.

(ii) The beam delivery system is composed of a variable attenuator, a beam homogeniser and a mask-projection system. The variable attenuator employs a partially reflective first plate (angled at various positions) reflecting part of the laser beam and a second plate that compensates for the parallel displacement of the beam caused by the first plate. The attenuator has a variable throughput from 10% up to 90% of the original beam. The beam homogenizer (Exitech Ltd., type EX-HS-700D) employs two lens arrays and a condenser lens. The condenser lens provides a laser spot of pre-determined size (13 × 13 mm²) at a certain distance with top-hat profiles on both axes (better than 2% uniformity). This homogenized beam coincides with a mask which is projected onto the sample via a combination of a field and projection lenses.

(iii) Finally, an X-Y-Z stage is in place in order to manipulate the sample and laser process different areas on it. The translational stage is operated via a controller with specialised software also being capable of externally triggering the laser. Thus, the whole operation is becoming fully automated. An appropriate mask and a ×5 objective lens set the laser spot delivered onto the samples to be a 2.5 × 2.5 mm² square spot.

The system has been designed in a way to be versatile and flexible, allowing for an extended degree of freedom of operation, thus providing the user with a wide spectrum of experimental parameters to vary from. These parameters are shown in **Figure 2**.

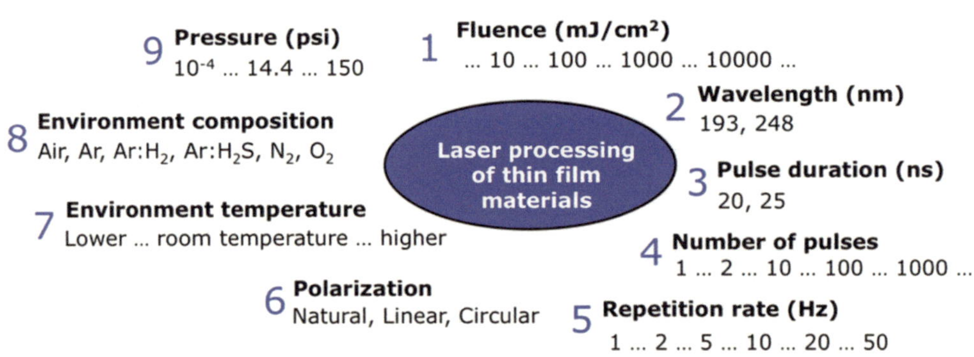

Figure 2. The nine experimental parameters of laser annealing system.

1. Fluence (mJ/cm²): The fluence of the laser, i.e. the energy density delivered to the sample, is adjusted by the variable attenuator. The maximum fluence may also vary depending on the size of the mask and the objective lens used to project the laser spot onto the sample's surface and of course from the excimer laser source: KrF laser is capable of providing a raw beam of 1200 mJ/pulse, whereas the ArF is capable of providing a raw beam of 700 mJ/pulse. Regarding the spot size, this is defined by the appropriate masks, on the mask station and the combination of field and projection lenses.

2. Wavelength (nm): The KrF laser emits light at 248 nm, while the ArF emits light deeper in the UV spectral range; at 193 nm. At this deep UV wavelength, most of the energy is absorbed by O_2 molecules in the air, which are then turned into ozone. To avoid the energy losses and thus the minimisation of the energy throughout the process in addition to avoiding health hazards due to the toxicity of ozone, the beam delivery path of the ArF laser is enclosed in sealed tubes, to allow for N_2 flow during the laser process.

3. Pulse duration (ns): Both excimer lasers are ultra-short pulsed UV lasers. The KrF has a pulse width of 25 ns, while the ArF has a pulse width of 20 ns.

4. Number of pulses: The applied number of pulses to the sample.

5. Repetition rate (Hz): The repetition rate is externally controlled and may vary from 1 to 50 Hz for both laser sources.

6. Polarization: While an excimer laser is usually producing completely randomly polarized light an optical filter that passes light of a specific polarization (polariser) may be used to block waves of other polarizations and convert a beam of light of undefined or mixed polarization into a beam with well-defined polarization state.

7. Environment temperature: The home-built system has the capability to have the sample sitting at −196°C (liquid nitrogen) or heated at elevated temperatures during the LA process.

8. Environment composition: The pressure cell, where the sample sits (shown in **Figure 1**), is supplied by a series of gases through appropriate pipelines connected to the cell. These gases are Ar, Ar:H_2, Ar:H_2S, N_2, and O_2. Depending on the application, the LA experiment may need to be conducted in an inert atmosphere (e.g., avoiding any oxidation occurring during the LA in air) or in a reacting atmosphere, where doping of the processed sample is seen necessary.

9. Pressure: In certain cases, low vacuum or high pressure application during LA (of any gaseous environment) is considered advantageous to avoid detrimental effects of ablation [17].

3. Sub-surface nanostructuring

3.1. Pulsed laser deposited nanocomposite AlN:Ag films

A robust nanoconstruction consisting of a narrow size distribution of PNPs embedded into a ceramic dielectric matrix emerges as the ideal configuration for applications such as authentication patterns and optical recording media specified for extreme conditions. To this end, AlN is an excellent candidate as a host matrix for PNPs due to its optical transparency, chemical inactivity, refractory character and high hardness. Ag is a characteristic example of noble metal that exhibits the strongest LSPR when formed into nanoparticles. However, silver's high diffusivity on one hand and the high melting point of AlN on the other hand makes the fabrication of AlN:Ag(PNPs) nanocomposites an unfeasible task: Ag is either out-diffused to the surface (upon thermal treatment) or is atomically dissolved into AlN during growth [55]. Such obstacles prevented the extended use of AlN in plasmonics. Herein, strategies for the successful growth of AlN:Ag(PNPs) films by LA-based micropatterning are discussed.

3.1.1. Growth and processing of materials

AlN:Ag nanocomposite films (70–100 nm thick) were grown on n-type c-Si(100) wafers *via* pulsed laser deposition (PLD) in an in-house-built high-vacuum system ($P_b < 5 \times 10^{-8}$ mbar) using the third harmonic of a Nd:YAG laser (6 ns pulse duration, fluence 30 J/cm^2 at 10 Hz repetition rate). More details on sample preparation can be found elsewhere [19, 20]. The resulting nanocomposite films consisted of an amorphous AlN matrix with embedded pure Ag nanospheres, as displayed in the transmission electron microscopy (TEM) image in the inset of **Figure 3**. The embedded nanospheres exhibited LSPR, spectrally located at around 420 nm [20].

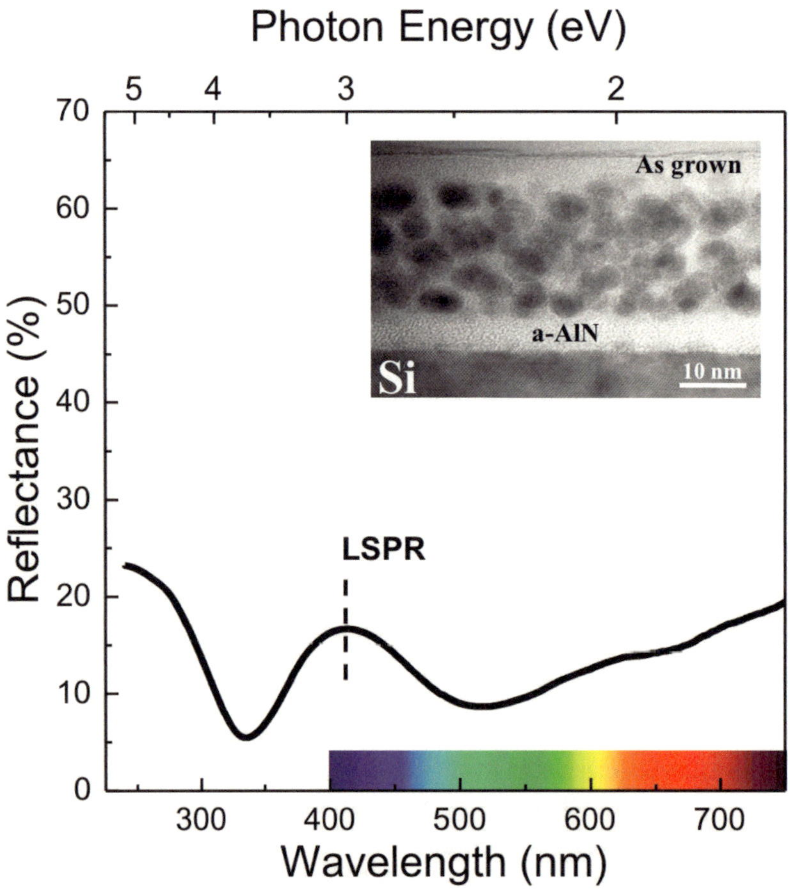

Figure 3. Optical reflectivity of AlN:Ag: as grown film. Inset depicts a cross section TEM image of the sample.

3.1.2. Laser processing

3.1.2.1. The effect of number of pulses

AlN:Ag films were processed with 1, 2 or 5 pulses at 200 mJ/cm^2 and 400 mJ/cm^2, i.e. just above the annealing threshold and just below the ablation threshold at 193 and 248 nm laser wavelengths. The pulse repetition rate in any case was 1 Hz, eliminating any possibility for transient thermal phenomena and thus allowing for each successive pulse to interact with a perfectly relaxed AlN:Ag film at room temperature. The optical reflectivity spectra are presented in **Figure 4**. In the case of the 193 nm wavelength at low fluence (200 mJ/cm^2) (**Figure 4a**), each successive pulse leads to a cumulative LSPR enhancement, in terms of the integral strength of the LSPR band, indicating that Ag diffusion results in shape and size improvement of the plasmonic nanoparticles; the Si substrate's E1 critical point at around 350 nm [56] is also gradually emerging with the number of pulses due to the film's transparency, because of the crystallization of AlN as we will discuss in more detail in the following paragraphs. On the contrary, at high fluence (400 mJ/cm^2) (**Figure 4b**), the evolution of the reflectivity spectra with the number of laser pulses exhibits severe changes. The effect of one pulse at 400 mJ/cm^2 to the plasmonic response is more pronounced compared to that of five pulses at 200 mJ/cm^2, indicating that the structural modifications taking place during the 193 nm LA process are better associated with the laser fluence than with the total laser energy delivered. This will be further examined in the next section where the effect of laser fluence is discussed. For two applied pulses, the LSPR band is substantially broader indicating an exceptionally broad size distribution of Ag nanoparticles that can only be explained by means of Ag outdiffusion. Finally, for five pulses the reflectivity spectrum is characteristic of a homogeneous Ag layer on top of the AlN:Ag film indicating that none of the remaining Ag is in the nanoparticle form.

3.1.2.2. The effect of laser wavelength

A major factor affecting the LA process is the laser wavelength, due to the different absorption coefficient of the amorphous AlN matrix and Ag. In the case of LA of AlN:Ag, using the 248 nm wavelength and fluence of 400 mJ/cm^2, the picture is quite more complex. At the first laser pulse (**Figure 4c**, red solid line), the pre-existing LSPR band is eliminated, possibly by dissolving of overheated, unstable Ag PNPs, whose constituent Ag atoms do not have the required diffusion time to reform into bigger Ag particles. Bigger particles, however, are observed after the second pulse resulting in intense LSPR peak (**Figure 4c**, green solid line). The lower photon energy at 248 nm and the fact that the pulse duration is shorter than the time required for Ostwald ripening in a cold environment [57, 58] (the process is indeed relatively cold because the 248 nm beam is not absorbed by the AlN matrix due to its high fundamental gap) allows for the Ostwald ripening process to take place in two steps: dissolution of particles for the first pulse and reforming of bigger particles close to the surface for the second pulse. Finally, the optical reflectivity spectrum for five pulses (**Figure 4c**, blue solid line) exhibits a broad LSPR band, resembling the broadening of the LSPR band for two applied pulses at 193 nm (**Figure 4b**, green solid line), but blue-shifted compared to the band for two pulses of 193 nm indicating a surrounding medium that is less optically dense than AlN (e.g. outdiffused Ag nanoparticles in air) [20].

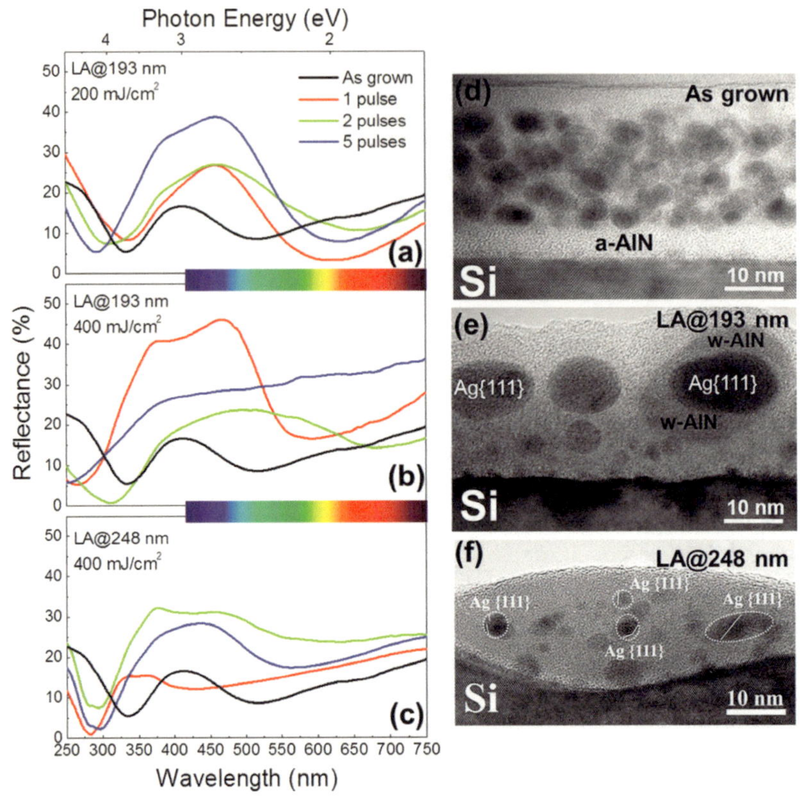

Figure 4. Optical reflectivity spectra for 1, 2 and 5 successive pulses for laser annealing using: (a) 193 nm wavelength, 200 mJ/cm^2 fluence, (b) 193 nm wavelength, 400 mJ/cm^2 fluence and (c) 248 nm wavelength, 400 mJ/cm^2 fluence. In all cases the black solid line, represents the as deposited film's optical response, which is shown for comparison. Cross-section TEM images of AlN:Ag/Si: (d) as-grown, (e) laser annealed with one pulse of 193 nm/400 mJ/cm^2 and (f) laser annealed with two pulses of 248 nm/400 mJ/cm^2.

Cross-sectional TEM images of the as-grown AlN:Ag and laser-treated areas (**Figure 4d–f**) reveal significant structural modifications for both wavelengths. The untreated films (**Figure 4d**) give no trace of crystalline AlN [59]. LA at 193 nm (**Figure 4e**) leads to agglomeration of Ag nanoparticles and formation of larger nanoparticles close to the surface of the film. This enlargement of Ag nanoparticles can be described as an Ostwald ripening phenomenon. Another important aspect of the 193 nm LA process is the partial crystallization of AlN into the wurtzite structure (w-AlN) around the enlarged nanoparticles (**Figure 4e**). Similar results occur after the 248 nm LA treatment (**Figure 4f**), although to a lesser extent due to the lower delivered photon energy, which is below the AlN's fundamental gap (6.2 eV). As a result, the laser beam interacts solely with the Ag nanoparticles. Consequently, the delivered energy to the substrate is higher than in the case of 193 nm LA, and thus the AlN/Si interface is significantly damaged. By the same token, the crystallization of the matrix is less likely to occur due to the lower effective temperature rise during the LA process.

To get an insight into the light absorption dynamics inside the film during LA, finite-difference time-domain (FDTD) simulations of a 60 nm thick nanostructured AlN:Ag film were performed, where Ag consists of 3 nm diameter inclusions at 15% volume filling ratio randomly distributed inside the AlN matrix [20]. Laser light, for both wavelengths (193 nm and/or 248 nm), is normally incident on the AlN:Ag film. Partially, it is absorbed by the film; partially, it is reflected and the rest is absorbed by the Si substrate. **Figure 5a** presents the absorption percentage per nm of length for the two wavelengths, where several configurations of randomly distributed Ag nanoparticles were averaged. The difference in the total film absorption, between the 193 and the 248 nm wavelengths, is small, with the 193 nm illumination being slightly more absorbed. The different standing wave patterns within the 60 nm film are also evident. Of great significance is the distribution of the absorbed light power among the different materials (**Figure 5b**). Considering that the Ag PNPs are dispersed at only 15% by volume, it is evident that they are the stronger absorbers in both cases. However, there is a significant deviation between the two annealing wavelengths: for 193 nm, a much more balanced absorption between AlN and Ag occurs, with minimal light energy reaching the Si substrate. In contrast, for 248 nm, most of the absorption occurs within the Ag inclusions, while at the same time, more light energy finds its way to the Si substrate causing significant damage. The absorption distribution maps for these two cases are shown in **Figure 5c**, considering a slice of one of the different randomized configurations. The fact that Ag nanoparticles act as absorption centres explains why AlN crystallization occurs only within a few nanometres from the nanoparticles surface. In addition, the more balanced absorption distribution in the film, as well as the minimal absorption by the Si substrate, qualitatively explains the superior annealing performance at 193 nm.

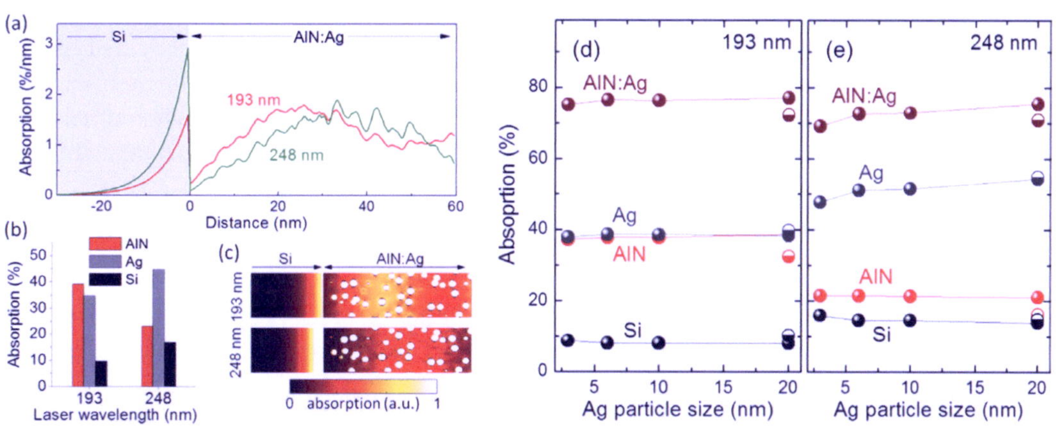

Figure 5. (a) Absorption distribution for a 60 nm thick AlN:Ag(15%vol)/Si film. (b) Absorption percentages for each constituent material within the AlN:Ag/Si structure. (c) Colour coded absorption distribution maps. (d) Absorption percentage within each material as a function of nanoparticle size, for 193 nm laser annealing wavelength and (e) Absorption percentage within each material as a function of nanoparticle size, for 248 nm laser annealing wavelength. The semi-open circles at 20 nm correspond to the nanoparticles having a non-absorbing 2.5 nm crystalline AlN ($n = 2.1$) layer around them.

Figures 5d and **5e** displays the cumulative absorption in each different material as a function of Ag particle size for the two annealing wavelengths. Ordered Ag NP arrays are considered here only for simplicity. Evidently, the 193 nm laser exhibits superior performance: unfluctuating performance for all particle sizes, more balanced absorption between matrix and inclusions, slightly stronger overall film absorption and less substrate absorption. Semi-open circles at 20 nm correspond to the nanoparticles having a non-absorbing 2.5 nm crystalline AlN ($n = 2.1$) shell. This results in a small decrease in the matrix absorption due to the higher fundamental energy gap of w-AlN. The effect of shell creation thus is towards the direction of limiting the film's overall absorption, although it does not seem to be large enough to qualify as a self-limiting factor for the annealing process.

3.1.2.3. The effect of laser fluence

The data presented in **Figure 5** demonstrate that the laser fluence is an essential parameter for the LA process. The optical reflectivity spectra of the laser annealed AlN:Ag vs. the laser fluence for both processing wavelengths and the corresponding quantitative results are presented in **Figure 6**. Increasing the fluence of the 193 nm wavelength, from 200 to 400 mJ/cm^2, induces a spectral shift concomitantly with a significant enhancement of plasmon resonance. Both effects are respectively associated with changes of the matrix's refractive index and the Ag nanoparticles size. The laser fluence decreased the bandwidth (full width at half maximum of the fitted LSPR band), indicating a narrower size distribution of Ag nanoparticles due to the matrix crystallization, as well as to Ag particle enlargement causing weaker electronic confinement. Eventually, at higher fluence (e.g. 600 mJ/cm^2), the reflectivity spectrum reveals the Si features, indicating that this fluence exceeds the ablation threshold.

On the other hand, for LA using two pulses at 248 nm, the picture is quite different (**Figure 6b**). At 300 mJ/cm^2, the reflectivity spectrum is similar to the one pulse at 248 nm and 400 mJ/cm^2 (**Figure 4c**), which is characterized by the absence of LSPR, indicating that the provided energy is not enough to cause ripening of Ag and only dissolution is observed. For fluences exceeding 400 mJ/cm^2, a consistent LSPR blue shift is observed due to only a possible reduction of the refractive index of the surrounding medium, supporting the Ag outdiffusion to the free surface and the formation of Ag nanoparticles surrounded by air.

3.1.3. Application in optical encoding of information

The combination of structural modifications and the variation of AlN:Ag plasmonic response, introduced by LA, produces strong reflectivity contrast, which has been exploited to encode spectrally sensitive patterns [19, 20]. Patterns such as 'ALN:AG', 'ALN', and 'AL' have been produced by LA onto films of AlN:Ag, AlN and Al, respectively, deposited either on Si or sapphire substrates, using either the 193 nm or the 248 nm beam, and their response (in transmittance and reflectance) to different light colours was investigated using optical microscopy [19, 20]. The reflectivity contrast $\Delta R(=R_{LA}-R_{untreated})$ between treated and untreated areas of an AlN:Ag sample demonstrates the level of contrast the human eye can resolve; the spectral variations of such ΔR contrasts for 193 and 248 nm LA are presented in **Figure 7b** for all cases of laser treated AlN:Ag deposited on opaque Si substrates. Depending on the processing wavelength and LA conditions exceptional flexibility in selecting the required encoding ΔR contrast over the whole visible spectral range can exist. In particular, the spectral selectivity

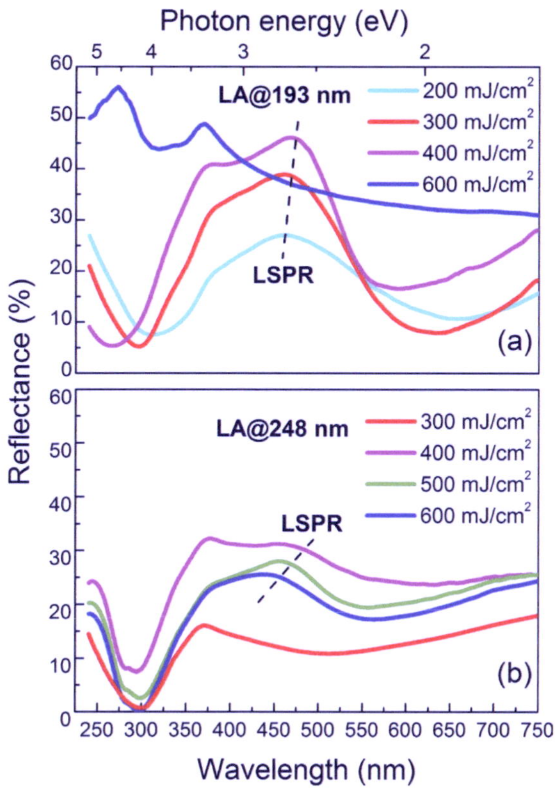

Figure 6. Optical reflectivity spectra with respect to laser fluence for LA at (a) 1 pulse at 193 nm and (b) 2 pulses at 248 nm.

is greater when large variations of ΔR are manifested in the visible spectrum. For example, for LA using one pulse of 400 mJ/cm² of 193 nm (**Figure 7b**, solid blue line), the ΔR ranges from 5% for red (650 nm) to 35% for blue (475 nm) providing very high spectral selectivity. These conditions are ideal for spectrally sensitive overt encoding. On the contrary, covert encoding requires more subtle spectral contrast ΔR in order for the pattern to be invisible under white light illumination and be revealed only under specific coloured illumination near the LSPR region. This can be done by, e.g., LA using two pulses of 500 mJ/cm² of 248 nm (**Figure 7b**, dashed magenta line).

The potential of plasmonic encoding has been better demonstrated by optical microscopy images, where the sample is illuminated by different coloured light (blue/468 nm, green/532 nm, red/655 nm) [20]. For this demonstration, AlN:Ag, AlN and Al films were grown on sapphire substrates in order to evaluate both the reflection and transmission reading [20]; the relevant optical microscopy images are presented in **Figure 8**. Images A1-D1 demonstrate covert encoding in transmission reading, facilitated by LA of an AlN:Ag film at 248 nm. The pattern, with a pixel size of 60 μm, is almost invisible under white light illumination (A1) and is fully uncovered only under blue illumination (D1) where the LSPR is manifested. Images A2-D2

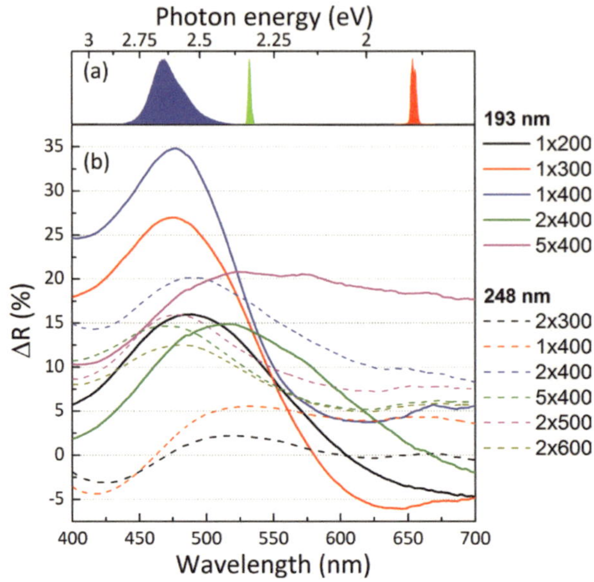

Figure 7. (a) Emission spectra of the three red-green-blue (RGB) light sources used for optical microscopy observations. (b) Spectral reflectivity contrasts (ΔR) for laser annealing of AlN:Ag on Si (opaque substrate) at 193 and 248 nm (purple and magenta lines, respectively) for various number of pulses and fluences demonstrating the flexibility in selecting the required encoding contrast.

and A3-D3 correspond to overt encoding in both transmission and reflection reading, realized by LA at 193 nm. The spectral sensitivity of the plasmonic patterns (A1-D1, A2-D2, A3-D3) contrasts the behaviour of patterns encoded by laser ablation in AlN (images A4-D4) and in Al (images A5-D5), which do exhibit similar contrasts for all colours, mostly due to scattering of light at the pixel edges. In addition to photosensitivity, AlN:Ag(PNPs) films have a unique combination of assets such as being harder than steel [20], chemically resistant against organic solvents [19, 20], almost having pitchless surface [19, 20] and are resistant to temperatures up to 600°C [20, 60].

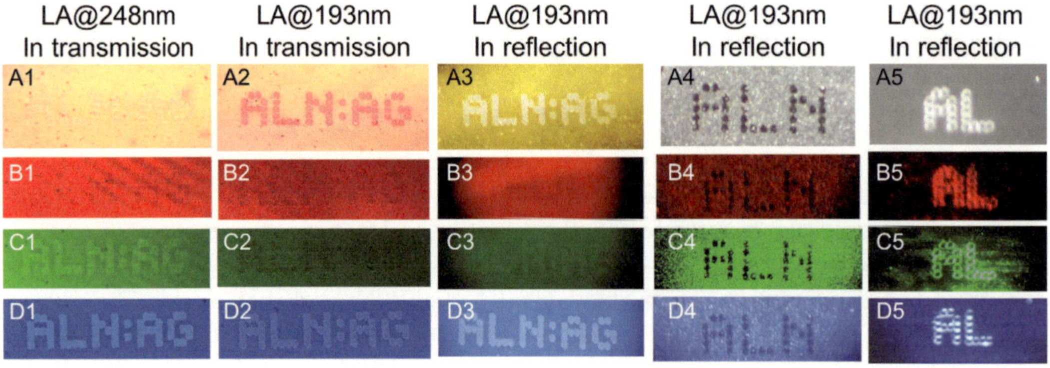

Figure 8. Optical microscopy images of AlN:Ag, AlN, Al on sapphire using A. white, B. red, C. Green and D. blue illumination, in transmission (1, 2) or reflection (3–5) mode on sapphire for 248 nm (1), and 193 nm (2–5) laser annealing, respectively [20].

3.2. Magnetron sputtered AlN/Ag multilayer thin films

In this section, an alternative pathway is demonstrated, namely, starting from non-plasmonic metal/dielectric multilayers, which are then transformed into plasmonic templates by single-shot UV LA leading to a subsurface nanoconstruction involving plasmonic Ag nanoparticles embedded in a hard and inert dielectric matrix. The subsurface encapsulation of Ag nanoparticles provides durability and long-term stability. The morphology of the final composite film depends primarily on the nanocrystalline character of the dielectric host and its thermal conductivity. The emergence of LSPR and its adjustability, depending on the applied fluence, is demonstrated.

3.2.1. Growth and processing of materials

Multilayers consisting of alternate thin layers of AlN and Ag were fabricated in an in-house built high vacuum system (base pressure 2×10^{-6} Pa) employing the reactive magnetron sputtering technique. The AlN crystal structure was varied in the two sets of AlN/Ag by applying two different power values to the Al target, 100 and 15 W. This variation lead to different categories of sputtered AlN [46]: at 100 W high crystalline wurtzite AlN (w-AlN) was obtained while at 15 W an amorphous AlN (a-AlN) was obtained. The produced samples were subjected to LA with an excimer ArF source. The effect of a single-pulse LA step onto the structural and optical properties was investigated as a function of fluence (400–700 mJ/cm^2) in an ambient atmosphere.

3.2.2. Laser processing

The heating dynamics involved in the multilayer structures during UV LA can be simulated/predicted performing optical and heat transport calculations. UV light propagation was solved via the FDTD method [61, 62], and a detailed map of the absorption profile in the multilayer and the substrate was obtained. The model system consists of a stratified multilayer structure of alternative dielectric (AlN) and metal layers (Ag) on a Si substrate, capped by AlN. The optical response of such a multilayer show high UV absorption (**Figure 9a**) due to silver's inter-band transitions at this spectral range. Under the proper experimental conditions, this high UV absorption was utilized to locally melt the structure. Due to the immiscibility of Ag with AlN, Ag diffusion will drive its transformation into a plasmonic nanocomposite with metallic PNPs embedded in a dielectric matrix (**Figure 9b**). The morphology of the nanocomposite and its optical response can be tuned by choosing materials and LA parameters such as: the nanocrystaline character of the dielectric which is strongly related to its thermal conductivity, the metal volume ratio and the total thickness of the multilayer.

The spatial absorption profile is convoluted with the temporal pulse profile (shown in **Figure 11b**, inset) and solved in the 1D heat transport equation [63]:

$$c(z)\rho(z)\partial_T T(z,t) = \partial_z\left[k(z)\partial_z T(z,t)\right] + Q(z,t) \tag{1}$$

where ∂_T, ∂_z denote partial derivatives with respect to time (t) and distance (z) from the top surface, respectively, and $Q(z,t) = \alpha(z)\varphi(t)$ is the laser heating source term defined by the product of the absorption spatial profile $\alpha(z)$ and the laser pulse temporal profile $\phi(t)$, with the total laser fluence given by:

$$f = \int_{-\infty}^{\infty} \varphi(t)dt \tag{2}$$

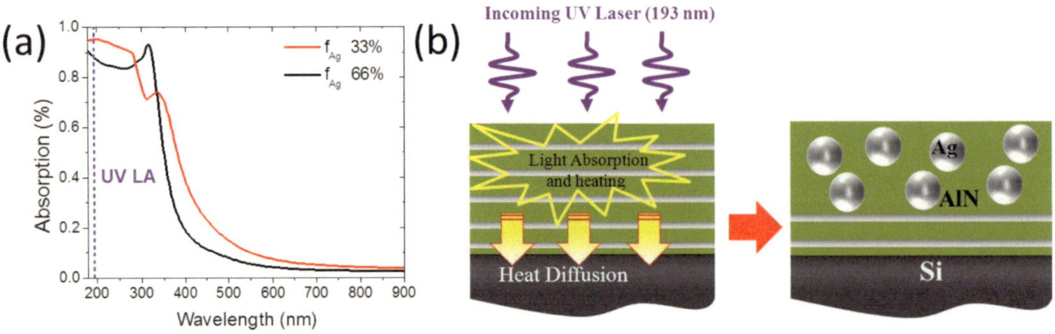

Figure 9. (a) Spectral absorption of a stratified metal (Ag) dielectric (AlN) structure illustrated in (b) for two metal volume ratios 33 and 66%. The high UV absorption can be utilized for the further modification of the nanocomposite. (b) Schematic illustration of the laser treatment of the multilayer structure. Under UV laser radiation, the absorbed energy is converted into heat, which is diffused by thermal conduction into the Si substrate. The result of the laser treatment is to totally transform the multilayer structure into a nanocomposite of embedded plasmonic metallic NPs inside a dielectric matrix.

c, ρ and k are the specific heat capacity, the mass density and the thermal conductivity, respectively. These three terms have a spatial dependence due to the interchange of different materials. Eq. (1) is time-integrated numerically (by a 4th order Runge-Kutta scheme) in order to calculate the explicit temperature transient. To do that, we assume that: (i) electronic and lattice temperatures are in equilibrium, (ii) radiation and convection losses are negligible, (iii) material properties (absorption, heat capacity, mass density and thermal conductivity) remain constant during heating.

Relevant recent experiments of a single UV laser annealing (LA) on a ceramic/metal (in particular AlN/Ag) multilayer film are discussed [17]. This LA step is capable of driving the subsurface modification of the metallic multilayers resulting into LSPR behaviour from PNPs that are formed and dispersed in a depth of several nanometre away from the surface. TEM observations from the as-deposited samples (**Figure 10a** and **b**) verified their multilayer character.

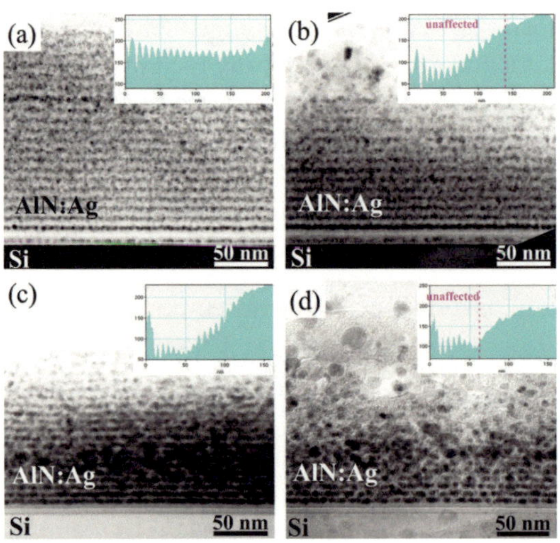

Figure 10. TEM cross sectional images of: (a) the as-deposited multilayer structure of w-AlN/Ag, (b) the same after LA, and (c) the as-deposited multilayer structure of a-AlN/Ag and (d) the same after LA. LA fluence for all relevant samples in this figure was set at 600 mJ/cm^2.

After LA treatment, the upper part of the multilayer was highly affected. The multilayer structure is destroyed and the enlarged NPs are fairly homogenously distributed inside the AlN matrix, retaining a high density in the case of a-AlN. On the contrary, in the case of w-AlN only, the top of the laser treated part of the sample is affected and is generally depleted of Ag, indicating a higher diffusivity of Ag in w-AlN compared to a-AlN. The w-AlN/Ag presents a destruction of periodicity affecting only the top six Ag layers, while two more layers were partially destroyed. The sample thickness that retained a stratified arrangement is indicated by a dashed line in the inset of **Figure 10b**, which depicts the contrast intensity profile along the film thickness. In the case of a-AlN/Ag, this structural re-arrangement is more extensive as shown in **Figure 10d**, and it affected the top 10 Ag layers, while again 2 additional layers were partially affected.

The above considerations make AlN's thermal conductivity a crucial parameter in the LA process. k_{AlN} is sensitive to the deposition method and conditions and can show large variations ranging from about 10 W/mK for amorphous AlN [64] to 285 W/mK for single crystal [65]. To explore its effect, the temperature transient is plotted (at the top of the multilayer) for several different values of the AlN thermal conductivity. A significant difference depending on k_{AlN} is found for the peak temperature rise when the laser pulse is on, but it quickly diminishes afterwards. The most important effect of k_{AlN} is shown in the spatial distribution of the peak temperature rise plotted in **Figure 11(c)**. In particular, for high values of k_{AlN}, the temperature profile is almost constant across the multilayer, even though most of the laser fluence is absorbed within the first 100 nm, as shown in **Figure 11(a)**, where

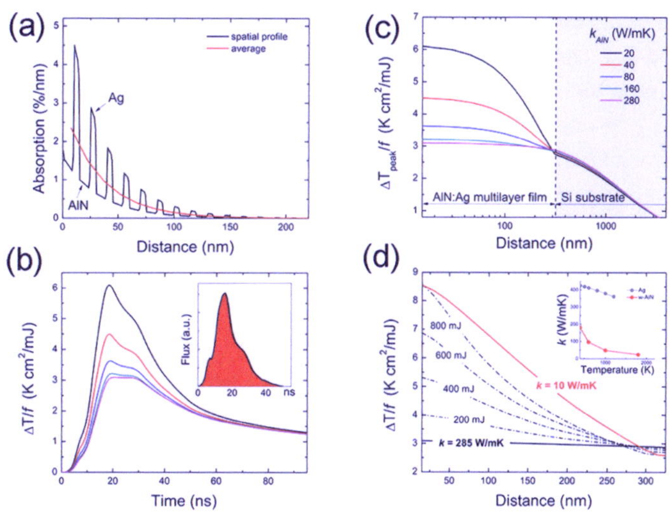

Figure 11. (a) UV laser (193 nm) absorption profile in our model structure (21 5nmAg/10nmAlN bilayers on Si), (b) normalised temperature transient at the top of the multilayer for various values of k_{AlN}. The inset shows the pulse profile of the excimer laser, (c) the normalized peak-temperature spatial distribution for the corresponding cases of (b). Large gradients across the film are obtained for low values of k_{AlN}, (d) peak transient temperature distribution as a function of distance from the film's top surface, at different incident laser fluence assuming a nonlinear thermal conductivity for the multilayer structure (w-AlN and Ag). Solid lines are for the linear case (zero fluence limit) for w-AlN (k = 285 W/mK) and for an ultra-low conductance a-AlN (k = 10 W/mK), both for a linear Ag thermal conductivity of 429 W/mK. Broken lines are for nonlinear w-AlN and Ag for different laser fluence. In the inset we plot the experimental temperature dependent thermal conductivities of Ag [50] and w-AlN [48] assumed in this work.

the spatial absorption profile within the multilayer is plotted. One can note that for low values of k_{AlN}, a significant temperature gradient is developed across the multilayer. Higher temperatures close to the surface and lower close to the substrate will result into different annealed structures, explaining our experimental results. It should be stressed here that up to now only the linear heat transport regime is studied in order to get insight into the temperature transients and gradients that can be developed and have not considered the thermodynamics of phase changes, e.g. melting and resolidification. Regarding radiation and convection losses from the top surface, a simple estimate suffices to show that they are of no consequence [20].

4. Surface nanostructuring

Plasmonic templates can be fabricated by LA of a thin metal film on a substrate, resulting into nanoparticle arrangements with many plasmonic applications as mentioned in Section 1 [11, 13, 14, 17, 19, 47–54]. The enabling property of these templates is the coherent surface plasmon oscillations excited by the incident light at specific frequencies, with high selectivity on the metal nanostructure size and shape. But a selective plasmon excitation is necessarily accompanied by selective light absorption, and thus by selective heating of the nanostructure. It is natural thus to allow this core property of the target application, to become itself one of the drivers in a multi-pulse laser induced self-assembly process. Besides heating, selective cooling is also an integral part of restructuring. But as the cooling rate depends on the geometric features of the nanostructure, a second driver for the self-assembly process is introduced. The available light absorption mechanisms in the case of nanostructured noble metals such as Ag, Au and Cu, can be generally split into (a) inter-band absorption from the metal's internal d-electrons, which is typically enabled at UV frequencies and in general is not size-dependent, and (b) resonant intra-band absorption, from the conduction electrons due to excitation of localized surface plasmon resonances (LSPR), which is strongly size and shape-selective. As an example, the overall absorption profile of an Ag nanoparticle (hemispherical shape) with diameter of 40 nm on top of a Si substrate is plotted in **Figure 12**, showing the spectral domains of the two absorption mechanisms. Utilization of these absorption modes combined with the size selectivity of cooling (~nanostructure surface/volume ratio) constitutes a toolbox for controlling the self-assembly of nanoparticles.

If after some laser treatment of a thin silver film on a Si substrate, Ag NPs with specific diameters have been formed, what will be the effect of a new incoming laser pulse? For this consideration, computer simulations can help explore the potential use of the various modes presented above. For computational simplicity a hexagonal lattice of Ag-hemispheres (40 nm in diameter) on a Si substrate with separation aspect ratios a/d = 1.2 (periodicity/diameter) have been assumed. First of all, **Figure 13a** shows the absorption per nm at different depths for this structure. The different modes relating to inter-band and intra-band absorptions in the nanoparticle, as well as the absorption in the substrate, are clearly shown. The absorption spatial distribution is then time-integrated with the appropriate laser temporal profile for each wavelength, depicted in **Figure 13c**. The experimentally acquired laser pulse

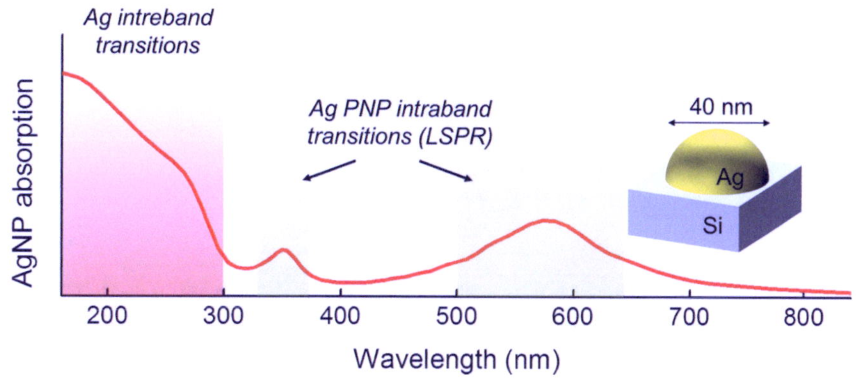

Figure 12. Simulated absorption spectrum of a 40 nm diameter Ag nanoparticle on top of a Si substrate. Inter-band absorption dominates the red shaded region, while LSPR driven absorption is dominating the green shaded regions (at ~350 nm the quadrupole 428 and ~580 nm the dipole contributions, respectively).

temporal profile from an excimer source was used for the 150–180nm range whereas the pulse temporal profile of a Nd:YAG source was used for the 300–900 nm range in a numerical solution of the numerically 1D heat diffusion equation presented in Eq. (1). Thus, we get the temperature transient profile at every point in the structure in degrees/mJ of laser fluence. In **Figure 13d**, the temperature transients for the two primary wavelengths used in recent experimental studies are presented. This way the mapping of the maximum temperature rise as a function of particle size, laser wavelength and laser fluence can be constructed. Bear in mind that the temperature transient follows the actual pulse temporal shape, but a significant transient tail also persists for longer times. The above discussion indicates that the processes discussed here are power and fluence dependent respectively, making the explicit time-dependent simulations essential. For a given laser fluence, direct information regarding particle heating can be extracted, and the probability of a nanoparticle melting or not can be assessed.

The global picture can then be revealed by considering how nanoparticle arrays of different sizes of nanoparticles behave to different wavelength illuminations. This is plotted in **Figure 13e**, showing the peak transient temperature, per unit of laser fluence, developed in the nanoparticle as a function of both the wavelength and the nanoparticle size. The dipole contribution is weaker than the contributions at shorter wavelengths (quadrupole LSPR and Si inter-band absorption), however, within a specific annealing plan one can always irradiate at one single wavelength, thus the strength of each band is of no importance other than defining the fluence that each laser wavelength should be used. Upon melting and resolidification, a random distribution of sizes will emerge. However, repeated laser pulses at a particular wavelength will lead to the melting of one particular size group and thus to its eventual extinction. The main concept behind this approach is, thus, repeated laser treatment, targeting different size groups with different laser wavelengths, driving the final size distribution towards a predetermined one. This process has been termed as 'modification of nanoparticles arrays by laser-induced self-assembly' (MONA-LISA) [47] and has been experimentally demonstrated.

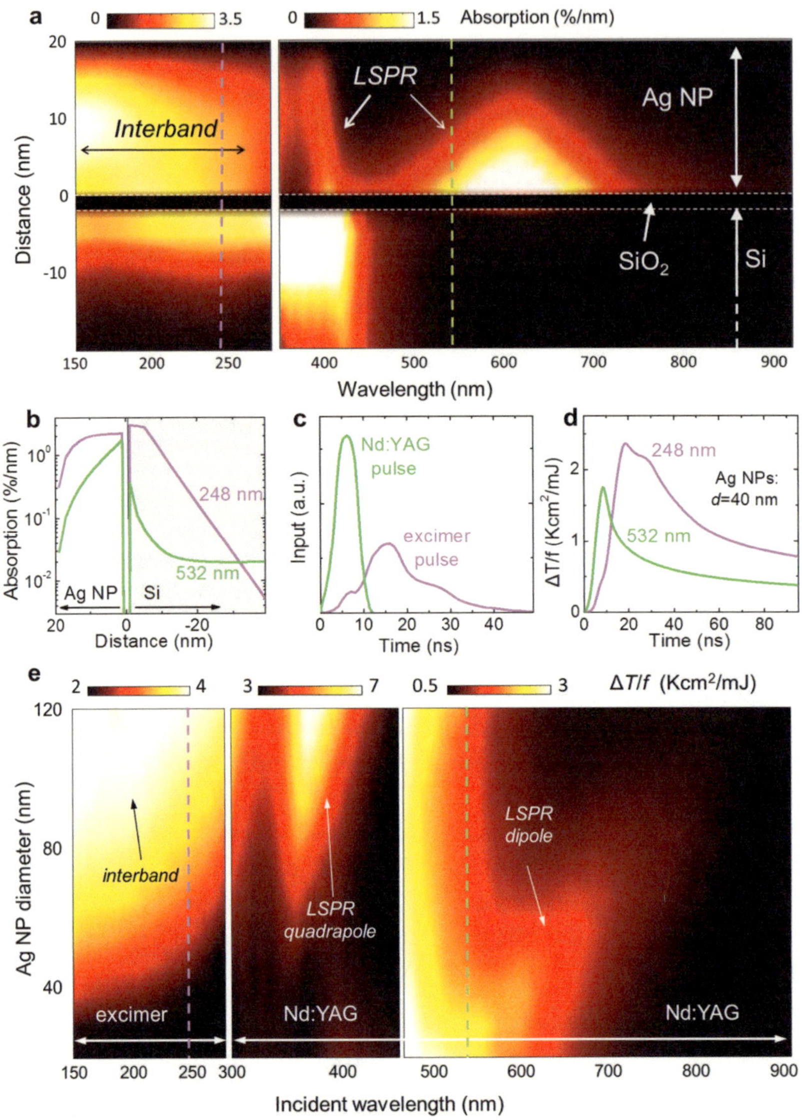

Figure 13. (a) Vertical absorption (% per nm) profile for the 40 nm diameter Ag nanoparticle. Indicated with the two dashed lines are the two primary wavelengths used (248 and 532 nm). (b) Absorption percentage per nm profile in the model structure for the two wavelengths used. (c) Temporal profile of the two different laser sources. (d) Corresponding temperature transients developed in the Ag NP (normalized to the laser fluence). (e) Global map showing the peak temperature developed in an Ag NP on top of Si versus wavelength and NP size.

As a first step, a single shot UV LA process was utilised, to initialize nanoparticle formation from a semi-continuous metal thin film, by laser induced self-assembly (LISA). This resulted into the restructuring of the Ag thin film into nanoparticle arrangements in macroscopic scales (several mm^2). The LISA process with a UV laser (248 nm) is strongly fluence-dependent as demonstrated by the SEM images of **Figure 14a**. The Ag NP size distributions are generally

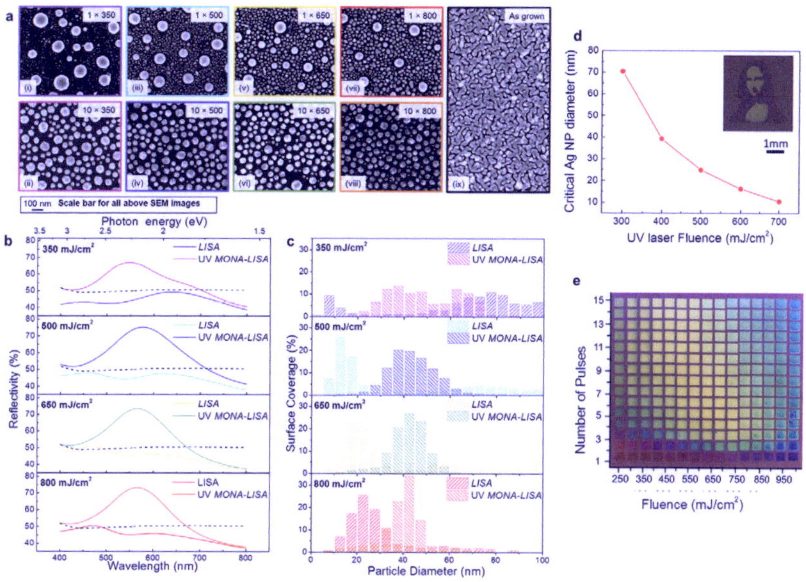

Figure 14. (a) SEM images of samples processed with LISA (1 pulse, 248 nm LA) and UV MONA-LISA (10 pulses, 248 nm) under various applied fluences. (b) The effect of LISA and UV MONA-LISA on the optical reflectivity spectra of a 10 nm Ag thin film under various applied fluences. (c) Surface coverage of the nanoparticles in relation to the particle diameter for all the cases of (a). (d) Critical nanoparticle diameter (the diameter value under which every particle remains practically cold during the UV LA process, hence these particles cannot be further manipulated upon irradiation) as a function of applied laser fluence. Inset depicts a portrait of Da Vinci's Mona Lisa: a microscopic pattern engraved on silicon wafer in a single pixel processing using series of photolithographic masks. (e) A digital photo of a grid of LISA / UV MONA-LISA spots of varying fluence (x-axis) and number of pulses (y-axis) produced on a 10 nm Ag thin film.

bimodal, consisting of one set of coarse and one set of fine NPs. Increasing the laser fluence from 350 to 800 mJ/cm^2 resulted into a decreasing difference between the two NP sets, as shown by the quantified SEM images (**Figure 14c**, purple, cyan, yellow and red lines). The modification of NP assemblies by multiple laser pulses (MONA-LISA process) involves the utilization of the selective optical absorption and the heat dissipation, as presented in the theoretical approach above. For this consideration, the initial assemblies of NPs were subjected to a series of UV and/ or VIS LA steps.

The case of UV MONA-LISA is first considered, i.e. processing by multiple UV (248 nm) laser pulses, and the effect of inter-band absorption is correlated with the size-selective heat dissipation. **Figure 14b** shows the optical reflectivity spectra of a 10 nm Ag film recorded after UV MONA-LISA applying 10 successive UV laser pulses at various fluences (magenta, blue, green and orange lines for 350, 500, 650 and 800 mJ/cm^2, respectively). Comparing these spectra with those from the previous LISA processing revealed that the consecutive pulses transform the two reflectivity peaks into one, indicating a gradual transformation of the bimodal distributions into unimodal ones. This was also confirmed by the corresponding SEM images and the size distribution histograms in **Figure 14a** and **c** (magenta, blue, green and orange bars), respectively. The reshaping of the particles is predominantly due to the disappearance of the larger particles. The physics behind that observation is based on the stronger heating of the larger NPs and faster heat dissipation from the finer NPs due to their higher effective surface (surface to volume ratio). As a result, the finer nanoparticles remain cold during UV MONA-LISA. To quantify this, photothermal calculations [47] were employed, in order to determine the equitemperature contour lines vs. nanoparticle diameter, from which the critical size under which the particles remain practically cold during the UV MONA-LISA. Considering the bulk Ag melting point of 960°C, the calculated values of the critical diameter vs. the laser fluence (248 nm) are displayed in **Figure 14d**. We remark that the critical diameter is highly dependent on the thermal conductivity of the substrate; less thermally conductive substrate would hinder heat dissipation. The UV MONA-LISA process is thus based on the ability of the larger particles to raise their temperature above their melting point, and consequently re-solidifying forming smaller particles with a stochastic size distribution, while the smaller ones remain below the melting threshold and therefore remain practically unaffected. As a result, successive pulses suppress the population of the coarse particles, gradually transforming them to finer particles via stochastic size redistribution. The wide range of size distributions achieved by LISA and UV MONA-LISA results in a vast colour palette of plasmonic responses as those presented in **Figure 14e** (real visual appearance of the Ag nanoparticle assemblies). Combining LISA and MONA-LISA process with projection masks enables the production of multicolour images of high spatial resolution and colour contrast. Such an example is presented in the inset of **Figure 14d**, depicting the portrait of Da-Vinci's famous 'Mona Lisa'.

The implementation of MONA-LISA employing exclusively UV pulses has the inherent limitation of the UV absorption being rather insensitive to nanoparticle size, and as a result, the size selectivity of the process is based exclusively on the size selectivity of the heat dissipation. Thus, only the large particles are affected by the successive laser pulses. Finer particles can be further manipulated by exploiting their selective optical absorption via LSPR using a visible wavelength, resulting in larger optical absorption and hence higher temperature rise, as predicted by the computational results presented in **Figure 13**. The optimal condition of narrowing the LSPR bandwidth at 532 nm was achieved by applying 500 pulses at 125 mJ/ cm^2, which shows that the VIS MONA-LISA is slower compared to the UV MONA-LISA that reached at steady state at 10 pulses. This is well expected considering the computational results of **Figure 13**, where a lower temperature rise is evident at visible wavelengths compared to UV processing.

Exploring the VIS MONA-LISA one can start from a UV MONA-LISA process (248 nm, 2 pulses at 350 mJ/cm^2) which demonstrates a double reflectance peak (due to bimodal nanostructuring) notably with no particular response close to 532 nm, as shown in **Figure 15a** (black line). Applying, VIS MONA-LISA to this area, with 500 pulses (i), 2000 pulses (ii) and 4000 pulses (iii) results in gradually re-adjusting the bimodal distribution to one with a peak very close to the laser processing wavelength (532 nm). The corresponding SEM images are shown in Ref. [47]. Analysis of these SEM images results in the histograms of **Figure 15b**. It is noteworthy that the VIS MONA-LISA probes particularly the finer Ag nanoparticles (<20 nm) confirming the computational analysis presented in **Figure 13**.

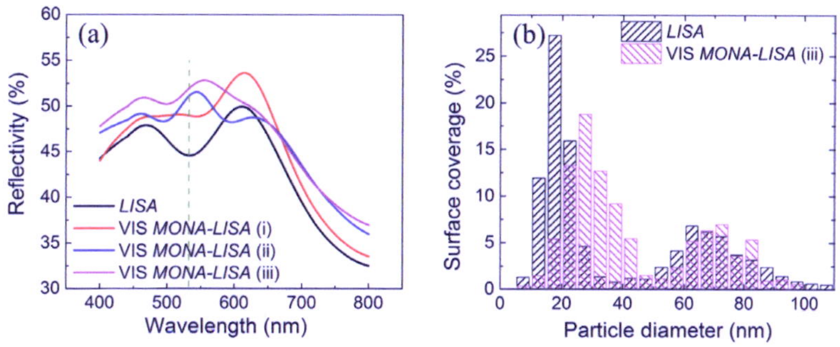

Figure 15. (a) Optical reflectivity spectra of LISA (2 pulse, 248 nm at 350 mJ/cm^2 - black line), used as a template for the subsequent Visible MONA-LISA (500 pulses (red line), 2000 pulses (blue line) and 4000 pulses (magenta line), 532 nm at 125 mJ/cm^2). The green dashed line sets the 532 nm mark. (b) Surface coverage of the nanoparticles in relation to the particle diameter for each laser treatment of (a).

5. Conclusions

Functional plasmonic templates can be fabricated by LA. In particular, two distinct cases have been discussed: laser nanostructuring of either ceramic/metal multilayers and composites or ultra-thin metal films results on sub-surface or surface plasmonic patterns, respectively, with many potential applications.

Theoretical investigations have highlighted the underline physics of laser-induced sub-surface and surface plasmonic nanostructuring: the absorption profile of the film and the temperature spatial distribution strongly depend on the structure parameters such as the metal volume ratio, the total thickness of the film and the thermal conductivity of the dielectric. By properly designing these parameters, combined with tuning the LA conditions (wavelength, number of pulses and fluence) one can arrive at plasmonic templates with predetermined morphology and optical response. The latter was demonstrated in experiments in which either AlN:Ag nanocomposites or multilayers of alternative AlN/Ag layers were subjected in UV LA leading to a sub-surface formation of PNPs sealed in an AlN matrix with a localized surface plasmon resonance response.

An extensive experimental and theoretical investigation has highlighted the photothermal processes involved in laser-induced surface plasmonic nanostructuring. Specifically, templates consisting of hemi-spherical Ag PNPs on Si substrate, with different arrangements and sizes, for both UV and optical irradiation have been explored. Different absorption mechanisms, which are size selective and enable geometric specificity, have been identified: (a) inter-band absorption taking place at the UV zone, probing the metal's internal d-electrons and (b) intra-band absorption taking place at the near-UV, and visible zone, probing the metal's free electrons at the specific spectral position of LSPR. In particular, by sequentially tuning the laser wavelength into resonance with different physical absorptions, we can selectively target the melting and re-solidification of different particle size groups. This process was termed: "modification of nanoparticle arrays by laser-induced self-assembly" (MONA-LISA).

Acknowledgements

N. Kalfagiannis acknowledges funding from People Programme (Marie Curie Actions) LASER-PLASMON of the European Union's Seventh Framework Programme (FP7/2007-2013) under REA grant agreement n° PIEF-GA-2012-330444. D.C. Koutsogeorgis acknowledges the School of Science and Technology at Nottingham Trent University for the approval of a sabbatical made possible via QR funds, and the Higher Education Funding Council for England (HEFCE) for providing the QR funds as a result of the Research Assessment Exercise 2008 (RAE2008). The research activities of E. Lidorikis and P. Patsalas that led to these results were co-financed by Hellenic Funds and by the European Regional Development Fund (ERDF) under the Hellenic National Strategic Reference Framework (NSRF) 2007–2013, according to contract no. 11ΣΥΝ-5-1280 of the Project 'Nano-Hybrid', within the Program 'Competitiveness and Entrepreneurship'.

Author details

Nikolaos Kalfagiannis[1]*, Demosthenes C. Koutsogeorgis[1], Elefterios Lidorikis[2] and Panos Patsalas[3]

*Address all correspondence to: nikolaos.kalfagiannis@ntu.ac.uk

1 School of Science and Technology, Nottingham Trent University, Nottingham, United Kingdom

2 Department of Materials Science and Engineering, University of Ioannina, Ioannina, Greece

3 Department of Physics, Aristotle University of Thessaloniki, Thessaloniki, Greece

References

[1] N. Lagos, M. M. Sigalas, E. Lidorikis. Theory of plasmonic near-field enhanced absorption in solar cells. Applied Physics Letters. 2011;**99**:063304.

[2] N. Kalfagiannis, P. G. Karagiannidis, C. Pitsalidis, N. T. Panagiotopoulos, C. Gravalidis, S. Kassavetis, P. Patsalas, S. Logothetidis. Plasmonic silver nanoparticles for improved organic solar cells. Solar Energy Materials and Solar Cells. 2012;**104**:165-174.

[3] H. A. Atwater, A. Polman. Plasmonics for improved photovoltaic devices. Nature Materials. 2010;**9**:205-213.

[4] M. J. Beliatis, S. J. Henley, S. Han, K. Gandhi, A. A. D. T. Adikaari, E. Stratakis, E. Kymakis, S. R. P. Silva. Organic solar cells with plasmonic layers formed by laser nanofabrication. Physical Chemistry Chemical Physics. 2013;**15**:8237-8244.

[5] H. Zoubos, L. E. Koutsokeras, D. F. Anagnostopoulos, E. Lidorikis, S. A. Kalogirou, A. R. Wildes, P. C. Kelires, P. Patsalas. Broadband optical absorption of amorphous carbon/

Ag nanocomposite films and its potential for solar harvesting applications. Solar Energy Materials and Solar Cells. 2013;**117**:350-356.

[6] P. Nikolaou, C. Mina, M. Constantinou, L.E. Koutsokeras, G. Constantinides, E. Lidorikis, A. Avgeropoulos, P.C. Kelires, P. Patsalas. Functionally graded poly(dimethylsiloxane)/silver nanocomposites with tailored broadband optical absorption. Thin Solid Films. 2015;**581**:14-19.

[7] E. Lidorikis. Modeling of enhanced absorption and Raman scattering caused by plasmonic nanoparticle near fields. Journal of Quantitative Spectroscopy and Radiative Transfer. 2012;**113**:2573-2584.

[8] M. Moskovits. Surface-enhanced spectroscopy. Reviews of Modern Physics. 1985;**57**:783-826.

[9] K. Kneipp, Y. Wang, H. Kneipp, L. T. Perelman, I. Itzkan, R. R. Dasari, M. S. Feld. Single molecule detection using surface-enhanced raman scattering (SERS). Physical Review Letters. 1997;**78**(9):1667.

[10] Y. Fang, N. H. Seong, D. D. Dlott. Measurement of the distribution of site enhancements in surface-enhanced Raman scattering. Science. 2008;**321:**388-392.

[11] S. J. Henley, J. D. Carey, S. R. P. Silva. Laser-nanostructured Ag films as substrates for surface-enhanced Raman spectroscopy. Applied Physics Letters. 2006;**88**:081904.

[12] S. J. Henley, S. R. P. Silva. Laser direct write of silver nanoparticles from solution onto glass substrates for surface-enhanced Raman spectroscopy. Applied Physics Letters. 2007;**91**:023107.

[13] K. Christou, I. Knorr, J. Ihlemann, H. Wackerbarth, V. Beushausen. Fabrication and characterization of homogeneous surface-enhanced Raman scattering substrates by single pulse UV-laser treatment of gold and silver films. Langmuir. 2010;**26**(23):18564-18569.

[14] C. H. Lin, L. Jiang, J. Zhou, H. Xiao, S. J. Chen, H. L Tsai. Laser-treated substrate with nanoparticles for surface-enhanced Raman scattering. Optics Letters. 2010;**35**(7):941-943.

[15] J. Chen, C. Glaus, R. Laforest, Q. Zhang, M. Yang, M. Gidding, M. J. Welch, Y. Xia. Gold nanocages as photothermal transducers for cancer treatment. Small. 2010;**6**:811-817.

[16] C. Ayala-Orozco, C. Urban, M. W. Knight, A. S. Urban, O. Neumann, S. W. Bishnoi, S. Mukherjee, A. M. Goodman, H. Charron, T. Mitchell, M. Shea, R. Roy, S. Nanda, R. Schiff, N. J. Halas, A. Joshi. Au nanomatryoshkas as efficient near-infrared photothermal transducers for cancer treatment: benchmarking against nanoshells. ACS Nano. 2014;**8**:6372-6381.

[17] A. Siozios, N. Kalfagiannis, D. V. Bellas, C. Bazioti, G.P. Dimitrakopulos, G. Vourlias, W.M. Cranton, E. Lidorikis, D. C. Koutsogeorgis, P. Patsalas. Sub-surface laser nanostructuring in stratified metal/dielectric media: a versatile platform towards flexible, durable and large-scale plasmonic writing. Nanotechnology. 2015;**26**:155301.

[18] C. Bazioti, G. P. Dimitrakopulos, T. Kehagias, P. Komninou, A. Siozios, E. Lidorikis, D. C. Koutsogeorgis, P. Patsalas. Influence of laser annealing on the structural properties of sputtered AlN: Ag plasmonic nanocomposites. Journal of Materials Science. 2014;**49**:3996-4006.

[19] A. Siozios, D. C. Koutsogeorgis, E. Lidorikis, G. P. Dimitrakopulos, T. Kehagias, H. Zoubos, P. Komninou, W. M. Cranton, C. Kosmidis, P. Patsalas. Optical encoding by plasmon-based patterning: hard and inorganic materials become photosensitive. Nano Letters. 2012;**12**:259-263.

[20] A. Siozios, D. C. Koutsogeorgis, E. Lidorikis, G. P. Dimitrakopulos, N. Pliatsikas, G. Vourlias, T. Kehagias, P. Komninou, W. Cranton, C. Kosmidis, P. Patsalas. Laser-matter interactions, phase changes and diffusion phenomena during laser annealing of plasmonic AlN: Ag templates and their applications in optical encoding. Journal of Physics D: Applied Physics. 2015;**48**:285306.

[21] C. F. Bohren, D. R. Huffman. Absorption and scattering of light by small particles. Wiley Interscience, New York; 1983.

[22] M. Rycenga, C. M. Cobley, J. Zeng, W. Li, C. H. Moran, Q. Zhang, D. Qin, Y. Xia. Controlling the synthesis and assembly of silver nanostructures for plasmonic applications. Chemical Reviews. 2011;**111**:3669-3712.

[23] S. A. Maier. Plasmonics: fundamentals and applications. Springer, New York; 2007.

[24] U. Kreibig, M. Vollmer. Optical properties of metal clusters. Springer, Berlin; 1995.

[25] N. J. Halas. Plasmonics: an emerging field fostered by Nano Letters. Nano Letters. 2010;**10**:3816-3822.

[26] M. L. Brongersma, V. M. Shalaev. The case for plasmonics. Science. 2010;**328**:440-441.

[27] E. Ozbay. Plasmonics: merging photonics and electronics at nanoscale dimensions. Science. 2006;**311**:189-193.

[28] K. S. Lee, M. A. J. El-Sayed. Gold and silver nanoparticles in sensing and imaging: sensitivity of plasmon response to size. Journal of Physical Chemistry B. 2006;**110**:19220-19225.

[29] M. M. Miller, A. A. Lazarides. Sensitivity of metal nanoparticle surface plasmon resonance to the dielectric environment. Journal of Physical Chemistry B. 2005;**109**:21556-21565.

[30] O. Kvitek, J. Siegel, V. Hnatowicz, V. Svorcik. Noble metal nanostructures: influence of structure and environment on their optical properties. Journal of Nanomaterials. 2013;**2013**:743684.

[31] M. R. Jones, K. D. Osberg, R. J. Macfarlane, M. R. Langille, C. A. Mirkin. Templated techniques for the synthesis and assembly of plasmonic nanostructures. Chemical Reviews. 2011;**111**:3736-3827.

[32] X. Lu, M. Rycenga, S. E. Skrabalak, B. J Wiley, Y. Xia. Chemical synthesis of novel plasmonic nanoparticles. Annual Reviews of Physical Chemistry. 2009;**60**:167-192.

[33] G. Walters, I. P. Parkin. The incorporation of noble metal nanoparticles into host matrix thin films: synthesis, characterisation and applications. Journal of Materials Chemistry. 2009;**19**:574-590.

[34] L. Armelao, D. Barreca, G. Bottaro, A. Gasparotto, S. Gross, C. Maragnob, E. Tondello. Recent trends on nanocomposites based on Cu, Ag and Au clusters: a closer look. Coordination Chemistry Reviews. 2006;**250**:1294-1314.

[35] J. H. Seo, J. H. Park, S. I. Kim, B. J. Park, Z. Ma, J. Choi, B. K. Ju. Nanopatterning by laser interference lithography: applications to optical devices. Journal of Nanoscience and Nanotechnology. 2014;**14**:1521-1532.

[36] U. Guler, J. C. Ndukaife, G. V. Naik, A. G. A. Nnanna, A. V. Kildishev, V. M. Shalaev, A. Boltasseva. Local heating with lithographically fabricated plasmonic titanium nitride nanoparticles. Nano Letters. 2013;**13**:6078-6083.

[37] G. Bracher, K. Schraml, M. Ossiander, S. Frédérick, J. J. Finley, M. Kaniber. Optical study of lithographically defined, subwavelength plasmonic wires and their coupling to embedded quantum emitters. Nanotechnology. 2014;**25**:075203.

[38] J. Toudert, D. Babonneau, S. Camelio, T. Girardeau, F. Yubero, J. P. Espinós, A. R. Gonzalez-Elipe. Using ion beams to tune the nanostructure and optical response of co-deposited Ag:BN thin films. Journal of Physics D: Applied Physics. 2007;**40**:4614-4620.

[39] V. A. Sivakov, K. Höflich, M. Becker, A. Berger, T. Stelzner, K. E. Elers, V. Pore, M. Ritala, S. H. Christiansen. Silver coated platinum core–shell nanostructures on etched Si nanowires: atomic layer deposition (ALD) processing and application in SERS. ChemPhysChem. 2010;**11**:1995-2000.

[40] Y. Lu, J. Y. Huang, C. Wang, S. Sun, J. Lou. Cold welding of ultrathin gold nanowires. Nature Nanotechnology. 2010;**5**:218-224.

[41] Y. K. Mishra, S. Mohapatra, D. Kabiraj, B. Mohanta, N. P. Lalla, J.C. Pivin, D. K. Avasthi. Synthesis and characterization of Ag nanoparticles in silica matrix by atom beam sputtering. Scripta Materialia. 2007;**56**:629-632.

[42] I. Tanahashi, Y. Manabe, T. Tohda, S. Sasaki, A. Nakamura. Optical nonlinearities of Au/SiO_2 composite thin films prepared by a sputtering method. Journal of Applied Physics. 1996;**79**:1244-1249.

[43] D. Lin, H. Tao, J. Trevino, J. P. Mondia, D. L. Kaplan, F. G. Omenetto, L. Dal Negro. Direct transfer of subwavelength plasmonic nanostructures on bioactive silk films. Advanced Materials. 2012;**24**:6088-6093.

[44] P. Nielsen, P. Morgen, A. C. Simonsen, O. Albrektsen. Hemispherical shell nanostructures from metal-stripped embossed alumina on aluminum templates. Journal of Physical Chemistry C. 2011;**115**:5552-5560.

[45] J. H. Park, P. Nagpal, K. M. McPeak, N. C. Lindquist, S. H. Oh, D. J. Norris. Fabrication of smooth patterned structures of refractory metals, semiconductors, and oxides via template stripping. ACS Applied Materials and Interfaces. 2013;**5**:9701-9708.

[46] P. Farzinpour, A. Sundar, K. D. Gilroy, Z. E. Eskin, R. A. Hughes, S. Neretina. Dynamic templating: a large area processing route for the assembly of periodic arrays of sub-micrometer and nanoscale structures. Nanoscale. 2013;**5**:1929-1938.

[47] N. Kalfagiannis, A. Siozios, D. V. Bellas, D. Toliopoulos, L. Bowen, N. Pliatsikas, W. M Cranton, C. Kosmidis, D. C. Koutsogeorgis, E. Lidorikis, P. Patsalas. Selective modification of nanoparticle arrays by laser-induced self assembly (MONA-LISA): putting control into bottom-up plasmonic nanostructuring. Nanoscale. 2016;**8**:8236-8244.

[48] M. J. Beliatis, S. J. Henley, S. R. P. Silva. Engineering the plasmon resonance of large area bimetallic nanoparticle films by laser nanostructuring for chemical sensors. Optics Letters. 2011;**36**(8):1362-1364.

[49] C. Favazza, R. Kalyanaraman, R. Sureshkumar. Robust nanopatterning by laser-induced dewetting of metal nanofilms. Nanotechnology. 2006;**17**:4229-4234.

[50] H. Krishna, N. Shirato, C. Favazza, R. Kalyanaraman. Pulsed laser induced self-organization by dewetting of metallic films. Journal of Materials Research. 2011;**26**(2):154-169.

[51] C. E. Rodrígueza, R. J. Peláeza, C. N. Afonsoa, S. Riedelb, P. Leidererb, D. Jimenez-Reyc, A. Climent-Fontc. Plasmonic response and transformation mechanism upon single laser exposure of metal discontinuous films. Applied Surface Science. 2014;**302**:32-36.

[52] F. Ruffino, A. Pugliara, E. Carria, C. Bongiorno, C. Spinella, M. G. Grimaldi. Formation of nanoparticles from laser irradiated Au thin film on SiO_2/Si: elucidating the Rayleigh-instability role. Materials Letters. 2012;**84**:27-30.

[53] F. Ruffino, E. Carria, S. Kimiagar, I. Crupi, F. Simone, M. G. Grimaldi. Formation and evolution of nanoscale metal structures on ITO surface by nanosecond laser irradiations of thin Au and Ag films. Science of Advanced Materials. 2012;**4**:708-718.

[54] J. Trice, D. Thomas, C. Favazza, R. Sureshkumar, R. Kalyanaraman. Pulsed-laser-induced dewetting in nanoscopic metal films: theory and experiments. Physical Review B. 2007;**75**:235439.

[55] Ch. E. Lekka, P. Patsalas, Ph. Komninou, G. A. Evangelakis. Electronic properties and bonding characteristics of AlN:Ag thin film nanocomposites. Journal of Applied Physics. 2011;**109**:054310.

[56] P. Lautenschlager, M. Garriga, L. Viña, M. Cardona. Temperature dependence of the dielectric function and interband critical points in silicon. Physical Review B. 1987;**36**:4821-4830.

[57] P. L. Redmond, A. J. Hallock, L. E. Brus. Electrochemical Ostwald ripening of colloidal Ag particles on conductive substrates. Nano Letters. 2005;**5**:131-135.

[58] A. Lotsari, G. P. Dimitrakopulos, T. Kehagias, P. Kavouras, H. Zoubos, L. E. Koutsokeras, P. Patsalas, Ph. Komninou. Structure, stability and mechanical performance of AlN:Ag nanocomposite films. Surface and Coatings Technology. 2010;**204**:1937-1941.

[59] G. M. Matenoglou, H. Zoubos, A. Lotsari, C.E. Lekka, Ph. Komninou, G. P. Dimitrakopulos, C. Kosmidis, G. A. Evangelakis, P. Patsalas. Metal-containing amorphous carbon (a-C:Ag) and AlN (AlN:Ag) metallo-dielectric nanocomposites. Thin Solid Films. 2009;**518**:1508-1511.

[60] A. Siozios, H. Zoubos, N. Pliatsikas, D. C. Koutsogeorgis, G. Vourlias, E. Pavlidou, W. Cranton, P. Patsalas. Growth and annealing strategies to control the microstructure of AlN:Ag nanocomposite films for plasmonic applications. Surface and Coatings Technology. 2014;**255**:28-36.

[61] J. D. Jackson. Classical electrodynamics. 3rd ed. Wiley, New York; 1999.

[62] E. Lidorikis, S. Egusa, J. D. Joannopoulos. Effective medium properties and photonic crystal superstructures of metallic nanoparticle arrays. Journal of Applied Physics. 2007;**101**:054304.

[63] R. Donald, L. E. S. Pitts. Schaum's outline of theory and problems of heat transfer. McGraw-Hill, New York; 1998.

[64] C. Duquenne, M. P. Besland, P. Y. Tessier, E. Gautron, Y. Scudeller, D. Averty. Thermal conductivity of aluminium nitride thin films prepared by reactive magnetron sputtering. Journal of Physics D: Applied Physics. 2012;**45**:015301.

[65] G. A. Slack, R. A. Tanzilli, R. O. Pohl, J. W. Vandersande. The intrinsic thermal conductivity of AlN. Journal of Physics and Chemistry of Solids. 1987;**46**:641-647.

Chapter 19

Fabrication of Plasmonic Crystalline Thin Film of Titanium Nitride (TiN) by Pulsed Laser Deposition with Third Harmonic of Nd:YAG Laser and Its Spectroscopic Analyses

Yasushi Oshikane

Additional information is available at the end of the chapter

http://dx.doi.org/10.5772/67765

Abstract

The author has been engaged in the development of a novel optical fiber probe using scanning near-field optical microscope (SNOM) with an efficient, plasmonic and asymmetric Metal-Insulator-Metal (MIM) structure at the probe tip. As a metallic layer, titanium nitride (TiN), one of the alternative plasmonic materials, is selected. A pulsed laser deposition (PLD) is used to fabricate the film by high-power Nd:YAG laser. The PLDed films have been analyzed by X-ray diffractometer (XRD), UV-Vis/NIR spectrophotometer, scanning electron microscope (SEM), and X-ray photoelectron spectroscopy (XPS). Though most of previous PLD studies of TiN film used a titanium target with reactive gases, the study presented in this chapter has significant features of (1) a hot pressed target of crystalline TiN powder and (2) third harmonic of injection-seeded Nd:YAG laser which have temporally smoothed Gaussian with a constant pulse energy. The very first PLD process has succeeded to fabricate flat and dense films of a few hundred nanometers. The TiN film, which lustered like gold, indicated two peaks at 36.7° (111) and 42.6° (200) in XRD patterns that correspond to crystal structure of TiN. An elementary analysis of the TiN film has carried out using XPS, and appropriate spectra with chemical shifts were observed.

Keywords: alternative plasmonic material, titanium nitride (TiN), plasmonic thin film, pulsed laser deposition (PLD), crystalline thin film, third harmonic of pulsed Nd:YAG laser, X-ray diffractometer (XRD), electromagnetic FEM simulation, X-ray photoelectron spectroscopy (XPS)

1. Introduction

In the 1980s, basic and applied research regarding surface plasmon, which is a compression wave excited at boundary between metallic and insulating thin films and propagates along the boundary, was started. Propagation characteristics of the wave depend on combined complex dielectric constant of the two media, which form the boundary. The surface plasmon, which is locally excited at the medium surface, has been widely used in (1) a highly sensitive detection mechanism for an infinitesimal amount of element or molecule, (2) formation of highly intense point light source at the apex of scanning probe used in scanning near-field optical microscopes (SNOMs), and (3) localized intense fields at the surface of single tiny metallic particle as a highly sensitive detector or a strong illuminator.

(1) Surface plasmon resonance caused by the total internal reflection inside of a transparent insulator covered with metallic thin layer and (2) surface plasmon polarities excited in subwavelength structure consisting metallic and insulating thin films need appropriate materials including free electrons which might be oscillated by external electromagnetic of light. And response of the electrons to the light field must show negative real component in complex dielectric constant. As a result, noble metals, i.e., gold (Au) and silver (Ag), have been widely used for plasmonic research. A silver thin film is known as the best plasmonic material in the visible region but transubstantiates easily. Also silver is soft and does not have excellent adhesive power to the glass surface. Gold thin film is chemically stable and soft and does not have negative real component of complex dielectric constant in the whole visible region. Surface plasmon and surface plasmon polariton have potentialities of the development in extremely localized and long-range propagation transmission medium for near future-interconnection system between the integrated electronic circuit and the optical communication circuit. In the actual electronic circuits in Ultra Large-Scale Integration (ULSIs), there is no device and waveguide based on silver but few tens of nanometer copper wirings only. But in CMOS devices, there are TiN thin layers as a copper diffusion barrier, and this film may act as excitation source and waveguide of plasmon or plasmon polariton waves when it has good crystallinity. TiN film is convenient for future optoelectronic circuits because of its compatibility with device fabrication process [1–11].

Titanium nitride (TiN) is one of the multirole compounds used in many technological areas. The application of TiN, which belongs to the early transition metal nitrides, is originated from their decent properties. Namely, they have high electrical and thermal conductivities, which are comparable to metals. Additionally, they are very hard with high melting points. Due to its good corrosion and wear resistances, TiN is used as a protective coating on industrial tools worked in inhospitable environment. With TiN coatings of about 2–10 μm thicknesses tool life is increased several fold generally. TiN also has important tasks in the semiconductor industry. It is used as a diffusion barrier layer in contact structures, both to prevent Cu and to provide an adhesive layer between dielectric material and Cu. Even though TiN is stable over a broad composition range, its structure and properties depend critically on its actual composition. Electrical resistivity of TiN depends strongly on its stoichiometry and morphology. The presence of oxygen and/or carbon in TiN takes on low hardness and high resistivity. One of the effective compounds is titanium oxynitride (TiON).

In recent research on alternative novel plasmonic materials, silicon, germanium, group III–V nitride semiconductors, arsenides, phosphides, and transparent conducive films are examined energetically. Transition metal nitrides, i.e., TiN, TaN, ZrN, HfN, attract increasing attention because of their golden luster, metallic properties in the Vis-NIR region, characteristics as nonstoichiometric and interstitial compound, and high free carrier densities.

As a result, the optical properties will be controlled by altering the composition of them. These materials are used as a diffusion barrier in damascene interconnect structure and gate metal in p-type/n-type transistor devices. In this work, I have been interested in TiN lately. One of the latest review papers described that this film has been formed widely and intensively by various techniques such as magnetron sputtering, cathode arc discharge, chemical vapor deposition, atomic layer deposition, pulsed laser deposition, ion beam-assisted deposition, high power impulse magnetron sputtering, etc. Achievements of the predecessors gave us various resultant TiN films with many different characteristics, which were quite sensitive to the experimental conditions.

Origin of the studies in disarray may connect closely with chemical composition of the films such as Ti_xN or TiN_x ($x \neq 1$). Also, when the chemical composition is nearly TiN_x ($x \sim 1$), its crystallinity was not enough or crystal grain size was very small. As a result, without taking our eyes off of plasmonics with TiN film, creation of extreme flat TiN film with high crystallinity has been a long-time consuming. Previous studies usually made a reaction field, which contains both Ti and N atoms, and the field was excited by a certain type of electrical discharge or laser-produced plasma. The resultant TiN film was deposited slowly [12–17].

In this study, pulsed laser deposition (PLD) by irradiation of ultraviolet laser pulses has been adopted into the thin-film formation. The reason is because (1) it is a clean process composed of pure solid target, clean substrate, and dense ultraviolet photons only, (2) it may be possible to form a film having a similar chemical composition, and (3) the deposited film thickness can be controlled with laser pulse number basically. In contrast to predecessors in thin film generation of TiN, PLD procedure in this experiment uses a hot pressed target of TiN crystal powder to create a highly crystalline TiN thin film, which is never formed [18–23].

Purposes of this study are (1) clarification of TiN film formation by PLD with the hot pressed TiN target, (2) promising samples culled from the PLD process will be used for basic experiments of surface plasmon resonance excitation, (3) metal-insulator-metal structure will be formed with the TiN film and the AlN film, which will be also deposited by the PLD method, and surface plasmon polarities will be excited in the multilayered structure.

2. First experience with commercial TiN films and instruments for fabrication and analyses of PLDed TiN thin film

At the beginning, since the author has no experience in generating and treating TiN film, several types of TiN samples for comparison were provided by courtesy of Mr. Akitaka Kashihara (Shinko Seiki). The samples were made by RF sputtering based on Ar/N_2 plasma, Ti target, and substrate (GS: S2112 glass slide, Matsunami Glass). Three samples of (a) TiN (50 nm)/GS, (b) SiO_2 (150 nm)/TiN (50 nm)/GS, and (c) TiN (200 nm)/SiO_2 (150 nm)/TiN (50 nm)/GS were produced experimentally with a view to asymmetric MIM structures. The thickness or duration of sputtering was calibrated with other samples.

And (d) a disparate TiN(50 nm)/QZ (Quartz) sample was also given by courtesy of Mr. Kaoru Hoshino (Parker S·N Kogyo). A quartz substrate covered with vacuum evaporated Ti thin film (50 nm) was treated by nitriding process with high temperature nitrogen gas (800°C, 1 hour), which is commonly applied to protection coating of industrial tools.

The X-ray diffraction (XRD) of these thin films was carried out using a Rigaku diffractometer (SmartLab) with a Cu Kα radiation operating at 20 kV voltage and 10 mA current. Unfortunately, most of them did not give effective XRD data. Only sample (c) had sharp three peaks corresponding to TiN(111), (200), and (222), which may be signals from outermost TiN (200 nm) layer and agreed well with previous papers. The thick TiN layer had luster of gold.

XRD data given by Mr. Hoshino showed that a thicker sample, which is the same as sample (d), has sharp peak of Ti_2N. Therefore, high temperature nitriding is good for tool protection but is not good for creation of plasmonic thin film.

As described herein below, samples (a)–(c) did not show good results in X-ray photoelectron spectroscopy (XPS). Therefore, the author chose the PLD method as a film deposition process for plasmonic TiN film. There is a commonly held view that PLD can deposit epitaxial thin film with stoichiometric composition in a clean atmosphere.

Deposition of TiN film was performed using a pulsed third harmonic (3ω) YAG laser system (λ = 355 nm, E_L = up to 350 mJ/pulse, pulse duration τ_L = 3.3 nsFWHM, Powerlite 9010, Continuum) as shown in **Figure 1(a)**. The pulse energy was monitored by a thermopile detector (30A-P-SH (AN/2), Ophir), and the pulse duration was observed by an universal streak camera (C5680-01, Hamamatsu Photonics) (**Figure 1(b)** and **(c)**) coupled with a cooled back-illuminated CCD camera (PIXIS1024B, Princeton Instruments) (**Figure 1(d)**). While most pulsed lasers used for PLD do not pay attention to their temporal pulse shape, and the shape is always considered a smoothed Gaussian, is not it? Actually, they are spiky, and the temporal shape changes with respect to each pulse. Therefore, the author decided to use an injection seeding by extremely narrow-spectrum (<5 kHz) IR fiber laser seeder (SI-2000, NP Photonics) at 1064 nm, and the resultant every pulse shape is maintained in smoothed Gaussian as shown in later. The resultant bandwidth of 3ω pulse was around 0.003 cm^{-1} and also each pulse energy kept constant. This may have the effect of both debris suppression and good surface flatness in the PLD process.

The pulsed laser beam was focused on the PLD target placed in center of the UHV chamber evacuated to 5×10^{-5} Pa by turbo-molecular pump (TMP, Varian) coupled with a dry scroll vacuum pump

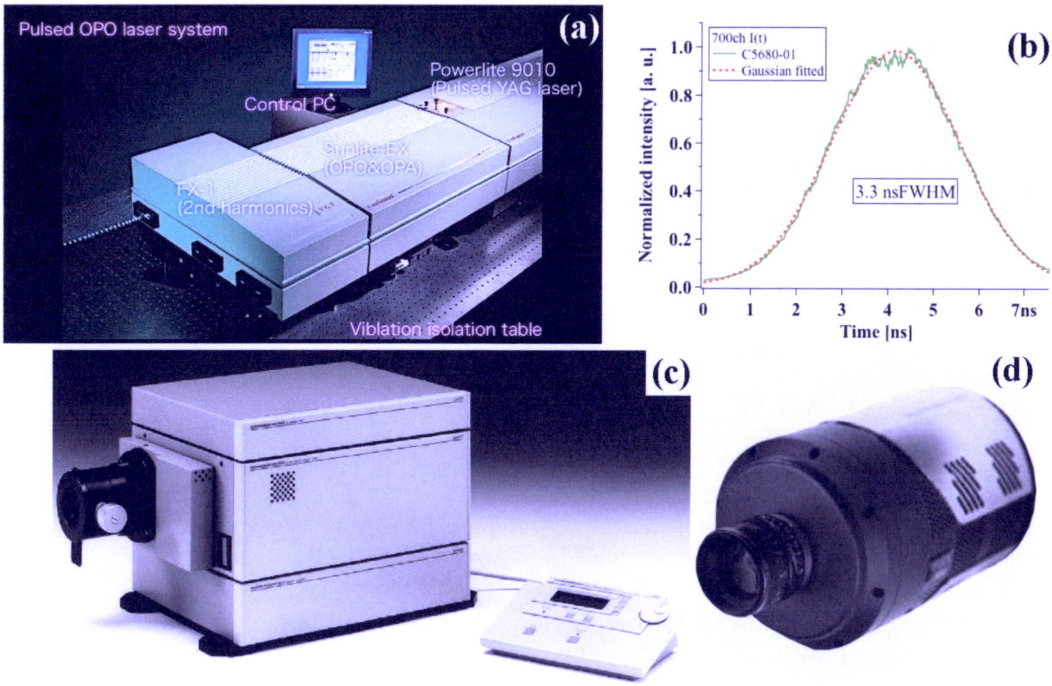

Figure 1. (a) An injection-seeded pulsed YAG laser (powerlite precision 9010) with OPO/OPA and doubler system (sunlite-EX & FX-1). (b) Typical temporal intensity profile of 355-nm pulse recorded by a streak camera. (c) A universal streak camera (C5680-01) and (d) a back-illuminated cooled CCD camera (PIXIS 1024B).

(SP, ANEST IWATA) as shown in **Figure 2**. The stoichiometric hot pressed TiN target ($\varphi = 20$ mm, $t = 5$ mm, three nines purity, filling rate = 88.6%, mass density = 4.65 g/cc, Toshima Manufacturing) was ablated at a pulse intensity ranging from 5.8×10^8 to 1.4×10^{10} W/cm^2 with pulse repetition rate of 10 Hz for several tens of minutes. For near future experiment for insulator film, a hot pressed AlN target ($\varphi = 20$ mm, $t = 5$ mm, three nines purity, filling rate = 99.4%, mass density = 3.24 g/cc, Toshima Manufacturing) was also prepared.

The glass substrate, kept around 5 cm away from the target without thermal control, was set through load lock chamber (TMP-pumped), which is positioned at right angle with laser optical axis, and temperature of target and substrate went along with the situation.

The PLD target hangs down from a vertically installed vacuum attachment rod, and the vertical position was set by a rack and pinion adjustment (±10 cm). The surface of PLD target makes a 360-degree turn by the rod. As a result, any position of the target was shot at any angle of incidence (typically 60°), and multilayer coating is available without exposure to the atmosphere when several different targets are installed. The deposition was carried out in the PLD chamber without introduction of nitrogen gas, and the chamber pressures were ranging from 10^{-4} to 5×10^{-4} Pa during deposition.

The 3ω laser pulse, which plane of polarization was adjusted to the vertical direction by a half-wave plate, was gradually focused via a AR-coated quartz plane-convex lens ($f = 750$ or 500 mm)

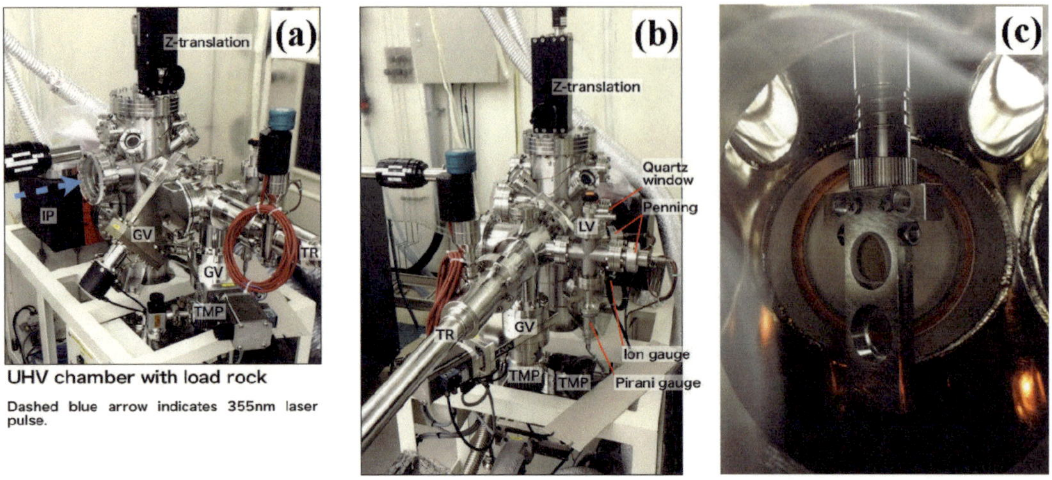

Figure 2. (a)(b) A UHV vacuum chamber system used in PLD experiments. (c) A photo of hot pressed TiN target installed in the center of UHV chamber.

struck the target surface through a homebuilt quartz Brewster window (θ_{inc} = 56°, t = 5 mm), as shown in **Figure 3**.

The deposition rate of the TiN films was investigated by measuring a height of artificially produced sub-wavelength step w/ and w/o mask during the PLD process. The step height was observed by both an optical surface profiler (NewView, Zygo) and a surface texture and contour measuring instruments (SURFCOM 1400D, Tokyo Seimitsu) as shown in **Figure 4(a)** and **(b)**, and it was determined to be close to 50 nm for a deposition period of 5 min with pulse energy of 340 mJ on target.

The substrate used in the PLD process was a glass slide as mentioned above and was cleaned by ultrasonic bath in the order of acetone, ethanol, and ultrapure water for 10 min each.

As a guide, time-integrated emission spectrum of the PLD plume ranging from 350 to 1050 nm was recorded by a diffraction grating spectrometer (USB4000, Ocean Optics) (**Figure 4(c)**) through a multimode optical fiber, in which one end faced to the chamber center. Scattered 3ω light was strongly attenuated by a UV-cut filter (MC UV SL-39, Kenko Tokina) for SLR cameras. Spectrometer wavelength calibration was carried out with 2ω (λ = 532 nm) and 3ω (λ = 355 nm) laser pulses, helium-neon laser (λ = 632.8 nm, Melles Griot), multiline argon ion laser (488 and 514.5 nm, Edmund Optics) and its argon lines (650–850 nm) from the discharge.

Spectral reflectance of the deposited TiN film was measured by an UV-Vis/NIR spectrophotometer (U-4000, Hitachi High-Technologies) (**Figure 5(a)**). Using a specified optional attachment, five-degree absolute and unpolarized specular reflectance of the sample were observed. A field of the measurement was 5 × 10 mm^2.

A one-time measurement for estimation of spectroscopic complex refractive index at the visible region was carried out by a spectroscopic ellipsometer (GES5E, SOPRA) with TiN film (50 nm) as shown in **Figure 5(b)**.

http://dx.doi.org/10.5772/67765

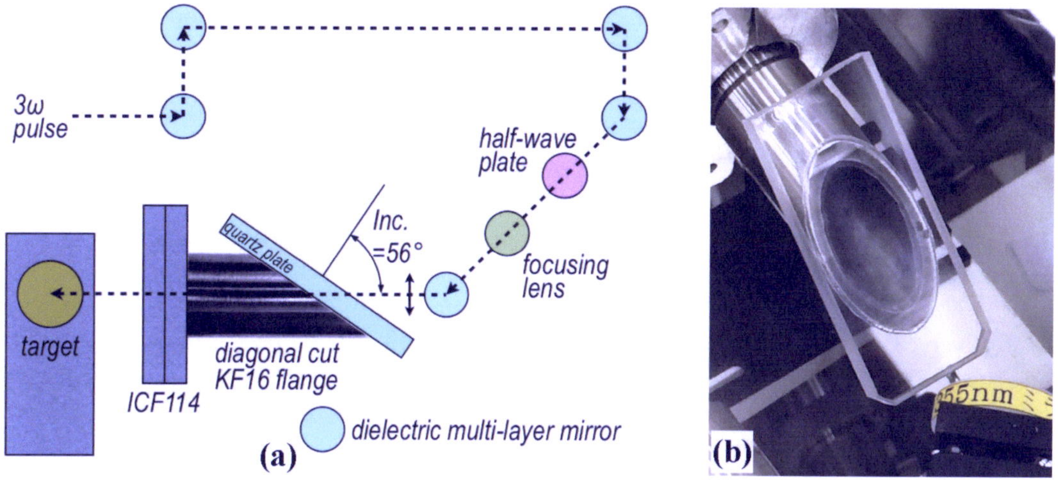

Figure 3. (a) Schematic diagram of optics used for guiding 3ω pulse into the UHV chamber center. (b) A photo of self-produced quartz Brewster window, which is effective with a half-wave plate.

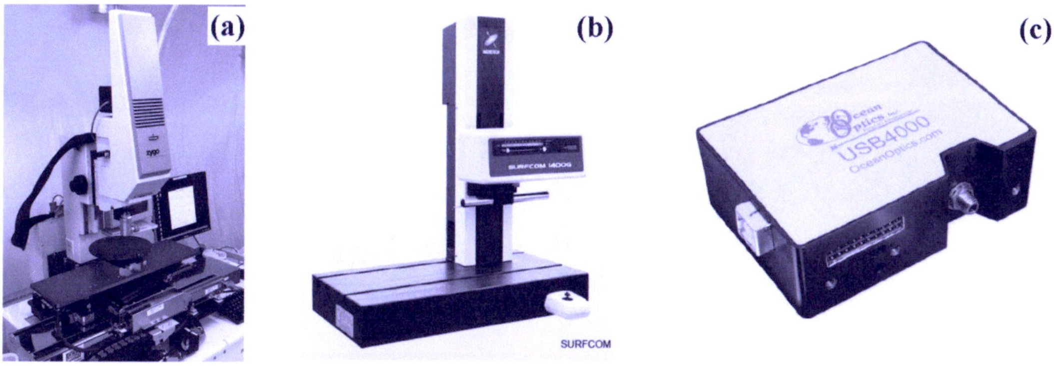

Figure 4. (a) An optical surface profiler by white-light interferometry (NewView). (b) A surface texture and contour measuring instrument (SURFCOM 1400D). (c) A fiber-type diffraction grating spectrometer (USB4000).

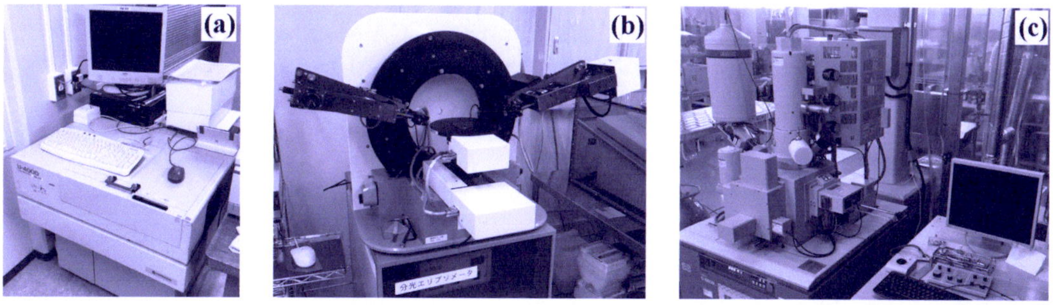

Figure 5. (a) A UV/Vis-NIR spectrophotometer (U-4000). (b) A spectroscopic ellipsometer (GES5E). (c) A scanning electron microscope (S-4800).

Position-dependent appearance, surface flatness, and existence of micro- and nanoparticles or debris were studied by a scanning electron microscope (SEM) (S-4800, Hitachi High-Technologies) (**Figure 5(c)**). Also, examinations of both a filling factor and growth process were performed with the SEM images.

Crystallinity of the PLD target was characterized by an X-ray diffractometer (XRD) (SmartLab, Rigaku) (**Figure 6(a)**) with Cu-Kα line (λ = 0.154 nm, E = 8.05 keV) before use. In this chapter, all the XRD patterns shown were measured by 2θ-ω scan with a crystal monochromator (Ge(220)×2). **Figure 13(a)** shows the XRD patterns of polycrystalline TiN. The diffraction peaks located at $2\theta = 33.7°$, 42.6, 61.8, 74.1 and 78.0° are corresponding to (111), (200), (220), (311), (222) of TiN, respectively. These peaks confirm that the PLD target is hot pressed polycrystalline TiN. But, an extra peak was found unexpectedly at $2\theta = 25°$, and this may be correspond to (101) of anatase-type titanium dioxide because the base powder may contain traces of oxygen. Deposition of TiN will be done with this target at the moment.

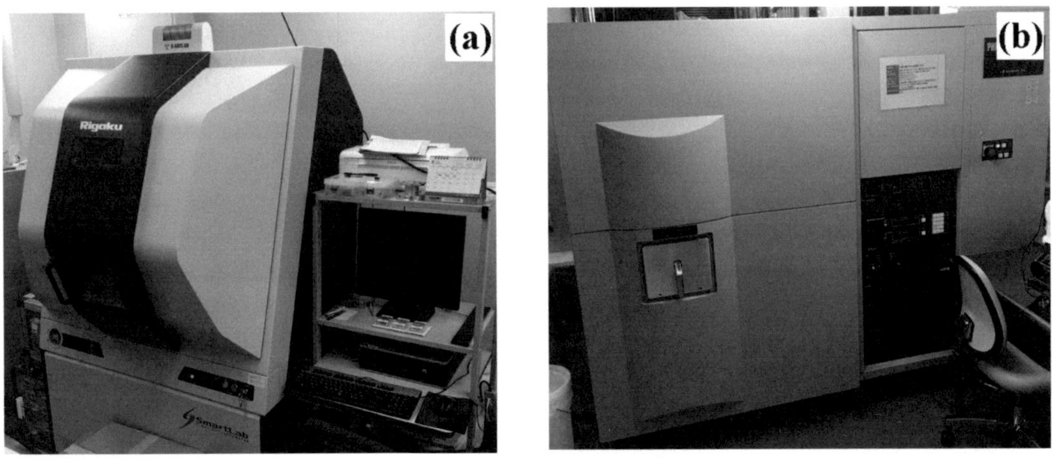

Figure 6. (a) X-ray diffractometer (SmartLab). (b) X-ray photoelectron spectroscopy system (PHI quantum2000 scanning ESCA microprobe).

Elemental composition, empirical formula, chemical state, and electronic state of the elements that exist within the TiN film was analyzed by a monochromatic X-ray photoelectron spectroscopy (XPS) system (PHI Quantum 2000 Scanning ESCA Microprobe, ULVAC-PHI) with Al Kα line (λ = 0.834 nm, E = 1.487 keV) as shown in **Figure 6(b)**.

3. Experimental results

3.1. Outline of the PLD experiment

Experimental condition is summarized in **Table 1**. The 3ω laser pulse is gradually focused by AR-coated flat-convex lens, and lens-target distance is adjusted to realize moderate intensity for laser irradiation. Basically, laser pulse energy is adjusted to maximum output.

http://dx.doi.org/10.5772/67765

Sample No.	focal length (flat-convex lens)[mm], coating	lens - target distance [mm]	E_{p_YAGout} [mJ/pulse] (3.3nsFWHM)	No. of 3ω mirror, R_{total}	No. of lens, T_{total}	terget - glass distance [mm]	No. of Brewster window, T_{total}, λ/2-plate	θ_{inc} of laser [deg]	PLD duration [sec], E_{total}[J]	Vacuum during PLD [Pa]	I_{Lave}[W/cm²], spot area [mm²], E_I[J/cm²/pulse]	outcome, appearance, analysis
1(⑤)	750 AR	415	360 (347.4 on T)	5 0.9995	1 0.98	50	1 0.99	60	600 2084.4	4.5E-4	5.26E+8 20 1.737	△ frosted
2(⑥)	750 AR	415	360 (347.4 on T)	5 0.9995	1 0.98	50	1 0.99	60	1200 4168.8	4.5E-4	5.26E+8 20 1.737	△ frosted
3(⑦)	750 AR	615	360 (347.4 on T)	5 0.9995	1 0.98	50	1 0.99	60	600 2084.4	4.5E-4	3.23E+9 3.26 10.656	◎ mirrored XPS
4(⑧)	750 AR	615	360 (347.4 on T)	5 0.9995	1 0.98	50	1 0.99	60	1200 4168.8	4.5E-4	3.23E+9 3.26 10.656	◎ mirrored XRD,XPS
5(⑨)	750 AR	515	360 (347.4 on T)	5 0.9995	1 0.98	50	1 0.99	60	1200 4168.8	4.5E-4	7.63E+8 9.82 3.538	△ frosted flake off
6(⑩)	750 AR	575	360 (347.4 on T)	5 0.9995	1 0.98	50	1 0.99	60	600 2084.4	4.5E-4	1.92E+9 5.48 6.34	◎ mirrored XRD,XPS
7(⑪)	750 AR	655	360 (347.4 on T)	5 0.9995	1 0.98	50	1 0.99	60	300+600 3126.6	4.5E-4	6.53E+9 1.612 21.55	mirrored step (50 nm)
8(⑫)	500 AR	455	360 (340 on T)	5 0.9995	1 0.98	70	1 0.99 #	60	2400 8160	4.5E-4	1.27E+10 0.814 41.77	◎ mirrored XPS
9(⑬)	500 AR	455	360 (340 on T)	5 0.9995	1 0.98	70	1 0.99 #	70	2400 8160	4.5E-4	8.66E+9 1.190 28.57	◎ mirrored XPS

 # half-waveplate(T=97.875%) was added for fine adjustment of polarization of laser pulse, and the resultant edge reflection of the Brewster window decreased.
⑤~⑬: serial number of the sample since the beginning of PLD experiment.

Table 1. Summary of the PLD experiment for the TiN thin film. Basically, double circles marked in rightmost column show acceptable results.

The plume flux ablated from the PLD target is so strong, and target-substrate distance is set more than 50 mm to suppress both sputtering and clacking of the deposited thin film. Incident angle of the laser pulse is set at 60° because a glass substrate is set at 90° from optical path of the laser beam. Duration of the PLD process is determined for thin film fabrication. Although base pressure of the main vacuum chamber is lower than 10^{-4} Pa, the load-lock system is used as a holder of a glass substrate, and the main and sub-chambers are connected. Then the PLD process is performed in a vacuum of 4.5×10^{-4} Pa. Laser irradiation intensity is adjusted by changing lens-target distance, and is realized from 5×10^{8} to 10^{10} W/cm². Rightmost column in Table 1 describes visual of the resultant film and X-ray analysis worked. In **Figure 7**, photos of well-deposited TiN films are shown, and adhesion of these films is quite good.

3.2. Reflectance spectrophotometry (U-4000)

By using a U-4000 spectrophotometer, unpolarized spectral reflectance at 5 degree was measured. As the digital photo shows, a mirrored PLD film of TiN had metallic cluster-like gold film, and the spectral reflectance appears to be the same as gold. Typical spectra reflectance curves are shown in **Figure 8**.

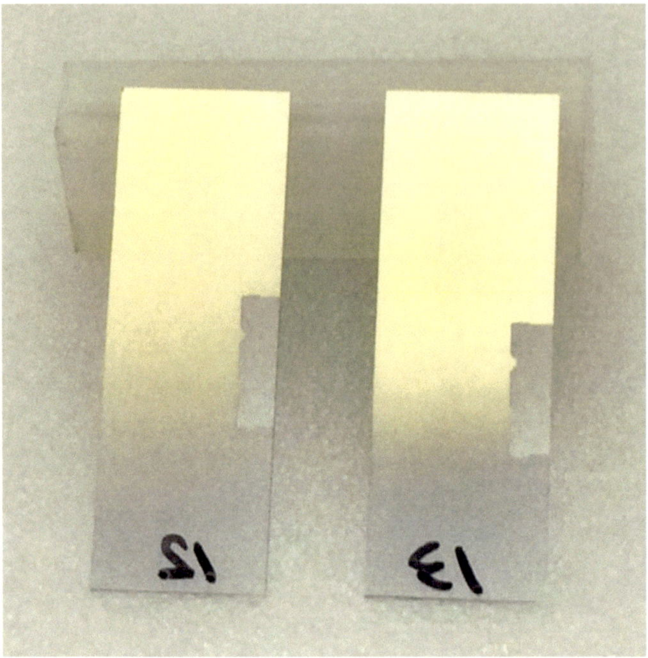

Figure 7. Photo of the PLDed TiN thin film (sample ⑫ & ⑬ in **Table 1**) on a glass slide.

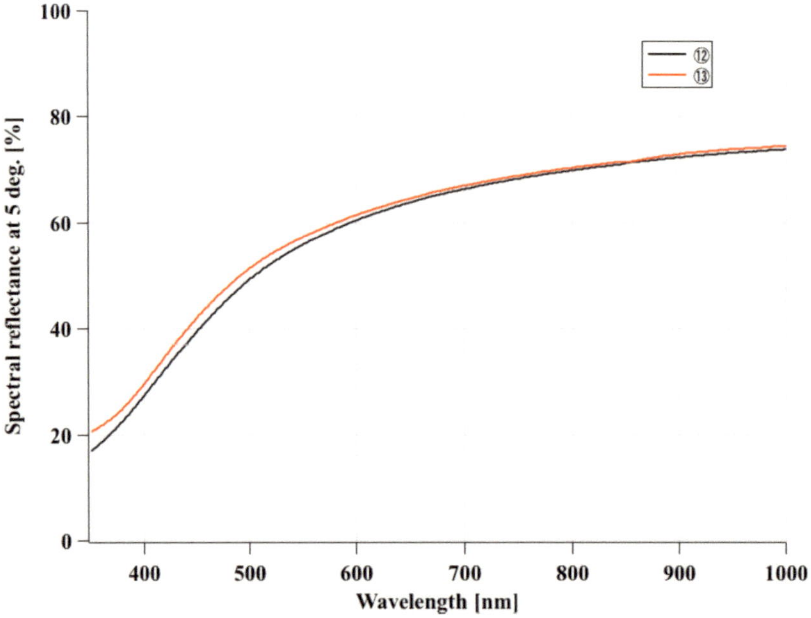

Figure 8. Unpolarized spectral reflectance of the PLDed TiN thin film (⑫ & ⑬ listed in **Table 1**).

3.3. Spectroscopic ellipsometry (GES5E)

Because of the instrument failure, spectroscopic ellipsometry for the TiN film was done only once before. Refer to the built-in database of refractive index, 50-nm-TiN film on a glass slide was measured by SE. Both the film thickness and the spectral complex refractive index were determined by the method of least squares. The resultant film thickness is almost same as 50 nm, and dependence of the index on light wavelength is also coincident with the database (**Figure 9**) [24–29].

3.4. Optical emission spectroscopy (USB4000)

Time-integrated optical emission spectrum of the PLD atmosphere or blight plume was measured by a fiber-type USB4000 grating spectrometer with a recorded wavelength range of 350–1000 nm, as shown in **Figure 10**. The input multimode optical fiber end was set at the right angle to one of the glass (Kovar) viewports installed on the experimental chamber. Through the viewport, light emission of the plume was collected at 45° from the PLD target normal.

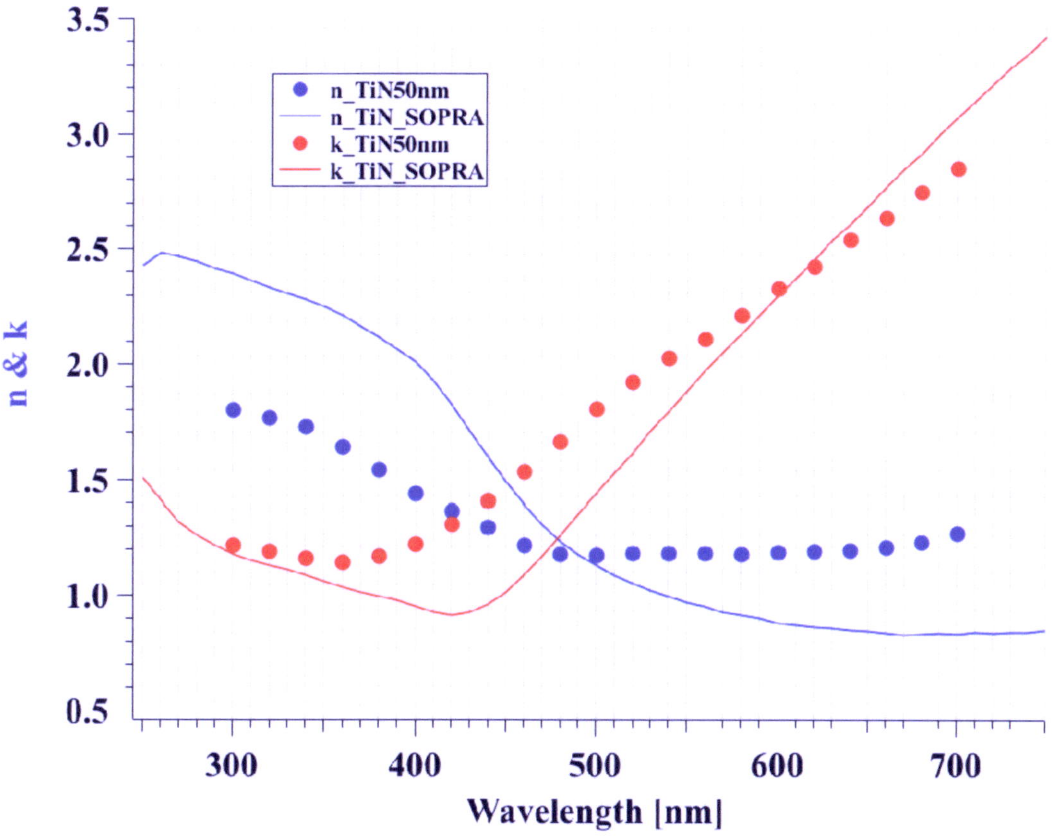

Figure 9. Complex refractive index curve of the PLDed TiN thin film determined by SE. Solid lines indicate a built-in reference data of TiN.

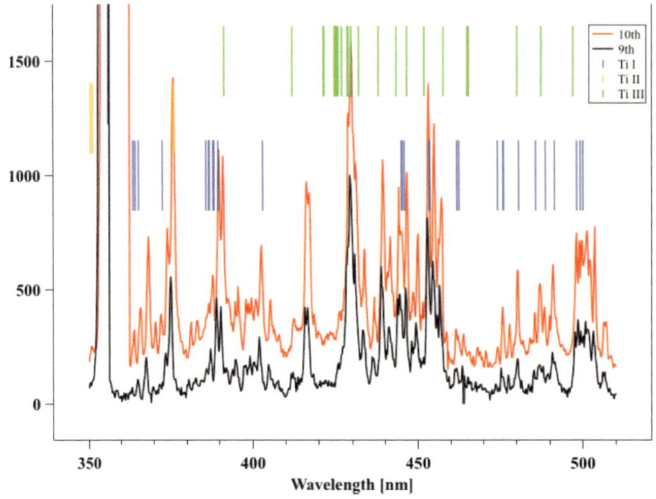

Figure 10. Time-integrated optical emission spectra of the samples ⑨ & ⑩ listed in **Table 1**. Colored markers of short line indicate line positions of both Ti atoms and Ti ions.

Because of low irradiation intensity of the pulsed UV laser light and non-use of discharges, neither multicharged titanium ion nor band emission of nitrogen was observed. In the UV region, emission of both Ti atom and singly ionized Ti atom was recorded.

3.5. White-light interferometry and stylus profilometry (NewView and Surfcom)

Estimation of a mean deposition rate of the PLD process was carried out by making a step on a film surface. After deposition of thin TiN layer, a part of the surface was masked, and additional deposition of TiN layer was executed for 5 min. The height of the handmade step was determined to be about 50 nm by a NewView interferometer, as shown in **Figure 11(a)**. Therefore, the mean deposition rate was estimated to be about 10 nm/min, and the PLD system will be able to control thickness of the film down to the sub-nanometer. The same step structure was also scanned by Surfcom, and the resultant height was same as that of NewView (**Figure 11(b)**).

3.6. Scanning electron microscopy (S-4800)

The surface of TiN thin film was observed by a scanning electron microscope as shown in **Figure 12**. Both (1) degradation of surface flatness of the film caused by shower of micro-nano debris presented in the PLD plume and (2) mass generation of Surface Plasmon Polariton (SPP) scatterer caused by buried debris become major problems in usual PLD process. At the beginning of this study, a glass slide was set within striking distance from the PLD target, and the resultant TiN film was powerfully influenced by spatial distribution of the PLD plume. Therefore, an extensive range of the film was observed by SEM, and the cross-sectional surface of peeling film was examined at high spatial resolution to find internal structure of the PLDed film. Then the central region of the film surface irradiated by very intense and hot

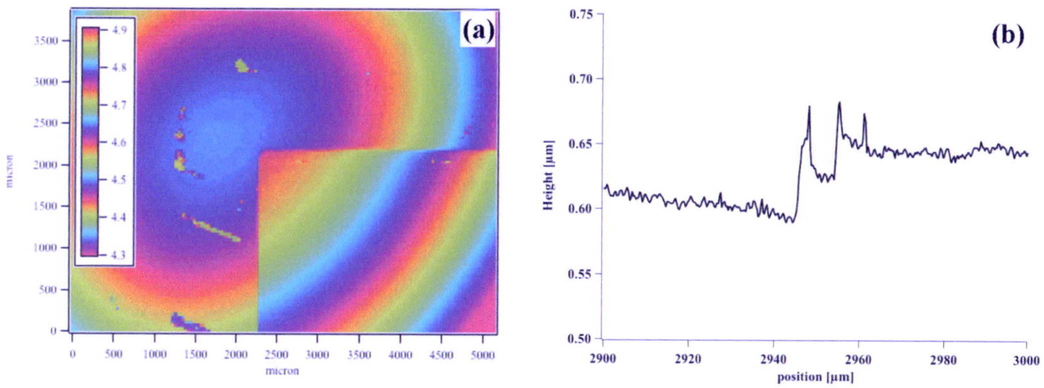

Figure 11. (a) Colored contour image, which was recorded by NewView, of the surface of PLDed sample (⑪ listed in **Table 1**) fabricated by two consecutive PLD processes. Square area (lower right) was masked during the second PLD process. Hence, the height is lower than that of upper left area. (b) Cross-sectional curve across the step structure recorded by Surfcom. Small dip at 2950 μm is no more than an unexpected hole.

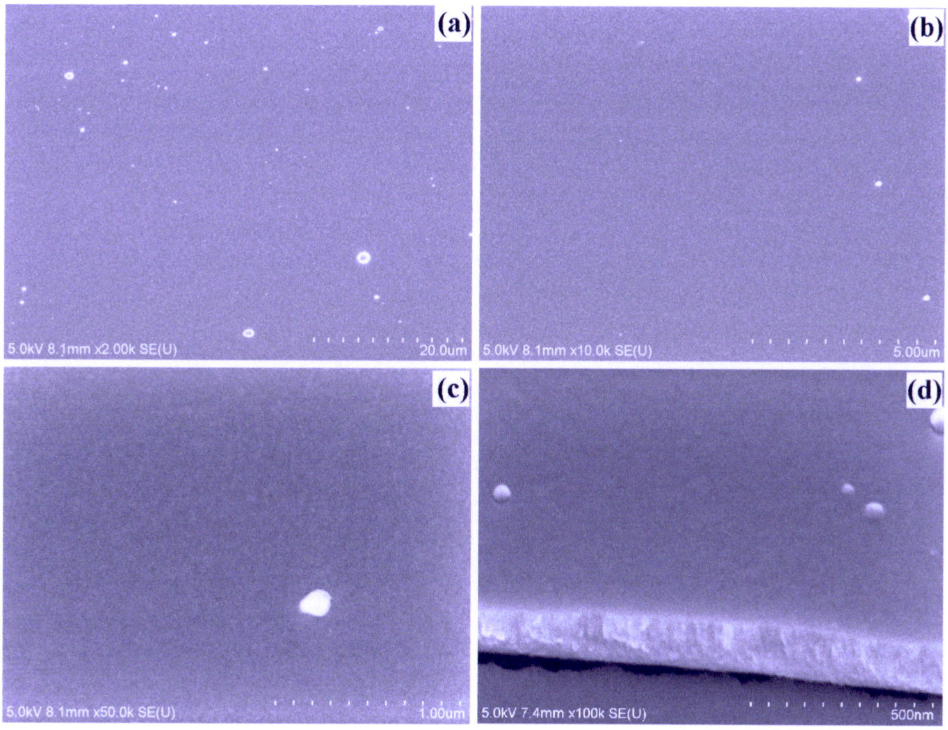

Figure 12. (a)–(c) SEM images of the PLDed TiN thin film (⑩ listed in **Table 1**). Debris is observed sparsely. (d) Cross-sectional SEM image of the PLDed thin film taken by oblique observation. Dense and flat structure of the film are observed clearly.

plume tarnished and cracked, and small pieces of peeling film existed. In the meantime, the film created by irradiation of fringe region of the plume had mirror completion, and the surface flatness was acceptable without dense flux of debris. Cross section of the peeling film was observed at maximum useful magnification (up to 100k), and an interior of the film seemed to be well filled without any apparent void. The cross-sectional surface had snaky and streaky structure in a longitudinal (surface normal) direction, and the growth of thin film may be governed by (111)-oriented growth. For this reason, out-of-plane XRD pattern of the film had a (111) peak only and in-plane pattern had no peaks.

3.7. X-ray diffractometry (SmartLab)

Application of TiN film for plasmonic phenomena is coupled with high crystallinity and large grain size in the film, and I decided to think a great deal about film quality before impending goal, i.e., excitation of SPP with TiN film. Crystallinity of the PLDed TiN film was examined by an X-ray diffractometer with a 2θ-ω scan (out-of-plane) method [30–32]. The PLD target was a hot pressed pellet of microcrystalline TiN powder and showed all the major peaks listed in well-known XRD databases as shown in blue curve of **Figure 13(a)**. Then TiN film, which was fabricated on a glass slide by plasma-assisted sputtering, was analyzed and only a weak (111) peak was observed. On the other hand, TiN film on a silicon (100) surface, which was fabricated by the same process, had no peak of TiN but Si(400) only, as shown in grey curve of **Figure 13(a)**. The PLDed TiN film had (111) and (200) peaks and had no peak in in-plane scan, as shown in **Figure 13(b)**. Therefore, highly (111)-oriented growth may act in a dominate fashion in the PLDed TiN film on a glass slide.

3.8. X-ray photoelectron spectroscopy (Quantum 2000)

Previous reports regarding TiN thin film fabrication, one of the major issues is an unexpected chemical composition of TiN and unwanted compound incorporation in the film. That is why

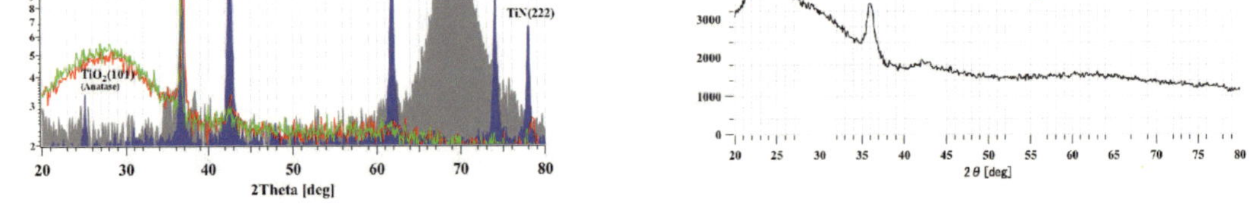

Figure 13. (a) XRD patterns of hot pressed PLD target and TiN film fabricated by plasma-assisted sputtering. (b) XRD pattern of the PLDed TiN thin film (⑩ listed in **Table 1**). The PLDed film is (111)-oriented, and the thickness is not enough to show higher order peaks.

http://dx.doi.org/10.5772/67765

an X-ray photoelectron spectroscopy was applied to the surface of TiN films fabricated by both plasma-assisted sputtering and PLD [33–38]. Every XPS analysis was carried out with X-ray beam at the right angle to the film surface. The survey spectrum was recorded in a binding energy of −8 to 1398 eV at low resolution ($\Delta E = 1$ eV), and high-resolution photoelectron spectra were measured for estimation of N(1s), O(1s), Ti(2s), and Ti(2p) as relevant atoms ($\Delta E = 0.05$ eV) as shown in **Figures 14** and **15**. Additionally, C(1s) was diagnosed for estimation of surface contamination and absolute energy reference for chemical shifts. The measurement took about 1 hour for each atom. All the XPS spectra were measured clearly, and the principal difference between the two films is noticeable on the spectrum of Ti(2p). The Ti(2p) spectrum has more peaks, which are strongly associated with a valuable work of Jaeger et al. [33]. Namely, peaks of Ti $2p^{3/2}$ and shake-up$^{3/2}$ are observed clearly. In the N(1s) spectrum, a peak originated from the N-Ti bond is much stronger than that of the O-Ti bond. In the O(1s) spectrum, a peak of the O-Ti bond is stronger than that of the O-N bond. The Ti(2s) spectrum is also obtained clearly with the PLDed film, and

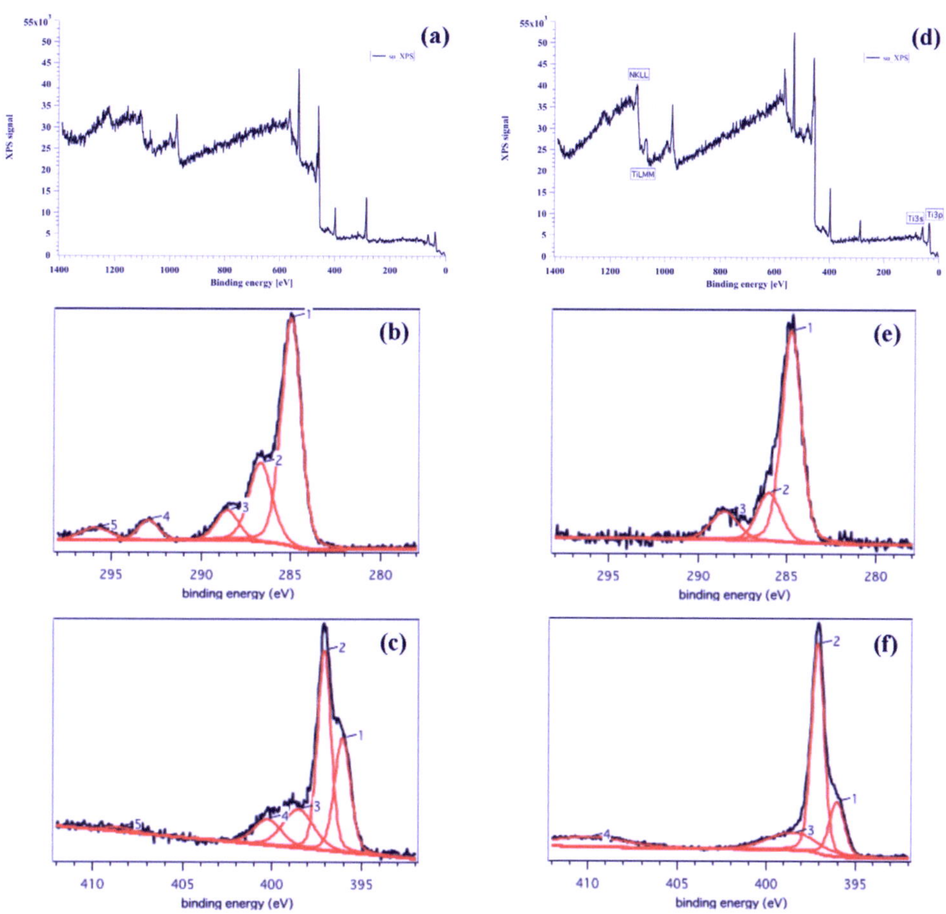

Figure 14. (a) Survey, (b) C(1s), (c) N(1s) spectrum of the sputtered TiN thin film recorded by XPS. (d) Survey, (e) C(1s), (f) N(1s) spectra of the PLDed TiN film (® listed in **Table 1**). Gray Gaussian curves are results of peak fitting by Igor Pro coupled with XPST 1.1 [39].

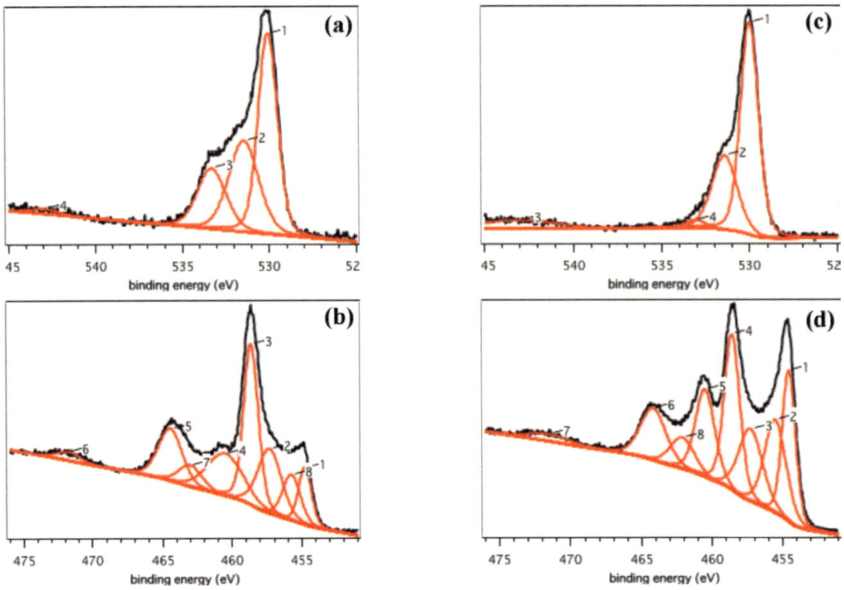

Figure 15. XPS spectrum of the sputtered TiN thin film (a) O(1s), (b) Ti(2p). XPS spectrum of the PLDed TiN thin film (⑩ listed in **Table 1**) (c) O(1s), (d) Ti(2p). In (d), two peaks originate from Ti-N combination are clearly observed, and the same two peaks do not exist in (b). Therefore, the PLDed TiN film may contain much Ti-N combinations.

broadened peak of first-order surface plasmon is also indicated. The high-resolution XPS spectra of N, O, and Ti are indicated simultaneously in **Figure 16** with peak positions referenced from NIST database. The N(1s) peak data agree well with Ti-N bonding. The O(1s) peak suggests the existence of titanium oxide in the PLDed film surface only. The Ti(2p) peaks seem to be close to peaks originated from the Ti-N bonding, but these are clearly influenced by oxides.

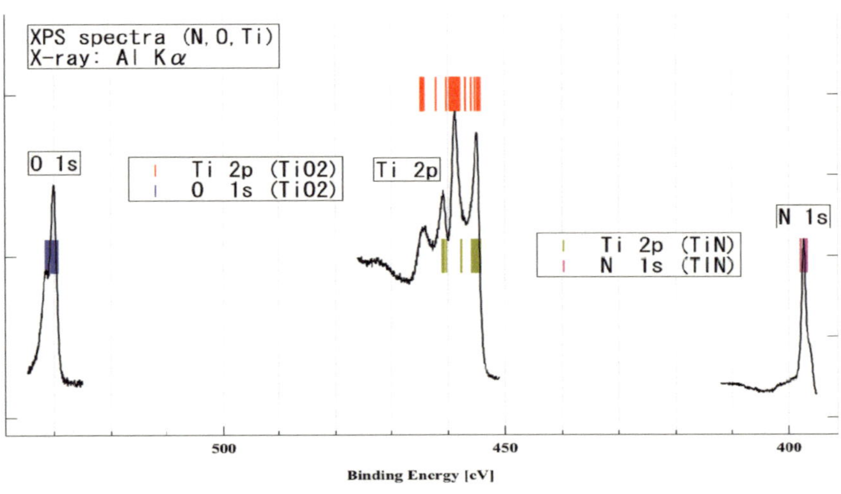

Figure 16. XPS spectra (N, O, Ti) of the PLDed TiN thin film (⑧ listed in **Table 1**) with reference peak positions (colored lines) related to TiN and TiO_2 combinations from the XPS database in the NIST homepage. TiN-based peaks are clearly observed, but TiO_2-based peaks are also exist because there might be considerable residual oxygen in the vacuum used for the PLD process.

4. FEM simulation

It is usually hard to observe excited surface plasmon polaritons in plasmonic experiments. So I have been performed numerical simulation concerning MIM structure with TiN as a metallic layer. A two-dimensional FEM model of MIM structure was designed as TiN(30 nm)/SiO_2(150 nm)/TiN(200 nm) layers on a glass substrate. And the multilayer was irradiated by *p*-polarized light by changing both incident angle and wavelength. As a well-known fact, the MIM structure could excite SPPs inside the insulating layer at small incident angles, and the wavelength of SPPs tends to be larger than the incident wavelength of light. Therefore, the model length in the *x*-direction is set at 2000 nm. The numerical simulation was done by an FEM-based electromagnetic simulation package, COMSOL Multiphysics and WaveOptics Module. In the calculation, the incident angle of light was changed from 0 to 75 degree at 15-degree interval. Increase in the incident angle caused (1) decrease in maximum at 550 nm and (2) blue shift in minimum at 670 nm (dip caused by SPPs) as shown in **Figure 17(a)**. With this result, the incident angle was fixed at 5 degree, and the dependence of minimum on the thickness of thinner TiN layer was simulated. The minimum fell to almost zero at TiN thickness of 20 nm. Calculated E_y map in the MIM structure is shown in **Figure 17(b)**. The incident wave is clearly indicated as a parallel pattern because no reflected wave exists. Inside the insulating layer (SiO_2), we can see intense and confined SPPs having extremely long space wavelength in the *x*-direction. So we could excite SPPs inside the MIM structure with TiN and SiO_2 but we need thinner-than-normal TiN layer for most efficient excitation of the SPPs. On the other hand, the thicker TiN layer must get thicker (>200 nm) to become perfect mirror layer.

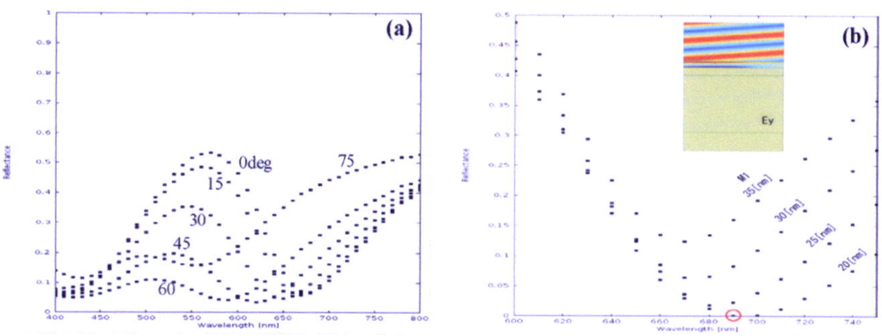

Figure 17. (a) Spectral reflectance curves of asymmetric MIM structure having TiN (30 nm)/SiO_2 (150 nm)/TiN (200 nm) layers on a quartz substrate depend on an incident angle of p-polarized light. (b) Minimum around 680 nm fell to almost zero at TiN thickness of 20 nm. The inset shows electric filed component *Ey* distribution inside the MIM structure. Top-striped area corresponds to vacuum layer just above the thin TiN layer, and most of the incident light converted into plasmon wave inside the MIM structure.

5. Conclusion

Thin film fabrication of titanium nitride has been studied by pulsed laser deposition with third harmonics of Nd:YAG laser. A metallic and adhesive thin film on a glass slide has been deposited at around 10^{10} [W/cm^2] of laser irradiation intensity. The mean deposition rate is around 10 nm/min. The PLDed TiN film has dense structure and is (111)-oriented by SEM and XRD analysis. Chemical composition analysis of the PLDed TiN film by XPS has good peaks related to Ti and N, but there are considerable influences of residual oxygen because of air-based vacuum. In the near future, a nitrogen-based high vacuum atmosphere will be realized, and XPS peaks of TiO_2 may be suppressed, and basic plasmonic experiments will be performed with the PLDed TiN film coupled with also PLDed AlN film.

Acknowledgements

The author expresses his deepest appreciation for overhaul and repair assistance to the relevant instruments from Center for Scientific Instrument Renovation and Manufacturing Support (CRM), Osaka University. The author also appreciates Dr. Jangwoo Kim (Staff Scientist, Postech) and Dr. Takahiro Yamada (Research Assistant Professor, Osaka University) for helping the XRD operation. The author gratefully acknowledges the technical support of the XPS operation by Mr. Kohei Hosoo (Master's Student, Osaka University) and Dr. Kenta Arima (Associate Professor, Osaka University). The author would like to acknowledge Dr. Katsuyoshi Endo (Professor, Osaka University) for his gratuitous conveyance of his spent- and well-preserved UHV chamber system. Several high power laser optics devices were provided by courtesy of Dr. Tomoyoshi Suenobu (Assistant Professor, Osaka University) and Dr. Mizuho Morita (Professor, Osaka University). Part of fabrication process for self-produced Brewster window was supported by Mr. Tadashi Hayashino (Technical Staff, Osaka University). Numerical electromagnetic simulation by COMSOL Multiphysics was done with technical support from Keisoku Engineering System Co., Ltd. and COMSOL Inc.

Author details

Yasushi Oshikane

Address all correspondence to: oshikane@prec.eng.osaka-u.ac.jp

Department of Precision Science & Technology and Applied Physics, Graduate School of Engineering, Osaka University, Suita, Osaka, Japan

References

[1] S. Rajak et al., Titanium nitrides as plasmonic materials in visible frequency range, Proc. of SPIE, Vol. 9654, 965412 (2015), doi: 10.1117/12.2181622.

[2] A. E. Khalifa et al., Plasmonic silicon solar cells using titanium nitride: a comparative study, J. Nanophotonics, Vol. 8, 084098 (2014).

[3] H. Mekawey et al., Dispersion analysis and engineering in TiN 2D plasmonic waveguides, Proc. of SPIE, Vol. 9371, 93711C (2015), doi: 10.1117/12.2076651.

[4] G. V. Naik et al., Oxides and nitrides as alternative plasmonic materials in the optical range, Opt. Mat. Exp., Vol. 1, No. 6, pp. 1090-1099 (2011).

[5] G. V. Naik et al., Alternative plasmonic materials: beyond gold and silver, Adv. Mat., Vol. 25, No. 24, pp. 3264-3294 (2013).

[6] P. Patsalas et al., Optical properties and plasmonic performance of titanium nitride, Mater., Vol. 8, pp. 3128-3154 (2015), doi:10.3390/ma8063128.

[7] V. Braic et al., The influence of deposition parameters on optical properties of titanium nitride thin films, Proc. of SPIE, Vol. 2461, pp. 597-599 (1995).

[8] N. Kinsey et al., Effective third-order nonlinearities in metallic refractory titanium nitride thin films, Opt Mater Express, Vol. 5, No. 11, pp. 2395-2403 (2015), doi: 10.1364/OME.5.002395.

[9] P. Patsalas et al., Optical, electronic, and transport properties of nanocrystalline titanium nitride thin films, J. Appl. Phys., Vol. 90, No. 9, pp. 4725-4734 (2001).

[10] Y. Zhong et al., Review of mid-infrared plasmonic materials, J. Nanophotonics, Vol. 9, 093791 (2015).

[11] F. Chiadini et al., Composite surface-plasmon-polariton waves guided by a thin metal layer sandwiched between a homogeneous isotropic dielectric material and a periodically multilayered isotropic dielectric material, J. Nanophotonics, Vol. 9, 093060 (2015).

[12] S. Fu et al., Impact of the basal material on the deposition titanium nitride thin films, Proc. of SPIE Vol. 9285, 928507, doi: 10.1117/12.2069017.

[13] S. Fu, The study of substrate material for deposited titanium nitride thin film and its influence, Proc. of SPIE, Vol. 9068, 90680H (2013), doi: 10.1117/12.2053929.

[14] J. A. Montes de Oca Valero et al., Low temperature, fast deposition of metallic titanium nitride films using plasma activated reactive evaporation, J. Vac. Sci. Technol., Vol. A 23, No. 3, pp. 393-400 (2005).

[15] S. Prayakarao et al., Gyroidal titanium nitride as nonmetallic metamaterial, Opt Mater Express, Vol. 5, No. 6, pp. 1316-1322 (2015), doi:10.1364/OME.5.001316

[16] H. Van Bui, Atomic layer deposition of TiN films: growth and electrical behavior down to sub-nanometer scale, PhD. Thesis – University of Twente, Enschede, the Netherlands (2013). ISBN: 978-90-365-3484-0, doi: 10.3990/1/9789036534840.

[17] J. M. Lackner et al., Pulsed laser deposition: a new technique for deposition of amorphous SiOx thin films, Surf and Coat Technol., Vol. 163-164, pp. 300-305 (2003).

[18] K. Obata et al., TiN growth by hybrid radical beam-PLD for Si barrier metal, Proc. of SPIE, Vol. 3933, pp.166-173 (2000).

[19] T. Ikegami et al., Effect of high voltage pulses on plasma plume in TiN deposition by PLD method, T. IEE Japan, Vol. 119-A, No. 6, pp. 860-865 (1999).

[20] H. Guo et al., Microstructures and properties of titanium nitride films prepared by pulsed laser deposition at different substrate temperatures, Appl. Surf Sci., Vol. 357, pp. 473-478 (2015).

[21] S. H. Kim et al., Structure and mechanical properties of titanium nitride thin films grown by reactive pulsed laser deposition, J Ceram Process Res., Vol. 10, No. 1, pp. 49-53 (2009).

[22] Z. P. Wang, et al., Growth of preferentially-oriented AlN films on amorphous substrate by pulsed laser deposition, Phys. Lett., Section A General Atom Solid State Phys., Vol. 375, No. 33, pp. 3007-3011 (2011).

[23] K. Jagannadham et al., Structural characteristics of AlN films deposited by pulsed laser deposition and reactive magnetron sputtering: a comparative study, J. Vac. Sci. Technol., Vol. A 16(5), pp. 2804-2815 (Sep/Oct 1998).

[24] S. Adachi, The Handbook on optical constants of metals, World Scientific Publishing Co. Pte. Ltd. 5 Toh Tuck Link, Singapore 596224. ISBN: 978-981-4405-94-2 (2012).

[25] E. Pascual et al., Surface reflectivity of TIN thin films measured by spectral ellipsometry, Surf Sci., Vol. 251/252, pp. 200-203 (1991).

[26] K. Postava et al., Optical characterization of TiN/SiO2(1000 nm)/Si system by spectroscopic ellipsometry and reflectometry, Appl. Surf. Sci., Vol. 175-176, pp. 270-275 (2001).

[27] J. Pflünger et al., Dielectric properties of TiCx, TiNx, VCx, and VNx from 1.5 to 40 eV determined by electron-energy-loss spectroscopy, Phys. Rev. B, Vol. 30, No. (3), pp. 1155-1163 (1984).

[28] K. Murai, Y. Oshikane et al., Design and fabrication of active spectral filter with metal-insulator-metal structure for visible light communication, Proc. SPIE, Vol. 8632, 863223 (2013).

[29] Y. Oshikane et al., 3D-FEM analysis of SPP excitation through nanoholes in asymmetric metal-insulator-metal structure at tip of circular truncated conical fiber, Proc. SPIE, Vol. 9163, 916317 (2014).

[30] J. I. Roh et al., Homo-epitaxial growth of 3C-SiC(100) thin films on SiC/Si substrate treated by chemical mechanical polishing, J. Korean Phys. Soc., Vol. 43, No. 1, pp. 96-101 (July 2003).

[31] S. S. Menon et al., Comparison of the Ti/TiN/AlCu/TiN stack with TiN/AlCu/Ti/TiN stack for application in ULSI metallization, Proc. of SPIE, Vol. 2875, pp. 322-333 (1996).

[32] K. Thamaphat et al., Phase characterization of TiO_2 powder by XRD and TEM, J. Kasetsart (Nat. Sci.), Vol. 42, pp. 357-361 (2008).

[33] D. Jaeger et al, J. Electron Spectrosc. Relat. Phenom., Vol. 185, pp. 523-534 (2012).

[34] N. C. Saha et al., Titanium nitride oxidation chemistry: an x-ray photoelectron spectroscopy study, J. Appl. Phys., Vol. 72, 1 (1992).

[35] N. Jiang et al., XPS study for reactively sputtered titanium nitride thin films deposited under different substrate bias, Physica B, Vol. 352, pp. 118-126 (2004).

[36] http://www.xpsfitting.com/2009/03/titanaium-nitride.html.

[37] http://xpssimplified.com/elements/titanium.php.

[38] F. H. Lu et al., XPS analyses of TiN films on Cu substrates after annealing in the controlled atmosphere, Thin Solid Films, Vol. 355-356, pp. 374-379 (1999).

[39] XPST 1.1 (X-ray photoelectron spectroscopy tools version 1.1 for Igor Pro) which is put on http://www.igorexchange.com/project/XPStools.